A

Comprehensive Introduction

to

DIFFERENTIAL GEOMETRY

VOLUME ONE

A
Comprehensive Introduction
to

DIFFERENTIAL GEOMETRY

VOLUME ONE
Third Edition

MICHAEL SPIVAK

PUBLISH OR PERISH, INC.

Houston, Texas   1999

Volume 1   ISBN   0-914098-70-5
Volume 2   ISBN   0-914098-71-3
Volume 3   ISBN   0-914098-72-1
Volume 4   ISBN   0-914098-73-X
Volume 5   ISBN   0-914098-74-8

Printed in the United States of America

ACKNOWLEDGEMENTS

I am greatly indebted to

Richard S. Palais

without his encouragement
these volumes would have remained
a short set of mimeographed notes

and

Donald E. Knuth

without his TEX program they would
never have become typeset books

PREFACE

The preface to the first edition, reprinted on the succeeding pages, excused this book's deficiencies on grounds that can hardly be justified now that these "notes" truly have become a book.

At one time I had optimistically planned to completely revise all this material for the momentous occasion, but I soon realized the futility of such an undertaking. As I examined these five volumes, written so many years ago, I could scarcely believe that I had once had the energy to learn so much material, or even recall how I had unearthed some of it.

So I have contented myself with the correction of errors brought to my attention by diligent readers, together with a few expository ameliorations; among these is the inclusion of a translation of Gauss' paper in Volume 2.

Aside from that, this third and final edition differs from the previous ones only in being typeset, and with figures redrawn. I have merely endeavored to typeset these books in a manner befitting a subject of such importance and beauty.

As a final note, it should be pointed out that since the first volumes of this series made their appearance in 1970, references in the text to "recent" results should be placed in context.

HOW THESE NOTES CAME TO BE

and how they did not come to be a book

For many years I have wanted to write the Great American Differential Geometry book. Today a dilemma confronts any one intent on penetrating the mysteries of differential geometry. On the one hand, one can consult numerous classical treatments of the subject in an attempt to form some idea how the concepts within it developed. Unfortunately, a modern mathematical education tends to make classical mathematical works inaccessible, particularly those in differential geometry. On the other hand, one can now find texts as modern in spirit, and as clean in exposition, as Bourbaki's Algebra. But a thorough study of these books usually leaves one unprepared to consult classical works, and entirely ignorant of the relationship between elegant modern constructions and their classical counterparts. Most students eventually find that this ignorance of the roots of the subject has its price -- no one denies that modern definitions are clear, elegant, and precise; it's just that it's impossible to comprehend how any one ever thought of them. And even after one does master a modern treatment of differential geometry, other modern treatments often appear simply to be about totally different subjects.

Of course, these remarks merely mean that no matter how well some of the present day texts achieve their objective, I nevertheless feel that an introduction to differential geometry ought to have quite different aims. There are two main premises on which these notes are based. The first premise is that it is absurdly inefficient to eschew the modern language of manifolds, bundles, forms, etc., which was developed precisely in order to rigorize the concepts of classical differential geometry. Rephrasing everything in more elementary terms involves incredible contortions which are not only unnecessary, but misleading. The work of Gauss, for example, which uses infinitesimals throughout, is most naturally rephrased in terms of differentials, even if it is possible to rewrite it in terms of derivatives. For this reason, the entire first volume of these notes is devoted to the theory of differentiable manifolds, the basic language of modern differential geometry. This language is compared whenever possible with the classical language, so that classical works can then be read.

The second premise for these notes is that in order for an introduction to differential geometry to expose the geometric aspect of the subject, an historical approach is necessary; there is no point in introducing the curvature tensor without explaining how it was invented and what it has to do with curvature. I personally felt that I could never acquire a satisfactory understanding of differentiable geometry until I read the original works. The second volume of these notes gives a detailed exposition of the fundamental papers of Gauss and Riemann. Gauss' work is now available in English (General Investigations of Curved Surfaces; Raven Press). There are also two English translations of Riemann's work, but I have provided a (very free) translation in the second volume.

Of course, I do not think that one should follow all the intricacies of the historical process, with its inevitable duplications and false leads. What is intended, rather, is a presentation of the subject along the lines which its development might have followed; as Bernard Morin said to me, there is no reason, in mathematics any more than in biology, why ontogeny must recapitulate phylogeny. When modern terminology finally is introduced, it should be as an outgrowth of this (mythical) historical development. And all the major approaches have to be presented, for they were all related to each other, and all still play an important role.

At this point I am reminded of a paper described in Littlewood's Mathematician's Miscellany. The paper began "The aim of this paper is to prove ..." and it transpired only much later that this aim was not achieved (the author hadn't claimed that it was). What I have outlined above is the content of a book the realization of whose basic plan and the incorporation of whose details would perhaps be impossible; what I have written is a second or third draft of a preliminary version of this book. I have had to restrict myself to what I could write and learn about within the present academic year, and all revisions and corrections have had to be made within this same period of time. Although I may some day be able to devote to its completion the time which such an undertaking deserves, at present I have no plans for this. Consequently, I would like to make these notes available now, despite their deficiencies, and with all the compromises I learned to make in the early hours of the morning.

These notes were written while I was teaching a year course in differential geometry at Brandeis University, during the academic year 1969-70. The course was taken by six juniors and seniors, and audited by a few graduate students. Most of them were familiar with the material in Calculus on Manifolds, which is essentially regarded as a prerequisite. More precisely, the complete prerequisites are advanced calculus using linear algebra and a basic knowledge of metric spaces. An acquaintance with topological spaces is even better, since it allows one to avoid the technical troubles which are sometimes relegated to the Problems, but I tried hard to make everything work without it.

The material in the present volume was covered in the first term, except for Chapter 10, which occupied the first couple of weeks of the second term, and Chapter 11, which was not covered in class at all. We found it necessary to take rest cures of nearly a week after completing Chapters 2, 3, and 7. The same material could easily be expanded to a full year course in manifold theory with a pace that few would describe as excessively leisurely. I am grateful to the class for keeping up with my accelerated pace, for otherwise the second half of these notes would not have been written. I am also extremely grateful to Richard Palais, whose expert knowledge saved me innumerable hours of labor.

Michael Spivak
Brandeis University
March, 1970

TABLE OF
CONTENTS

Although the chapters are not divided into sections,
the listing for each chapter gives some indication
which topics are treated, and on what pages.

CHAPTER 4. TENSORS

CHAPTER 5. VECTOR FIELDS AND DIFFERENTIAL EQUATIONS

CHAPTER 6. INTEGRAL MANIFOLDS

CHAPTER 7. DIFFERENTIAL FORMS

CHAPTER 8. INTEGRATION

CHAPTER 9. RIEMANNIAN METRICS

CHAPTER 10. LIE GROUPS

CHAPTER 11. EXCURSION IN THE REALM OF ALGEBRAIC TOPOLOGY

APPENDIX A

A
Comprehensive Introduction
to
DIFFERENTIAL GEOMETRY

VOLUME ONE

CHAPTER 1

MANIFOLDS

The nicest example of a metric space is **Euclidean n-space** $\mathbb{R}^n$, consisting of all n-tuples $x = (x^1, \ldots, x^n)$ with each $x^i \in \mathbb{R}$, where $\mathbb{R}$ is the set of real numbers. Whenever we speak of $\mathbb{R}^n$ as a metric space, we shall assume that it has the "usual metric"

$$d(x, y) = \sqrt{\sum_{i=1}^{n}(y^i - x^i)^2} \, ,$$

unless another metric is explicitly suggested. For $n = 0$ we will interpret $\mathbb{R}^0$ as the single point $0 \in \mathbb{R}$.

A manifold is supposed to be "locally" like one of these exemplary metric spaces $\mathbb{R}^n$. To be precise, a **manifold** is a metric space M with the following property:

> If $x \in M$, then there is some neighborhood U of x and some integer $n \geq 0$ such that U is homeomorphic to $\mathbb{R}^n$.

The simplest example of a manifold is, of course, just $\mathbb{R}^n$ itself; for each $x \in \mathbb{R}^n$ we can take U to be all of $\mathbb{R}^n$. Clearly, $\mathbb{R}^n$ supplied with an equivalent metric (one which makes it homeomorphic to $\mathbb{R}^n$ with the usual metric), is also a manifold. Indeed, a hasty recollection of the definition shows that anything homeomorphic to a manifold is also a manifold—the specific metric with which M is endowed plays almost no role, and we shall almost never mention it.

[If you know anything about topological spaces, you can replace "metric space" by "topological space" in our definition; this new definition allows some pathological creatures which are not metrizable and which fail to have other properties one might carelessly assume must be possessed by spaces which are locally so nice. Appendix A contains remarks, supplementing various chapters, which should be consulted if one allows a manifold to be non-metrizable.]

The second simplest example of a manifold is an open ball in $\mathbb{R}^n$; in this case we can take U to be the entire open ball since an open ball in $\mathbb{R}^n$ is homeomorphic to $\mathbb{R}^n$. This example immediately suggests the next: any open subset V of $\mathbb{R}^n$ is a manifold—for each $x \in V$ we can choose U to be some open ball with $x \in U \subset V$. Exercising a mathematician's penchant for generalization,

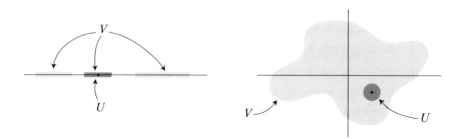

we immediately announce a proposition whose proof is left to the reader: An open subset of a manifold is also a manifold (called, quite naturally, an **open submanifold** of the original manifold).

The open subsets of $\mathbb{R}^n$ already provide many different examples of manifolds (just how many is the subject of Problem 24), though by no means all. Before proceeding to examine other examples, which constitute most of this chapter, some preliminary remarks need to be made.

If x is a point of a manifold M, and U is a neighborhood of x (U contains · some open set V with $x \in V$) which is homeomorphic to $\mathbb{R}^n$ by a homeomorphism $\phi : U \to \mathbb{R}^n$, then $\phi(V) \subset \mathbb{R}^n$ is an open set containing $\phi(x)$. Conse-

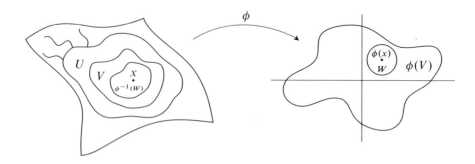

quently, there is an open ball W with $\phi(x) \in W \subset \phi(V)$. Thus $x \in \phi^{-1}(W) \subset V \subset U$. Since $\phi : V \to \mathbb{R}^n$ is continuous, the set $\phi^{-1}(W)$ is open *in* V, and thus open in M; it is, of course, homeomorphic to W, and thus to $\mathbb{R}^n$. This complicated little argument just shows that we can always choose the neighborhood U in our definition to be an open neighborhood.

With a little thought, it begins to appear that, in fact, U *must* be open. But to prove this, we need the following theorem, stated here without proof.*

1. THEOREM. If $U \subset \mathbb{R}^n$ is open and $f : U \to \mathbb{R}^n$ is one-one and continuous, then $f(U) \subset \mathbb{R}^n$ is open. (It follows that $f(V)$ is open for any open $V \subset U$, so f^{-1} is continuous, and f is a homeomorphism.)

Theorem 1 is called "Invariance of Domain", for it implies that the property of being a "domain" (a connected open set) is invariant under one-one continuous maps into $\mathbb{R}^n$. The proof that the neighborhood U in our definition must be open is a simple deduction from Invariance of Domain, left to the reader as an easy exercise (it is also easy to see that if Theorem 1 were false, then there would be an example where the U in our definition was not open).

We next turn our attention to the integer n appearing in our definition. Notice that n may depend on the point x. For example, if $M \subset \mathbb{R}^3$ is

$$M = \{(x, y, z) : z = 0\} \cup \{(x, y, z) : x = 0 \text{ and } z = 1\}$$
$$= M_1 \cup M_2,$$

then we can choose $n = 2$ for points in M_1 and $n = 1$ for points in M_2. This

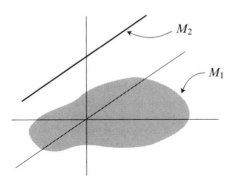

example, by the way, is an unnecessarily complicated device for producing one manifold from two. In general, given M_1 and M_2, with metrics d_1 and d_2, we can first replace each d_i with an equivalent metric $\bar{d}_i$ such that $\bar{d}_i(x, y) < 1$ for all $x, y \in M_i$; for example, we can define

$$\bar{d}_i = \frac{d_i}{1 + d_i} \quad \text{or} \quad \bar{d}_i = \min(d_i, 1).$$

---

*All proofs require some amount of machinery. The quickest routes are probably provided by Vick, *Homology Theory* and Massey, *Singular Homology Theory*. An old-fashioned, but pleasantly geometric, treatment may be found in Newman, *Topology of Plane Sets*.

Then we can define a metric d on $M = M_1 \cup M_2$ by

$$d(x, y) = \begin{cases} \bar{d}_i(x, y) & \text{if there is some } i \text{ such that } x, y \in M_i \\ 1 & \text{otherwise} \end{cases}$$

(we assume that M_1 and M_2 are disjoint; if not, they can be replaced by new sets which are). In the new space M, both M_1 and M_2 are open sets. If M_1 and M_2 are manifolds, M is clearly a manifold also. This construction can be applied to any number of spaces—even uncountably many; the resulting metric space is called the **disjoint union** of the metric spaces M_i. A disjoint union of manifolds is a manifold. In particular, since a space with one point is a manifold, so is any discrete space M, defined by the metric

$$d(x, y) = \begin{cases} 0 & \text{if } x = y \\ 1 & \text{if } x \neq y. \end{cases}$$

Although different n's may be required at different points of a manifold M, it would seem that only one n can work at a given point $x \in M$. For the proof of this intuitively obvious assertion we have recourse once again to Invariance of Domain. As a first step, we note that $\mathbb{R}^n$ is not homeomorphic to $\mathbb{R}^m$ when $n \neq m$, for if $n > m$, then there is a one-one continuous map from $\mathbb{R}^m$ into a non-open subset of $\mathbb{R}^n$. The further deduction, that the n of our definition is unique at each $x \in M$, is left to the reader. This unique n is called the **dimension of M at x**. A manifold has **dimension n** or is **n-dimensional** or is an **n-manifold** if it has dimension n at each point. It is convenient to refer to the manifold M as M^n when we want to indicate that M has dimension n.

Consider once more a discrete space, which is a 0-dimensional manifold. The only compact subsets of such a space are finite subsets. Consequently, an uncountable discrete space is not σ-compact (it cannot be written as a countable union of compact subsets). The same phenomenon occurs with higher-dimensional manifolds, as we see by taking a disjoint union of uncountably many manifolds homeomorphic to $\mathbb{R}^n$. In these examples, however, the manifold is not connected. We will often need to know that this is the only way in which σ-compactness can fail to hold.

2. THEOREM. If X is a connected, locally compact metric space, then X is σ-compact.

PROOF. For each $x \in X$ consider those numbers $r > 0$ such that the closed ball

$$\{y \in X : d(x, y) \leq r\}$$

is a compact set (there is at least one such $r > 0$, since X is locally compact). The set of all such $r > 0$ is an interval. If, for some x, this set includes all $r > 0$, then X is σ-compact, since

$$X = \bigcup_{n=1}^{\infty} \{y \in X : d(x, y) \le n\}.$$

If not, then for each $x \in X$ define $r(x)$ to be one-half the least upper bound of all such r.

The triangle inequality implies that

$$\{y \in X : d(x_1, y) \le r\} \subset \{y \in X : d(x_2, y) \le r + d(x_1, x_2)\},$$

so that

$$\{y \in X : d(x_1, y) \le r - d(x_1, x_2)\} \subset \{y \in X : d(x_2, y) \le r\},$$

which implies that

(1) $$r(x_1) \ge r(x_2) - \frac{1}{2}d(x_1, x_2).$$

Interchanging x_1 and x_2 gives

(2) $$|r(x_1) - r(x_2)| \le \frac{1}{2}d(x_1, x_2),$$

so the function $r : X \to \mathbb{R}$ is continuous. This has the following important consequence. Suppose $A \subset X$ is compact. Let A' be the union of all closed balls of radius $r(y)$ and center y, for all $y \in A$. Then A' is also compact. The proof is as follows.

Let $z_1, z_2, z_3, \ldots$ be a sequence in A'. For each i there is a $y_i \in A$ such that z_i is in the ball of radius $r(y_i)$ with center y_i. Since A is compact, some subsequence of the y_i, which we might as well assume is the sequence itself, converges to some point $y \in A$. Now the closed ball B of radius $\frac{3}{2}r(y)$ and

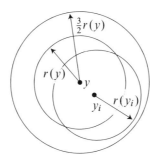

center y is compact. Since $y_i \to y$ and since the function r is continuous, eventually the closed balls

$$\{y \in X : d(y, y_i) \leq r(y_i)\}$$

are contained in B. So the sequence z_i is eventually in the compact set B, and consequently some subsequence converges. Moreover, the limit point is actually in the closed ball of radius $r(y)$ and center y (Problem 10). Thus A' is compact.

Now let $x_0 \in X$ and consider the compact sets

$$A_1 = \{x_0\}$$
$$A_{n+1} = A_n'.$$

Their union A is clearly open. It is also closed. To see this, suppose that x is a point in the closure of A. Then there is some $y \in A$ with $d(x, y) < \frac{2}{3}r(x)$.

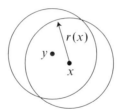

By (1),

$$r(y) \geq r(x) - \frac{1}{2}d(x, y)$$
$$> r(x) - \frac{1}{2} \cdot \frac{2}{3}r(x) = \frac{2}{3}r(x)$$
$$> d(x, y).$$

This shows that if $y \in A_n$, then $x \in A_n'$, so $x \in A$.

Since X is connected, and $A \neq \emptyset$ is open and closed, it must be that $X = A$, which is σ-compact. ❖

After this hassle with point-set topology, we present the long-promised examples of manifolds. The only connected 1-manifolds are the line $\mathbb{R}$ and the circle, or 1-dimensional sphere, S^1, defined by

$$S^1 = \{x \in \mathbb{R}^2 : d(x, 0) = 1\}.$$

The function $f: (0, 2\pi) \rightarrow S^1$ defined by $f(\theta) = (\cos\theta, \sin\theta)$ is a homeomorphism; it is even continuous, though not one-one, on $[0, 2\pi]$. We will often denote the point $(\cos\theta, \sin\theta) \in S^1$ simply by $\theta \in [0, 2\pi]$. (Of course, it is always necessary to check that use of this notation is valid.) The function $g: (-\pi, \pi) \rightarrow S^1$, defined by the same formula, is also a homeomorphism; together with f it shows that S^1 is indeed a manifold.

There is another way to prove this, better suited to generalization. The projection P from the point $(0, 1)$ onto the line $\mathbb{R} \times \{-1\} \subset \mathbb{R} \times \mathbb{R}$, illustrated in

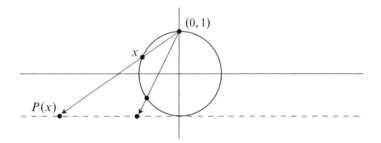

the above diagram, is a homeomorphism of $S^1 - \{(0, 1)\}$ onto $\mathbb{R} \times \{-1\}$: this is proved most simply by calculating $P: S^1 - \{(0, 1)\} \rightarrow \mathbb{R} \times \{-1\}$ explicitly. The point $(0, 1)$ may be taken care of similarly, by projecting onto $\mathbb{R} \times \{1\}$, or it suffices to note that S^1 is "homogeneous"—there is a homeomorphism taking any point into any other (namely, an appropriate rotation of $\mathbb{R}^2$). Considerations similar to these now show that the ***n*-sphere**

$$S^n = \{x \in \mathbb{R}^{n+1} : d(x, 0) = 1\}$$

is an n-manifold. The 2-sphere S^2, commonly known as "the sphere", is our first example of a compact 2-manifold or **surface**.

From these few manifolds we can already construct many others by noting that if M_i are manifolds of dimension n_i $(i = 1, 2)$, then $M_1 \times M_2$ is an $(n_1 + n_2)$-manifold. In particular

$$\underbrace{S^1 \times \cdots \times S^1}_{n \text{ times}}$$

is called the n-torus, while $S^1 \times S^1$ is commonly called "the torus". It is obviously homeomorphic to a subset of $\mathbb{R}^4$, and it is also homeomorphic to a certain subset of $\mathbb{R}^3$ which is what most people have in mind when they speak of "the torus": This subset may be obtained by revolving the circle

$$\{(0, y, z) \in \mathbb{R}^3 : (y-1)^2 + z^2 = 1/4\}$$

around the z-axis. The same construction may be applied to any 1-manifold

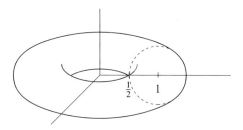

contained in $\{(0, y, z) \in \mathbb{R}^3 : y > 0\}$. The resulting surface, called a **surface of revolution**, has components homeomorphic either to the torus or to the cylinder $S^1 \times \mathbb{R}$, the latter of which is also homeomorphic to the annulus, the region of the plane contained between two concentric circles.

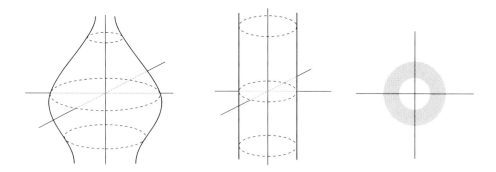

The next simplest compact 2-manifold is the 2-holed torus. To provide a more

explicit description of the 2-holed torus, it is easiest to begin with a "handle", a space homeomorphic to a torus with a hole cut out; more precisely, we throw away all the points on one side of a certain circle, which remains in our handle, and which will be referred to as the boundary of the handle. The 2-holed

torus may be obtained by piecing two of these together; it is also described as

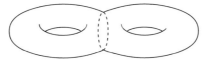

the disjoint union of two handles with corresponding points on the boundaries "identified".

The *n*-holed torus may be obtained by repeated applications of this procedure. It is homeomorphic to the space obtained by starting with the disjoint

union of *n* handles and a sphere with *n* holes, and then identifying points on the boundary of the ith handle with corresponding points on the ith boundary piece of the sphere.

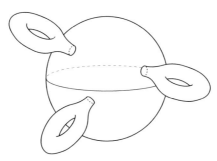

There is one 2-manifold of which most budding mathematicians make the acquaintance when they still know more about paper and paste than about metric spaces—the famous *Möbius strip*, which you "make" by giving a strip of paper a half twist before pasting its ends together. This can be described

 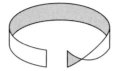 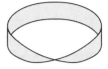

analytically as the image in $\mathbb{R}^3$ of the function $f: [0, 2\pi] \times (-1, 1) \to \mathbb{R}^3$ defined by

$$f(\theta, t) = \left(2\cos\theta + t\cos\tfrac{\theta}{2}\cos\theta, \; 2\sin\theta + t\cos\tfrac{\theta}{2}\sin\theta, \; t\sin\tfrac{\theta}{2}\right).$$

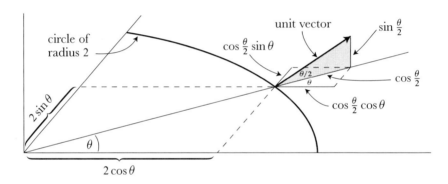

If we define f on $[0, 2\pi] \times [-1, 1]$ instead, we obtain the Möbius strip with a boundary; as investigation of the paper model will show, this boundary is homeomorphic to a circle, not to two disjoint circles. With our recently introduced terminology, the Möbius strip can also be described as $[0, 1] \times (-1, 1)$ with $(0, t)$ and $(1, -t)$ "identified".

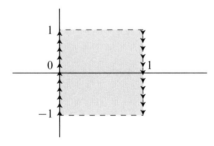

We have not yet had to make precise this notion of "identification", but our next example will force the issue. We wish to identify each point $x \in S^2$ with its antipodal point $-x \in S^2$. The space which results, the **projective plane**, $\mathbb{P}^2$, is a lot harder to visualize than previous examples; indeed, there is no subset of $\mathbb{R}^3$ which represents it adequately.

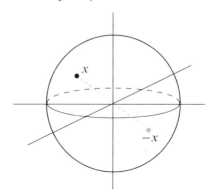

The precise definition of $\mathbb{P}^2$ uses the same trick that mathematicians always use when they want two things which are not equal to be equal. The *points* of $\mathbb{P}^2$ are defined to be the sets $\{p, -p\}$ for $p \in S^2$. We will denote this set by $[p] \in \mathbb{P}^2$, so that $[-p] = [p]$. We thus have a map $f : S^2 \to \mathbb{P}^2$ given by $f(p) = [p]$, for which $f(p) = f(q)$ implies $p = \pm q$. We will postpone for a while the problem of defining the metric giving the distance between two points $[p]$ and $[q]$, but we can easily say what the open sets will turn out to be (and this is all you need to know in order to check that $\mathbb{P}^2$ is a surface). A subset $U \subset \mathbb{P}^2$ will be open if and only if $f^{-1}(U) \subset S^2$ is open. This just means that the open sets of $\mathbb{P}^2$ are of the form $f(V)$ where $V \subset S^2$ is an open set with the additional important property that if it contains p it also contains $-p$.

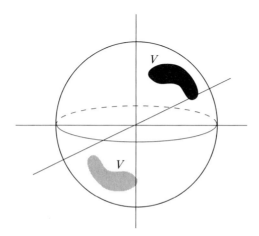

In exactly the same way, we could have defined the points of the Möbius strip M to be

$$\text{all points } (s,t) \in (0,1) \times (-1,1)$$

together with

$$\text{all sets } \{(0,t),(1,-t)\}, \quad \text{denoted by } [(0,t)] \text{ or } [(1,-t)].$$

There is a map $f : [0,1] \times (-1,1) \to M$ given by

$$f((s,t)) = \begin{cases} (s,t) & \text{if } s \neq 0,1 \\ [(s,t)] & \text{if } s = 0 \text{ or } 1, \end{cases}$$

and $U \subset M$ is open if and only if $f^{-1}(U) \subset [0,1] \times (-1,1)$ is open, so that the open sets of M are of the form $f(V)$ where V is open and contains $(s,-t)$ whenever it contains (s,t) for $s = 0$ or 1.

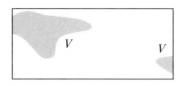

To get an idea of what $\mathbb{P}^2$ looks like, we can make things easier for ourselves by first throwing away all points of S^2 below the (x,y)-plane, since they are identified with points above the (x,y)-plane anyway. This leaves the upper hemisphere (including the bounding circle), which is homeomorphic to the disc

$$D^2 = \{x \in \mathbb{R}^2 : d(x,0) \leq 1\},$$

and we must identify each $p \in S^1$ with $-p \in S^1$. Squaring things off a bit, this is the same as identifying points on the sides of a square according to the scheme shown below (points on sides with the same label are identified in such a way that the heads of the arrows are identified with each other). The dotted lines in this picture are the key to understanding $\mathbb{P}^2$. If we distort the

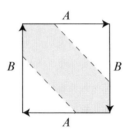

region between them a bit we see that the front part of *B* followed by the back part of *A*, at the upper left, is to be identified with the same thing at the lower right, in reverse direction; in other words, we obtain a Möbius strip with

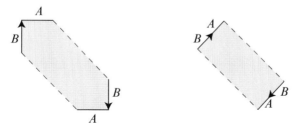

a boundary (namely, the dotted line, which is a single circle). If this Möbius strip is removed, we are left with two pieces which can be rearranged to form something homeomorphic to a disc. The projective plane is thus obtained from

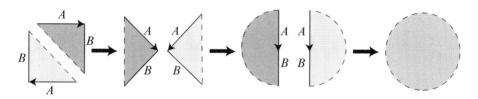

the disjoint union of a disc and a Möbius strip with a boundary, by identifying points on the boundary and points on the boundary of the disc, both of which are circles. Thus to make a model of $\mathbb{P}^2$ we just have to sew a circular piece of cloth and a cloth Möbius strip together along their edges. Unfortunately, a little experimentation will convince you that this cannot be done (without having the two pieces of cloth pass through each other).

The subset of $\mathbb{R}^3$ obtained as the union of the Möbius strip and a disc, although not homeomorphic to $\mathbb{P}^2$, can still be described mathematically in terms

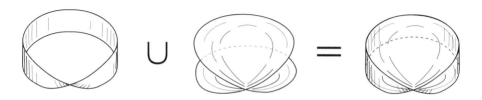

of $\mathbb{P}^2$. There is clearly a continuous function $f \colon \mathbb{P}^2 \to \mathbb{R}^3$ whose image is this subset; moreover, although f is not one-one, it is **locally one-one**, that is, every point $p \in \mathbb{P}^2$ has a neighborhood U on which f is one-one. Such a function f

is called a **topological immersion** (the single word "immersion" has a more specialized meaning, explained in Chapter 2). We can thus say that $\mathbb{P}^2$ can be topologically immersed in $\mathbb{R}^3$, although not topologically imbedded (there is no homeomorphism f from $\mathbb{P}^2$ to a subset of $\mathbb{R}^3$). In $\mathbb{R}^4$, however, with an extra dimension to play around with, the disc can be added so as not to intersect the Möbius strip.

Another topological immersion of $\mathbb{P}^2$ in $\mathbb{R}^3$ can be obtained by first immersing the Möbius strip so that its boundary circle lies in a plane; this can be done in the following way. The figures below show that the Möbius strip may be obtained from an annulus by identifying opposite points of the inner circle. (This is also obvious from the fact that the Möbius strip is the projective plane with a disc removed.) This inner circle can be replaced by a quadrilateral. When the

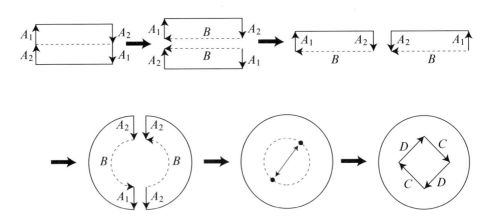

resulting figure is drawn up into 3-space and the appropriate identifications are made we obtain the "cross-cap". The cross-cap together with the disc at the bottom is a topologically immersed $\mathbb{P}^2$.

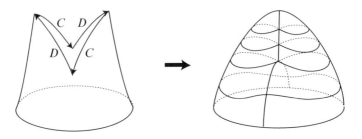

The one gap in the preceding discussion is the definition of a metric for $\mathbb{P}^2$. The missing metric can be supplied by an appeal to Problem 3-1, which will later be used quite often, and which the reader should peruse sometime before reading Chapter 3. Roughly speaking, it shows that things like $\mathbb{P}^2$, which ought to be manifolds, are. (Those who know about topological spaces will recognize it as a disguised case of the Urysohn Metrization Theorem.) For the present, however, we will obtain our metric by a trick that simultaneously provides an imbedding of $\mathbb{P}^2$ in $\mathbb{R}^4$. Consider the function $f : S^2 \to \mathbb{R}^4$ defined by

$$f(x, y, z) = (yz, xz, xy, x^2 + 2y^2 + 3z^2).$$

Clearly $f(p) = f(-p)$. We maintain that $f(p) = f(q)$ implies that $p = \pm q$. To prove this, suppose that $f(x, y, z) = f(a, b, c)$. We have, first of all

(1)
$$\begin{aligned} yz &= bc \\ xz &= ac \\ xy &= ab. \end{aligned}$$

If $a, b, c \neq 0$, this leads to

(2)
$$\begin{aligned} y &= \frac{bx}{a} \\ z &= \frac{cx}{a}. \end{aligned}$$

Now

$$\begin{aligned} (x + y + z)^2 &= x^2 + y^2 + z^2 + 2(xy + xz + yz) \\ &= 1 + 2(xy + xz + yz), \end{aligned}$$

so we also have
$$(x + y + z)^2 = (a + b + c)^2,$$

hence

(3)
$$a + b + c = \pm(x + y + z).$$

Using (2), this gives

$$a + b + c = \pm x \left(1 + \frac{b}{a} + \frac{c}{a}\right) = \pm x \left(\frac{a + b + c}{a}\right),$$

so $x = \pm a$. Similarly, we obtain $y = \pm b$, $z = \pm b$, with the same sign (which comes from (3)) holding for all three equations. In this case we have proved our contention without even using the fourth coordinate of f. Now suppose $a = 0$. If $x \neq 0$, then (1) would immediately give $y = z = 0$, so that

$$(x, y, z) = (\pm 1, 0, 0).$$

But $y = z = 0$ implies (by (1) again) that $bc = 0$, so $b = 0$ or $c = 0$ and

$$(a, b, c) = (0, \pm 1, 0) \quad \text{or} \quad (0, 0, \pm 1).$$

These equations clearly contradict

$$x^2 + 2y^2 + 3z^2 = a^2 + 2b^2 + 3c^2.$$

Thus $x = 0$ also, and we have

(4) $\quad yz = bc$

(5) $\quad 2y^2 + 3z^2 = 2b^2 + 3c^2$

(6) $\quad \begin{aligned} y^2 + z^2 &= 1 \\ b^2 + c^2 &= 1. \end{aligned}$

But (6) implies that

$$2y^2 + 3z^2 = 2y^2 + 3(1 - y^2)$$
$$= 3 - y^2,$$

and similarly for b and c, so (5) gives

$$3 - y^2 = 3 - b^2$$

(7) $$y = \pm b.$$

Now (4) gives

(8) $$z = \pm c$$

(this holds even if $y = b = 0$, since then $z, c = \pm 1$). Clearly, (4) also shows that the same sign holds in (7) and (8), which completes the proof.

Since $f(p) = f(q)$ precisely when $p = \pm q$, we can define $\bar{f} : \mathbb{P}^2 \to \mathbb{R}^4$ by

$$\bar{f}([p]) = f(p).$$

This map is one-one and we can use it to define the metric in $\mathbb{P}^2$:

$$\bar{d}([p], [q]) = d\big(\bar{f}([p]), \bar{f}([q])\big) = d\big(f(p), f(q)\big).$$

Then one can check that the open sets are indeed the ones described above.

By the way, the map $g \colon \mathbb{P}^2 \to \mathbb{R}^3$ defined by the first 3 components of f,

$$g([x, y, z]) = (yz, xz, xy)$$

is a topological immersion of $\mathbb{P}^2$ in $\mathbb{R}^3$. The image in $\mathbb{R}^3$ is Steiner's "Roman surface".

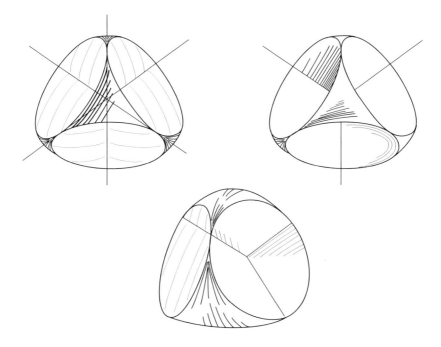

With the new surface $\mathbb{P}^2$ at our disposal, we can create other surfaces in the same way as the n-holed torus. For example, to a handle we can attach a projective space with a hole cut out, or, what amounts to the same thing, a Möbius strip. The closest we can come to picturing this is by drawing a cross-cap sticking on a torus. We can also join together a pair of projective planes with holes cut out, which amounts to sewing two Möbius strips together along their boundary. Although this can be pictured as two cross-caps joined together, it has a nicer, and famous, representation. Consider the surface obtained from the square with identifications indicated below; it may also be obtained from the cylinder $[0, 1] \times S^1$ by identifying $(0, x) \in [0, 1] \times S^1$ with $(1, x')$, where x' is the reflection of x through a fixed diameter of the circle. Notice that the identifications on the square force P_1, P_2, P_3, and P_4 to be identified, so that the set $\{P_1, P_2, P_3, P_4\}$ is a single point of our new space. The dotted lines

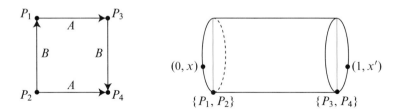

below, dividing the sides into thirds, form a single circle, which separates the surface into two parts, one of which is shaded.

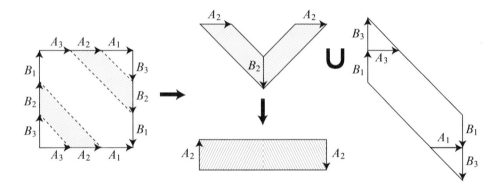

Rearrangement of the two parts shows that this surface is precisely two Möbius strips with corresponding points on their boundary identified. The description in terms of $[0, 1] \times S^1$ immediately suggests an immersion of the surface. Turning one end of the cylinder around and pushing it through itself orients the left-hand boundary so that $(0, x)$ is directly opposite $(1, x')$, to which it can then be joined, forming the "Klein bottle".

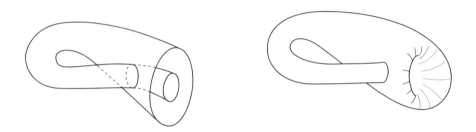

Examples of higher-dimensional manifolds will not be treated in nearly such detail, but, in addition to the family of n-manifolds S^n, we will mention the related family of "projective spaces". **Projective n-space** $\mathbb{P}^n$ is defined as the collection of all sets $\{p, -p\}$ for $p \in S^n$. The description of the open sets in $\mathbb{P}^n$ is precisely analogous to the description for $\mathbb{P}^2$. Although these spaces seem to form a family as regular as the family S^n, we will see later that the spaces $\mathbb{P}^n$ for even n differ in a very important way from the same spaces for odd n.

One further definition is needed to complete this introduction to manifolds. We have already discussed some spaces which are not manifolds only because they have a "boundary", for example, the Möbius strip and the disc. Points on these "boundaries" do not have neighborhoods homeomorphic to $\mathbb{R}^n$, but they do have neighborhoods homeomorphic to an important subset of $\mathbb{R}^n$. The (**closed**) **half-space** $\mathbb{H}^n$ is defined by

$$\mathbb{H}^n = \{(x^1, \ldots, x^n) \in \mathbb{R}^n : x^n \geq 0\}.$$

A **manifold-with-boundary** is a metric space M with the following property:

> If $x \in M$, then there is some neighborhood U of x and some integer $n \geq 0$ such that U is homeomorphic to either $\mathbb{R}^n$ or $\mathbb{H}^n$.

A point in a manifold-with-boundary cannot have a neighborhood homeomorphic to both $\mathbb{R}^n$ and $\mathbb{H}^n$ (Invariance of Domain again); we can therefore distinguish those points $x \in M$ having a neighborhood homeomorphic to $\mathbb{H}^n$. The set of all such x is called the **boundary** of M and is denoted by ∂M. If M is actually a manifold, then $\partial M = \emptyset$. Notice that if M is a subset of $\mathbb{R}^n$, then ∂M is not necessarily the same as the boundary of M in the old sense (defined for any subset of $\mathbb{R}^n$); indeed, if M is a manifold-with-boundary of dimension $< n$, then all points of M will be boundary points of M.

If manifolds-with-boundary are studied as frequently as manifolds, it becomes bothersome to use this long designation. Often, the word "manifold" is used for "manifold-with-boundary". A manifold in our sense is then called "non-bounded"; a non-bounded compact manifold is called a "closed manifold". We will stick to the other terminology, but will sometimes use "bounded manifold" instead of "manifold-with-boundary".

PROBLEMS

1. Show that if d is a metric on X, then both $\bar{d} = d/(1 + d)$ and $\bar{d} = \min(1, d)$ are also metrics and that they are equivalent to d (i.e., the identity map $1\colon (X, d) \to (X, \bar{d})$ is a homeomorphism).

2. If (X_i, d_i) are metric spaces, for $i \in I$, with metrics $d_i < 1$, and $X_i \cap X_j = \emptyset$ for $i \neq j$, then (X, d) is a metric space, where $X = \bigcup_i X_i$, and $d(x, y) = d_i(x, y)$ if $x, y \in X_i$ for some i, while $d(x, y) = 1$ otherwise. Each X_i is an open subset of X, and Y is homeomorphic to X if and only if $Y = \bigcup_i Y_i$ where the Y_i are disjoint open sets and Y_i is homeomorphic to X_i for each i. The space (X, d) (or any space homeomorphic to it) is called the **disjoint union** of the spaces X_i.

3. (a) Every manifold is locally compact.
(b) Every manifold is locally pathwise connected, and a connected manifold is pathwise connected.
(c) A connected manifold is arcwise connected. (A path is a continuous image of $[0, 1]$, but an arc is a *one-one* continuous image. A difficult theorem states that every path contains an arc between its end points, but a direct proof of arcwise-connectedness can be given for manifolds.)

4. A space X is called locally connected if for each $x \in X$ it is the case that every neighborhood of x contains a connected neighborhood.

(a) Connectedness does not imply local connectedness.
(b) An open subset of a locally connected space is locally connected.
(c) X is locally connected if and only if components of open sets are open, so every neighborhood of a point in a locally connected space contains an *open* connected neighborhood.
(d) A locally connected space is homeomorphic to the disjoint union of its components.
(e) Every manifold is locally connected, and consequently homeomorphic to the disjoint union of its components, which are open submanifolds.

5. (a) The neighborhood U in our definition of a manifold is always open.
(b) The integer n in our definition is unique for each x.

6. (a) A subset of an n-manifold is an n-manifold if and only if it is open.
(b) If M is connected, then the dimension of M at x is the same for all $x \in M$.

7. (a) If $U \subset \mathbb{R}$ is an interval and $f\colon U \to \mathbb{R}$ is continuous and one-one, then f is either increasing or decreasing.

(b) The image $f(U)$ is open.

(c) The map f is a homeomorphism.

8. For this problem, assume

    (1) (The Generalized Jordan Curve Theorem) If $A \subset \mathbb{R}^n$ is homeomorphic to S^{n-1}, then $\mathbb{R}^n - A$ has 2 components, and A is the boundary of each.

    (2) If $B \subset \mathbb{R}^n$ is homeomorphic to $D^n = \{x \in \mathbb{R}^n : d(x, 0) \leq 1\}$, then $\mathbb{R}^n - B$ is connected.

(a) One component of $\mathbb{R}^n - A$ (the "outside of A") is unbounded, and the other (the "inside of A") is bounded.

(b) If $U \subset \mathbb{R}^n$ is open, $A \subset U$ is homeomorphic to S^{n-1} and $f : U \to \mathbb{R}^n$ is one-one and continuous (so that f is a homeomorphism on A), then $f(\text{inside of } A) = \text{inside of } f(A)$. (First prove $\subset$.)

(c) Prove Invariance of Domain.

9. (a) Give an elementary proof that $\mathbb{R}^1$ is not homeomorphic to $\mathbb{R}^n$ for $n > 1$.

(b) Prove directly from the Generalized Jordan Curve Theorem that $\mathbb{R}^m$ is not homeomorphic to $\mathbb{R}^n$ for $m \neq n$.

10. In the proof of Theorem 2, show that the limit of a convergent subsequence of the z_i is actually in the closed ball of radius $r(y)$ and center y.

11. Every connected manifold (which is a metric space) has a countable base for its topology, and a countable dense subset.

12. (a) Compute the composition $f = S^1 - \{(0, 1)\} \overset{P}{\longrightarrow} \mathbb{R}^1 \times \{-1\} \to \mathbb{R}^1$ explicitly for the map P on page 7, and show that it is a homeomorphism.

(b) Do the same for $f : S^{n-1} - \{(0, \ldots, 0, 1)\} \to \mathbb{R}^{n-1}$.

13. (a) The text describes the open subsets of $\mathbb{P}^2$ as sets of the form $f(V)$, where $V \subset S^2$ is open and contains $-p$ whenever it contains p. Show that this last condition is actually unnecessary.

(b) The analogous condition *is* necessary for the Möbius strip, which is discussed immediately afterwards. Explain how the two cases differ.

14. (a) Check that the metric defined for $\mathbb{P}^2$ gives the open sets described in the text.

(b) Check that $\mathbb{P}^2$ is a surface.

15. (a) Show that $\mathbb{P}^1$ is homeomorphic to S^1.

(b) Since we can consider $S^{n-1} \subset S^n$, and since antipodal points in S^{n-1} are still antipodal when considered as points in S^n, we can consider $\mathbb{P}^{n-1} \subset \mathbb{P}^n$ in an obvious way. Show that $\mathbb{P}^n - \mathbb{P}^{n-1}$ is homeomorphic to interior $D^n = \{x \in \mathbb{R}^n : d(x,0) < 1\}$.

16. A classical theorem of topology states that every compact surface other than S^2 is obtained by gluing together a certain number of tori and projective spaces, and that all compact surfaces-with-boundary are obtained from these by cutting out a finite number of discs. To which of these "standard" surfaces are the following homeomorphic?

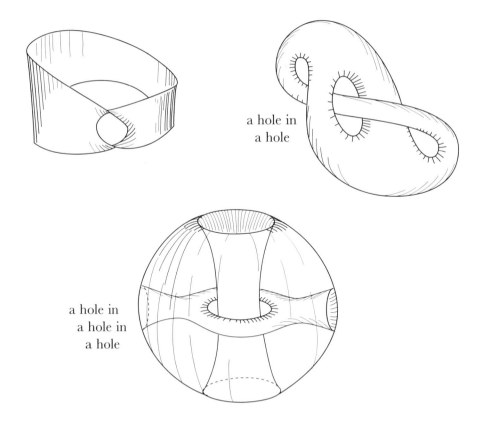

a hole in
a hole

a hole in
a hole in
a hole

17. Let $C \subset \mathbb{R} \subset \mathbb{R}^2$ be the Cantor set. Show that $\mathbb{R}^2 - C$ is homeomorphic to the surface shown at the top of the next page.

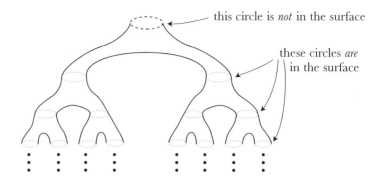

this circle is *not* in the surface

these circles *are* in the surface

18. A locally compact (but non-compact) space X "has one end" if for every compact $C \subset X$ there is a compact K such that $C \subset K \subset X$ and $X - K$ is connected.

(a) $\mathbb{R}^n$ has one end if $n > 1$, but not if $n = 1$.

(b) $\mathbb{R}^n - \{0\}$ does not "have one end" so $\mathbb{R}^n - \{0\}$ is not homeomorphic to $\mathbb{R}^m$.

19. This problem is a sequel to the previous one; it will be used in Problem 24. An **end** of X is a function ε which assigns to each compact subset $C \subset X$ a non-empty component $\varepsilon(C)$ of $X - C$, in such a way that $C_1 \subset C_2$ implies $\varepsilon(C_2) \subset \varepsilon(C_1)$.

(a) If $C \subset \mathbb{R}$ is compact, then $\mathbb{R} - C$ has exactly 2 unbounded components, the "left" component containing all numbers < some N, the "right" one containing all numbers > some N. If ε is an end of $\mathbb{R}$, show that $\varepsilon(C)$ is either always the "left" component of $\mathbb{R} - C$, or always the "right" one. Thus $\mathbb{R}$ has 2 ends.

(b) Show that $\mathbb{R}^n$ has only one end ε for $n > 1$. More generally, X has exactly one end ε if and only if X "has one end" in the sense of Problem 18.

(c) This part requires some knowledge of topological spaces. Let $\mathcal{E}(X)$ be the set of all ends of a connected, locally connected, locally compact Hausdorff space X. Define a topology on $X \cup \mathcal{E}(X)$ by choosing as neighborhoods $N_C(\varepsilon_0)$ of an end ε_0 the sets

$$N_C(\varepsilon_0) = \varepsilon_0(C) \cup \{\text{ends } \varepsilon : \varepsilon(C) = \varepsilon_0(C)\},$$

for all compact C. Show that $X \cup \mathcal{E}(X)$ is a compact Hausdorff space. What is $\mathbb{R} \cup \mathcal{E}(\mathbb{R})$, and $\mathbb{R}^n \cup \mathcal{E}(\mathbb{R}^n)$ for $n > 1$?

20. Consider the following three surfaces.

(A) The infinite-holed torus: ⬡⬡⬡⬡ • • •

(B) The doubly infinite-holed torus: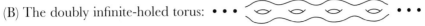

(C) The infinite jail cell window:

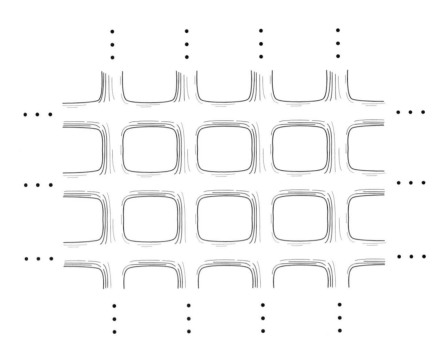

(a) Surfaces (A) and (C) have one end, while surface (B) does not.

(b) Surfaces (A) and (C) are homeomorphic! *Hint*: The region cut out by the lines in the picture below is a cylinder, which occurs at the left of (A). Now draw in two more lines enclosing more holes, and consider the region between the two pairs.

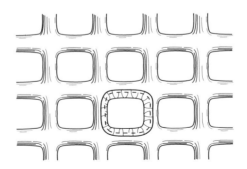

21. (a) The three open subsets of $\mathbb{R}^2$ shown below are homeomorphic.

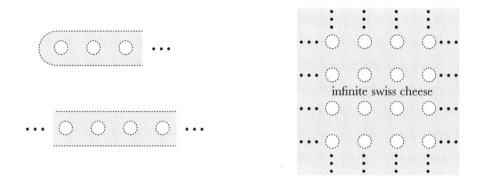

infinite swiss cheese

(b) The points *inside* the three surfaces of Problem 20 are homeomorphic.

22. (a) Every open subset of $\mathbb{R}$ is homeomorphic to the disjoint union of intervals.
(b) There are only countably many non-homeomorphic open subsets of $\mathbb{R}$.

23. For the purposes of this problem we will use a consequence of the Urysohn Metrization Theorem, that for any connected manifold M, there is a homeomorphism f from M to a subset of the countable product $\mathbb{R} \times \mathbb{R} \times \cdots$.

(a) If M is a connected non-compact manifold, then there is a continuous function $f \colon M \to \mathbb{R}$ such that f "goes to ∞ at ∞", i.e., if $\{x_n\}$ is a sequence which is eventually in the complement of every compact set, then $f(x_n) \to \infty$. (Compare with Problem 2-30.)
(b) Given a homeomorphism $f \colon M \to \mathbb{R} \times \mathbb{R} \times \cdots$ and a $g \colon M \to \mathbb{R}$ which goes to ∞ at ∞, define $\bar{f} \colon M \to \mathbb{R} \times (\mathbb{R} \times \mathbb{R} \times \cdots)$ by $\bar{f}(x) = (g(x), f(x))$. Show that $\bar{f}(M)$ is closed.
(c) There are at most $\mathfrak{c}$ non-homeomorphic connected manifolds (where $\mathfrak{c} = 2^{\aleph_0}$ is the cardinality of $\mathbb{R}$).

24. (a) It is possible for $\mathbb{R}^2 - A$ and $\mathbb{R}^2 - B$ to be homeomorphic even though A and B are non-homeomorphic closed subsets.
(b) If $A \subset \mathbb{R}^2$ is closed and totally disconnected (the only components of A are points), then $\mathcal{E}(\mathbb{R}^2 - A)$ is homeomorphic to A. Hence $\mathbb{R}^2 - A$ and $\mathbb{R}^2 - B$ are non-homeomorphic if A and B are non-homeomorphic closed totally disconnected sets.
(c) The **derived set** A' of A is the set of all non-isolated points. We define $A^{(n)}$ inductively by $A^{(1)} = A'$ and $A^{(n+1)} = (A^{(n)})'$. For each n there is a subset A_n of $\mathbb{R}$ such that $A_n{}^{(n)}$ consists of one point.

*(d) There are $\mathfrak{c}$ non-homeomorphic closed totally disconnected subsets of $\mathbb{R}^2$. *Hint*: Let C be the Cantor set, and $c_1 < c_2 < c_3 < \cdots$ a sequence of points in C. For each sequence $n_1 < n_2 < \cdots$, one can add a set A_{n_i} such that its n_i^{th} derived set is $\{c_i\}$.

(e) There are $\mathfrak{c}$ non-homeomorphic connected open subsets of $\mathbb{R}^2$.

25. (a) A manifold-with-boundary could be defined as a metric space M with the property that for each $x \in M$ there is a neighborhood U of x and an integer $n \geq 0$ such that U is homeomorphic to an open subset of $\mathbb{H}^n$.

(b) If M is a manifold-with-boundary, then ∂M is a closed subset of M and ∂M and $M - \partial M$ are manifolds.

(c) If C_i, $i \in I$ are the components of ∂M, and $I' \subset I$, then $M - \bigcup_{i \in I'} C_i$ is a manifold-with-boundary.

26. If $M \subset \mathbb{R}^n$ is a closed set and an n-dimensional manifold-with-boundary, then the topological boundary of M, as a subset of $\mathbb{R}^n$, is ∂M. This is not necessarily true if M is not a closed subset.

27. (a) Every point (a, b, c) on Steiner's surface satisfies $b^2 c^2 + a^2 c^2 + a^2 b^2 = abc$.

(b) If (a, b, c) satisfies this equation and $0 \neq D = \sqrt{b^2 c^2 + a^2 c^2 + a^2 b^2}$, then (a, b, c) is on Steiner's surface. *Hint*: Let $x = bc/D$, etc.

(c) The set $\{(a, b, c) \in \mathbb{R}^3 : b^2 c^2 + a^2 c^2 + a^2 b^2 = abc\}$ is the union of the Steiner surface and of the portions $(-\infty, -1/2)$ and $(1/2, \infty)$ of each axis.

CHAPTER 2

DIFFERENTIABLE STRUCTURES

We are now ready to apply analysis to the study of manifolds. The necessary tools of "advanced calculus", which the reader should bring along freshly sharpened, are contained in Chapters 2 and 3 of *Calculus on Manifolds*. We will use freely the notation and results of these chapters, *including* some problems, notably 2-9, 2-15, 2-25, 2-26, 2-29, 3-32, and 3-35; however, we will denote the identity map from $\mathbb{R}^n$ to $\mathbb{R}^n$ by I, rather than by π (which will be used often enough in other contexts), so that $I^i(x) = x^i$.

On a general manifold M the notion of a continuous function $f\colon M \to \mathbb{R}$ makes sense, but the notion of a differentiable function $f\colon M \to \mathbb{R}$ does not. This is the case despite the fact that M is locally like $\mathbb{R}^n$, where differentiability of functions can be defined. If $U \subset M$ is an open set and we choose a homeomorphism $\phi\colon U \to \mathbb{R}^n$, it would seem reasonable to define f to be differentiable on U if $f \circ \phi^{-1}\colon \mathbb{R}^n \to \mathbb{R}$ is differentiable. Unfortunately, if $\psi\colon V \to \mathbb{R}^n$ is another homeomorphism, and $U \cap V \neq \emptyset$, then it is not necessarily true that $f \circ \psi^{-1}\colon \mathbb{R}^n \to \mathbb{R}$ is also differentiable. Indeed, since

$$f \circ \psi^{-1} = f \circ \phi^{-1} \circ (\phi \circ \psi^{-1}),$$

we can expect $f \circ \psi^{-1}$ to be differentiable for all f which make $f \circ \phi^{-1}$ differentiable only if $\phi \circ \psi^{-1}\colon \mathbb{R}^n \to \mathbb{R}^n$ is differentiable. This is certainly not always

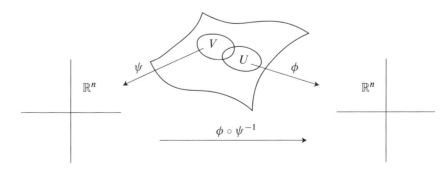

the case; for example, one need merely choose ϕ to be $h \circ \psi$, where $h \colon \mathbb{R}^n \to \mathbb{R}^n$ is a homeomorphism that is not differentiable.

If we insist on defining differentiable functions on any manifold, there is no way out of this impasse. It is necessary to adorn our manifolds with a little additional structure, the precise nature of which is suggested by the previous discussion.

Among all possible homeomorphisms from $U \subset M$ onto $\mathbb{R}^n$, we wish to select a certain collection with the property that $\phi \circ \psi^{-1}$ is differentiable whenever ϕ, ψ are in the collection. This is precisely what we shall do, but a few refinements will be introduced along the way.

First of all, we will be interested almost exclusively in functions $f \colon \mathbb{R}^n \to \mathbb{R}^n$ which are C^∞ (that is, each component function f^i possesses continuous partial derivatives of all orders); sometimes we will use the words "differentiable" or "smooth" to mean C^∞.

Moreover, instead of considering homeomorphisms from open subsets U of M onto $\mathbb{R}^n$, it will suffice to consider homeomorphisms $x \colon U \to x(U) \subset \mathbb{R}^n$ onto open subsets of $\mathbb{R}^n$.

The use of the letters x, y, etc., for these homeomorphisms, henceforth adhered to almost religiously, is meant to encourage the casual confusion of a point $p \in M$ with $x(p) \in \mathbb{R}^n$, which has "coordinates" $x^1(p), \ldots, x^n(p)$. The only time this notation will be confusing (and it will be) is when we are referring to the manifold $\mathbb{R}^n$, where it is hard not to lapse back into the practice of denoting points by x and y. We will often mention the pair (x, U), instead of x alone, just to provide a convenient name for the domain of x.

If U and V are open subsets of M, two homeomorphisms $x \colon U \to x(U) \subset \mathbb{R}^n$ and $y \colon V \to y(V) \subset \mathbb{R}^n$ are called C^∞-**related** if the maps

$$y \circ x^{-1} \colon x(U \cap V) \to y(U \cap V)$$
$$x \circ y^{-1} \colon y(U \cap V) \to x(U \cap V)$$

are C^∞. This make sense, since $x(U \cap V)$ and $y(U \cap V)$ are open subsets of $\mathbb{R}^n$. Also, it makes sense, and is automatically true, if $U \cap V = \emptyset$.

A family of mutually C^∞-related homeomorphisms whose domains cover M is called an **atlas** for M. A particular member (x, U) of an atlas $\mathcal{A}$ is called a **chart** (for the atlas $\mathcal{A}$), or a **coordinate system** on U, for the obvious reason that it provides a way of assigning "coordinates" to points on U, namely, the coordinates $x^1(p), \ldots, x^n(p)$ to the point $p \in U$.

We can even imagine a mesh of coordinate lines on U, by considering the inverse images under x of lines in $\mathbb{R}^n$ parallel to one of the axes.

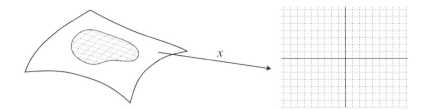

The simplest example of a manifold together with an atlas consists of $\mathbb{R}^n$ with an atlas $\mathcal{A}$ of only one map, the identity $I: \mathbb{R}^n \to \mathbb{R}^n$. We can easily make the atlas bigger; if U and V are homeomorphic open subsets of $\mathbb{R}^n$, we can adjoin any homeomorphism $x: U \to V$ with the property that x and x^{-1} are C^∞. Indeed, we can adjoin as many such x's as we like—it is easy to check that they are all C^∞-related to each other. The advantage of this bigger atlas $\mathcal{U}$ is that the single word "chart", when applied to this atlas, denotes something which must be described in cumbersome language if one can refer only to $\mathcal{A}$. Aside from this, $\mathcal{U}$ differs only superficially from $\mathcal{A}$; one can easily construct $\mathcal{U}$ from $\mathcal{A}$ (and one would be foolish not to do so once and for all). What has just been said for the atlas $\{I\}$ applies to any atlas:

1. LEMMA. If $\mathcal{A}$ is an atlas of C^∞-related charts on M, then $\mathcal{A}$ is contained in a unique maximal atlas $\mathcal{A}'$ for M.

PROOF. Let $\mathcal{A}'$ be the set of all charts y which are C^∞-related to all charts $x \in \mathcal{A}$. It is easy to check that all charts in $\mathcal{A}'$ are C^∞-related, so $\mathcal{A}'$ is an atlas, and it is clearly the unique maximal atlas containing $\mathcal{A}$. ❖

We now define a C^∞ **manifold** (or **differentiable manifold**, or **smooth manifold**) to be a pair $(M, \mathcal{A})$, where $\mathcal{A}$ is a *maximal* atlas for M. Thus, about the simplest example of a C^∞ manifold is $(\mathbb{R}^n, \mathcal{U})$, where $\mathcal{U}$ (the "usual C^∞-structure for $\mathbb{R}^n$") is the maximal atlas containing $\{I\}$. Another example is $(\mathbb{R}, \mathcal{V})$ where $\mathcal{V}$ contains the homeomorphism $x \mapsto x^3$, whose inverse is *not* C^∞, together with all charts C^∞-related to it. Although $(\mathbb{R}, \mathcal{U})$ and $(\mathbb{R}, \mathcal{V})$ are not the same, there is a one-one onto function $f: \mathbb{R} \to \mathbb{R}$ such that

$$x \in \mathcal{U} \text{ if and only if } x \circ f \in \mathcal{V},$$

namely, the obvious map $f(x) = x^3$. Thus $(\mathbb{R}, \mathcal{U})$ and $(\mathbb{R}, \mathcal{V})$ are the sort of structures one would want to call "isomorphic". The term actually used is "diffeomorphic": two C^∞ manifolds $(M, \mathcal{A})$ and $(N, \mathcal{B})$ are **diffeomorphic** if there is a one-one onto function $f: M \to N$ such that ·

$x \in \mathcal{B}$ if and only if $x \circ f \in \mathcal{A}$.

The map f is called a **diffeomorphism**, and f^{-1} is clearly a diffeomorphism also. If we had not required our atlases to be maximal, the definition of diffeomorphism would have had to be more complicated.

Normally, of course, we will suppress mention of the atlas for a differentiable manifold, and speak elliptically of "the differentiable manifold M"; the atlas for M is sometimes referred to as the *differentiable structure* for M. It will always be understood that $\mathbb{R}^n$ refers to the pair $(\mathbb{R}^n, \mathcal{U})$.

It is easy to see that a diffeomorphism must be continuous. Consequently, its inverse must also be continuous, so that a diffeomorphism is automatically a homeomorphism. This raises the natural question whether, conversely, two homeomorphic manifolds are necessarily diffeomorphic. Later (Problem 9-24) we will be able to prove easily that $\mathbb{R}$ with any atlas is diffeomorphic to $(\mathbb{R}, \mathcal{U})$. A proof of the corresponding assertion for $\mathbb{R}^2$ is much harder, the proof for $\mathbb{R}^3$ would certainly be too difficult for inclusion here, and the proof of the essential uniqueness of C^∞ structures on $\mathbb{R}^n$ for $n \geq 5$ requires very difficult techniques from topology.

In the case of spheres, the projections P_1 and P_2 from the points $(0, \ldots, 0, 1)$ and $(0, \ldots, 0, -1)$ of S^{n-1} are easily seen to be C^∞-related. They therefore determine an atlas—the "usual C^∞ structure for S^{n-1}". This atlas may also be described in terms of the $2n$ homeomorphisms

$$f_i: S^{n-1} \cap \{x \in \mathbb{R}^n : x^i > 0\} \to \mathbb{R}^{n-1}$$
$$g_i: S^{n-1} \cap \{x \in \mathbb{R}^n : x^i < 0\} \to \mathbb{R}^{n-1}$$

defined by $f_i(x) = g_i(x) = (x^1, \ldots, x^{i-1}, x^{i+1}, \ldots, x^n)$, which are C^∞-related to P_1 and P_2. There are, up to diffeomorphism, unique differentiable structures on S^n for $n \leq 6$. But there are 28 diffeomorphism classes of differentiable structures on S^7, and over 16 million on S^{31}. However, we shall not come close to proving these assertions, which are part of the field called "differential topology", rather then differential geometry. (Perhaps most astonishing of all is the quite recent discovery that $\mathbb{R}^4$ has a differentiable structure that is not diffeomorphic to the usual differentiable structure!)

Other examples of differentiable manifolds will be given soon, but we can already describe a differentiable structure $\mathcal{A}'$ on any open submanifold N of a differentiable manifold $(M, \mathcal{A})$; the atlas $\mathcal{A}'$ consists of all (x, U) in $\mathcal{A}$ with $U \subset N$.

Just as diffeomorphisms are analogues for C^{∞} manifolds of homeomorphisms, there are analogues of continuous maps. A function $f: M \to N$ is called **differentiable** if for every coordinate system (x, U) for M and (y, V) for N, the map $y \circ f \circ x^{-1}: \mathbb{R}^n \to \mathbb{R}^m$ is differentiable. More particularly, f

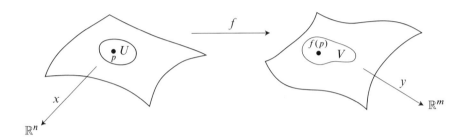

is called **differentiable at** $p \in M$ if $y \circ f \circ x^{-1}$ is differentiable at $x(p)$ for coordinate systems (x, U) and (y, V) with $p \in U$ and $f(p) \in V$. If this is true for one pair of coordinate systems, it is easily seen to be true for any other pair. We can thus define differentiability of f on any open subset $M' \subset M$; as one would suspect, this coincides with differentiability of the restricted map $f|M': M' \to N$. Clearly, a differentiable map is continuous.

A differentiable function $f: M \to \mathbb{R}$ refers, of course, to the usual differentiable structure on $\mathbb{R}$, and hence f is differentiable if and only if $f \circ x^{-1}$ is differentiable for each chart x. It is easy to see that

(1) a function $f: \mathbb{R}^n \to \mathbb{R}^m$ is differentiable as a map between C^{∞} manifolds if and only if it is differentiable in the usual sense;

(2) a function $f: M \to \mathbb{R}^m$ is differentiable if and only if each $f^i: M \to \mathbb{R}^m$ is differentiable;

(3) a coordinate system (x, U) is a diffeomorphism from U to $x(U)$;

(4) a function $f: M \to N$ is differentiable if and only if each $y^i \circ f$ is differentiable for each coordinate system y of N;

(5) a differentiable function $f: M \to N$ is a diffeomorphism if and only if f is one-one onto and $f^{-1}: N \to M$ is differentiable.

The differentiable structures on many manifolds are designed to make certain functions differentiable. Consider first the product $M_1 \times M_2$ of two differentiable manifolds M_i, and the two "projections" $\pi_i \colon M_1 \times M_2 \to M_i$ defined by $\pi_i(p_1, p_2) = p_i$. It is easy to define a differentiable structure on $M_1 \times M_2$ which makes each π_i differentiable. For each pair (x_i, U_i) of coordinate systems on M_i, we construct the homeomorphism

$$x_1 \times x_2 \colon U_1 \times U_2 \to \mathbb{R}^{n_1 + n_2}$$

defined by

$$x_1 \times x_2(p_1, p_2) = (x_1(p_1), x_2(p_2)), \quad \text{i.e.,} \quad x_1 \times x_2 = (x_1 \circ \pi_1, x_2 \circ \pi_2).$$

Then we extend this atlas to a maximal one.

Similarly, there is a differentiable structure on $\mathbb{P}^n$ which makes the map $f \colon S^n \to \mathbb{P}^n$ (defined by $f(p) = [p] = \{p, -p\}$) differentiable. Consider any coordinate system (x, U) for S^n, where U does *not* contain $-p$ if it contains p, so that $f|U$ is one-one. The map $x \circ (f|U)^{-1}$ is a homeomorphism on $f(U) \subset \mathbb{P}^n$, and any two such are C^∞-related. The collection of these homeomorphisms can then be extended to a maximal atlas.

To obtain differentiable structures on other surfaces, we first note that a C^∞ manifold-with-boundary can be defined in an obvious way. It is only necessary to know when a map $f \colon \mathbb{H}^n \to \mathbb{R}^n$ is to be considered differentiable; we call f **differentiable** when it can be extended to a differentiable function on an open neighborhood of $\mathbb{H}^n$. A "handle" is then a C^∞ manifold-with-boundary.

A differentiable structure on the 2-holed torus can be obtained by "matching" the differentiable structure on two handles. The details involved in this process are reserved for Problem 14.

To deal with C^∞ functions effectively, one needs to know that there are lots of them. The existence of C^∞ functions on a manifold depends on the existence of C^∞ functions on $\mathbb{R}^n$ which are 0 outside of a compact set. We briefly recall here the necessary facts about such C^∞ functions (c.f. *Calculus on Manifolds*, pg. 29).

(1) The function $h: \mathbb{R} \to \mathbb{R}$ defined by

$$h(x) = \begin{cases} e^{-1/x^2} & x \neq 0 \\ 0 & x = 0 \end{cases}$$

is C^∞, and $h^{(n)}(0) = 0$ for all n.

(2) The function $j: \mathbb{R} \to \mathbb{R}$ defined by

$$j(x) = \begin{cases} e^{-(x-1)^{-2}} \cdot e^{-(x+1)^{-2}} & x \in (-1, 1) \\ 0 & x \notin (-1, 1) \end{cases}$$

is C^∞.

Similarly, there is a C^∞ function $k: \mathbb{R} \to \mathbb{R}$ which is positive on $(0, \delta)$ and 0 elsewhere.

(3) The function $l: \mathbb{R} \to \mathbb{R}$ defined by

$$l(x) = \left(\int_0^x k \right) \Big/ \left(\int_0^\delta k \right)$$

is C^∞; it is 0 for $x \leq 0$, increasing on $(0, \delta)$, and 1 for $x \geq \delta$.

(4) The function $g: \mathbb{R}^n \to \mathbb{R}$ defined by

$$g(x) = j(x^1/\varepsilon) \cdots j(x^n/\varepsilon)$$

is C^∞; it is positive on $(-\varepsilon, \varepsilon) \times \cdots \times (-\varepsilon, \varepsilon)$ and 0 elsewhere.

On a C^∞ manifold M we can now produce many non-constant C^∞ functions. The closure $\overline{\{x : f(x) \neq 0\}}$ is called the **support** of f, and denoted simply by support f (or sometimes supp f).

2. LEMMA. Let $C \subset U \subset M$ with C compact and U open. Then there is a C^∞ function $f: M \to [0, 1]$ such that $f = 1$ on C and support $f \subset U$. (Compare *Case 2* of the proof of Theorem 15.)

PROOF. For each $p \in C$, choose a coordinate system (x, V) with $\overline{V} \subset U$ and $x(p) = 0$. Then $x(V) \supset (-\varepsilon, \varepsilon) \times \cdots \times (-\varepsilon, \varepsilon)$ for some $\varepsilon > 0$. The function $g \circ x$ (where g is defined in (4)) is C^∞ on V. Clearly it remains C^∞ if we extend

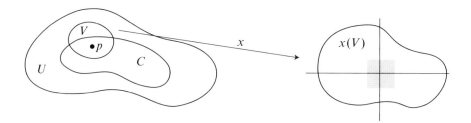

it to be 0 outside of V. Let f_p be the extended function. The function f_p can be constructed for each p, and is positive on a neighborhood of p whose closure is contained in U. Since C is compact, finitely many such neighborhoods cover C, and the sum, $f_{p_1} + \cdots + f_{p_n}$, of the corresponding functions has support $\subset U$. On C it is positive, so on C it is $\geq \delta$ for some $\delta > 0$. Let $f = l \circ (f_{p_1} + \cdots + f_{p_n})$, where l is defined in (3). ❖

By the way, we could have defined C^r manifolds for each $r \geq 1$, not just for "$r = \infty$". (A function $f \colon \mathbb{R}^n \to \mathbb{R}$ is C^r if it has continuous partial derivatives up to order r). A "C^0 function" is just a continuous function, so a C^0 manifold is just a manifold in the sense of Chapter 1. We can also define *analytic* manifolds (a function $f \colon \mathbb{R}^n \to \mathbb{R}$ is analytic at $a \in \mathbb{R}^n$ if f can be expressed as a power series in the $(x^i - a^i)$ which converges in some neighborhood of a). The symbol C^ω stands for analytic, and it is convenient to agree that $r < \infty < \omega$ for each integer $r \geq 0$. If $\alpha < \beta$, then the charts of a maximal C^β atlas are all C^α-related, but this atlas can always be extended to a *bigger* atlas of C^α-related charts, as in Lemma 1. Thus, a C^β structure on M can always be extended to a C^α structure in a unique way; the smaller structure is the "stronger" one, the C^0 structure (consisting of all homeomorphisms $x \colon U \to \mathbb{R}^n$) being the largest. The converse of this trivial remark is a hard theorem: For $\alpha \geq 1$, every C^α structure contains a C^β structure for each $\beta > \alpha$; it is not unique, of course, but it is unique up to diffeomorphism. This will not be proved here.* In fact, C^α manifolds for $\alpha \neq \infty$ will hardly ever be mentioned again. One remark is in order now; the proof of Lemma 2 produces an appropriate C^α function f on a C^α manifold, for $0 \leq \alpha \leq \infty$. Of course, for $\alpha = \omega$ the proof

---

* For a proof see Munkres, *Elementary Differential Topology*.

fails completely (and the result is false—an analytic function which is 0 on an open set is 0 everywhere).

With differentiable functions now at our disposal, it is fitting that we begin differentiating them. What we shall define are the partial derivatives of a differentiable function $f: M \to \mathbb{R}$, with respect to a coordinate system (x, U). At this point classical notation for partial derivatives is systematically introduced, so it is worth recalling a logical notation for the partial derivatives of a function $f: \mathbb{R}^n \to \mathbb{R}$. We denote by $D_i f(a)$ the number

$$\lim_{h \to 0} \frac{f(a^1, \ldots, a^i + h, \ldots, a^n) - f(a)}{h}.$$

The Chain Rule states that if $g: \mathbb{R}^m \to \mathbb{R}^n$ and $f: \mathbb{R}^n \to \mathbb{R}$, then

$$D_j(f \circ g)(a) = \sum_{i=1}^{n} D_i f(g(a)) \cdot D_j g^i(a).$$

Now, for a function $f: M \to \mathbb{R}$ and a coordinate system (x, U) we define

$$\frac{\partial f}{\partial x^i}(p) = \left. \frac{\partial f}{\partial x^i} \right|_p = D_i(f \circ x^{-1})(x(p)),$$

(or simply $\dfrac{\partial f}{\partial x^i} = D_i(f \circ x^{-1}) \circ x$, as an equation between functions). If we define the curve $c_i: (-\varepsilon, \varepsilon) \to M$ by

$$c_i(h) = x^{-1}(x(p) + (0, \ldots, h, \ldots, 0)),$$

then this partial derivative is just

$$\lim_{h \to 0} \frac{f(c_i(h)) - f(p)}{h},$$

so it measures the rate change of f along the curve c_i; in fact it is just $(f \circ c_i)'(0)$. Notice that

$$\frac{\partial x^i}{\partial x^j}(p) = \delta_j^i = \begin{cases} 1 & \text{if } i = j \\ 0 & \text{if } i \neq j. \end{cases}$$

If x happens to be the identity map of $\mathbb{R}^n$, then $D_i f(p) = \partial f / \partial x^i(p)$, which is the classical symbol for this partial derivative.

Another classical instance of this notation, often not completely clarified, is the use of the symbols $\partial/\partial r$ and $\partial/\partial\theta$ in connection with "polar coordinates". On the subset A of $\mathbb{R}^2$ defined by

$$A = \mathbb{R}^2 - \{(x, y) \in \mathbb{R}^2 : y = 0 \text{ and } x \geq 0\}$$
$$= \mathbb{R}^2 - L$$

we can introduce a "coordinate system" $P \colon A \to \mathbb{R}^2$ by

$$P(x, y) = (r(x, y), \theta(x, y)),$$

where $r(x, y) = \sqrt{x^2 + y^2}$ and $\theta(x, y)$ is the unique number in $(0, 2\pi)$ with

$$x = r(x, y) \cos\theta(x, y)$$
$$y = r(x, y) \sin\theta(x, y).$$

This really is a coordinate system on A in our sense, with its image being the set $\{r : r > 0\} \times (0, 2\pi)$. (Of course, the polar coordinate system is often

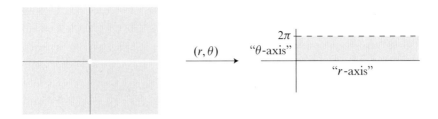

not restricted to the set A. One can delete any ray other than L if $\theta(x, y)$ is restricted to lie in the appropriate interval $(\theta_0, \theta_0 + 2\pi)$; many results are essentially independent of which line is deleted, and this sometimes justifies the sloppiness involved in the definition of the polar coordinate system.)

We have really defined P as an inverse function, whose inverse P^{-1} is defined simply by

$$P^{-1}(r, \theta) = (r\cos\theta, r\sin\theta).$$

From this formula we can compute $\partial f/\partial r$ explicitly:

$$(f \circ P^{-1})(r, \theta) = f(r \cos \theta, r \sin \theta),$$

so

$$\frac{\partial f}{\partial r}(x, y) = D_1(f \circ P^{-1})(P(x, y))$$

$$= D_1 f(P^{-1}(P(x, y)) \cdot D_1[P^{-1}]^1(P(x, y))$$
$$+ D_2 f(P^{-1}(P(x, y)) \cdot D_1[P^{-1}]^2(P(x, y))$$
by the Chain Rule
$$= D_1 f(x, y) \cdot \cos \theta(x, y) + D_2 f(x, y) \cdot \sin \theta(x, y).$$

This formula just gives the value of the directional derivative of f at (x, y), along a unit vector $v = (\cos \theta(x, y), \sin \theta(x, y))$ pointing outwards from the origin to (x, y). This is to be expected, because c_1, the inverse image under P

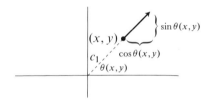

of a curve along the "r-axis", is just a line in this direction.

A similar computation gives

$$\frac{\partial f}{\partial \theta}(x, y) = D_1 f(x, y)[-r(x, y) \sin \theta(x, y)] + D_2 f(x, y)[r(x, y) \cos \theta(x, y)].$$

The vector $w = (-\sin \theta(x, y), \cos \theta(x, y))$ is perpendicular to v, and thus the direction, at the point (x, y), of the curve c_2 which is the inverse image under P of a curve along the "θ-axis". The factor $r(x, y)$ appears because this curve

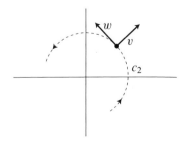

goes around a circle of that radius as θ goes from 0 to 2π, so it is going $r(x, y)$ times as fast as it should go in order to be used to compute the directional derivative of f in the direction w. Note that $\partial f/\partial\theta$ is independent of which line is deleted from the plane in order to define the function θ unambiguously.

Using the notation $\partial f/\partial x$ for $D_1 f$, etc., and suppressing the argument (x, y) everywhere (thus writing an equation about functions), we can write the above equations as

$$\frac{\partial f}{\partial r} = \frac{\partial f}{\partial x}\cos\theta + \frac{\partial f}{\partial y}\sin\theta$$

$$\frac{\partial f}{\partial \theta} = \frac{\partial f}{\partial x}(-r\sin\theta) + \frac{\partial f}{\partial y}r\cos\theta.$$

In particular, these formulas also tell us what $\partial x/\partial r$ etc., are, where (x, y) denotes the identity coordinate system of $\mathbb{R}^2$. We have $\partial x/\partial r = \cos\theta$, etc., so our formulas can be put in the form

$$\frac{\partial f}{\partial r} = \frac{\partial f}{\partial x}\frac{\partial x}{\partial r} + \frac{\partial f}{\partial y}\frac{\partial y}{\partial r}$$

$$\frac{\partial f}{\partial \theta} = \frac{\partial f}{\partial x}\frac{\partial x}{\partial \theta} + \frac{\partial f}{\partial y}\frac{\partial y}{\partial \theta}.$$

In classical notation, the Chain Rule would always be written in this way. It is a pleasure to report that henceforth this may always be done:

3. PROPOSITION. If (x, U) and (y, V) are coordinate systems on M, and $f : M \to \mathbb{R}$ is differentiable, then on $U \cap V$ we have

(1)
$$\frac{\partial f}{\partial y^i} = \sum_{j=1}^{n} \frac{\partial f}{\partial x^j}\frac{\partial x^j}{\partial y^i}.$$

PROOF. It's the Chain Rule, of course, if you just keep your cool:

$$\frac{\partial f}{\partial y^i}(p) = D_i(f \circ y^{-1})(y(p))$$

$$= D_i([f \circ x^{-1}] \circ [x \circ y^{-1}])(y(p))$$

$$= \sum_{j=1}^{n} D_j(f \circ x^{-1})([x \circ y^{-1}](y(p))) \cdot D_i[x \circ y^{-1}]^j(y(p))$$

$$= \sum_{j=1}^{n} D_j(f \circ x^{-1})(x(p)) \cdot D_i[x^j \circ y^{-1}](y(p))$$

$$= \sum_{j=1}^{n} \frac{\partial f}{\partial x^j}(p) \cdot \frac{\partial x^j}{\partial y^i}(p). \ \clubsuit$$

At this point we could introduce the "Einstein summation convention". Notice that the summation in this formula occurs for the index j, which appears both "above" (in $\partial x^j / \partial y^i$) and "below" (in $\partial f / \partial x^j$). There are scads of formulas in which this happens, often with hoards of indices being summed over, and the convention is to omit the $\sum$ sign completely—double indices (which by luck, the nature of things, and felicitous choice of notation, almost always occur above and below) being summed over. I won't use this notation because whenever I do, I soon forget I'm supposed to be summing, and because by doing things "right", one can avoid what Élie Cartan has called the "debauch of indices".

We will often write formula (1) in the form

$$\frac{\partial}{\partial y^i} = \sum_{j=1}^{n} \frac{\partial x^j}{\partial y^i} \frac{\partial}{\partial x^j};$$

here $\partial / \partial y^i$ is considered as an operator taking the function f to $\partial f / \partial y^i$. The operator taking f to $\partial f / \partial y^i (p)$ is denoted by

$$\frac{\partial}{\partial y^i}\bigg|_p ; \quad \text{thus} \quad \frac{\partial}{\partial y^i}\bigg|_p = \sum_{j=1}^{n} \frac{\partial x^j}{\partial y^i}(p) \frac{\partial}{\partial x^j}\bigg|_p .$$

For later use we record a property of $\ell = \partial / \partial x^i|_p$: it is a "point-derivation".

4. PROPOSITION. For any differentiable $f, g : M \to \mathbb{R}$, and any coordinate system (x, U) with $p \in U$, the operator $\ell = \partial / \partial x^i|_p$ satisfies

$$\ell(fg) = f(p)\ell(g) + \ell(f)g(p).$$

PROOF. Left to the reader. $\clubsuit$

If (x, U) and (x', U') are two coordinate systems on M, the $n \times n$ matrix

$$\left(\frac{\partial x'^i}{\partial x^j}(p) \right)$$

is just the Jacobian matrix of $x' \circ x^{-1}$ at $x(p)$. It is non-singular; in fact, its inverse is clearly

$$\left(\frac{\partial x^i}{\partial x'^j}(p)\right).$$

Now if $f \colon M^n \to N^m$ is C^∞ and (y, V) is a coordinate system around $f(p)$, the rank of the $m \times n$ matrix

$$\left(\frac{\partial(y^i \circ f)}{\partial x^j}(p)\right)$$

clearly does not depend on the coordinate system (x, U) or (y, V). It is called the **rank of** f **at** p. The point p is called a **critical point** of f if the rank of f at p is $< m$ (the dimension of the image N); if p is not a critical point of f, it is called a **regular point** of f. If p is a critical point of f, the value $f(p)$ is called a **critical value** of f. Other points in N are **regular values**; thus $q \in N$ is a regular value if and only if p is a regular point of f for every $p \in f^{-1}(q)$. This is true, in particular, if $q \notin f(M)$—a non-value of f is still a "regular value".

If $f \colon \mathbb{R} \to \mathbb{R}$, then x is a critical point of f if and only if $f'(x) = 0$. It is possible for all points of the interval $[a, b]$ to be critical points, although this can happen only if f is constant on $[a, b]$. If $f \colon \mathbb{R}^2 \to \mathbb{R}$ has all points as critical values, then $D_1 f = D_2 f = 0$ everywhere, so f is again constant. On the other hand, a function $f \colon \mathbb{R}^2 \to \mathbb{R}^2$ may have all points as critical points without being constant, for example, $f(x, y) = x$. In this case, however, the image $f(\mathbb{R}^2) = \mathbb{R} \times \{0\} \subset \mathbb{R}^2$ is still a "small" subset of $\mathbb{R}^2$. The most important theorem about critical points generalizes this fact. To state it, we will need some terminology.

Recall that a set $A \subset \mathbb{R}^n$ has "measure zero" if for every $\varepsilon > 0$ there is a sequence $B_1, B_2, B_3, \ldots$ of (closed or open) rectangles with

$$A \subset \bigcup_{n=1}^{\infty} B_n$$

and

$$\sum_{n=1}^{\infty} v(B_n) < \varepsilon,$$

where $v(B_n)$ is the volume of B_n. We want to define the same concept for a subset of a manifold. To do this we need a lemma, which in turn depends on a lemma from *Calculus on Manifolds*, which we merely state.

5. LEMMA. Let $A \subset \mathbb{R}^n$ be a rectangle and let $f \colon A \to \mathbb{R}^n$ be a function such that $|D_j f^i| \leq K$ on A for $i, j = 1, \ldots, n$. Then

$$|f(x) - f(y)| \leq n^2 K |x - y|$$

for all $x, y \in A$.

6. LEMMA. If $f \colon \mathbb{R}^n \to \mathbb{R}^n$ is C^1 and $A \subset \mathbb{R}^n$ has measure 0, then $f(A)$ has measure 0.

PROOF. We can assume that A is contained in a compact set C (since $\mathbb{R}^n$ is a countable union of compact sets). Lemma 5 implies that there is some K such that

$$|f(x) - f(y)| \leq n^2 K |x - y|$$

for all $x, y \in C$. Thus f takes rectangles of diameter d into sets of diameter $\leq n^2 K d$. This clearly implies that $f(A)$ has measure 0 if A does. ❖

A subset A of a C^∞ n-manifold M has **measure zero** if there is a sequence of charts (x_i, U_i), with $A \subset \bigcup_i U_i$, such that each set $x_i(A \cap U_i) \subset \mathbb{R}^n$ has measure 0. Using Lemma 6, it is easy to see that if $A \subset M$ has measure 0, then $x(A \cap U) \subset \mathbb{R}^n$ has measure 0 for any coordinate system (x, U). Conversely, if this condition is satisfied and M is connected, or has only countably many components, then it follows easily from Theorem 1-2 that A has measure 0. (But if M is the disjoint union of uncountably many copies of $\mathbb{R}$, and A consists of one point from each component, then A does not have measure 0). Lemma 6 thus implies another result:

7. COROLLARY. If $f \colon M \to N$ is a C^1 function between two n-manifolds and $A \subset M$ has measure 0, then $f(A) \subset N$ has measure 0.

PROOF. There is a sequence of charts (x_i, U_i) with $A \subset \bigcup_i U_i$ and each set $x_i(A \cap U_i)$ of measure 0. If (y, V) is a chart on N, then $f(A) \cap V = \bigcup_i f(A \cap U_i) \cap V$. Each set

$$y(f(A \cap U_i) \cap V) = y \circ f \circ x^{-1}(x(A \cap U_i))$$

has measure 0, by Lemma 6. Thus $y(f(A) \cap V)$ has measure 0. Since $f(\bigcup_i U_i)$ is contained in the union of at most countably many components of N, it follows that $f(A)$ has measure 0. ❖

8. THEOREM (SARD'S THEOREM). If $f : M \to N$ is a C^1 map between n-manifolds, and M has at most countably many components, then the critical values of f form a set of measure 0 in N.

PROOF. It clearly suffices to consider the case where M and N are $\mathbb{R}^n$. But this case is just Theorem 3.14 of *Calculus on Manifolds.* ❖

The stronger version of Sard's Theorem, which we will never use (except once, in Problem 8-24), states* that the critical values of a C^k map $f : M^n \to N^m$ are a set of measure 0 if $k \geq 1 + \max(n - m, 0)$. Theorem 8 is the easy case, and the case $m > n$ is the trivial case (Problem 20). Although Theorem 8 will be very important later on, for the present we are more interested in knowing what the image of $f : M \to N$ looks like locally, in terms of the rank k of f at $p \in M$. More exact information can be given when f actually has rank k in a neighborhood of p. It should be noted that f must have rank $\geq k$ in some neighborhood of p, because some $k \times k$ submatrix of $(\partial(y^i \circ f)/\partial x^i)$ has non-zero determinant at p, and hence in a neighborhood of p.

9. THEOREM. (1) If $f : M^n \to N^m$ has rank k at p, then there is some coordinate system (x, U) around p and some coordinate system (y, V) around $f(p)$ with $y \circ f \circ x^{-1}$ in the form

$$y \circ f \circ x^{-1}(a^1, \ldots, a^n) = (a^1, \ldots, a^k, \psi^{k+1}(a), \ldots, \psi^m(a)).$$

Moreover, given any coordinate system y, the appropriate coordinate system on N can be obtained merely by permuting the component functions of y.

(2) If f has rank k in a neighborhood of p, then there are coordinate systems (x, U) and (y, V) such that

$$y \circ f \circ x^{-1}(a^1, \ldots, a^n) = (a^1, \ldots, a^k, 0, \ldots, 0).$$

Remark: The special case $M = \mathbb{R}^n$, $N = \mathbb{R}^m$ is equivalent to the general theorem, which gives only local results. If y is the identity of $\mathbb{R}^m$, part (1) says that by first performing a diffeomorphism on $\mathbb{R}^n$, and then permuting the coordinates in $\mathbb{R}^m$, we can insure that f keeps the first k components of a point fixed. These diffeomorphisms on $\mathbb{R}^n$ and $\mathbb{R}^m$ are clearly necessary, since f may not even be one-one on $\mathbb{R}^k \times \{0\} \subset \mathbb{R}^n$, and its image could, for example, contain only points with first coordinate 0.

---

*For a proof, see Milnor, *Topology From the Differentiable Viewpoint* or Sternberg, *Lectures on Differential Geometry.*

In part (2) we must clearly allow more leeway in the choice of y, since $f(\mathbb{R}^n)$ may not be contained in any k-dimensional subspace of $\mathbb{R}^n$.

PROOF. (1) Choose some coordinate system u around p. By a permutation of the coordinate functions u^i and y^i we can arrange that

$$(1) \qquad \det\left(\frac{\partial(y^\alpha \circ f)}{\partial u^\beta}(p)\right) \neq 0 \qquad \alpha, \beta = 1, \ldots, k.$$

Define

$$\begin{aligned} x^\alpha &= y^\alpha \circ f & \alpha &= 1, \ldots, k \\ x^r &= u^r & r &= k+1, \ldots, n. \end{aligned}$$

Condition (1) implies that

$$(2) \qquad \det\left(\frac{\partial x^i}{\partial u^j}(p)\right) = \det\left(\begin{array}{c|c} \boxed{\dfrac{\partial(y^\alpha \circ f)}{\partial u^\beta}} & \diagdown \\ \hline \mathbf{0} & \begin{smallmatrix} 1 \\ & \ddots \\ & & 1 \end{smallmatrix} \end{array}\right) \neq 0.$$

This shows that $x = (x \circ u^{-1}) \circ u$ is a coordinate system in some neighborhood of p, since (2) and the Inverse Function Theorem show that $x \circ u^{-1}$ is a diffeomorphism in a neighborhood of $u(p)$. Now

$$q = x^{-1}(a^1, \ldots, a^n) \quad \text{means} \quad x(q) = (a^1, \ldots, a^n),$$
$$\text{hence} \quad x^i(q) = a^i,$$
$$\text{hence} \quad \begin{cases} y^\alpha \circ f(q) = a^\alpha & \alpha = 1, \ldots, k \\ u^r(q) = a^r & r = k+1, \ldots, n, \end{cases}$$

so

$$y \circ f \circ x^{-1}(a^1, \ldots, a^n) = y \circ f(q) \qquad \text{for } q = x^{-1}(a^1, \ldots, a^n)$$
$$= (a^1, \ldots, a^k, \underline{\quad}).$$

(2) Choose coordinate systems x and v so that $v \circ f \circ x^{-1}$ has the form in (1). Since rank $f = k$ in a neighborhood of p, the lower square in the matrix

$$\left(\frac{\partial(v^i \circ f)}{\partial x^j}\right) = \left(\begin{array}{c|c} \begin{smallmatrix} 1 \\ & \ddots \\ & & 1 \end{smallmatrix} & \mathbf{0} \\ \hline \diagup & \begin{smallmatrix} D_{k+1}\psi^{k+1} \\ & \ddots \\ & & D_m\psi^m \end{smallmatrix} \end{array}\right)$$

must vanish in a neighborhood of p. Thus we can write

$$\psi^r(a) = \bar{\psi}^r(a^1, \ldots, a^k) \qquad r = k+1, \ldots, m.$$

Define

$$y^\alpha = v^\alpha$$
$$y^r = v^r - \bar{\psi}^r \circ (v^1, \ldots, v^k).$$

Since

(3) $\quad y \circ v^{-1}(b^1, \ldots, b^m) = y(q) \qquad$ for $v(q) = (b^1, \ldots, b^m)$

$\qquad = (b^1, \ldots, b^k, \; b^{k+1} - \bar{\psi}^{k+1}(b^1, \ldots, b^k), \; \ldots, \; b^m - \bar{\psi}^m(b^1, \ldots, b^k)),$

the Jacobian matrix

$$\left(\frac{\partial y^i}{\partial v^j}\right) = \begin{pmatrix} \boxed{\begin{smallmatrix} 1 & & \\ & \ddots & \\ & & 1 \end{smallmatrix}} & \text{\Large 0} \\ \text{\Large X} & \boxed{\begin{smallmatrix} 1 & & \\ & \ddots & \\ & & 1 \end{smallmatrix}} \end{pmatrix}$$

has non-zero determinant, so y is a coordinate system in a neighborhood of $f(p)$. Moreover,

$y \circ f \circ x^{-1}(a^1, \ldots, a^n)$

$\qquad = y \circ v^{-1} \circ v \circ f \circ x^{-1}(a^1, \ldots, a^n)$

$\qquad = y \circ v^{-1}(a^1, \ldots, a^k, \psi^{k+1}(a), \ldots, \psi^m(a))$

$= (a^1, \ldots, a^k, \; \psi^{k+1}(a) - \bar{\psi}^{k+1}(a^1, \ldots, a^k), \; \ldots, \; \psi^m(a) - \bar{\psi}^m(a^1, \ldots, a^k))$

$\qquad\qquad$ by (3)

$= (a^1, \ldots, a^k, 0, \ldots, 0).$ ❖

Theorem 9 acquires a special form when the rank of f is n or m:

10. THEOREM. (1) If $m \leq n$ and $f: M^n \to N^m$ has rank m at p, then for any coordinate system (y, V) around $f(p)$, there is some coordinate system (x, U) around p with

$$y \circ f \circ x^{-1}(a^1, \ldots, a^n) = (a^1, \ldots, a^m).$$

(2) If $n \leq m$ and $f: M^n \to N^m$ has rank n at p, then for any coordinate system (x, U) around p, there is a coordinate system (y, V) around $f(p)$ with
$$y \circ f \circ x^{-1}(a^1, \ldots, a^n) = (a^1, \ldots, a^n, 0, \ldots, 0).$$

PROOF. (1) This is practically a special case of (1) in Theorem 9; it is only necessary to observe that when $k = m$, it is clearly unnecessary, in the proof of this case, to permute the y^i in order to arrange that
$$\det\left(\frac{\partial(y^\alpha \circ f)}{\partial u^\beta}(p)\right) \neq 0 \qquad \alpha, \beta = 1, \ldots, m;$$
only the u^i need be permuted.

(2) Since the rank of f at any point must be $\leq n$, the rank of f equals n in some neighborhood of p. It is convenient to think of the case $M = \mathbb{R}^n$ and $N = \mathbb{R}^m$ and produce the coordinate system y for $\mathbb{R}^m$ when we are given the identity coordinate system for $\mathbb{R}^n$. Part (2) of Theorem 9 yields coordinate systems ϕ for $\mathbb{R}^n$ and ψ for $\mathbb{R}^m$ such that
$$\psi \circ f \circ \phi^{-1}(a^1, \ldots, a^n) = (a^1, \ldots, a^n, 0, \ldots, 0).$$
Even if we do not perform ϕ^{-1} first, the map f still takes $\mathbb{R}^n$ into the subset

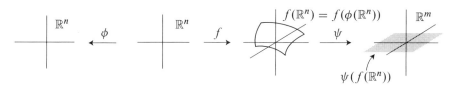

$f(\mathbb{R}^n)$ which ψ takes to $\mathbb{R}^n \times \{0\} \subset \mathbb{R}^m$—the points of $\mathbb{R}^n$ just get moved to the wrong place in $\mathbb{R}^n \times \{0\}$. This can be corrected by another map on $\mathbb{R}^m$. Define λ by
$$\lambda(b^1, \ldots, b^m) = (\phi^{-1}(b^1, \ldots, b^n), b^{n+1}, \ldots, b^m).$$
Then
$$\lambda \circ \psi \circ f(a^1, \ldots, a^n) = \lambda \circ \psi \circ f \circ \phi^{-1}(b^1, \ldots, b^n)$$
$$\text{for } (b^1, \ldots, b^n) = \phi(a)$$
$$= \lambda(b^1, \ldots, b^n, 0, \ldots, 0)$$
$$= (\phi^{-1}(b^1, \ldots, b^n), 0, \ldots, 0)$$
$$= (a^1, \ldots, a^n, 0, \ldots, 0),$$
so $\lambda \circ \psi$ is the desired y. If we are given a coordinate system x on $\mathbb{R}^n$ other than the identify, we just define
$$\lambda(b^1, \ldots, b^m) = (x(\phi^{-1}(b^1, \ldots, b^n)), b^{n+1}, \ldots, b^m);$$
it is easily checked that $y = \lambda \circ \psi$ is now the desired y. ❖

Although p is a regular point of f in case (1) of Theorem 10 and a critical point in case (2) (if $n < m$), it is case (2) which most interests us. A differentiable function $f\colon M^n \to N^m$ is called an **immersion** if the rank of f is n, the dimension of the domain M, at all points of M. Of course, it is necessary that $m \geq n$, and it is clear from Theorem 10(2) that an immersion is locally one-one (so it is a topological immersion, as defined in Chapter 1). On the other hand, a differentiable map f need not be an immersion even if it is globally one-one. The simplest example is the function $f\colon \mathbb{R} \to \mathbb{R}$ defined by $f(x) = x^3$, with $f'(0) = 0$. Another example is

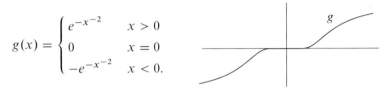

$$g(x) = \begin{cases} e^{-x^{-2}} & x > 0 \\ 0 & x = 0 \\ -e^{-x^{-2}} & x < 0. \end{cases}$$

A more illuminating example is the function $h\colon \mathbb{R} \to \mathbb{R}^2$ defined by

$$h(x) = (g(x), |g(x)|);$$

although its image is the graph of a non-differentiable function, the curve itself manages to be differentiable by slowing down to velocity 0 at the point $(0,0)$. One can easily define a similar curve whose image looks like the picture below.

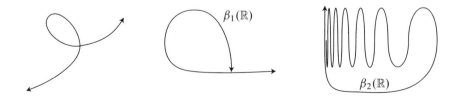

Three immersions of $\mathbb{R}$ in $\mathbb{R}^2$ are shown below. Although the second and third immersions β_1 and β_2 are one-one, their images are not homeomorphic to $\mathbb{R}$. Of course, even if the one-one immersion $f: P \to M$ is not a homeo-morphism onto its image, there is certainly some metric and some differentiable structure on $f(P)$ which makes the inclusion map $i: f(P) \to M$ an immersion. In general, a subset $M_1 \subset M$, with a differentiable structure (not nec-essarily compatible with the metric M_1 inherits as a subset of M), is called an **immersed submanifold** of M if the inclusion map $i: M_1 \to M$ is an immersion. The following picture, indicating the image of an immersion $\beta_3: \mathbb{R} \to S^1 \times S^1$,

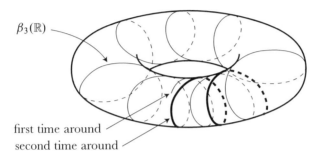

shows that M_1 may even be a dense subset of M.

Despite these complications, if M_1 is a k-dimensional immersed submanifold of M^n and U_1 is a neighborhood *in* M_1 of a point $p \in M_1$, then there is a coordinate system (y, V) of M around p, such that

$$U_1 \cap V = \{q \in M : y^{k+1}(q) = \cdots = y^n(q) = 0\};$$

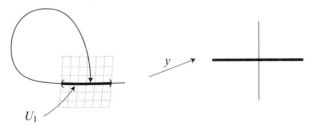

this is an immediate consequence of Theorem 10(2), with $f = i$. Thus, if $g: M_1 \to N$ is C^∞ (considered as a function on the manifold M_1) in a neigh-borhood of a point $p \in M_1$, then there is a C^∞ function $\tilde{g}$ on a neighborhood $V \subset M$ of p such that $g = \tilde{g} \circ i$ on $V \cap M_1$—we can define

$$\tilde{g}(q) = g(q'), \qquad \text{where} \quad \begin{cases} y^\alpha(q') = y^\alpha(q) & \alpha = 1, \ldots, k \\ y^r(q') = 0 & r = k+1, \ldots, n. \end{cases}$$

On the other hand, even if g is C^∞ on all of M_1 we may not be able to define $\tilde{g}$ on M. For example, this cannot be done if g is one of the functions $\beta_i^{-1} : \beta_i(M) \to \mathbb{R}$.

One other complication arises with immersed submanifolds. If $M_1 \subset M$ is an immersed submanifold, and $f : P \to M$ is a C^∞ function with $f(P) \subset M_1$, it is not necessarily true that f is C^∞ when considered as a map into M_1, with its C^∞ structure. The following figure shows that f might not even be continuous

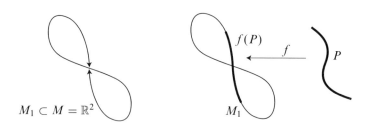

as a map into M_1. Actually, this is the only thing that can go wrong:

11. PROPOSITION. If $M_1 \subset M$ is an immersed manifold, $f : P \to M$ is a C^∞ function with $f(P) \subset M_1$, and f is continuous considered as a map into M_1, then f is also C^∞ considered as a map into M_1.

PROOF. Let $i : M_1 \to M$ be the inclusion map. We want to show that $i^{-1} \circ f$ is C^∞ if it is continuous. Given $p \in P$, choose a coordinate system (y, V) for M around $f(p)$ such that

$$U_1 = \{q \in V : y^{k+1}(q) = \cdots = y^n(q) = 0\}$$

is a neighborhood of $f(p)$ in M_1 and $(y^1|U_1, \ldots, y^k|U_1)$ is a coordinate system of M_1 on U_1.

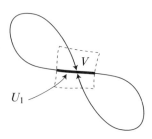

By assumption, $i^{-1} \circ f$ is continuous, so

$$f^{-1} \circ i(\text{open set}) \text{ is an open set.}$$

Since U_1 is open in M_1, this means that $f^{-1}(U_1) \subset P$ is open. Thus f takes some neighborhood of $p \in P$ into U_1. Since all $y^j \circ f$ are C^∞, and $y^1, \ldots, y^k$ are a coordinate system on U_1, the function f is C^∞ considered as a map into M_1. ❖

Most of these difficulties disappear when we consider one-one immersions $f : P \to M$ which are homeomorphisms onto their image. Such an immersion is called an **imbedding** ("embedding" for the English). An immersed submanifold $M_1 \subset M$ is called simply a (C^∞) **submanifold** of M if the inclusion map $i : M_1 \to M$ is an imbedding; it is called a **closed submanifold** of M if M_1 is also a closed subset of M.

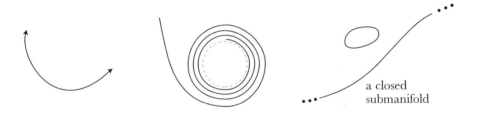

a closed
submanifold

There is one way of getting submanifolds which is very important, and gives the sphere $S^{n-1} \subset \mathbb{R}^n - \{0\} \subset \mathbb{R}^n$, defined as $\{x : |x|^2 = 1\}$, as a special case.

12. PROPOSITION. If $f : M^n \to N$ has constant rank k on a neighborhood of $f^{-1}(y)$, then $f^{-1}(y)$ is a closed submanifold of M of dimension $n - k$ (or is empty). In particular, if y is a regular value of $f : M^n \to N^m$, then $f^{-1}(y)$ is an $(n - m)$-dimensional submanifold of M (or is empty).

PROOF. Left to the reader. ❖

It is to be hoped that however abstract the notion of C^∞ manifolds may appear, submanifolds of $\mathbb{R}^N$ will seem like fairly concrete objects. Now it turns out that *every* (connected) C^∞ manifold can be imbedded in some $\mathbb{R}^N$, so that manifolds can be pictured as subsets of Euclidean space (though this picture is not always the most useful one). We will prove this fact only for compact manifolds, but we first develop some of the machinery which would be used in the general case, since we will need it later on anyway. Unfortunately, there are many definitions and theorems involved.

If $\mathcal{O}$ is a cover of a space M, a cover $\mathcal{O}'$ of M is a **refinement** of $\mathcal{O}$ (or "refines $\mathcal{O}$") if for every U in $\mathcal{O}'$ there is some V in $\mathcal{O}$ with $U \subset V$ (the sets of $\mathcal{O}'$ are "smaller" than those of $\mathcal{O}$)—a subcover is a very special case of a refining cover. A cover $\mathcal{O}$ is called **locally finite** if every $p \in M$ has a neighborhood W which intersects only finitely many sets in $\mathcal{O}$.

13. THEOREM. If $\mathcal{O}$ is an open cover of a manifold M, then there is an open cover $\mathcal{O}'$ of M which is locally finite and which refines $\mathcal{O}$. Moreover, we can choose all members of $\mathcal{O}'$ to be open sets diffeomorphic to $\mathbb{R}^n$.

PROOF. We can obviously assume that M is connected. By Theorem 1.2, there are compact sets $C_1, C_2, C_3, \ldots$ with $M = C_1 \cup C_2 \cup C_3 \cup \cdots$. Clearly C_1 has an open neighborhood U_1 with compact closure. Then $\overline{U_1} \cup C_2$ has an open neighborhood U_2 with compact closure. Continuing in this way, we obtain open sets U_i, with $\overline{U_i}$ compact and $\overline{U_i} \subset U_{i+1}$, whose union contains all C_i, and hence is M. Let $U_{-1} = U_0 = \emptyset$.

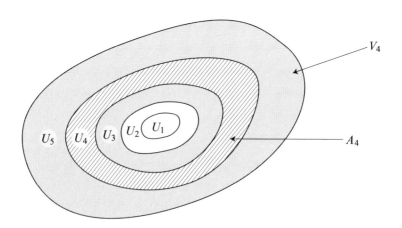

Now M is the union for $i > 1$ of the "annular" regions $A_i = \overline{U_i} - U_{i-1}$. Since each A_i is compact, we can obviously cover A_i by a finite number of open sets, each contained in some member of $\mathcal{O}$, and each contained in $V_i = U_{i+1} - U_{i-2}$. We can also choose these open sets to be diffeomorphic to $\mathbb{R}^n$. In this way we obtain a cover $\mathcal{O}'$ which refines $\mathcal{O}$ and which is locally finite, since a point in U_i is not in V_j for $j \geq 2 + i$. ❖

Notice that if $\mathcal{O}$ is an open locally finite cover of a space M and $C \subset M$ is compact, then C intersects only finitely many members of $\mathcal{O}$. This shows that an open locally finite cover of a connected manifold must be countable (like the cover constructed in the proof of Theorem 13).

14. THEOREM (THE SHRINKING LEMMA). Let $\mathcal{O}$ be an open locally finite cover of a manifold M. Then it is possible to choose, for each U in $\mathcal{O}$, an open set U' with $\overline{U'} \subset U$ in such a way that the collection of all U' is also an open cover of M.

PROOF. We can clearly assume that M is connected. Let $\mathcal{O} = \{U_1, U_2, U_3, \dots\}$. Then

$$C_1 = U_1 - (U_2 \cup U_3 \cup \cdots)$$

is a closed set contained in U_1, and $M = C_1 \cup U_2 \cup U_3 \cup \cdots$. Let U_1' be an open set with $C_1 \subset U_1' \subset \overline{U_1'} \subset U_1$. Now

$$C_2 = U_2 - (U_1' \cup U_3 \cup \cdots)$$

is a closed set contained in U_2, and $M = U_1' \cup C_2 \cup U_3 \cup \cdots$. Let U_2' be an open set with $C_2 \subset U_2' \subset \overline{U_2'} \subset U_2$. Continue in this way.

For any $p \in M$ there is a largest n with $p \in U_n$, because $\mathcal{O}$ is locally finite. Now

$$p \in U_1' \cup U_2' \cup \cdots \cup U_n' \cup \left(U_{n+1} \cup U_{n+2} \cup \cdots\right);$$

it follows that

$$p \in U_1' \cup U_2' \cup \cdots,$$

since replacing U_{n+i} by U_{n+i}' cannot possibly eliminate p. ❖

15. THEOREM. Let $\mathcal{O}$ be an open locally finite cover of a manifold M. Then there is a collection of C^∞ functions $\phi_U : M \to [0, 1]$, one for each U in $\mathcal{O}$, such that

(1) support $\phi_U \subset U$ for each U,

(2) $\sum_U \phi_U(p) = 1$ for all $p \in M$ (this sum is really a finite sum in some neighborhood of p, by (1)).

PROOF. *Case 1. Each U in $\mathcal{O}$ has compact closure.* Choose the U' as in Theorem 14. Apply Lemma 2 to $\overline{U'} \subset U \subset M$ to obtain a C^∞ function $\psi_U : M \to [0, 1]$ which is 1 on $\overline{U'}$ and has support $\subset U$. Since the U' cover M, clearly

$$\sum_{U \in \mathcal{O}} \psi_U > 0 \quad \text{everywhere.}$$

Define

$$\phi_U = \frac{\psi_U}{\displaystyle\sum_{U \in \mathcal{O}} \psi_U}.$$

Case 2. General case. This case can be proved in the same way, provided that Lemma 2 is true for $C \subset U \subset M$ with C closed (but not necessarily compact) and U open. But this is a consequence of *Case 1*:

For each $p \in C$ choose an open set $U_p \subset U$ with compact closure. Cover $M - C$ with open sets V_α having compact closure and contained in $M - C$. The open cover $\{U_p, V_\alpha\}$ has an open locally finite refinement $\mathcal{O}$ to which *Case 1* applies. Let

$$f = \sum_{U \in \mathcal{O}'} \phi_U, \qquad \text{where} \quad \mathcal{O}' = \{U \in \mathcal{O} : U \subset U_p \text{ for some } p\}.$$

This sum is C^∞, since it is a finite sum in a neighborhood of each point. Since $\sum_U \phi_U(p) = 1$ for all p, and $\phi_U(p) = 0$ when $U \subset V_\alpha$, clearly $f(p) = 1$ for all $p \in C$. Using the fact that $\mathcal{O}$ is locally finite, it is easy to see that support $f \subset U$. ❖

16. COROLLARY. If $\mathcal{O}$ is any open cover of a manifold M, then there is a collection of C^∞ functions $\phi_i : M \to [0, 1]$ such that

(1) the collection of sets $\{p : \phi_i(p) \neq 0\}$ is locally finite,

(2) $\sum_i \phi_i(p) = 1$ for all $p \in M$,

(3) for each i there is a $U \in \mathcal{O}$ such that support $\phi_i \subset U$.

(A collection $\{\phi_i : M \to [0, 1]\}$ satisfying (1) and (2) is called a **partition of unity**; if it satisfies (3), it is called **subordinate to** $\mathcal{O}$.)

It is now fairly easy to prove the last theorem of this chapter.

17. THEOREM. If M^n is a compact C^∞ manifold, then there is an imbedding $f : M \to \mathbb{R}^N$ for some N.

PROOF. There are a finite number of coordinate systems $(x_1, U_1), \ldots, (x_k, U_k)$ with $M = U_1 \cup \cdots \cup U_k$. Choose U_i' as in Theorem 14, and functions $\psi_i \colon M \to [0,1]$ which are 1 on $\overline{U_i'}$ and have support $\subset U_i$. Define $f \colon M \to \mathbb{R}^N$, where $N = nk + k$, by

$$ f = (\psi_1 \cdot x_1, \ldots, \psi_k \cdot x_k, \psi_1, \ldots, \psi_k). $$

This is an immersion, because any point p is in U_i' for some i, and on U_i', where $\psi_i = 1$, the $N \times n$ Jacobian matrix

$$ \left(\frac{\partial f^\alpha}{\partial x_i^\beta} \right) \quad \text{contains the } n \times n \text{ matrix} \quad \left(\frac{\partial x_i^\alpha}{\partial x_i^\beta} \right) = I. $$

It is also one-one. For suppose that $f(p) = f(q)$. There is some i such that $p \in U_i'$. Then $\psi_i(p) = 1$, so also $\psi_i(q) = 1$. This shows that we must have $q \in U_i$. Moreover,

$$ \psi_i \cdot x_i(p) = \psi_i \cdot x_i(q), $$

so $p = q$, since x_i is one-one on U_i. ❖

Problem 3-33 shows that, in fact, we can always choose $N = 2n + 1$.

PROBLEMS

1. (a) Show that being C^∞-related is *not* an equivalence relation.
(b) In the proof of Lemma 1, show that all charts in $\mathcal{A}'$ are C^∞-related, as claimed.

2. (a) If M is a metric space together with a collection of homeomorphisms $x \colon U \to \mathbb{R}^n$ whose domains cover M and which are C^∞-related, show that the n at each point is unique *without* using Invariance of Domain.
(b) Show similarly that ∂M is well-defined for a C^∞ manifold-with-boundary M.

3. (a) All C^∞ functions are continuous, and the composition of C^∞ functions is C^∞.
(b) A function $f \colon M \to N$ is C^∞ if and only if $g \circ f$ is C^∞ for every C^∞ function $g \colon N \to \mathbb{R}$.

4. How many distinct C^∞ structures are there on $\mathbb{R}$? (There is only one up to diffeomorphism; that is not the question being asked.)

5. (a) If $N \subset M$ is open and $\mathcal{A}'$ consists of all (x, U) in $\mathcal{A}$ with $U \subset N$, show that $\mathcal{A}'$ is maximal for N if $\mathcal{A}$ is maximal for M.

(b) Show that $\mathcal{A}'$ can also be described as the set of all $(x|V \cap N, V \cap N)$ for (x, V) in $\mathcal{A}$.

(c) Show that the inclusion $i : N \to M$ is C^∞, and that $\mathcal{A}'$ is the unique atlas with this property.

6. Check that the two projections P_1 and P_2 on S^{n-1} are C^∞ related to the $2n$ homeomorphisms f_i and g_i.

7. (a) If M is a connected C^∞ manifold and $p, q \in M$, then there is a C^∞ curve $c : [0, 1] \to M$ with $c(0) = p$ and $c(1) = q$.

(b) It is even possible to choose c to be one-one.

8. (a) Show that $(M_1 \times M_2) \times M_3$ is diffeomorphic to $M_1 \times (M_2 \times M_3)$ and that $M_1 \times M_2$ is diffeomorphic to $M_2 \times M_1$.

(b) The differentiable structure on $M_1 \times M_2$ makes the "slice" maps

$$p_1 \mapsto (p_1, \bar{p}_2)$$
$$p_2 \mapsto (\bar{p}_1, p_2)$$

of $M_1, M_2 \to M_1 \times M_2$ differentiable for all $\bar{p}_1 \in M_1, \bar{p}_2 \in M_2$.

(c) More generally, a map $f : N \to M_1 \times M_2$ is C^∞ if and only if the compositions $\pi_1 \circ f : N \to M_1$ and $\pi_2 \circ f : N \to M_2$ are C^∞. Moreover, the C^∞ structure we have defined for $M_1 \times M_2$ is the only one with this property.

(d) If $f_i : N \to M_i$ are C^∞ $(i = 1, 2)$, can one determine the rank of $(f_1, f_2) : N \to M_1 \times M_2$ at p in terms of the ranks of f_i at p? For $f_i : N_i \to M_i$, show that $f_1 \times f_2 : N_1 \times N_2 \to M_1 \times M_2$, defined by $f_1 \times f_2(p_1, p_2) = (f_1(p_1), f_2(p_2))$, is C^∞ and determine its rank in terms of the ranks of f_i.

9. Let $g : S^n \to \mathbb{P}^n$ be the map $p \mapsto [p]$. Show that $f : \mathbb{P}^n \to M$ is C^∞ if and only if $f \circ g : S^n \to M$ is C^∞. Compare the rank of f and the rank of $f \circ g$.

10. (a) If $U \subset \mathbb{R}^n$ is open and $f : U \to \mathbb{R}$ is locally C^∞ (every point has a neighborhood on which f is C^∞), then f is C^∞. (Obvious.)

(b) If $f : \mathbb{H}^n \to \mathbb{R}$ is locally C^∞, then f is C^∞, i.e., f can be extended to a C^∞ function on a neighborhood of $\mathbb{H}^n$. (Not so obvious.)

11. If $f : \mathbb{H}^n \to \mathbb{R}$ has two extensions g, h to C^∞ functions in a neighborhood of $\mathbb{H}^n$, then $D_j g$ and $D_j h$ are the same at points of $\mathbb{R}^{n-1} \times \{0\}$ (so we can speak of $D_j f$ at these points).

12. If M is a C^∞ manifold-with-boundary, then there is a unique C^∞ structure on ∂M such that the inclusion map $i : \partial M \to M$ is an imbedding.

13. (a) Let $U \subset M^n$ be an open set such that boundary U is an $(n-1)$-dimensional (differentiable) submanifold. Show that $\overline{U}$ is an n-dimensional manifold-with-boundary. (It is well to bear in mind the following example: if $U = \{x \in \mathbb{R}^n : d(x, 0) < 1$ or $1 < d(x, 0) < 2\}$, then $\overline{U}$ is a manifold-with-boundary, but $\partial \overline{U} \neq$ boundary U.)

(b) Consider the figure shown below. This figure may be extended by putting

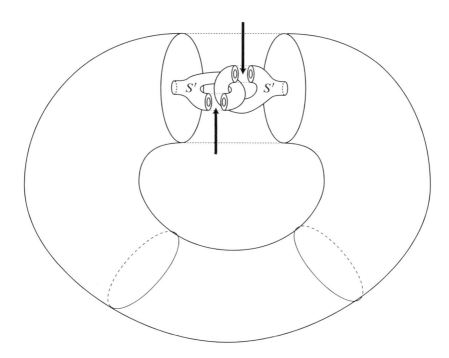

smaller copies of the two parts of S' into the regions indicated by arrows, and then repeating this construction indefinitely. The closure S of the final resulting figure is known as *Alexander's Horned Sphere*. Show that S is homeomorphic to S^2. (*Hint*: The additional points in the closure are homeomorphic to the Cantor set.) If U is the unbounded component of $\mathbb{R}^3 - S$, then $S =$ boundary U, but $\overline{U}$ is not a 2-dimensional manifold-with-boundary, so part (a) is true only for differentiable submanifolds.

14. (a) There is a map $f : \mathbb{R}^2 \to \mathbb{R}^2$ such that

(1) $f(x, 0) = (x, 0)$ for all x,

(2) $f(x, y) \subset \mathbb{H}^2$ for $y \geq 0$,

(3) $f(x, y) \subset \mathbb{R}^2 - \mathbb{H}^2$ for $y < 0$,

(4) f restricted to the upper half-plane or the lower half-plane is C^∞, but f itself is not C^∞.

(b) Suppose M and N are C^∞ manifolds-with-boundary and $f \colon \partial M \to \partial N$ is a diffeomorphism. Let $P = M \cup_f N$ be obtained from the disjoint union of M and N by identifying $x \in \partial M$ with $f(x) \in \partial N$. If (x, U) is a coordinate system around $p \in \partial M$ and (y, V) a coordinate system around $f(p)$, with $f(U \cap \partial M) = V \cap \partial N$, and $(y \circ f)|U \cap \partial M = x|U \cap \partial M$, we can define a homeomorphism from $U \cup V \subset P$ to $\mathbb{R}^n$ by sending U to $\mathbb{H}^n$ by x and V to the lower half-plane by the reflection of y. Show that this procedure does *not*

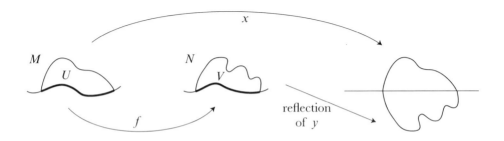

define a C^∞ structure on P.

(c) Now suppose that there is a neighborhood U of ∂M in M and a diffeomorphism $\alpha \colon U \to \partial M \times [0, 1)$, such that $\alpha(p) = (p, 0)$ for all $p \in \partial M$, and a similar diffeomorphism $\beta \colon V \to \partial N \times [0, 1)$. (We will be able to prove later that such diffeomorphisms always exist). Show that there is a unique C^∞ structure

on P such that the inclusions of M and N are C^∞ and such that the map from $U \cup V$ to $\partial M \times (-1, 1)$ induced by α and β is a diffeomorphism.

(d) By using two different pairs (α, β), define two different C^∞ structures on $\mathbb{R}^2$, considered as the union of two copies of $\mathbb{H}^2$ with corresponding points on $\partial \mathbb{H}^2$ identified. Show that the resulting C^∞ manifolds are diffeomorphic, but that the diffeomorphism *cannot* be chosen arbitrarily close to the identity map.

15. (a) Find a C^∞ structure on $\mathbb{H}^1 \times \mathbb{H}^1$ which makes the inclusion into $\mathbb{R}^2$ a C^∞ map. Can the inclusion be an imbedding? Are the projections on each factor C^∞ maps?

(b) If M and N are manifolds-with-boundary, construct a C^∞ structure on $M \times N$ such that all the "slice maps" (defined in Problem 8) are C^∞.

16. Show that the function $f \colon \mathbb{R} \to \mathbb{R}$ defined by

$$f(x) = \begin{cases} e^{-1/x} & x > 0 \\ 0 & x \leq 0 \end{cases}$$

is C^∞ (the formula e^{-1/x^2} is used just to get a function which is > 0 for $x < 0$, and $e^{-1/|x|}$ could be used just as well).

17. Lemma 2 (as addended by the proof of Theorem 15) shows that if C_1 and C_2 are disjoint closed subsets of M, then there is a C^∞ function $f \colon M \to [0,1]$ such that $C_1 \subset f^{-1}(0)$ and $C_2 \subset f^{-1}(1)$. Actually, we can even find f with $C_1 = f^{-1}(0)$ and $C_2 = f^{-1}(1)$. The proof turns out to be quite easy, once you know the trick.

(a) It suffices to find, for any closed $C \subset M$, a C^∞ function f with $C = f^{-1}(0)$.

(b) Let $\{U_i\}$ be a countable cover of $M - C$, where each U_i is of the form

$$U_i = x^{-1}(\{a \in \mathbb{R}^n : |a| < 1\})$$

for some coordinate system x taking an open subset of $M - C$ onto $\mathbb{R}^n$. Let $f_i \colon M \to [0,1]$ be a C^∞ function with $f_i > 0$ on U_i and $f_i = 0$ on $M - U_i$. Functions like

$$\frac{\partial f_i}{\partial x^j}, \quad \frac{\partial^2 f_i}{\partial x^j \partial x^k}, \quad \cdots$$

will be called mixed partials of f_i, of order $1, 2, \ldots$. Let

$$\alpha_i = \text{sup of all mixed partials of } f_1, \ldots, f_i \text{ of all orders } \leq i.$$

Show that

$$f = \sum_{i=1}^{\infty} \frac{f_i}{\alpha_i 2^i}$$

is C^∞, and $C = f^{-1}(0)$.

18. Consider the coordinate system (y^1, y^2) for $\mathbb{R}^2$ defined by

$$y^1(a, b) = a$$
$$y^2(a, b) = a + b.$$

(a) Compute $\partial f / \partial y^1(a, b)$ from the definition.

(b) Also compute it from Proposition 3 (to find $\partial I^i / \partial y^j$, write each I^i in terms of y^1 and y^2).

Notice that $\partial f / \partial y^1 \neq \partial f / \partial I^1$ even though $y^1 = I^1$; the operator $\partial / \partial y^i$ depends on y and i, not just on y^i.

19. Compute the "Laplacian"

$$\frac{\partial^2}{\partial x^2} + \frac{\partial^2}{\partial y^2}$$

in terms of polar coordinates. (First compute $\partial / \partial x$ in terms of $\partial / \partial r$ and $\partial / \partial \theta$; then compute $\partial^2 / \partial x^2$ from this). *Answer:* $\frac{1}{r} [\frac{\partial}{\partial r} (r \frac{\partial}{\partial r}) + \frac{\partial}{\partial \theta} (\frac{1}{r} \frac{\partial}{\partial \theta})]$.

20. If $f: M^n \to N^m$ is C^1 and $m > n$, then $f(M)$ has measure 0 (provided that M has only countably many components).

21. The following pictures show, for $n = 1, 2$, and 3, a subdivision of $[0, 1] \times [0, 1]$ into 2^{2n} squares, $A_{n,1}, \ldots, A_{n,2^{2n}}$; square $A_{n,k}$ is labeled simply k. The numbering is determined by the following conditions:

(a) The lower left square is $A_{n,1}$.

(b) The upper left square is $A_{n,2^{2n}}$.

(c) Squares $A_{n,k}$ and $A_{n,k+1}$ have a common side.

(d) Squares $A_{n,4l+1}, A_{n,4l+2}, A_{n,4l+3}, A_{n,4l+4}$ are contained in $A_{n-1,l+1}$.

4	3
1	2

16	13	12	11
15	14	9	10
2	3	8	7
1	4	5	6

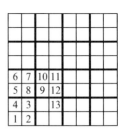

Define $f: [0, 1] \to [0, 1] \times [0, 1]$ by the condition

$$f(t) \in A_{n,k} \quad \text{for all} \quad \frac{k-1}{2^{2n}} \leq t \leq \frac{k}{2^{2n}}.$$

Show that f is continuous, onto $[0, 1] \times [0, 1]$, and not one-one.

22. For $p/2^n \in [0, 1]$, define $f(p/2^n) \in \mathbb{R}^2$ as shown below.

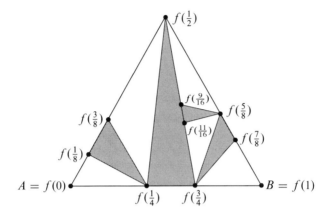

(a) Show that f is uniformly continuous, so that it has a continuous extension $g \colon [0, 1] \to \mathbb{R}^2$. Show that g is one-one, and that its image will not have measure 0 if the shaded triangles are chosen correctly.

(b) Consider the homeomorphic image of S^1 obtained by adding, below the image of g, a semi-circle with diameter the line segment AB. What does the inside of this curve look like?

23. Let $c \colon [0, 1] \to \mathbb{R}^n$ be continuous. For each partition $P = \{t_0, \ldots, t_k\}$ of $[0, 1]$, define

$$\ell(c, P) = \sum_{i=1}^{k} d(c(t_i), c(t_{i-1})).$$

The curve c is **rectifiable** if $\{\ell(c, P)\}$ is bounded above (with **length** equal to $\sup\{\ell(c, P)\}$). Show that the image of a rectifiable curve has measure 0.

24. (a) If M is a C^∞ manifold, a set $M_1 \subset M$ can be made into a k-dimensional submanifold of M if and only if around each point in M_1 there is a coordinate system (x, U) on M such that $M_1 \cap U = \{p \colon x^{k+1}(p) = \cdots = x^n(p) = 0\}$.

(b) The subset M_1 can be made into a closed submanifold if and only if such coordinate systems exist around every point of M.

25. The set $\{(x, |x|) \colon x \in \mathbb{R}\}$ is not the image of any immersion of $\mathbb{R}$ into $\mathbb{R}^2$.

26. (a) If $U \subset \mathbb{R}^k$ is open and $f \colon U \to \mathbb{R}^{n-k}$ is C^∞, then the **graph of** $f = \{(p, f(p)) \in \mathbb{R}^n \colon p \in U\}$ is a submanifold of $\mathbb{R}^n$.

(b) Every submanifold of $\mathbb{R}^n$ is locally of this form, after renumbering coordinates. (Neither Theorem 9 nor 10 is quite strong enough. You will need the implicit function theorem (*Calculus on Manifolds*, pg. 41). Theorem 10 is essentially Theorem 2-13 of *Calculus on Manifolds*; comparison with the implicit function theorem will show how some information has been allowed to escape.)

27. (a) An immersion from one n-manifold to another is an open map (the image of an open set is open).
(b) If M and N are n-manifolds with M compact and N connected, and $f: M \to N$ is an immersion, then f is onto.

28. Prove Proposition 12: If $f: M^n \to N$ has constant rank k on a neighborhood of $f^{-1}(y)$, then $f^{-1}(y)$ is a (closed) submanifold of M of dimension $n - k$ (or is empty).

29. Let $f: \mathbb{P}^2 \to \mathbb{R}^3$ be the map

$$g([x, y, z]) = (yz, xz, xy)$$

defined in Chapter 1, whose image is the Steiner surface. Show that g fails to be an immersion at 6 points (the image points are the points at distance $\pm 1/2$ on each axis). There *is* a way of immersing $\mathbb{P}^2$ in $\mathbb{R}^3$, known as Boy's Surface. See Hilbert and Cohn-Vossen, *Geometry and the Imagination*, pp. 317–321.

30. A continuous function $f: X \to Y$ is **proper** if $f^{-1}(C)$ is compact for every compact $C \subset Y$. The **limit set** $L(f)$ of f is the set of all $y \in Y$ such that $y = \lim f(x_n)$ for some sequence $x_1, x_2, x_3, \ldots \in X$ with no convergent subsequence.

(a) $L(f) = \emptyset$ if and only if f is proper.
(b) $f(X) \subset Y$ is closed if and only if $L(f) \subset f(X)$.
(c) There is a continuous $f: \mathbb{R} \to \mathbb{R}^2$ with $f(\mathbb{R})$ closed, but $L(f) \neq \emptyset$.
(d) A one-one continuous function $f: X \to Y$ is a homeomorphism (onto its image) if and only if $L(f) \cap f(Y) = \emptyset$.
(e) A submanifold $M_1 \subset M$ is a closed submanifold if and only if the inclusion map $i: M_1 \to M$ is proper.
(f) If M is a manifold, there is a proper map $f: M \to \mathbb{R}$; the function f can be made C^∞ if M is a C^∞ manifold.

31. (a) Find a cover of $[0, 1]$ which is not locally finite but which is "point-finite": every point of $[0, 1]$ is in only finitely many members of the cover.
(b) Prove the Shrinking Lemma when the cover $\mathcal{O}$ is point-finite and countable (notice that local-finiteness is not really used).
(c) Prove the Shrinking Lemma when $\mathcal{O}$ is a (not necessarily countable) point-finite cover of any space. (You will need Zorn's Lemma; consider collections $\mathcal{C}$

of pairs (U, U') where $U \in \mathcal{O}$, $U' \subset U$, and the union of all U' for $(U, U') \in \mathcal{C}$, together with all other $U \in \mathcal{O}$ covers the space.)

32. (a) If $M_1 \subset M$ is a closed submanifold, $U \supset M_1$ is any neighborhood, and $f : M_1 \to \mathbb{R}$ is C^∞, then there is a C^∞ function $\tilde{f} : M \to \mathbb{R}$ with $\tilde{f} = f$ on M_1, and with support $\tilde{f} \subset U$.
(b) This is false if $M = \mathbb{R}$ and $M_1 = (0, 1)$.
(c) This is false if $\mathbb{R}$ is replaced by a disconnected manifold N.

Remark: It is also false if $M = \mathbb{R}^2, M_1 = N = S^1$, and $f =$ identity; in fact, in this case, f has no continuous extension to a map from $\mathbb{R}^2$ to S^1, but the proof requires some topology. However, f can always be extended to a C^∞ function in a neighborhood of M_1 (extend locally, and use partitions of unity).

33. (a) The set of all non-singular $n \times n$ matrices with real entries is called $GL(n, \mathbb{R})$, the **general linear group**. It is a C^∞ manifold, since it is an open subset of $\mathbb{R}^{n^2}$. The **special linear group** $SL(n, \mathbb{R})$, or **unimodular group**, is the subgroup of all matrices with det $= 1$. Using the formula for $D(\det)$ in *Calculus on Manifolds*, pg. 24, show that $SL(n, \mathbb{R})$ is a closed submanifold of $GL(n, \mathbb{R})$ of dimension $n^2 - 1$.
(b) The symmetric $n \times n$ matrices may be thought of as $\mathbb{R}^{n(n+1)/2}$. Define $\psi : GL(n, \mathbb{R}) \to$ (symmetric matrices) by $\psi(A) = A \cdot A^t$, where A^t is the transpose of A. The subgroup $\psi^{-1}(I)$ of $GL(n, \mathbb{R})$ is called the **orthogonal group** $O(n)$. Show that $A \in O(n)$ if and only if the rows [or columns] of A are orthonormal.
(c) Show that $O(n)$ is compact.
(d) For any $A \in GL(n, \mathbb{R})$, define $R_A : GL(n, \mathbb{R}) \to GL(n, \mathbb{R})$ by $R_A(B) = BA$. Show that R_A is a diffeomorphism, and that $\psi \circ R_A = \psi$ for all $A \in O(n)$. By applying the chain rule, show that for $A \in O(n)$ the matrix

$$\left(\frac{\partial \psi^{ij}}{\partial x^{kl}} (A) \right) \text{ has the same rank as } \left(\frac{\partial \psi^{ij}}{\partial x^{kl}} (I) \right).$$

(Here x^{kl} are the coordinate functions in $\mathbb{R}^{n^2}$, and ψ^{ij} the $n(n+1)/2$ component functions of ψ.) Conclude from Proposition 12 that $O(n)$ is a submanifold of $GL(n, \mathbb{R})$.
(e) Using the formula

$$\psi^{ij}(A) = \sum_k a_{ik} a_{jk} \qquad (A = (a_{ij})),$$

show that

$$\frac{\partial \psi^{ij}}{\partial x^{kl}}(A) = \begin{cases} a_{jl} & k = i \neq j \\ a_{il} & k = j \neq i \\ 2a_{il} & k = i = j \\ 0 & \text{otherwise.} \end{cases}$$

Show that the rank of this matrix is $n(n + 1)/2$ at I (and hence at A for all $A \in O(n)$.) Conclude that $O(n)$ has dimension $n(n - 1)/2$.

(f) Show that $\det A = \pm 1$ for all $A \in O(n)$. The group $O(n) \cap SL(n, \mathbb{R})$ is called the **special orthogonal group** $SO(n)$, or the **rotation group** $R(n)$.

34. Let $M(m, n)$ denote the set of all $m \times n$ matrices, and $M(m, n; k)$ the set of all $m \times n$ matrices of rank k.

(a) For every $X_0 \in M(m, n; k)$ there are permutation matrices P and Q such that

$$PX_0Q = \begin{pmatrix} A_0 & B_0 \\ C_0 & D_0 \end{pmatrix}, \qquad \text{where } A_0 \text{ is } k \times k \text{ and non-singular.}$$

(b) There is some $\varepsilon > 0$ such that A is non-singular whenever all entries of $A - A_0$ are $< \varepsilon$.

(c) If

$$PXQ = \begin{pmatrix} A & B \\ C & D \end{pmatrix}$$

where the entries of $A - A_0$ are $< \varepsilon$, then X has rank k if and only if $D = CA^{-1}B$. *Hint*: If I_k denotes the $k \times k$ identity matrix, then

$$\begin{pmatrix} I_k & 0 \\ X & I_{p-k} \end{pmatrix} \begin{pmatrix} A & B \\ C & D \end{pmatrix} = \begin{pmatrix} A & B \\ XA + C & XB + D \end{pmatrix}.$$

(d) $M(m, n; k) \subset M(m, n)$ is a submanifold of dimension $k(m + n - k)$ for all $k \leq m, n$.

CHAPTER 3

THE TANGENT BUNDLE

A point $v \in \mathbb{R}^n$ is frequently pictured as an arrow from 0 to v. But there are many situations where we would like to picture this same arrow as starting

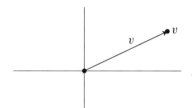

at a different point $p \in \mathbb{R}^n$:

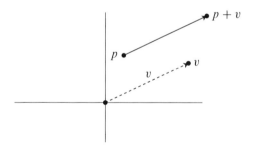

For example, suppose $c \colon \mathbb{R} \to \mathbb{R}^n$ is a differentiable curve. Then $c'(t) = (c^{1\prime}(t), \ldots, c^{n\prime}(t))$ is just a point of $\mathbb{R}^n$, but the line between $c(t)$ and $c(t) + c'(t)$ is tangent to the curve, and the "velocity vector" or "tangent vector" $c'(t)$ of the curve c is customarily pictured as the arrow from $c(t)$ to $c(t) + c'(t)$.

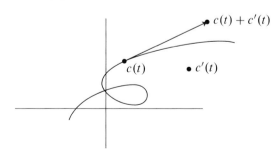

To give this picture mathematical substance, we simply describe the "arrow" from p to $p + v$ by the pair (p, v). The set of all such pairs is just $\mathbb{R}^n \times \mathbb{R}^n$, which we will also denote by $T\mathbb{R}^n$, the "tangent space of $\mathbb{R}^n$"; elements of $T\mathbb{R}^n$ are called "tangent vectors" of $\mathbb{R}^n$. We will often denote $(p, v) \in T\mathbb{R}^n$ by v_p ("the vector v at p"); in conformity with this notation, we will denote the set of all (p, v) for $v \in \mathbb{R}^n$ by $\mathbb{R}^n{}_p$. At times, it is more convenient to denote a member of $T\mathbb{R}^n$ by a single letter, like v. To recover the first member of a pair $v \in T\mathbb{R}^n$, we define the "projection" map $\pi \colon \mathbb{R}^n \times \mathbb{R}^n \to \mathbb{R}^n$ by $\pi(a, b) = a$. For any tangent vector v, the point $\pi(v)$ is "where it's at".

The set $\pi^{-1}(p)$ may be pictured as all arrows starting at p. Alternately,

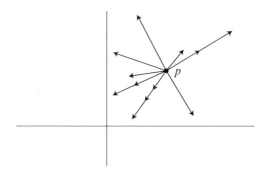

it can be pictured more geometrically as a particular subset of $\mathbb{R}^n \times \mathbb{R}^n$, the one visualizable case occurring when $n = 1$. This picture gives rise to some

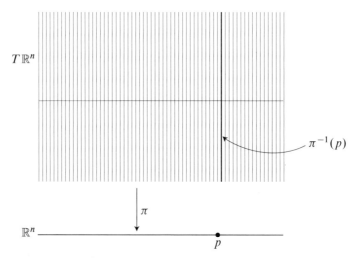

terminology—we call $\pi^{-1}(p)$ the **fibre over** p. This fibre can be made into a vector space in an obvious way: we define

$$(p, v) \oplus (p, w) = (p, v + w)$$
$$a \bullet (p, v) = (p, a \cdot v).$$

(The operations $\oplus$ and $\bullet$ should really be thought of as defined on

$$\bigcup_{p \in \mathbb{R}^n} \pi^{-1}(p) \times \pi^{-1}(p), \qquad \text{and} \qquad \mathbb{R} \times T\mathbb{R}^n, \quad \text{respectively.}$$

Usually we will just use ordinary $+$ and $\cdot$ instead of $\oplus$ and $\bullet$.)

If $f: \mathbb{R}^n \to \mathbb{R}^m$ is a differentiable map, and $p \in \mathbb{R}^n$, then the linear transformation $Df(p): \mathbb{R}^n \to \mathbb{R}^m$ may be used to produce a linear map from $\mathbb{R}^n{}_p \to \mathbb{R}^m{}_{f(p)}$ defined by

$$v_p \mapsto [Df(p)(v)]_{f(p)}.$$

This map, whose apparently anomalous features will soon be justified, is denoted by f_{*p}; the symbol f_* denotes the map $f_*: T\mathbb{R}^n \to T\mathbb{R}^m$ which is the union of all f_{*p}. Since $f_{*p}(v)$ is defined to be a vector $\in \mathbb{R}^m{}_{f(p)}$, the following diagram "commutes" (the two possible compositions from $T\mathbb{R}^n$ to $\mathbb{R}^m$ are equal),

$$
\begin{array}{ccc}
T\mathbb{R}^n & \xrightarrow{\ f_*\ } & T\mathbb{R}^m \\
{\scriptstyle \pi}\downarrow & & \downarrow{\scriptstyle \pi} \\
\mathbb{R}^n & \xrightarrow{\ f\ } & \mathbb{R}^m
\end{array}
\qquad \pi \circ f_* = f \circ \pi.
$$

Thus, f_* has the map f, as well as all maps $Df(p)$, built into it.

This is not the only reason for defining f_* in this particular way, however. Suppose that $g: \mathbb{R}^m \to \mathbb{R}^k$ is another differentiable function, so that, by the chain rule,

(1) $$D(g \circ f)(p) = Dg(f(p)) \circ Df(p).$$

By our definition,

$$g_*\left([Df(p)(v)]_{f(p)}\right) = \left(Dg(f(p))(Df(p)(v))\right)_{g(f(p))}.$$

This looks horribly complicated, but, using (1), it can be written

$$g_*(f_*(v_p)) = (g \circ f)_*(v_p);$$

thus we have

$$g_* \circ f_* = (g \circ f)_*.$$

This relation would clearly fall apart completely if $f_*(v_p)$ were not in $\mathbb{R}^m{}_{f(p)}$; with our present definition of f_*, it is merely an elegant restatement of the chain rule.

Henceforth, we will state almost all concepts about Jacobian matrices, like rank or singularity, in terms of f_*, rather than Df. The "tangent vector" of a curve $c: \mathbb{R} \to \mathbb{R}^n$ can be defined in terms of this concept, also. The **tangent vector of** c **at** t may be defined as

$$c'(t)_{c(t)} \in \mathbb{R}^n{}_{c(t)}.$$

[If c happens to be of the form

$$c(t) = (t, f(t)) \text{ for } f: \mathbb{R} \to \mathbb{R}$$

then

$$c'(t)_{c(t)} = (1, f'(t))_{c(t)};$$

this vector lies along the tangent line to the graph of f at $(t, f(t))$.] Notice that the tangent vector of c at t is the same as

$$c_*(1_t) = [Dc(t)(1)]_{c(t)} = (c^{1\prime}(t), \ldots, c^{n\prime}(t))_{c(t)},$$

where $1_t = (t, 1)$ is the "unit" tangent vector of $\mathbb{R}$ at t.

If $g: \mathbb{R}^n \to \mathbb{R}^m$ is differentiable, then $g \circ c$ is a curve in $\mathbb{R}^m$. The tangent vector of $g \circ c$ at t is

$$(g \circ c)_*(1_t) = g_*(c_*(1_t))$$
$$= g_*(\text{tangent vector of } c \text{ at } t).$$

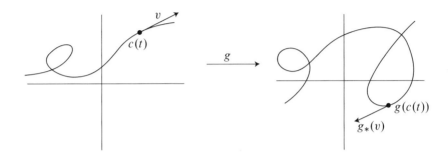

Consider now an n-dimensional manifold M and an imbedding $i : M \to \mathbb{R}^N$. Suppose we take a coordinate system (x, U) around p. Then $i \circ x^{-1}$ is a map from $\mathbb{R}^n$ to $\mathbb{R}^N$ with rank n. Consequently, $(i \circ x^{-1})_*(\mathbb{R}^n{}_{x(p)})$ is an n-dimensional subspace of $\mathbb{R}^N{}_{i(p)}$. This subspace doesn't depend on the coordinate

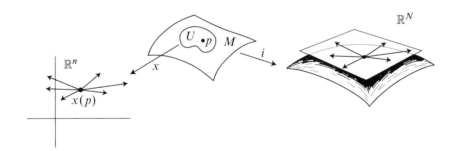

system x, for if y is another coordinate system, then

$$(i \circ y^{-1})_* = (i \circ x^{-1} \circ x \circ y^{-1})_*$$
$$= (i \circ x^{-1})_* \circ (x \circ y^{-1})_*$$

and

$$(x \circ y^{-1})_{*y(p)} : \mathbb{R}^n{}_{y(p)} \to \mathbb{R}^n{}_{x(p)}$$

is an isomorphism (with inverse $(y \circ x^{-1})_{*x(p)}$).

There is another way to see this, which justifies the picture we have drawn. If $c\colon (-\varepsilon, \varepsilon) \to \mathbb{R}^n$ is a curve with $c(0) = x(p)$, then $\alpha = i \circ x^{-1} \circ c$ is a curve in $\mathbb{R}^N$ which lies in $i(M)$, and every differentiable curve in $i(M)$ is of this

form (Proof?). Now

$$\alpha_*(1_0) = (i \circ x^{-1})_* \circ c_*(1_0),$$

so the tangent vector of every α is in $(i \circ x^{-1})_*\big(\mathbb{R}^n_{x(p)}\big)$. Moreover, every vector in this subspace is the tangent vector of some α, since every vector in $\mathbb{R}^n_{x(p)}$ is the tangent vector of some curve c. Thus, our n-dimensional subspace is just the set of all tangent vectors at $i(p)$ to differentiable curves in $i(M)$. We will denote this n-dimensional subspace by $(M, i)_p$.

We now want to look at the (disjoint) union

$$T(M, i) = \bigcup_{p \in M} (M, i)_p \subset i(M) \times \mathbb{R}^N \subset T\mathbb{R}^N.$$

We can define a "projection" map

$$\pi\colon T(M, i) \to M$$

by

$$\pi(v) = p \quad \text{if} \quad v \in (M, i)_p.$$

As in the case of $T\mathbb{R}^n$, each "fibre" $\pi^{-1}(p)$ has a vector space structure also. Beyond this we have to look a little more carefully at some specific examples.

Consider first the manifold $M = S^1$ and the inclusion $i\colon S^1 \to \mathbb{R}^2$. The curve $c(\theta) = (\cos\theta, \sin\theta)$ passes through every point of S^1, and

$$c'(\theta) = (-\sin\theta, \cos\theta) \neq 0.$$

For each $p = (\cos\theta, \sin\theta) \in S^1$, let $u_p = (-\sin\theta, \cos\theta)_p$ (it clearly doesn't matter which of the infinitely many possible θ's we choose). Then $(S^1, i)_p$ consists of all multiples of the vector u_p. We can therefore define a homeomorphism

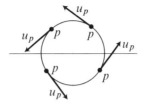

$f_1\colon T(S^1, i) \to S^1 \times \mathbb{R}^1$ by $f_1(\lambda u_p) = (p, \lambda)$, which makes the following diagram commute.

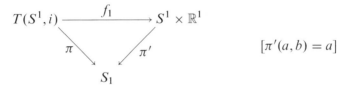

$$[\pi'(a, b) = a]$$

If we define the "fibres" of π' to be the sets $\pi'^{-1}(p)$, then each fibre has a vector space structure in a natural way. Commutativity of the diagram means that f_1 takes fibres into fibres; clearly f_1 restricted to a fibre is a linear isomorphism onto the image.

Now consider the manifold $M = S^2$ and the inclusion $i\colon S^2 \subset \mathbb{R}^3$. In this case there is *no* map $f_2\colon T(S^2, i) \to S^2 \times \mathbb{R}^2$ with the properties of the map f_1. If there were, then, for a fixed vector $v \neq 0$ in $\mathbb{R}^2$, the set of vectors

$$\{f_2^{-1}(v_p) : p \in S^2\}$$

would be a collection of non-zero tangent vectors, one at each point of S^2, which varied continuously. It is a well-known (hard) theorem of topology that this is impossible (you can't comb the hair on a sphere).

There is another example where we can prove that no appropriate homeomorphism $T(M, i) \to M \times \mathbb{R}^2$ exists, without appealing to a hard theorem of topology. The map i will just be the inclusion $M \to \mathbb{R}^3$ where M is a Möbius strip, to be precise, the particular subset of $\mathbb{R}^3$ defined in Chapter 1—M is the image of the map $f : [0, 2\pi] \times (-1, 1) \to \mathbb{R}^3$ defined by

$$f(\theta, t) = \left(2\cos\theta + t\cos\tfrac{\theta}{2}\cos\theta, \ 2\sin\theta + t\cos\tfrac{\theta}{2}\sin\theta, \ t\sin\tfrac{\theta}{2}\right).$$

At each point $p = (2\cos\theta, 2\sin\theta, 0)$ of M, the vector

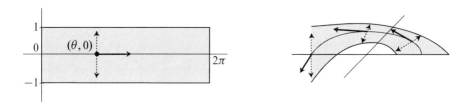

$$v_p = (-2\sin\theta, 2\cos\theta, 0)_p = f_*\big((1, 0)_{(\theta, 0)}\big)$$

is a tangent vector. The same is true for all multiples of $f_*\big((0, 1)_{(\theta, 0)}\big)$, shown as dashed arrows in the picture. Notice that

$$f_*\big((0, 1)_{(0,0)}\big) = [Df(0, 0)(0, 1)]_{(2,0,0)}$$

$$= \left[\frac{\partial f}{\partial t}(0, 0)\right]_{(2,0,0)} = (1, 0, 0)_{(2,0,0)},$$

while

$$f_*\big((0, 1)_{(2\pi, 0)}\big) = \left[\frac{\partial f}{\partial t}(2\pi, 0)\right]_{(2,0,0)} = (-1, 0, 0)_{(2,0,0)}.$$

This means that we can never pick non-zero dashed vectors *continuously* on the set of all points $(2\cos\theta, 2\sin\theta, 0)$: If we could, then each vector would be

$$f_*\big((0, \lambda(\theta))_{(\theta, 0)}\big)$$

for some continuous function $\lambda : [0, 2\pi] \to \mathbb{R}$. This function would have to be non-zero everywhere and also satisfy $\lambda(2\pi) = -\lambda(0)$, which it can't (by an easy theorem of topology). The impossibility of choosing non-zero dashed vectors continuously clearly shows that there is no way to map $T(M, i)$, fibre by fibre, homeomorphically onto $M \times \mathbb{R}^2$. We thus have another case where $T(M, i)$ does not "look like" a product $M \times \mathbb{R}^n$.

For any imbedding $i: M \to \mathbb{R}^N$, however, the structure of $T(M, i)$ is always simple *locally*: if (x, U) is a coordinate system on M, then $\pi^{-1}(U)$, the part of $T(M, i)$ over U, can always be mapped, fibre by fibre, homeomorphically onto $U \times \mathbb{R}^n$. In fact, for each $p \in U$, the fibre

$$(M, i)_p \quad \text{equals} \quad (i \circ x^{-1})_{*x(p)} \left(\mathbb{R}^n{}_{x(p)} \right) = m_p \left(\mathbb{R}^n{}_{x(p)} \right),$$

where the abbreviation m_p has been introduced temporarily; we can therefore define

$$f: \pi^{-1}(U) \to U \times \mathbb{R}^n$$

by

$$f\left(m_p(v_{x(p)}) \right) = (p, v).$$

In standard jargon, $T(M, i)$ is "locally trivial". This additional feature qualifies $T(M, i)$ to be included among an extremely important class of structures:

An **n-dimensional vector bundle** (or **n-plane bundle**) is a five-tuple

$$\xi = (E, \pi, B, \oplus, \odot),$$

where

(1) E and B are spaces (the "total space" and "base space" of ξ, respectively),

(2) $\pi: E \to B$ is a continuous map *onto* B,

(3) $\oplus$ and $\odot$ are maps

$$\oplus: \bigcup_{p \in B} \pi^{-1}(p) \times \pi^{-1}(p) \to E, \qquad \odot: \mathbb{R} \times E \to E,$$

with $\oplus \left(\pi^{-1}(p) \times \pi^{-1}(p) \right) \subset \pi^{-1}(p)$ and $\odot \left(\mathbb{R} \times \pi^{-1}(p) \right) \subset \pi^{-1}(p)$, which make each fibre $\pi^{-1}(p)$ into an n-dimensional vector space over $\mathbb{R}$,

such that the following "local triviality" condition is satisfied:

For each $p \in B$, there is a neighborhood U of p and a homeomorphism $t: \pi^{-1}(U) \to U \times \mathbb{R}^n$ which is a vector space isomorphism from each $\pi^{-1}(q)$ onto $q \times \mathbb{R}^n$, for all $q \in U$.

Because this local triviality condition really is a local condition, each bundle $\xi = (E, \pi, B, \oplus, \odot)$ automatically gives rise to a bundle $\xi | A$ over any subset $A \subset B$; to be precise,

$$\xi | A = \left(\pi^{-1}(A), \ \pi | \pi^{-1}(A), \ A, \ \oplus | \bigcup_{p \in A} \pi^{-1}(p) \times \pi^{-1}(p), \ \odot | \mathbb{R} \times \pi^{-1}(A) \right).$$

Notation as cumbersome as all this invites abuse, and we shall usually refer simply to a bundle $\pi \colon E \to B$, or even denote the bundle by E alone. For vectors $v, w \in \pi^{-1}(p)$ and $a \in \mathbb{R}$, we will denote $\oplus(v, w)$ and $\odot(a, v)$ by $v + w$, and $a \cdot v$ or av, respectively.

The simplest example of an n-plane bundle is just $X \times \mathbb{R}^n$ with $\pi \colon X \times \mathbb{R}^n \to X$ the projection on the first factor, and the obvious vector space structure on each fibre. This is called the **trivial** n-plane bundle over X and will be denoted by $\varepsilon^n(X)$. The "tangent bundle" $T\mathbb{R}^n$ is just $\varepsilon^n(\mathbb{R}^n)$.

The bundle $T(S^1, i)$ considered before is equivalent to $\varepsilon^1(S^1)$. Equivalence is here a technical term: Two vector bundles $\xi_1 = \pi_1 \colon E_1 \to B$ and $\xi_2 = \pi_2 \colon E_2 \to B$ are **equivalent** ($\xi_1 \simeq \xi_2$) if there is a homeomorphism $h \colon E_1 \to E_2$ which takes each fibre $\pi_1^{-1}(p)$ isomorphically onto $\pi_2^{-1}(p)$. The map h is called an **equivalence**. A bundle equivalent to $\varepsilon^n(B)$ is called **trivial**. (The local triviality condition for a bundle ξ just says that $\xi | U$ is trivial for some neighborhood U of p.)

The bundles $T(S^2, i)$ and $T(M, i)$ are not trivial, but there is an even simpler example of a non-trivial bundle. The Möbius strip *itself* (not $T(M, i)$) can be considered as a 1-dimensional vector bundle over S^1, for M can be obtained from $[0, 1] \times \mathbb{R}$ by identifying $(0, a)$ with $(1, -a)$, while S^1 can be obtained from

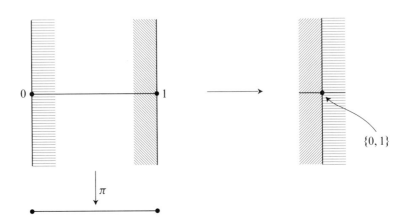

[0, 1] by identifying 0 with 1; the map π is defined by $\pi(t, a) = t$ for $0 < t < 1$ and $\pi(\{(0, a), (1, -a)\}) = \{0, 1\}$. The diagram above illustrates local triviality near the point $\{0, 1\}$ of S^1. Suppose that $s : S^1 \to M$ is a continuous function with $\pi \circ s = $ identity of M (such a function is called a **section**). Such a map

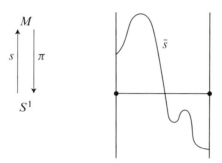

corresponds to a continuous function $\bar{s} : [0, 1] \to \mathbb{R}$ with $\bar{s}(0) = -\bar{s}(1)$. Since $\bar{s}$ must be 0 somewhere, the section s must be 0 somewhere (that is, $s(\theta) \in \pi^{-1}(\theta)$ must be the 0 vector for some $\theta \in S^1$). This surely shows that M is not a trivial bundle.

An equivalence is obviously the analogue of an isomorphism. The analogue of a homomorphism is the following.* A **bundle map** from ξ_1 to ξ_2 is a pair of continuous maps $(\tilde{f}, f)$, with $\tilde{f} : E_1 \to E_2$ and $f : B_1 \to B_2$, such that

(1) the following diagram commutes

$$
\begin{array}{ccc}
E_1 & \xrightarrow{\ \tilde{f}\ } & E_2 \\
{\scriptstyle \pi_1}\downarrow & & \downarrow{\scriptstyle \pi_2} \\
B_1 & \xrightarrow{\ f\ } & B_2 \ ,
\end{array}
$$

(2) $\tilde{f} : \pi_1^{-1}(p) \to \pi_2^{-1}(f(p))$ is a linear map.

The pair (f_*, f) is a bundle map from $T\mathbb{R}^k$ to $T\mathbb{R}^l$ for any differentiable $f : \mathbb{R}^k \to \mathbb{R}^l$. If $M^n \subset \mathbb{R}^k$ and $N^m \subset \mathbb{R}^l$ are submanifolds, $i : M \to \mathbb{R}^k$ and $j : N \to \mathbb{R}^l$ are the inclusions, and the map f satisfies $f(M) \subset N$, then f_*

---

*There are actually several possible choices, depending on whether one is considering all bundles at once, fixed bundles over various spaces, or a fixed base space with varying bundles. Thus f may be restricted to be an isomorphism on fibres and f to be the identity, or a homeomorphism. The relations between some of these cases are considered in the problems.

takes $T(M,i)$ to $T(N,j)$; to see this, just remember that $v \in T(M,i)_p$ is the tangent vector of a curve c in M, so $f_*(v)$ is the tangent vector of the curve $f \circ c$ in N, and consequently $f_*(v) \in T(N,j)$. In this way we obtain a bundle map from $T(M,i)$ to $T(N,j)$. Actually, it would have sufficed to begin with a C^∞ function $f \colon M \to N$, since f can be extended to $\mathbb{R}^k$ locally. In fact, this construction could be generalized much further, to the case where i and j are merely imbeddings of two abstract manifolds M and N, and $f \colon M \to N$ is C^∞; we just consider the function $j \circ f \circ i^{-1} \colon i(M) \to i(N)$ and extend it locally to $\mathbb{R}^k$. The case which we want to examine most carefully is the simplest: where $M = N$ and f is the identity, while i and j are two imbeddings of M in $\mathbb{R}^k$ and $\mathbb{R}^l$, respectively. Elements of $T(M,i)_p$ are of the form $(i \circ x^{-1})_*(w)$

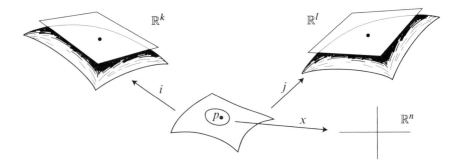

for $w \in \mathbb{R}^n{}_{x(p)}$, while elements of $T(M,j)_p$ are of the form $(j \circ x^{-1})_*(w)$ for $w \in \mathbb{R}^n{}_{x(p)}$. If we map

$$(i \circ x^{-1})_*(w) \mapsto (j \circ x^{-1})_*(w)$$

we obtain a bundle map from $T(M,i)|U$ to $T(M,j)|U$, which is obviously an equivalence. The map $(M,i)_p \to (M,j)_p$ induced on fibres is independent of the coordinate system x, for if (y,V) is another coordinate system, then

$$(i \circ y^{-1})_*(w) = (i \circ x^{-1})_*\big((x \circ y^{-1})_*(w)\big)$$

$$\big\updownarrow \qquad\qquad \big\updownarrow$$

$$(j \circ y^{-1})_*(w) = (j \circ x^{-1})_*\big((x \circ y^{-1})_*(w)\big).$$

We can therefore put all these maps together, and obtain an *equivalence* from $T(M,i)$ to $T(M,j)$. In other words, the dependence of $T(M,i)$ on i is almost illusory; we could abbreviate $T(M,i)$ to TM, if we agreed that TM really denotes an equivalence class of bundles, rather than one bundle. That is the sort of thing an algebraist might do, and it is undoubtedly ugly. What we would like to do is to get a single bundle for each M, in some natural way, which has all the properties any one of these particular bundles $T(M, i)$ has. Can we do this? Yes, we can. When we do, $T\mathbb{R}^n$ will be different from our old definition (namely, $\varepsilon^n(\mathbb{R}^n)$), and so will f_* for $f: \mathbb{R}^n \to \mathbb{R}^m$, so in stating our result precisely we will write "old f_*" when necessary.

1. THEOREM. It is possible to assign to each n-manifold M an n-plane bundle TM over M, and to each C^∞ map $f: M \to N$ a bundle map (f_*, f), such that:

(1) If $1: M \to M$ is the identity, then $1_*: TM \to TM$ is the identity. If $g: N \to P$, then $(g \circ f)_* = g_* \circ f_*$.

(2) There are equivalences $t^n: T\mathbb{R}^n \to \varepsilon^n(\mathbb{R}^n)$ such that for every C^∞ function $f: \mathbb{R}^n \to \mathbb{R}^m$ the following commutes.

$$
\begin{array}{ccc}
T\mathbb{R}^n & \xrightarrow{\ f_*\ } & T\mathbb{R}^m \\
\Big\downarrow{\scriptstyle t^n} & & \Big\downarrow{\scriptstyle t^m} \\
\varepsilon^n(\mathbb{R}^n) & \xrightarrow{\text{old } f_*} & \varepsilon^m(\mathbb{R}^m)
\end{array}
$$

(3) If $U \subset M$ is an open submanifold, then TU is equivalent to $(TM)|U$, and for $f: M \to N$ the map $(f|U)_*: TU \to TN$ is just the restriction of f_*. More precisely, there is an equivalence $TU \simeq (TM)|U$ such that the following diagrams commute, where $i: U \to M$ is the inclusion.*

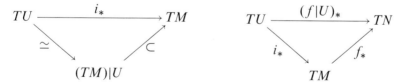

PROOF. The construction of TM is an ingenious, though quite natural, subterfuge. We will obtain a single bundle for TM, but the *elements* of TM will each be large equivalence classes.

---

*When using the notation f_*, it must be understood that the symbol "f" really refers to a triple (f, M, N) where $f: M \to N$. The identity map 1 of U to itself and the inclusion map $i: U \to M$ have to be considered as different, since the maps $1_*: TU \to TU$ and $i_*: TU \to TM$ are certainly different (they map TU into two different sets).

The construction is much easier to understand if we first imagine that we *already had* our bundles TM. Then if (x, U) is a coordinate system, we would have a map $x_*: TU \to T(x(U))$, and this would be an equivalence (with inverse $(x^{-1})_*$). Since TU should be essentially $(TM)|U$, and $T(x(U))$ should essentially be $x(U) \times \mathbb{R}^n$, a point $e \in \pi^{-1}(p)$ would be taken by x_* to some $(x(p), v)$. Here v is just an element of $\mathbb{R}^n$ (and every v would occur, since x_* maps $\pi^{-1}(p)$ isomorphically onto $\{p\} \times \mathbb{R}^n$). If y is another coordinate system, then $y_*(e)$ would be $(y(p), w)$ for some $w \in \mathbb{R}^n$. We can easily figure out what the relationship between v and w would be; since $(x(p), v)$ is taken to $(y(p), w)$ by $y_* \circ x_*^{-1} = (y \circ x^{-1})_*$, and $(y \circ x^{-1})_*$ is supposed to be the old $(y \circ x^{-1})_*$, we would have

(a) $$w = D(y \circ x^{-1})(x(p))(v).$$

This condition makes perfect sense without any mention of bundles. It is the clue which enables us to now define TM.

If x and y are coordinate systems whose domains contain p, and $v, w \in \mathbb{R}^n$, we *define*

$$(x, v) \underset{p}{\sim} (y, w) \qquad \text{if (a) is satisfied.}$$

It is easy to check (using the chain rule) that $\underset{p}{\sim}$ is an equivalence relation; the equivalence class of (x, v) will be denoted by $[x, v]_p$. These equivalence classes will be called tangent vectors at p, and TM is defined to be the set of all tangent vectors at all points $p \in M$; the map π takes $\underset{p}{\sim}$ equivalence classes to p. We define a vector space structure on $\pi^{-1}(p)$ by the formulas

$$[x, v]_p + [x, w]_p = [x, v + w]_p$$
$$a \cdot [x, v]_p = [x, a \cdot v]_p;$$

this definition is independent of the particular coordinate system x or y, because $D(y \circ x^{-1})(x(p))$ is an isomorphism from $\mathbb{R}^n$ to $\mathbb{R}^n$.

Our definition of TM provides a one-one onto map

(b) $$t_x: \pi^{-1}(U) \to U \times \mathbb{R}^n, \quad \text{namely} \quad [x, v]_q \mapsto (q, v).$$

We want this to be a homeomorphism, so we want $t_x^{-1}(A)$ to be open for every open $A \subset U \times \mathbb{R}^n$, and thus we want any union of such sets to be open. There is a metric with exactly these sets as open sets, but it is a little ticklish to produce, so we leave this one part of the proof to Problem 1.

We now have a bundle $\pi: TM \to M$. We will denote the fibre $\pi^{-1}(p)$ by M_p, in conformity with the notation $\mathbb{R}^n{}_p$, though TM_p might be better. If

$f: M \to N$, and (x, U) and (y, V) are coordinate systems around p and $f(p)$, respectively, we define

(c) $$f_*([x, v]_p) = [y, D(y \circ f \circ x^{-1})(x(p))(v)]_{f(p)}.$$

Of course, it must be checked that this definition is independent of x and y (the chain rule again).

Condition (1) of our theorem is obvious.

To prove (2), we define t^n to be t_I, where I is the identity map of $\mathbb{R}^n$ and t_x is defined in (b); it is trivial, though perhaps confusing to the novice, to prove commutativity of the diagram.

Condition (3) is practically obvious also. In fact, the fibre of TU over $p \in U$ is almost exactly the same as the fibre of TM over p; the only difference is that each equivalence class for M contains some extra members, since in M there are more coordinate systems around p than there are in $U \subset M$. ❖

Henceforth, the bundle $\pi: TM \to M$ will be called the **tangent bundle** of M. If $i: M \to \mathbb{R}^k$ is an imbedding, then TM is equivalent to $T(M, i)$. In fact, if (x, U) is a coordinate system around p, and I is the identity coordinate system of $\mathbb{R}^k$, then

$$i_*([x, v]_p) = [I, D(i \circ x^{-1})(x(p))(v)]_{i(p)} \qquad \text{by (c)}$$

$$t^n = t_I \quad \Big\downarrow$$

$$(i(p), D(i \circ x^{-1})(x(p))(v)) \in (M, i)_p;$$

the composition $t^n i_*$ is easily seen to be an equivalence. But $T(M, i)$ will play no further role in this story—the abstract substitute TM will always be used instead.

Having succeeded in producing a bundle over each M, which is equivalent to $T(M, i)$, we next ask how fortuitous this was. Can one find other bundles with the same properties? The answer is yes, and we proceed to define two different such bundles.

For the first example, we consider curves $c: (-\varepsilon, \varepsilon) \to M$, each defined on some interval around 0, with $c(0) = p$. If (x, U) is a coordinate system around p, we define

$$c_1 \underset{p}{\approx} c_2 \quad \text{if and only if:} \qquad \begin{array}{l} x \circ c_1 \text{ and } x \circ c_2, \text{ mapping } \mathbb{R} \text{ to } \mathbb{R}^n, \\ \text{have the same derivative at 0.} \end{array}$$

The equivalence classes, for all $p \in M$, will be the elements of our new bundle, $T'M$. For $f: M \to N$ there is a map $f_\sharp$ taking the $\underset{p}{\approx}$ equivalence class of c to the $\widetilde{\widetilde{f(p)}}$ equivalence class of $f \circ c$. Without bothering to check details, we can already see that this example is "really the same" as TM —

$$[x, v]_p \quad \text{corresponds to:} \qquad \begin{array}{l} \text{the } \underset{p}{\approx} \text{ equivalence class of } x^{-1} \circ \gamma, \\ \text{where } \gamma \text{ is a curve in } \mathbb{R}^n \text{ with } \gamma'(0) = v; \end{array}$$

under this correspondence, $f_\sharp$ corresponds to f_*.

In the second example, things are not so simple. We define a tangent vector at p to be a linear operator ℓ which operates on all C^∞ functions f and which is a "derivation at p":

$$\ell(fg) = f(p)\ell(g) + g(p)\ell(f).$$

We have already seen that the operators $\ell = \partial/\partial x^i\big|_p$ have this property. For these operators, clearly $\ell(f) = \ell(g)$ if $f = g$ in a neighborhood of p. This condition is actually true for any derivation ℓ. For, suppose that $f = 0$ in a neighborhood of p. There is a C^∞ function $h \colon M \to \mathbb{R}$ with $h(p) = 1$ and support $h \subset f^{-1}(0)$. Then

$$0 = \ell(0) = \ell(fh) = f(0)\ell(h) + h(0)\ell(f) = 0 + \ell(f).$$

Thus, if $f = g$ in a neighborhood of 0, then $0 = \ell(f - g) = \ell(f) - \ell(g)$. If f is defined only in a neighborhood of p, we may use this trick to define $\ell(f)$: choose h to be 1 on a neighborhood of p, with support $h \subset f^{-1}(0)$, and define $\ell(f)$ as $\ell(fh)$.

The set of all such operators is a vector space, but it is not *a priori* clear what its dimension is. This comes out of the following.

2. **LEMMA.** Let f be a C^∞ function in a convex open neighborhood U of 0 in $\mathbb{R}^n$, with $f(0) = 0$. Then there are C^∞ functions $g_i \colon U \to \mathbb{R}$ with

(1) $f(x^1, \ldots, x^n) = \sum_{i=1}^n x^i g_i(x^1, \ldots, x^n)$ for $x \in U$,

(2) $g_i(0) = D_i f(0)$.

(The second condition actually follows from the first.)

PROOF. For $x \in U$, let $h_x(t) = f(tx)$; this is defined for $0 \le t \le 1$, since U is convex. Then

$$f(x) = f(x) - f(0) = \int_0^1 h_x'(t)\, dt = \int_0^1 \sum_{i=1}^n D_i f(tx) \cdot x^i\, dt.$$

Therefore we can let $g(x) = \int_0^1 D_i f(tx)\, dt.$ ❖

3. THEOREM. The set of all linear derivations at $p \in M^n$ is an n-dimensional vector space. In fact, if (x, U) is a coordinate system around p, then

$$\left. \frac{\partial}{\partial x^1} \right|_p, \dots, \left. \frac{\partial}{\partial x^n} \right|_p$$

span this vector space, and any derivation ℓ can be written

$$\ell = \sum_{i=1}^{n} \ell(x^i) \cdot \left. \frac{\partial}{\partial x^i} \right|_p,$$

(so ℓ is determined by the numbers $\ell(x^i)$).

PROOF. Notice that

$$\ell(l) = \ell(1 \cdot 1) = 1 \cdot \ell(1) + 1 \cdot \ell(1),$$

so $\ell(1) = 0$. Hence $\ell(c) = c \cdot \ell(1) = 0$ for any constant function c on U. Consider the case where $M = \mathbb{R}^n$ and $p = 0$. Assume U is convex. Given f on U, choose g_i as in Lemma 2, for the function $f - f(0)$. Then

$$\ell(f) = \ell(f - f(0)) = \ell\left(\sum_{i=1}^{n} I^i g_i \right) \quad \begin{matrix} (I^i \text{ denotes the } i^{\text{th}} \\ \text{coordinate function}) \end{matrix}$$

$$= \sum_{i=1}^{n} [\ell(I^i) g_i(0) + I^i(0) \ell(g_i)]$$

$$= \sum_{i=1}^{n} \ell(I^i) \frac{\partial f}{\partial I^i}(0) + 0.$$

This shows that $\partial/\partial I^i|_0$ span the vector space; they are clearly linearly independent. It is a simple exercise to use the coordinate system x to transfer this result from $\mathbb{R}^n$ to M. ❖

From Theorem 3 we can see that, once again, a bundle constructed from all derivations at all points of M is "really the same" as TM. We can let

$$\ell = \sum_{i=1}^{n} a^i \left. \frac{\partial}{\partial x^i} \right|_p \quad \text{correspond to} \quad [x, a]_p;$$

the formula

$$\left. \frac{\partial}{\partial x^i} \right|_p = \sum_{j=1}^{n} \frac{\partial y^j}{\partial x^i}(p) \left. \frac{\partial}{\partial y^j} \right|_p,$$

derived in Chapter 2, shows that

$$\sum_{i=1}^{n} a^i \left.\frac{\partial}{\partial x^i}\right|_p = \sum_{i=1}^{n} b^i \left.\frac{\partial}{\partial y^i}\right|_p \quad \text{if and only if} \quad b^j = \sum_{i=1}^{n} a^i \frac{\partial y^j}{\partial x^i}(p),$$

and this is precisely the equation which says that $(x, a) \underset{p}{\sim} (y, b)$. It is easily checked that under this correspondence, the map which corresponds to f_* can be defined as follows:

$$[f_*(\ell)](g) = \ell(g \circ f).$$

Notice that if x denotes the identity coordinate system on $\mathbb{R}^n$, then $\sum_{i=1}^{n} a^i \left.\frac{\partial}{\partial x^i}\right|_p$ corresponds to a_p when we identify $T\mathbb{R}^n$ with $\varepsilon^n(\mathbb{R}^n)$.

We will usually make no distinction whatsoever between a tangent vector $v \in M_p$ and the linear derivation it corresponds to, that is, between $[x, a]_p$ and

$$\sum_{i=1}^{n} a^i \left.\frac{\partial}{\partial x^i}\right|_p ;$$

consequently, we will not hesitate to write $v(f)$ for a differentiable function f defined in a neighborhood of p. In fact, a tangent vector is often most easily described by telling what derivation it corresponds to, and the map f_* is often most easily analyzed from the relation

$$(f_* v)(g) = v(g \circ f).$$

It is customary to denote the identity coordinate system on $\mathbb{R}^1$ by t, and to write

$$\left.\frac{d}{dt}\right|_{t_0} \quad \text{for} \quad \left.\frac{\partial}{\partial t}\right|_{t_0} ;$$

this is a basis for $\mathbb{R}_{t_0}$. If $c\colon \mathbb{R} \to M$ is a differentiable curve, then

$$c_* \left(\left.\frac{d}{dt}\right|_{t_0} \right) \in M_{c(t_0)}$$

is called the **tangent vector to** c **at** t_0. We will denote it by the suggestive symbol

$$\left.\frac{dc}{dt}\right|_{t_0} .$$

This symbol will be subjected to the standard abuses one finds (unexplained) in calculus textbooks: the symbol

$$\frac{dc}{dt} \quad \text{will often stand for} \quad \frac{dc}{dt}\bigg|_t ,$$

the subscript "t" now denoting a particular number $t \in \mathbb{R}$, as well as the identity coordinate system.

As you might well expect, it is no accident that our second and third examples turned out to be "really the same" as TM. There is a general theorem that all "reasonable" examples will have this property, but it is a little delicate to state, and quite a mess to prove, so it has been quarantined in an Addendum to this chapter.

The tangent bundle TM of a C^∞ manifold has a little more structure than an arbitrary n-plane bundle. Since TM locally looks like $U \times \mathbb{R}^n$, clearly TM is itself a manifold; there is, moreover, a natural way to put a C^∞ structure on TM. If $x\colon U \to \mathbb{R}^n$ is a chart on M, then every element $v \in (TM)|U$ is uniquely of the form

$$v = \sum_{i=1}^n a^i \frac{\partial}{\partial x^i}\bigg|_p , \qquad p = \pi(v).$$

Let us denote a^i by $\dot{x}^i(v)$. Then the map

$$v \mapsto (x^1(\pi(v)), \ldots, x^n(\pi(v)), \dot{x}^1(v), \ldots, \dot{x}^n(v)) \in \mathbb{R}^{2n}$$

is a homeomorphism from $(TM)|U$ to $x(U) \times \mathbb{R}^n$. This map, $(x \circ \pi, \dot{x})$, is simply the map x_* when we identify TU with $U \times \mathbb{R}^n$ in the standard way. If (y, V) is another coordinate system, and

$$v = \sum_{j=1}^n b^j \frac{\partial}{\partial y^j}\bigg|_p ,$$

then, as we have already seen,

$$b^j = \sum_{i=1}^n a^i \frac{\partial y^j}{\partial x^i}(p) = \sum_{i=1}^n a^i D_i(y^j \circ x^{-1})(x(p)).$$

This shows that if $(t, a) = (t^1, \ldots, t^n, a^1, \ldots, a^n) \in \mathbb{R}^{2n}$, then

$$y_* \circ (x_*)^{-1}(t, a)$$
$$= \left(y \circ x^{-1}(t), \ \sum_{i=1}^n a^i D_i(y^1 \circ x^{-1})(t), \ \ldots, \ \sum_{i=1}^n a^i D_i(y^n \circ x^{-1})(t) \right).$$

This expression shows that $y_* \circ (x_*)^{-1}$ is C^∞.

We thus have a collection of C^∞-related charts on TM, which can be extended to a maximal atlas.

With this C^∞ structure, the local trivializations x_* are C^∞. In general, a vector bundle $\pi: E \to B$ is called a C^∞ vector bundle if E and B are C^∞ manifolds and there are C^∞ local trivializations in a neighborhood of each point. It follows that $\pi: E \to B$ is C^∞.

Recall that a **section** of a bundle $\pi: E \to B$ is a continuous function $s: B \to E$ such that $\pi \circ s = $ identity of B; for C^∞ vector bundles we can also speak of C^∞ sections. A section of TM is called a **vector field** on M; for submanifolds M of $\mathbb{R}^n$, a vector field may be pictured as a continuous selection of arrows tangent to M. The theorem that you can't comb the hair on a sphere just states that

there is no vector field on S^2 which is everywhere non-zero. We have shown that there do not exist two vector fields on the Möbius strip which are everywhere linearly independent.

Vector fields are customarily denoted by symbols like X, Y, or Z, and the vector $X(p)$ is often denoted by X_p (sometimes X may be used to denote a single vector, in some M_p). If we think of TM as the set of derivations, then for any coordinate system (x, U), we have

$$X(p) = \sum_{i=1}^{n} a^i(p) \left.\frac{\partial}{\partial x^i}\right|_p \qquad \text{for all } p \in U.$$

The functions a^i are continuous or C^∞ if and only if $X: U \to TM$ is continuous or C^∞.

If X and Y are two vector fields, we define a new vector field $X + Y$ by

$$(X + Y)(p) = X(p) + Y(p).$$

Similarly, if $f: M \to \mathbb{R}$, we define the vector field fX by

$$(fX)(p) = f(p)X(p).$$

Clearly $X + Y$ and fX are C^∞ if X, Y, and f are C^∞. On U we can write

$$X = \sum_{i=1}^{n} a^i \frac{\partial}{\partial x^i},$$

the symbol $\partial/\partial x^i$ now denoting the vector field

$$p \mapsto \frac{\partial}{\partial x^i}\bigg|_p.$$

If $f: M \to \mathbb{R}$ is a C^∞ function, and X is a vector field, then we can define a new *function* $\bar{X}(f): M \to \mathbb{R}$ by letting X operate on f at each point:

$$\bar{X}(f)(p) = X_p(f).$$

It is not hard to check that if X is a C^∞ vector field, then $\bar{X}(f)$ is C^∞ for every C^∞ function f; indeed, if locally

$$X(p) = \sum_{i=1}^{n} a^i(p) \frac{\partial}{\partial x^i}\bigg|_p,$$

then

$$\bar{X}(f) = \sum_{i=1}^{n} a^i \frac{\partial f}{\partial x^i},$$

which is a sum of products of C^∞ functions. Conversely, if $\bar{X}(f)$ is C^∞ for *every* C^∞ function f, then X is a C^∞ vector field (since $\bar{X}(x^i) = a^i$).

Let $\mathcal{F}$ denote the set of all C^∞ functions on M. We have just seen that a C^∞ vector field X gives rise to a function $\bar{X}: \mathcal{F} \to \mathcal{F}$. Clearly,

$$\bar{X}(f + g) = \bar{X}(f) + \bar{X}(g)$$
$$\bar{X}(fg) = f\bar{X}(g) + g\bar{X}(f);$$

thus $\bar{X}$ is a "derivation" of the ring $\mathcal{F}$. Often, a C^∞ vector field X is identified with the derivation $\bar{X}$. The reason for this is that if $A: \mathcal{F} \to \mathcal{F}$ is any derivation, then $A = \bar{X}$ for a unique C^∞ vector field X. In fact, we clearly must define

$$X_p(f) = A(f)(p),$$

and the operator X_p thus defined is a derivation at p.

The tangent bundle is the true beginning of the study of differentiable manifolds, and you should not read further until you grok it.* The next few chapters constitute a detailed study of this bundle. One basic theme in all these chapters is that any structure one can put on a vector space leads to a structure on any vector bundle, in particular on the tangent bundle of a manifold. For the present, we will discuss just one new concept about manifolds, which arises in this very way from the notion of "orientation" in a vector space.

The non-singular linear maps $f \colon V \to V$ from a finite dimensional vector space to itself fall into two groups, those with $\det f > 0$, and those with $\det f < 0$; linear transformations in the first group are called **orientation preserving** and the others are called **orientation reversing**. A simple example of the latter is the map $f \colon \mathbb{R}^n \to \mathbb{R}^n$ defined by $f(x) = (x^1, \ldots, x^{n-1}, -x^n)$ (reflection in the hyperplane $x^n = 0$). There is no way to pass continuously between these two groups: if we identify linear maps $\mathbb{R}^n \to \mathbb{R}^n$ with $n \times n$ matrices, and thus with $\mathbb{R}^{n^2}$, then the orientation preserving and orientation reversing maps are disjoint open subsets of the set of all non-singular maps (those with $\det \neq 0$). The terminology "orientation preserving" is a bit strange, since we have not yet defined anything called "orientation", which is being preserved. The problem becomes more acute if we want to define orientation preserving isomorphisms between two different (but isomorphic) vector spaces V and W; this clearly makes no sense unless we supply V and W with more structure.

To provide this extra structure, we note that two ordered bases $(v_1, \ldots, v_n)$ and $(v'_1, \ldots, v'_n)$ for V determine an isomorphism $f \colon V \to V$ with $f(v_i) = v'_i$; the matrix $A = (a_{ij})$ of f is given by the equations

$$v'_i = \sum_{j=1}^n a_{ji} v_j.$$

We call $(v_1, \ldots, v_n)$ and $(v'_1, \ldots, v'_n)$ **equally oriented** if $\det A > 0$ (i.e., if f is orientation preserving) and **oppositely oriented** if $\det A < 0$.

The relation of being equally oriented is clearly an equivalence relation, dividing the collection of all ordered bases into just two equivalence classes. Either of these two equivalence classes is called an **orientation** for V. The class to which $(v_1, \ldots, v_n)$ belongs will be denoted by $[v_1, \ldots, v_n]$, so that if μ is an orientation of V, then $(v_1, \ldots, v_n) \in \mu$ if and only if $[v_1, \ldots, v_n] = \mu$. If μ denotes one

---

*A cult word of the sixties, "grok" was coined, purportedly as a word from the Martian language, by Robert A. Heinlein in his pop science fiction novel *Stranger in a Strange Land*. Its sense is nicely conveyed by the definition in *The American Heritage Dictionary*: "To understand profoundly through intuition or empathy".

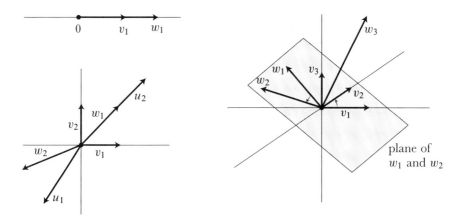

Examples of equally oriented ordered bases in $\mathbb{R}$, $\mathbb{R}^2$, and $\mathbb{R}^3$.

orientation of V, the other will be denoted by $-\mu$, and the orientation $[e_1, \ldots, e_n]$ for $\mathbb{R}^n$ will be called the "standard orientation".

Now if (V, μ) and (W, ν) are two n-dimensional vector spaces, together with orientations, an isomorphism $f : V \to W$ is called **orientation preserving (with respect to μ and ν)** if $[f(v_1), \ldots, f(v_n)] = \nu$ whenever $[v_1, \ldots, v_n] = \mu$; if this holds for any one $(v_1, \ldots, v_n)$, it clearly holds for all.

For the trivial bundle $\varepsilon^n(X) = X \times \mathbb{R}^n$ we can put the "standard orientation" $[(x, e_1), \ldots, (x, e_n)]$ on each fibre $\{x\} \times \mathbb{R}^n$. If $f : \varepsilon^n(X) \to \varepsilon^n(X)$ is an equivalence, and X is connected, then f is either orientation preserving or orientation reversing on each fibre, for if we define the functions $a_{ij} : X \to \mathbb{R}$ by

$$f(x, e_i) = \sum_{j=1}^{n} a_{ji}(x) \cdot (x, e_j),$$

then $\det(a_{ij}) : X \to \mathbb{R}$ is continuous and never 0. If $\pi : E \to B$ is a nontrivial n-plane bundle, an **orientation** μ of E is defined to be a collection of orientations μ_p for $\pi^{-1}(p)$ which satisfy the following "compatibility condition" for any open connected set $U \subset B$:

If $t : \pi^{-1}(U) \to U \times \mathbb{R}^n$ is an equivalence, and the fibres of $U \times \mathbb{R}^n$ are given the standard orientation, then t is either orientation preserving or orientation reversing on all fibres.

Notice that if this condition is satisfied for a certain t, and $t' : \pi^{-1}(U) \to U \times \mathbb{R}^n$ is another equivalence, then t' automatically satisfies the same condition, since

$t' \circ t^{-1} \colon U \times \mathbb{R}^n \to U \times \mathbb{R}^n$ is an equivalence. This shows that the orientations μ_p define an orientation of E if the compatibility condition holds for a collection of sets U which cover B.

If a bundle E has orientation $\mu = \{\mu_p\}$, it has another orientation $-\mu = \{-\mu_p\}$, but not every bundle has an orientation. For example, the Möbius strip, considered as a 1-dimensional bundle over S^1, has no orientation. For, although the Möbius strip has no non-zero section, we can pick two vectors from each fibre so that the totality A looks like two sections. For example, we can let A be $[0,1] \times \{-1,1\}$ with $(0,a)$ identified with $(1,-a)$; then A just looks like the boundary of the Möbius strip obtained from $[0,1] \times [-1,1]$. If

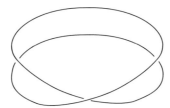

we had compatible orientations μ_p, we could define a section $s \colon S^1 \to M$ by choosing $s(p)$ to be the unique vector $s(p) \in A \cap \pi^{-1}(p)$ with $[s(p)] = \mu_p$.

A bundle is called **orientable** if it has an orientation, and **non-orientable** otherwise; an **oriented bundle** is just a pair (ξ, μ) where μ is an orientation for ξ. This definition can be applied, in particular, to the tangent bundle TM of a C^∞ manifold M. In this case, we call M itself **orientable** or **non-orientable** depending on whether TM is orientable or non-orientable; an orientation of TM is also called an **orientation** of M, and an **oriented manifold** is a pair (M, μ) where μ is an orientation for TM.

The manifold $\mathbb{R}^n$ is orientable, since $T\mathbb{R}^n \simeq \varepsilon^n(\mathbb{R}^n)$, on which we have the standard orientation. The sphere $S^{n-1} \subset \mathbb{R}^n$ is also orientable. To see this we

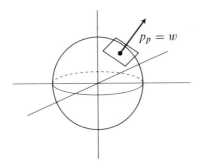

note that for each $p \in S^{n-1}$ the vector $w = p_p \in \varepsilon^n(\mathbb{R}^n) \simeq T\mathbb{R}^n$ is *not* in $i_*(S^{n-1}{}_p) \in T\mathbb{R}^n{}_p$ (Problem 21), so for $v_1, \dots, v_{n-1} \in S^{n-1}{}_p$ we can define $(v_1, \dots, v_{n-1}) \in \mu_p$ if and only if $(w, i_*(v_1), \dots, i_*(v_{n-1}))$ is in the standard orientation of $\mathbb{R}^n{}_p$. The orientation $\mu = \{\mu_p : p \in S^{n-1}\}$ thus defined is called the "standard orientation" of S^{n-1}.

The torus $S^1 \times S^1$ is another example of an orientable manifold. This can be seen by noting that for any two manifolds M_1 and M_2 the fibre $(M_1 \times M_2)_p$ of $T(M_1 \times M_2)$ can be written as $V_{1p} \oplus V_{2p}$ where $(\pi_i)_* : V_{ip} \to (M_i)_p$ is an isomorphism and the subspaces V_{ip} vary continuously (Problem 26). Since TS^1 is trivial, this shows that $T(S^1 \times S^1)$ is also trivial, and consequently orientable. Any n-holed torus is also orientable—the proof is presented in Problem 16, which also discusses the tangent bundle of a manifold-with-boundary.

The Möbius strip M is the simplest example of a non-orientable 2-manifold. For the imbedding of M considered previously we have already seen that on the

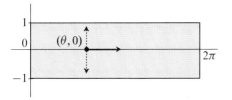
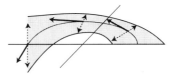

subset $S = \{(2\cos\theta, 2\sin\theta, 0)\} \subset M$ there are continuously varying vectors v_p, but that it is impossible to choose continuously from among the dashed vectors $w_p = f_*((0,1)_{(\theta,0)})$ and their negatives. If we had orientations μ_p for $p \in S$, then we could simply choose w_p if $[v_p, w_p] = \mu_p$ and $-w_p$ otherwise.

The projective plane $\mathbb{P}^2$ must be non-orientable also, since it contains the Möbius strip (for any orientable bundle $\xi = \pi : E \to B$, the restriction $\xi | B'$ to any subset $B' \subset B$ is also orientable). Non-orientability of $\mathbb{P}^2$ can be seen in another way, by considering the "antipodal map" $A : S^2 \to S^2$ defined by $A(p) = -p$. This map is just the restriction of a linear map $\bar{A} : \mathbb{R}^3 \to \mathbb{R}^3$ defined by the same formula. The map $A_* : S^2{}_p \to S^2{}_{A(p)}$ is just $(p, v) \mapsto (\bar{A}(p), \bar{A}(v))$, when $S^2{}_p$ is identified with a subspace of $\{p\} \times \mathbb{R}^3$. The map $\bar{A}$ is orientation reversing, so if $v_i = (p, u_i) \in S^2{}_p$, the bases

$$(u_1, u_2, p) \quad \text{and} \quad (\bar{A}(u_1), \bar{A}(u_2), \bar{A}(p))$$

are oppositely oriented. This shows that if μ is the standard orientation of S^2 and $[v_1, v_2] \in \mu_p$, then $[A_*v_1, A_*v_2] \in -\mu_{A(p)}$. Thus the map $A : S^2 \to S^2$ is "orientation reversing" (the notion of an **orientation preserving** or **orientation reversing** map $f: M \to N$ makes sense for any imbedding f of one oriented manifold into another oriented manifold of the same dimension). From this fact it follows easily that $\mathbb{P}^2$ is not orientable: If $\mathbb{P}^2$ had an orientation $\nu = \{\nu_{[p]}\}$ and $g: S^2 \to \mathbb{P}^2$ is the map $p \mapsto [p]$, then we could define an orientation $\{\bar{\mu}_p\}$ on S^n by requiring g to be orientation preserving; the map A would then be orientation preserving with respect to $\bar{\mu}$, which is impossible, since $\bar{\mu} = \mu$ or $-\mu$.

For projective 3-space $\mathbb{P}^3$ the situation is just the opposite. In this case, the antipodal map $A: S^3 \to S^3$ *is* orientation preserving. If $g: S^3 \to \mathbb{P}^3$ is the map $p \mapsto [p]$, we obviously can define orientations ν_p for $\mathbb{P}^3$ by requiring g to be orientation preserving. In general, these same arguments show that $\mathbb{P}^n$ is orientable for n odd and non-orientable for n even.

There is a more "elementary" definition of orientability, which does not use the tangent bundle of M at all. According to this definition, M is orientable if there is a subset $\mathcal{A}'$ of the atlas $\mathcal{A}$ for M such that

(1) the domains of all $(x, U) \in \mathcal{A}'$ cover M,
(2) for all (x, U) and $(y, V) \in \mathcal{A}'$,

$$\det\left(\frac{\partial y^i}{\partial x^j}\right) > 0 \quad \text{on} \quad U \cap V.$$

An orientation μ of TM allows us to distinguish the subset $\mathcal{A}'$ as the collection of all (x, U) for which $x_*: TM|U \to T(x(U)) \simeq x(U) \times \mathbb{R}^n$ is orientation preserving (when $x(U) \times \mathbb{R}^n$ is given the standard orientation). Condition (2) holds, because it is just the condition that $(y \circ x^{-1})_*: T(x(U)) \to T(x(U))$ is orientation preserving. Conversely, given $\mathcal{A}'$ we can orient the fibres of $TM|U$ in such a way that x_* is orientation preserving, and obtain an orientation of TM. Although our original definition is easier to picture geometrically, the determinant condition will be very important later on.

ADDENDUM
EQUIVALENCE OF TANGENT BUNDLES

The fact that all reasonable candidates for the tangent bundle of M turn out to be essentially the same is stated precisely as follows.

4. THEOREM*. If we have a bundle $T'M$ over M for each M, and a bundle map $(f_\sharp, f)$ for each C^∞ map $f : M \to N$ satisfying

(1) of Theorem 1,

(2) of Theorem 1, for certain equivalences t'^n,

(3) of Theorem 1, for certain equivalences $T'U \simeq (T'M)|U$,

then there are equivalences

$$e_M : TM \to T'M$$

such that the following diagram commutes for every C^∞ map $f : M \to N$.

$$
\begin{array}{ccc}
TM & \xrightarrow{\ f_* \ } & TN \\
e_M \downarrow & & \downarrow e_N \\
T'M & \xrightarrow{\ f_\sharp \ } & T'N
\end{array}
$$

PROOF. The details of this proof are so horrible that you should probably skip it (and you should definitely quit when you get bogged down); the welcome symbol ❖ occurs quite a ways on. Nevertheless, the idea behind the proof is simple enough. If (x, U) is a chart on M, then both $(TM)|U$ and $(T'M)|U$ "look like" $x(U) \times \mathbb{R}^n$, so there ought to be a map taking the fibres of one to the fibres of the other. What we have to hope is that our conditions on TM and $T'M$ make them "look alike" in a sufficiently strong way for this idea to really work out. Those who have been through this sort of rigamarole before know (i.e., have faith) that it's going to work out; those for whom this sort of proof is a new experience should find it painful and instructive.

---

*Functorites will notice that Theorems 1 and 4 say that there is, up to natural equivalence, a unique functor from the category of C^∞ manifolds and C^∞ maps to the category of bundles and bundle maps which is naturally equivalent to $(\varepsilon^n, \text{old } f_*)$ on Euclidean spaces, and to the restriction of the functor on open submanifolds.

Let (x, U) be a coordinate system on M. Then we have the following string of equivalences. Two of them, which are denoted by the same symbol $\simeq$, are the equivalences mentioned in condition (3). Let α_x denote the composition $\alpha_x = (t^n | x(U)) \circ \simeq \circ x_* \circ (\simeq)^{-1}$.

$$(TM)|U \xleftarrow{\ \simeq\ } TU \xrightarrow{\ x_*\ } T(x(U)) \xrightarrow{\ \simeq\ } (T\mathbb{R}^n)|x(U) \xrightarrow{\ t^n|x(U)\ } \varepsilon^n(\mathbb{R}^n)|x(U)$$

$$\alpha_x$$

Similarly, using equivalence $\simeq'$ for T', we can define β_x.

$$(T'M)|U \xleftarrow{\ \simeq'\ } T'U \xrightarrow{\ x_\sharp\ } T'(x(U)) \xrightarrow{\ \simeq'\ } (T\mathbb{R}^n)|x(U) \xrightarrow{\ t'^n|x(U)\ } \varepsilon^n(\mathbb{R}^n)|x(U)$$

$$\beta_x$$

Then

$$\beta_x^{-1} \circ \alpha_x : (TM)|U \to (T'M)|U$$

is an equivalence, so it takes the fibre of TM over p isomorphically to the fibre of $T'M$ over p for each $p \in U$. Our main task is to show that this isomorphism between the fibres over p is independent of the coordinate system (x, U). This will be done in three stages.

(I) Suppose $V \subset U$ is open and $y = x|V$. We will need to name all the inclusion maps

$$i : U \to M$$
$$\bar{\imath} : V \to M$$
$$j : V \to U$$
$$k : y(V) \to x(U).$$

To compare α_x and α_y, consider the following diagram.

(1)

$$
\begin{array}{ccccccccc}
(TM)|U & \xleftarrow{\simeq} & TU & \xrightarrow{x_*} & T(x(U)) & \xrightarrow{\simeq} & (T\mathbb{R}^n)|x(U) & \xrightarrow{t^n|x(U)} & \varepsilon^n(\mathbb{R}^n)|x(U) \\
\uparrow{\scriptstyle\subset} & \textcircled{1} & \uparrow{\scriptstyle j_*} & \textcircled{2} & \uparrow{\scriptstyle k_*} & \textcircled{3} & \uparrow{\scriptstyle\subset} & \textcircled{4} & \uparrow{\scriptstyle\subset} \\
(TM)|V & \xleftarrow{\simeq} & TV & \xrightarrow{y_*} & T(y(V)) & \xrightarrow{\simeq} & (T\mathbb{R}^n)|y(V) & \xrightarrow{t^n|y(V)} & \varepsilon^n(\mathbb{R}^n)|y(V)
\end{array}
$$

Each of the four squares in this diagram commutes. To see this for square ①, we enlarge it, as shown below. The two triangles on the left commute by condition (3) for *TM*, and the one on the right commutes because $i \circ j = \bar{i}$.

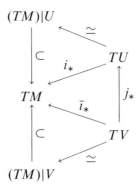

Square ② commutes because $k \circ y = x \circ j$. Square ③ commutes for the same reason as square ①; the inclusions $x(U) \to \mathbb{R}^n$ and $y(V) \to \mathbb{R}^n$ come into play. Square ④ obviously commutes. Chasing through diagram (**1**) now shows that the following commutes.

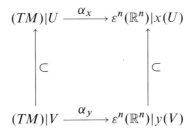

This means that for $p \in V$, the isomorphism α_y between the fibres over p is the same as α_x. Clearly the same is true for β_x and β_y, since our proof used only properties (1), (2), and (3), not the explicit construction of *TM*. Thus $\beta_y^{-1} \circ \alpha_y = \beta_x^{-1} \circ \alpha_x$ on the fibres over p, for every $p \in V$.

(II) We now need a Lemma which applies to both *TM* and $T'M$. Again, it will be proved for *TM* (where it is actually obvious), using only properties (1), (2), and (3), so that it is also true for $T'M$.

LEMMA. If $A \subset \mathbb{R}^n$ and $B \subset \mathbb{R}^m$ are open, and $f \colon A \to B$ is C^∞, then the following diagram commutes.

$$
\begin{array}{ccccc}
TA & \overset{\simeq}{\longrightarrow} & (T\mathbb{R}^n)|A & \overset{t^n|A}{\longrightarrow} & \varepsilon^n(\mathbb{R})^n|A \\
\downarrow{\scriptstyle f_*} & & & & \downarrow{\scriptstyle \text{old } f_*} \\
TB & \overset{\simeq}{\longrightarrow} & (T\mathbb{R}^m)|B & \overset{t^m|B}{\longrightarrow} & \varepsilon^m(\mathbb{R})^m|B
\end{array}
$$

PROOF. *Case 1.* There is a map $\bar{f} \colon \mathbb{R}^n \to \mathbb{R}^m$ with $\bar{f} = f$ on A. Consider the following diagram, where $i \colon A \to \mathbb{R}^n$ and $j \colon B \to \mathbb{R}^m$ are the inclusion maps.

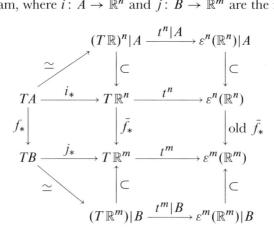

Everything in this diagram obviously commutes. This implies that the two compositions

$$
TA \overset{\simeq}{\longrightarrow} (T\mathbb{R}^n)|A \overset{t^n|A}{\longrightarrow} \varepsilon^n(\mathbb{R}^n)|A \overset{\subset}{\longrightarrow} \varepsilon^n(\mathbb{R}^n) \overset{\text{old } \bar{f}_*}{\longrightarrow} \varepsilon^m(\mathbb{R}^m)
$$

and

$$
TA \overset{f_*}{\longrightarrow} TB \overset{\simeq}{\longrightarrow} (T\mathbb{R}^m)|B \overset{t^m|B}{\longrightarrow} \varepsilon^m(\mathbb{R}^m)|B \overset{\subset}{\longrightarrow} \varepsilon^m(\mathbb{R}^m)
$$

are equal and this proves the Lemma in Case 1, since the maps "old $\bar{f}_*$" and "old f_*" are equal on A.

Case 2. General case. For each $p \in A$, we want to show that two maps are the same on the fibre over p. Now there is a map $\bar{f} \colon \mathbb{R}^n \to \mathbb{R}^m$ with $\bar{f} = f$ on an open set A', where $p \in A' \subset A$. We then have the following diagram, where every $\simeq$ comes from the fact that some set is an open submanifold of another, and $i: A' \to A$ is the inclusion map.

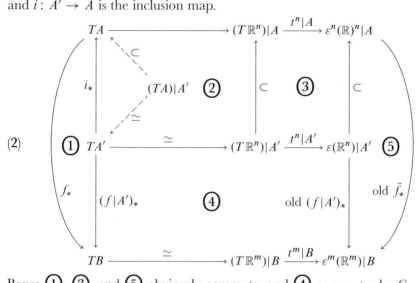

(2)

Boxes ①, ③, and ⑤ obviously commute, and ④ commutes by *Case 1*. To see that square ② (which has a triangle within it) commutes, we imbed it in a larger diagram, in which $j: A \to \mathbb{R}^n$ is the inclusion map, and other maps have also been named, for ease of reference.

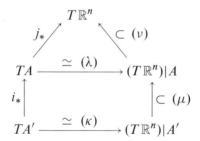

To prove that $\lambda \circ i_* = \mu \circ \kappa$, it suffices to prove that

$$\nu \circ \lambda \circ i_* = \nu \circ \mu \circ \kappa,$$

since ν is one-one. Thus it suffices to prove $j_* \circ i_* = \nu \circ \mu \circ \kappa$, which amounts to proving commutativity of the following diagram.

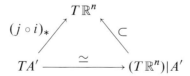

Since $j \circ i$ is just the inclusion of A' in $\mathbb{R}^n$, this does commute.

Commutativity of diagram **(2)** shows that the composition

$$TA \xrightarrow{f_*} TB \xrightarrow{\simeq} (T\mathbb{R}^m)|B \xrightarrow{t^m|B} \varepsilon^m(\mathbb{R}^m)|B$$

coincides, *on the subset* $(TA)|A'$, with the composition

$$TA \xrightarrow{\simeq} (T\mathbb{R}^n)|A \xrightarrow{t^n|A} \varepsilon^n(\mathbb{R}^n)|A \xrightarrow{\text{old } \bar{f}_*} \varepsilon^m(\mathbb{R}^m)|B,$$

and on A' we can replace "old $\bar{f}_*$" by "old f_*". In other words, the two compositions are equal in a neighborhood of any $p \in A$, and are thus equal, which proves the Lemma.

(III) Now suppose (x, U) and (y, V) are any two coordinate systems with $p \in U \cap V$. To prove that $\beta_y^{-1} \circ \alpha_y$ and $\beta_x^{-1} \circ \alpha_y$ induce the same isomorphism on the fibre of TM at p, we can assume without loss of generality that $U = V$, because part (I) applies to x and $x|U \cap V$, as well as to y and $y|U \cap V$.

Assuming $U = V$, we have the following diagram.

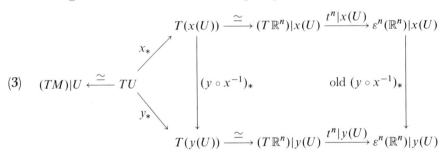

The triangle obviously commutes, and the rectangle commutes by part (II). Diagram **(3)** thus shows that

$$\alpha_y = \text{old } (y \circ x^{-1})_* \circ \alpha_x.$$

Exactly the same result holds for T':

$$\beta_y = \text{old } (y \circ x^{-1})_* \circ \beta_x.$$

The desired result $\beta_y^{-1} \circ \alpha_y = \beta_x^{-1} \circ \alpha_x$ follows immediately.

Now that we have a well-defined bundle map $TM \to T'M$ (the union of all $\beta_x^{-1} \circ \alpha_x$), it is clearly an equivalence e_M. The proof that $e_N \circ f_* = f_\sharp \circ e_M$ is left as a masochistic exercise for the reader. ❖

PROBLEMS

1. Let M be any set, and $\{(x_i, U_i)\}$ a sequence of one-one functions $x_i : U_i \to \mathbb{R}^n$ with $U_i \subset M$ and $x(U_i)$ open in $\mathbb{R}^n$, such that each

$$x_j \circ x_i^{-1} : x_i(U_i \cap U_j) \to x_j(U_i \cap U_j)$$

is continuous. It would seem that M ought to have a metric which makes each U_i open and each x_i a homeomorphism. Actually, this is not quite true:

(a) Let $M = \mathbb{R} \cup \{*\}$, where $* \notin \mathbb{R}$. Let $U_1 = \mathbb{R}$ and $x_1 : U_1 \to \mathbb{R}$ be the identity, and let $U_2 = \mathbb{R} - \{0\} \cup \{*\}$, with $x_2 : U_2 \to \mathbb{R}$ defined by

$$x_2(a) = a, \quad a \neq 0, *$$
$$x_2(*) = 0.$$

Show that there is no metric on M of the required sort, by showing that every neighborhood of 0 would have to intersect every neighborhood of $*$. Nevertheless, we can find on M a pseudometric ρ (a function $\rho : M \times M \to \mathbb{R}$ with all properties for a metric except that $\rho(p, q)$ may be 0 for $p \neq q$) such that ρ is a metric on each U_i and each x_i is a homeomorphism:

(b) If $A \subset \mathbb{R}^n$ is open, then there is a sequence $A_1, A_2, A_3, \ldots$ of open subsets of A such that every open subset of A is a union of certain A_i's.

(c) There is a sequence of continuous functions $f_i : A \to [0, 1]$, with support $f_i \subset A$, which "separates points and closed sets": if C is closed and $p \in A - C$, then there is some f_i with $f_i(p) \notin \overline{f_i(A \cap C)}$. *Hint:* First arrange in a sequence all pairs (A_i, A_j) of part (b) with $\overline{A_i} \subset A_j$.

(d) Let $f_{i,j}$, $j = 1, 2, 3, \ldots$ be such a sequence for each open set $x_i(U_i)$. Define $g_{i,j} : M \to [0, 1]$ by

$$g_{i,j}(p) = \begin{cases} f_{i,j}(p) & p \in U_i \\ 0 & p \notin U_i. \end{cases}$$

Arrange all $g_{i,j}$ in a single sequence $G_1, G_2, G_3, \ldots$, let d be a bounded metric on $\mathbb{R}$, and define ρ on M by

$$\rho(p, q) = \sum_{i=1}^{\infty} \frac{1}{2^i} d\big(G_i(p), G_i(q)\big).$$

Show that ρ is the required pseudometric.

(e) Suppose that for every $p, q \in M$ there is a U_i and U_j with $p \in U_i$ and $q \in U_j$ and open sets $B_i \subset x_i(U_i)$ and $B_j \subset x_j(U_j)$ so that $p \in x_i^{-1}(B_i)$, $q \in x_j^{-1}(B_j)$, and $x_i^{-1}(B_i) \cap x_j^{-1}(B_j) = \emptyset$. Show that ρ is actually a metric on M.

2. (a) Suppose (x, U) and (y, V) are two coordinate systems, giving rise to two maps on TM,

$$t_x : \pi^{-1}(U) \to U \times \mathbb{R}^n, \quad [x, v]_q \mapsto (q, v),$$
$$t_y : \pi^{-1}(V) \to V \times \mathbb{R}^n, \quad [y, w]_q \mapsto (q, w).$$

Show that in $\pi^{-1}(U \cap V)$ the sets of the form $t_x{}^{-1}(A)$ for $A \subset U \times \mathbb{R}^n$ open are exactly the sets of the form $t_y{}^{-1}(B)$ for $B \subset V \times \mathbb{R}^n$ open.
(b) Show that if there is a metric on TM such that t_{x_i} is a homeomorphism for a collection (x_i, U_i) with $M = \bigcup_i U_i$, then all t_x are homeomorphisms.
(c) Conclude from Problem 1 that there is a metric on TM which makes each t_x a homeomorphism.

3. Show that in the definition of an equivalence it suffices to assume that the map $E_1 \to E_2$ is continuous. (To prove the inverse continuous, note that locally it is just a map $U \times \mathbb{R}^n \to U \times \mathbb{R}^n$).

4. Show that in the definition of a bundle map, continuity of $f : B_1 \to B_2$ follows automatically from continuity of $\tilde{f} : E_1 \to E_2$.

5. A **weak equivalence** between two bundles over the same base space B is a bundle map $(\tilde{f}, f)$ where $\tilde{f}$ is an isomorphism on each fibre, and f is a homeomorphism of B onto itself. Find two inequivalent, but weakly equivalent, bundles over the following base spaces:

(i) the disjoint union of two circles,

(ii) a figure eight ,

(iii) the torus.

6. Given a bundle map $(\tilde{f}, f)$, show that $\tilde{f} = g \circ h$ where g and h are continuous maps such that h takes fibres linearly to fibres, while g is an isomorphism on each fibre.

7. (a) Show that for any bundle $\pi : E \to B$, the map $s : B \to E$ with $s(p)$ the 0 vector of $\pi^{-1}(p)$ is a section.
(b) Show that an n-plane bundle ξ is trivial if and only if there are n sections $s_1, \ldots, s_n$ which are everywhere linearly independent, i.e., $s_1(p), \ldots, s_n(p) \in \pi^{-1}(p)$ are linearly independent for all $p \in B$.
(c) Show that locally every n-plane bundle has n linearly independent sections.

8. (a) Check that $\underset{p}{\sim}$ is an equivalence relation on the set of pairs (x, v).
(b) Check that the definition of f_* is independent of the coordinate systems x and y which are used.
(c) Check the remaining details in Theorem 1.

9. (a) Show that the correspondence between TM and equivalence classes of curves under which $[x, v]_p$ corresponds to the $\underset{p}{\approx}$ equivalence class of $x^{-1} \circ \gamma$, for γ a curve in $\mathbb{R}^n$ with $\gamma'(0) = v$, makes f_* correspond to $f_\#$.

(b) Show that under the correspondence $[x, a]_p \mapsto \sum_i a^i \partial/\partial x^i \big|_p$, the map f_* can be defined by

$$[f_*(\ell)](g) = \ell(g \circ f).$$

10. If V is a finite dimensional vector space over $\mathbb{R}$, define a C^∞ structure on V and a homeomorphism from $V \times V$ to TV which is independent of choice of bases. As in the case of $\mathbb{R}^n$, for $v, w \in V$ we will denote by $v_w \in V_w$ the vector corresponding to (w, v).

11. If $g \colon \mathbb{R} \to \mathbb{R}$ is C^∞ show that

$$g(x) = g(0) + g'(0)x + x^2 h(x)$$

for some C^∞ function $h \colon \mathbb{R} \to \mathbb{R}$.

12. (a) Let $\mathcal{F}_p$ be the set of all C^∞ functions $f \colon M \to \mathbb{R}$ with $f(p) = 0$, and let $\ell \colon \mathcal{F}_p \to \mathbb{R}$ be a linear operator with $\ell(fg) = 0$ for all $f, g \in \mathcal{F}_p$. Show that ℓ has a unique extension to a derivation.

(b) Let W be the vector subspace of $\mathcal{F}_p$ generated by all products fg for $f, g \in \mathcal{F}_p$. Show that the vector space of all derivations at p is isomorphic to the dual space $(\mathcal{F}_p/W)^*$.

(c) Since $(\mathcal{F}_p/W)^*$ has dimension $n = $ dimension of M, the same must be true of $\mathcal{F}_p/W$. If x is a coordinate system with $x(p) = 0$, show that $x^1 + W, \ldots,$ $x^n + W$ is a basis for $\mathcal{F}_p/W$ (use Lemma 2). The situation is quite different for C^1 functions, as the next problem shows.

13. (a) Let V be the vector space of all C^1 functions $f \colon \mathbb{R} \to \mathbb{R}$ with $f(0) = 0$, and let W be the subspace generated by all products. Show that $\lim_{x \to 0} f(x)/x^2$ exists for all $f \in W$.

(b) For $0 < \varepsilon < 1$, let

$$f_\varepsilon(x) = \begin{cases} x^{1+\varepsilon} & x \geq 0 \\ 0 & x \leq 0. \end{cases}$$

Show that all f_ε are in V, and that they represent linearly independent elements of V/W.

(c) Conclude that $(V/W)^*$ has dimension $\mathfrak{c}^{\mathfrak{c}} = 2^{\mathfrak{c}}$.

14. If $f \colon M \to N$ and f_* is the 0 map on each fibre, then f is constant on each component of M.

15. (a) A map $f: M \to N$ is an immersion if and only if f_* is one-one on each fibre of TM. More generally, the rank of f at $p \in M$ is the rank of the linear transformation $f_*: M_p \to N_{f(p)}$.

(b) If $f \circ g = f$, where g is a diffeomorphism, then the rank of $f \circ g$ at a equals the rank of f at $g(a)$. (Compare with Problem 2-33(d).)

16. (a) If M is a manifold-with-boundary, the tangent bundle TM is defined exactly as for M; elements of M_p are $\underset{p}{\sim}$ equivalence classes of pairs (x, v). Although x takes a neighborhood of $p \in \partial M$ onto $\mathbb{H}^n$, rather than $\mathbb{R}^n$, the vectors v still run through $\mathbb{R}^n$, so M_p still has tangent vectors "pointing in all directions". If $p \in \partial M$ and $x: U \to \mathbb{H}^n$ is a coordinate system around p, then

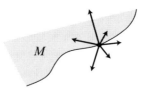

$x_*^{-1}(\mathbb{R}^{n-1}{}_{x(p)}) \subset M_p$ is a subspace. Show that this subspace does not depend on the choice of x; in fact, it is $i_*(\partial M)_p$, where $i: \partial M \to M$ is the inclusion.

(b) Let $a \in \mathbb{R}^{n-1} \times \{0\} \subset \mathbb{H}^n$. A tangent vector in $\mathbb{H}^n{}_a$ is said to point "inward" if, under the identification of $T\mathbb{H}^n$ with $\varepsilon^n(\mathbb{H}^n)$, the vector is (a, v) where $v^n > 0$. A vector $v \in M_p$ which is not in $i_*(\partial M)_p$ is said to point "inward" if

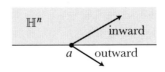

$x_*(v) \in \mathbb{H}^n{}_{x(p)}$ points inward. Show that this definition does not depend on the coordinate system x.

(c) Show that if M has an orientation μ, then ∂M has a unique orientation $\partial \mu$ such that $[v_1, \ldots, v_{n-1}] = (\partial \mu)_p$ if and only if $[w, i_* v_1, \ldots, i_* v_{n-1}] = \mu_p$ for every outward pointing $w \in M_p$.

(d) If μ is the usual orientation of $\mathbb{H}^n$, show that $\partial \mu$ is $(-1)^n$ times the usual orientation of $\mathbb{R}^{n-1} = \partial \mathbb{H}^n$. (The reason for this choice will become clear in Chapter 8.)

(e) Suppose we are in the setup of Problem 2-14. Define $g: \partial M \times [0, 1) \to \partial N \times [0, 1)$ by $g(p, t) = (f(p), t)$. Show that TP is obtained from $TM \cup TN$

by identifying

$$v \in (\partial M)_p \quad \text{with} \quad (\beta^{-1})_* g_* \alpha_*(v) \in (\partial N)_{f(p)}.$$

(f) If M and N have orientations μ and ν and $f\colon (\partial M, \partial \mu) \to (\partial N, \partial \nu)$ is orientation-*reversing*, show that P has an orientation which agrees with μ and ν on $M \subset P$ and $N \subset P$.

(g) Suppose M is S^2 with two holes cut out, and N is $[0, 1] \times S^1$. Let f be a diffeomorphism from M to N which is orientation preserving on one copy of S^1 and orientation reversing on the other. What is the resulting manifold P?

17. Show that $T\mathbb{P}^2$ is homeomorphic to the space obtained from $T(S^2, i)$ by identifying $(p, v) \in (S^2, i)_p$ with $(-p, -v) \in (S^2, i)_{-p}$.

18. Although there is no everywhere non-zero vector field on S^2, there is one on $S^2 - \{(0, 0, 1)\}$, which is diffeomorphic to $\mathbb{R}^2$. Show that such a vector field can be picked so that near $(0, 0, 1)$ the vector field looks like the following picture (a "magnetic dipole"):

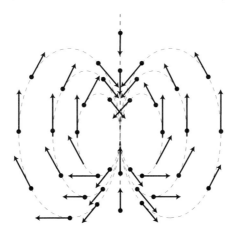

19. Suppose we have a "multiplication" map $(a, b) \mapsto a \cdot b$ from $\mathbb{R}^n \times \mathbb{R}^n$ to $\mathbb{R}^n$ that makes $\mathbb{R}^n$ into a (non-associative) division algebra. That is,

$$
\begin{aligned}
(a_1 + a_2) \cdot b &= a_1 \cdot b + a_2 \cdot b \\
a \cdot (b_1 + b_2) &= a \cdot b_1 + a \cdot b_2 \\
\lambda(a \cdot b) &= (\lambda a) \cdot b = a \cdot (\lambda b) \quad \text{for } \lambda \in \mathbb{R} \\
a \cdot (1, 0, \dots, 0) &= a
\end{aligned}
$$

and there are no zero divisors:

$$a, b \neq 0 \implies ab \neq 0.$$

(For example, for $n = 1$, we can use ordinary multiplication, and for $n = 2$ we can use "complex multiplication", $(a, b) \cdot (c, d) = (ac - bd, ad + bc)$.) Let $e_1, \ldots, e_n$ be the standard basis of $\mathbb{R}^n$.

(a) Every point in S^{n-1} is $a \cdot e_1$ for a unique $a \in \mathbb{R}^n$.

(b) If $a \neq 0$, then $a \cdot e_1, \ldots, a \cdot e_n$ are linearly independent.

(c) If $p = a \cdot e_1 \in S^{n-1}$, then the projection of $a \cdot e_2, \ldots, a \cdot e_n$ on $(S^{n-1}, i)_p$ are linearly independent.

(d) Multiplication by a is continuous.

(e) TS^{n-1} is trivial.

(f) $T\mathbb{P}^{n-1}$ is trivial.

The tangent bundles TS^3 and TS^7 are both trivial. Multiplications with the required properties on $\mathbb{R}^4$ and $\mathbb{R}^8$ are provided by the "quaternions" and "Cayley numbers", respectively; the quaternions are not commutative and the Cayley numbers are not even associative. It is a classical theorem that the reals, complexes, and quaternions are the only associative examples. For a simple proof, see R. S. Palais, *The Classification of Real Division Algebras*, Amer. Math. Monthly **75** (1968), 366–368. J. F. Adams has proved, using methods of algebraic topology, that $n = 1, 2, 4,$ or 8.

[Incidentally, non-existence of zero divisors immediately implies that for $a \neq 0$ there is some b with $ab = (1, 0, \ldots, 0)$ and b' with $b'a = (1, 0, \ldots, 0)$. If the multiplication is associative it follows easily that $b = b'$, so that we always have multiplicative inverses. Conversely, this condition implies that there are no zero divisors if the multiplication is associative; otherwise it suffices to assume the existence of a *unique* b with $a \cdot b = b \cdot a = (1, 0, \ldots, 0)$.]

20. (a) Consider the space obtained from $[0, 1] \times \mathbb{R}^n$ by identifying $(0, v)$ with $(1, Tv)$, where $T : \mathbb{R}^n \to \mathbb{R}^n$ is a vector space isomorphism. Show that this can be made into the total space of a vector bundle over S^1 (a generalized Möbius strip).

(b) Show that the resulting bundle is orientable if and only if T is orientation preserving.

21. Show that for $p \in S^2$, the vector $p_p \in \mathbb{R}^3{}_p$ is not in $i_*(S^2{}_p)$ by showing that the inner product $\langle p, c'(0) \rangle = 0$ for all curves c with $c(0) = p$ and $|c(t)| = 1$ for all t. (Recall that

$$\langle f, g \rangle'(t) = \langle f'(t)^{\mathbf{t}}, g(t) \rangle + \langle f(t), g'(t)^{\mathbf{t}} \rangle,$$

where $^{\mathbf{t}}$ denotes the transpose; see *Calculus on Manifolds*, pg. 23.)

22. Let M be a C^∞ manifold. Suppose that $(TM)|A$ is trivial whenever $A \subset M$ is homeomorphic to S^1. Show that M is orientable. *Hint:* An arc c from

$p_0 \in M$ to $p \in M$ is contained in some such A so $(TM)|c$ is trivial. Thus one can "transport" the orientation of M_{p_0} to M_p. It must be checked that this is independent of the choice of c. First consider pairs c, c' which meet only at p_0 and p. The general, possibly quite messy, case can be treated by breaking up c into small pieces contained in coordinate neighborhoods.

Remark: Using results from the Addendum to Chapter 9, together with Problem 29, we can conclude that a neighborhood of some $S^1 \subset M$ is non-orientable if M is non-orientable.

The next two problems deal with important constructions associated with vector bundles.

23. (a) Suppose $\xi = \pi\colon E \to X$ is a bundle and $f\colon Y \to X$ is a continuous map. Let $E' \subset Y \times E$ be the set of all (y, e) with $f(y) = \pi(e)$, define $\pi'\colon E' \to Y$ by $\pi'(y, e) = y$, and define $\tilde{f}\colon E' \to E$ by $\tilde{f}(y, e) = e$. A vector space structure can be defined on

$$\pi'^{-1}(y) = \{(y, e)\colon e \in \pi^{-1}(f(y))\}$$

by using the vector space structure on $\pi^{-1}(f(y))$. Show that $\pi'\colon E' \to Y$ is a bundle, and $(\tilde{f}, f)$ a bundle map which is an isomorphism on each fibre. This bundle is denoted by $f^*(\xi)$, and is called the bundle **induced** (from ξ) by f.
(b) Suppose we have another bundle $\xi'' = \pi''\colon E'' \to Y$ and a bundle map $(\tilde{\tilde{f}}, f)$ from ξ'' to ξ which is an isomorphism on each fibre. Show that $\xi'' \simeq \xi' = f^*(\xi)$. *Hint*: Map $e \in E''$ to $(\pi''(e), \tilde{\tilde{f}}(e)) \in E'$.
(c) If $g\colon Z \to Y$, then $(f \circ g)^*(\xi) \simeq g^*(f^*(\xi))$.
(d) If $A \subset X$ and $i\colon A \to X$ is the inclusion map, then $i^*(\xi) \simeq \xi|A$.
(e) If ξ is orientable, then $f^*(\xi)$ is also orientable.
(f) Give an example where ξ is non-orientable, but $f^*(\xi)$ is orientable.
(g) Let $\xi = \pi\colon E \to B$ be a vector bundle. Since $\pi\colon E \to B$ is a continuous map from a space to the base space B of ξ, the symbol $\pi^*(\xi)$ makes sense. Show that if ξ is not orientable, then $\pi^*(\xi)$ is not orientable.

24. (a) Given an n-plane bundle $\xi = \pi\colon E \to B$ and an m-plane bundle $\eta = \pi'\colon E' \to B$, let $E'' \subset E \times E'$ be the set of all pairs (e, e') with $\pi(e) = \pi'(e')$. Let $\pi''(e, e') = \pi(e) = \pi'(e')$. Show that $\pi''\colon E'' \to B$ is an $(n + m)$-plane bundle. It is called the **Whitney sum** $\xi \oplus \eta$ of ξ and η; the fibre of $\xi \oplus \eta$ over p is the direct sum $\pi^{-1}(p) \oplus \pi'^{-1}(p)$.
(b) If $f\colon Y \to B$, show that $f^*(\xi \oplus \eta) \simeq f^*(\xi) \oplus f^*(\eta)$.

(c) Given bundles $\xi_i = \pi_i : E_i \to B_i$, define $\pi : E_1 \times E_2 \to B_1 \times B_2$ by $\pi(e_1, e_2) = (\pi_1(e_1), \pi_2(e_2))$. Show that this is a bundle $\xi_1 \times \xi_2$ over $B_1 \times B_2$.
(d) If $\Delta : B \to B \times B$ is the "diagonal map", $\Delta(x) = (x, x)$, show that $\xi \oplus \eta \simeq \Delta^*(\xi \times \eta)$.
(e) If ξ and η are orientable, show that $\xi \oplus \eta$ is orientable.
(f) If ξ is orientable, and η is non-orientable, show that $\xi \oplus \eta$ is also non-orientable.
(g) Define a "natural" orientation on $V \oplus V$ for any vector space V, and use this to show that $\xi \oplus \xi$ is always orientable.
(h) If X is a "figure eight" (c.f. Problem 5), find two non-orientable 1-plane bundles ξ and η over X such that $\xi \oplus \eta$ is also non-orientable.

25. (a) If $\pi : E \to M$ is a C^∞ vector bundle, then π_* has maximal rank at each point, and each fibre $\pi^{-1}(p)$ is a C^∞ submanifold of E.
(b) The 0-section of E is a submanifold, carried diffeomorphically onto B by π.

26. (a) If M and N are C^∞ manifolds, and π_M [or π_N] $: M \times N \to M$ [or N] is the projection on M [or N], then $T(M \times N) \simeq \pi_M^*(TM) \oplus \pi_N^*(TN)$.
(b) If M and N are orientable, then $M \times N$ is orientable.
(c) If $M \times N$ is orientable, then both M and N are orientable.

27. Show that the Jacobian matrix of $y_* \circ (x_*)^{-1}$ is of the form

$$\begin{pmatrix} D_j y^i \circ x^{-1} & 0 \\ \times & D_j y^i \circ x^{-1} \end{pmatrix}.$$

This shows that the manifold TM is *always orientable*, i.e., the bundle $T(TM)$ is orientable. (Here is a more conceptual formulation: for $v \in TM$, the orientation for $(TM)_v$ can be defined as

$$\left[\frac{\partial}{\partial(x^1 \circ \pi)} \bigg|_v, \dots, \frac{\partial}{\partial(x^n \circ \pi)} \bigg|_v, \frac{\partial}{\partial \dot{x}^1} \bigg|_v, \dots, \frac{\partial}{\partial \dot{x}^n} \bigg|_v \right];$$

the form of $y_* \circ (x_*)^{-1}$ shows that this orientation is independent of the choice of x.) A different proof that TM is orientable is given in Problem 29.

28. (a) Let (x, U) be a coordinate system on M with $x(p) = 0$ and let $v \in M_p$ be $\sum_{i=1}^n a^i \, \partial/\partial x^i \big|_p$. Consider the curve c in TM defined by

$$c(t) = v + t \frac{\partial}{\partial x^i} \bigg|_p.$$

Show that

$$\frac{dc}{dt}(0) = \frac{\partial}{\partial \dot{x}^i}\bigg|_v .$$

(b) Find a curve whose tangent vector at 0 is $\partial/\partial(x^i \circ \pi)\big|_v$.

29. This problem requires some familiarity with the notion of exact sequences

(c.f. Chapter 11). A sequence of bundle maps $E_1 \xrightarrow{\tilde{f}} E_2 \xrightarrow{\tilde{g}} E_3$ with $f = g =$ identity of B is **exact** if at each fibre it is exact as a sequence of vector space maps.

(a) If $\xi = \pi\colon E \to B$ is a C^∞ vector bundle, show that there is an exact sequence

$$0 \to \pi^*(\xi) \to TE \to \pi^*(TB) \to 0.$$

Hint: (1) An element of the total space of $\pi^*(\xi)$ is a pair of points in the same fibre, which determines a tangent vector of the fibre. (2) Map $X \in (TE)_e$ to $(e, \pi_* X)$.

(b) If $0 \to E_1 \to E_2 \to E_3 \to 0$ is exact, then each bundle E_i is orientable if the other two are.

(c) $T(TM)$ is always orientable.

(d) If $\pi\colon E \to M$ is not orientable, then the manifold E is not orientable. (This is why the proof that the Möbius strip is a non-orientable manifold is so similar to the proof that the Möbius bundle over S^1 is not orientable.)

The next two Problems contain more information about the groups introduced in Problem 2-33. In addition to being used in Problem 32, this information will all be important in Chapter 10.

30. (a) Let $p_0 \in S^{n-1}$ be the point $(0, \ldots, 0, 1)$. For $n \geq 2$ define $f\colon SO(n) \to S^{n-1}$ by $f(A) = A(p_0)$. Show that f is continuous and open. Show that $f^{-1}(p_0)$ is homeomorphic to $SO(n-1)$, and then show that $f^{-1}(p)$ is homeomorphic to $SO(n-1)$ for all $p \in S^{n-1}$.

(b) $SO(1)$ is a point, so it is connected. Using part (a), and induction on n, prove that $SO(n)$ is connected for all $n \geq 1$.

(c) Show that $O(n)$ has exactly two components.

31. (a) If $T\colon \mathbb{R}^n \to \mathbb{R}^n$ is a linear transformation, $T^*\colon \mathbb{R}^n \to \mathbb{R}^n$, the **adjoint** of T, is defined by $\langle T^*v, w \rangle = \langle v, Tw \rangle$ (for each v, the map $w \mapsto \langle v, Tw \rangle$ is linear, so it is $w \mapsto \langle T^*v, w \rangle$ for a unique T^*v). If A is the matrix of T with respect to the usual basis, show that the matrix of T^* is the transpose A^t.

(b) A linear transformation $T\colon \mathbb{R}^n \to \mathbb{R}^n$ is **self-adjoint** if $T = T^*$, so that $\langle Tv, w \rangle = \langle v, Tw \rangle$ for all $v, w \in \mathbb{R}^n$. If A is the matrix of T with respect to the standard basis, then T is self-adjoint if and only if A is symmetric, $A^{\mathbf{t}} = A$. It is a standard theorem that a symmetric A can be written as CDC^{-1} for some diagonal matrix D (for an analytic proof, see *Calculus on Manifolds*, pg. 122). Show that C can be chosen orthogonal, by showing that eigenvectors for distinct eigenvalues are orthogonal.

(c) A self-adjoint T (or the corresponding symmetric A) is called **positive semi-definite** if $\langle Tv, v \rangle \geq 0$ for all $v \in \mathbb{R}^n$, and **positive definite** if $\langle Tv, v \rangle > 0$ for all $v \neq 0$. Show that a positive definite A is non-singular. *Hint*: Use the Schwarz inequality.

(d) Show that $A^{\mathbf{t}} \cdot A$ is always positive semi-definite.

(e) Show that a positive semi-definite A can be written as $A = B^2$ for some B. (Remember that A is symmetric.)

(f) Show that every $A \in \mathrm{GL}(n, \mathbb{R})$ can be written uniquely as $A = A_1 \cdot A_2$ where $A_1 \in O(n)$ and A_2 is positive definite. *Hint*: Consider $A^{\mathbf{t}} \cdot A$, and use part (e).

(g) The matrices A_1 and A_2 are continuous functions of A. *Hint*: If $A^{(n)} \to A$ and $A^{(n)} = A^{(n)}{}_1 \cdot A^{(n)}{}_2$, then some subsequence of $\{A^{(n)}{}_1\}$ converges.

(h) $\mathrm{GL}(n, \mathbb{R})$ is homeomorphic to $O(n) \times \mathbb{R}^{n(n+1)/2}$ and has exactly two components, $\{A\colon \det A > 0\}$ and $\{A\colon \det A < 0\}$. (Notice that this also gives us another way of finding the dimension of $O(n)$.)

32. Two continuous functions $f_0, f_1 \colon X \to Y$ are called **homotopic** if there is a continuous function $H\colon X \times [0, 1] \to Y$ such that

$$f_i(x) = H(x, i) \qquad i = 0, 1.$$

The functions $H_t \colon X \to Y$ defined by $H_t(x) = H(x, t)$ may be thought of as a path of functions from $H_0 = f_0$ to $H_1 = f_1$. The map H is called a **homotopy** between f_0 and f_1.

   The notation $f\colon (X, A) \to (Y, B)$, for $A \subset X$ and $B \subset Y$, means that $f\colon X \to Y$ and $f(A) \subset B$. We call $f_0, f_1 \colon (X, A) \to (Y, B)$ **homotopic (as maps from (X, A) to (Y, B))** if there is an H as above such that each $H_t \colon (X, A) \to (Y, B)$.

(a) If $A\colon [0, 1] \to \mathrm{GL}(n, \mathbb{R})$ is continuous and $H\colon \mathbb{R}^n \times [0, 1] \to \mathbb{R}^n$ is defined by $H(x, t) = A(t)(x)$, show that H is continuous, so that H_0 and H_1 are homotopic as maps from $(\mathbb{R}^n, \mathbb{R}^n - \{0\})$ to $(\mathbb{R}^n, \mathbb{R}^n - \{0\})$. Conclude that a non-singular linear transformation $T\colon (\mathbb{R}^n, \mathbb{R}^n - \{0\}) \to (\mathbb{R}^n, \mathbb{R}^n - \{0\})$ with $\det T > 0$ is homotopic to the identity map.

(b) Suppose $f\colon \mathbb{R}^n \to \mathbb{R}^n$ is C^∞ and $f(0) = 0$, while $f(\mathbb{R}^n - \{0\}) \subset \mathbb{R}^n - \{0\}$. If $Df(0)$ is non-singular, show that $f\colon (\mathbb{R}^n, \mathbb{R}^n - \{0\}) \to (\mathbb{R}^n, \mathbb{R}^n - \{0\})$ is homotopic to $Df(0): (\mathbb{R}^n, \mathbb{R}^n - \{0\}) \to (\mathbb{R}^n, \mathbb{R}^n - \{0\})$. *Hint:* Define $H(x, t) = f(tx)$ for $0 < t \le 1$ and $H(x, 0) = Df(0)(x)$. To prove continuity at points $(x, 0)$, use Lemma 2.

(c) Let U be a neighborhood of $0 \in \mathbb{R}^n$ and $f: U \to \mathbb{R}^n$ a homeomorphism with $f(0) = 0$. Let $B_r \subset V$ be the open ball with center 0 and radius r, and let $h: \mathbb{R}^n \to B_r$ be the homeomorphism

$$h(x) = \left(\frac{2r}{\pi} \arctan |x| \right) x;$$

then

$$f \circ h: (\mathbb{R}^n, \mathbb{R}^n - \{0\}) \to (\mathbb{R}^n, \mathbb{R}^n - \{0\}).$$

We will say that f is **orientation preserving at** 0 if $f \circ h$ is homotopic to the identity map $1: (\mathbb{R}^n, \mathbb{R}^n - \{0\}) \to (\mathbb{R}^n, \mathbb{R}^n - \{0\})$. Check that this does not depend on the choice of $B_r \subset V$.

(d) For $p \in \mathbb{R}^n$, let $T_p: \mathbb{R}^n \to \mathbb{R}^n$ be $T_p(q) = p + q$. If $f: U \to V$ is a homeomorphism, where $U, V \subset \mathbb{R}^n$ are open, we will say that f is **orientation preserving at** p if $T_{-f(p)} \circ f \circ T_p$ is orientation preserving at 0. Show that if M is orientable, then there is a collection $\mathcal{C}$ of charts whose domains cover M such that for every (x, U) and (y, V) in $\mathcal{C}$, the map $y \circ x^{-1}$ is orientation preserving at $x(p)$ for all $p \in U \cap V$.

(e) Notice that the condition on $y \circ x^{-1}$ in part (d) makes sense even if $y \circ x^{-1}$ is not differentiable. Thus, if M is any (not necessarily differentiable) manifold, we can define M to be **orientable** if there is a collection $\mathcal{C}$ of homeomorphisms $x: U \to \mathbb{R}^n$ whose domains cover M, such that $\mathcal{C}$ satisfies the condition in part (d). To prove that this definition agrees with the old one we need a fact from algebraic topology: If $f: \mathbb{R}^n \to \mathbb{R}^n$ is a homeomorphism with $f(0) = 0$ and $T: \mathbb{R}^n \to \mathbb{R}^n$ is $T(x^1, \dots, x^n) = (x^1, \dots, x^{n-1}, -x^n)$, then precisely one of f and $T \circ f$ is orientation preserving at 0. Assuming this result, show that if M has such a collection $\mathcal{C}$ of homeomorphisms, then for any C^∞ structure on M the tangent bundle TM is orientable.

33. Let $M^n \subset \mathbb{R}^N$ be a C^∞ n-dimensional submanifold. By a **chord** of M we mean a point of $\mathbb{R}^N$ of the form $p - q$ for $p, q \in M$.

(a) Prove that if $N > 2n + 1$, then there is a vector $v \in S^{N-1}$ such that

    (i) no chord of M is parallel to v,

    (ii) no tangent plane M_p contains v.

Hint: Consider certain maps from appropriate open subsets of $M \times M$ and TM to S^{N-1}.

(b) Let $\mathbb{R}^{N-1} \subset \mathbb{R}^N$ be the subspace perpendicular to v, and $\pi : \mathbb{R}^N \to \mathbb{R}^{N-1}$ the corresponding projection. Show that $\pi | M$ is a one-one immersion. In particular, if M is compact, then $\pi | M$ is an imbedding.

(c) Every compact C^∞ n-dimensional manifold can be imbedded in $\mathbb{R}^{2n+1}$.

Note: This is the easy case of Whitney's classical theorem, which gives the same result even for non-compact manifolds (H. Whitney, *Differentiable manifolds*, Ann. of Math. **37** (1935), 645–680). Proofs may be found in Auslander and MacKenzie, *Introduction to Differentiable Manifolds* and Sternberg, *Lectures on Differential Geometry*. In Munkres, *Elementary Differential Topology*, there is a different sort of argument to prove that a not-necessarily-compact n-manifold M can be imbedded in some $\mathbb{R}^N$ (in fact, with $N = (n+1)^2$). Then we may show that M imbeds in $\mathbb{R}^{2n+1}$ using essentially the argument above, together with the existence of a proper map $f : M \to \mathbb{R}$, given by Problem 2-30 (compare Guillemin and Pollack, *Differential Topology*). A much harder result of Whitney shows that M^n can actually be imbedded in $\mathbb{R}^{2n}$ (H. Whitney, *The self-intersections of a smooth n-manifold in 2n-space*, Ann. of Math. **45** (1944), 220–246).

CHAPTER 4

TENSORS

A ll the constructions on vector bundles carried out in this chapter have a common feature. In each case, we replace each fibre $\pi^{-1}(p)$ by some other vector space, and then fit all these new vector spaces together to form a new vector bundle over the same base space.

The simplest case arises when we replace each fibre V by its dual space V^*. Recall that V^* denotes the vector space of all linear functions $\lambda \colon V \to \mathbb{R}$. If $f \colon V \to W$ is a linear transformation, then there is a linear transformation $f^* \colon W^* \to V^*$ defined by

$$(f^*\lambda)(v) = \lambda(fv).$$

It is clear that if $1_V \colon V \to V$ is the identity, then 1_V^* is the identity map of V^* and if $g \colon U \to V$, then $(f \circ g)^* = g^* \circ f^*$. These simple remarks already show that f^* is an isomorphism if $f \colon V \to W$ is, for $(f^{-1} \circ f)^* = 1_V^*$ and $(f \circ f^{-1})^* = 1_W^*$.

The dimension of V^* is the same as that of V, for finite dimensional V. In fact, if $v_1, \ldots, v_n$ is a basis for V, then the elements $v^*_i \in V^*$, defined by

$$v^*_i(v_j) = \delta^i_j,$$

are easily checked to be a basis for V^*. The linear function v^*_i depends on the entire set $v_1, \ldots, v_n$, not just on v_i alone, and the isomorphism from V to V^* obtained by sending v_i to v^*_i is *not* independent of the choice of basis (consider what happens if v_1 is replaced by $2v_1$).

On the other hand, if $v \in V$, we can define $v^{**} \in V^{**} = (V^*)^*$ unambiguously by

$$v^{**}(\lambda) = \lambda(v) \qquad \text{for every } \lambda \in V^*.$$

If $v^{**}(\lambda) = 0$ for every $\lambda \in V^*$, then $\lambda(v) = 0$ for all $\lambda \in V^*$, which implies that

$v = 0$. Thus the map $v \mapsto v^{**}$ is an isomorphism from V to V^{**}. It is called the **natural isomorphism** from V to V^{**}.

(Problem 6 gives a precise meaning to the word "natural", formulated only after the term had long been in use. Once the meaning is made precise, we can prove that there is *no* natural isomorphism from V to V^{*}.)

Now let $\xi = \pi \colon E \to B$ be any vector bundle. Let

$$E' = \bigcup_{p \in B} [\pi^{-1}(p)]^{*},$$

and define the function $\pi' \colon E' \to B$ to take each $[\pi^{-1}(p)]^{*}$ to p. If $U \subset B$, and $t \colon \pi^{-1}(U) \to U \times \mathbb{R}^{n}$ is a trivialization, then we can define a function

$$t' \colon \pi'^{-1}(U) \to U \times (\mathbb{R}^{n})^{*}$$

in the obvious way: since the map t restricted to a fibre,

$$t_{p} \colon \pi^{-1}(p) \to \{p\} \times \mathbb{R}^{n},$$

is an isomorphism, it gives us an isomorphism

$$(t_{p}^{*})^{-1} \colon [\pi^{-1}(p)]^{*} \to \{p\} \times (\mathbb{R}^{n})^{*}.$$

We can make $\pi' \colon E' \to B$ into a vector bundle, the **dual bundle** ξ^{*} of ξ, by requiring that all such t' be local trivializations. (We first pick an isomorphism from $(\mathbb{R}^{n})^{*}$ to $\mathbb{R}^{n}$, once and for all.)

At first it might appear that $\xi^{*} \simeq \xi$, since each $\pi^{-1}(p)$ is isomorphic to $\pi'^{-1}(p)$. However, this is true merely because the two vector spaces have the same dimension. The lack of a natural isomorphism from V to V^{*} prevents us from constructing an equivalence between ξ^{*} and ξ. Actually, we will see later that in "most" cases ξ^{*} *is* equivalent to ξ; for the present, readers may ponder this question for themselves. In contrast, the bundle $\xi^{**} = (\xi^{*})^{*}$ is *always* equivalent to ξ. We construct the equivalence by mapping the fibre V of ξ over p to the fibre V^{**} of ξ^{**} over p by the natural isomorphism. If you think about how ξ^* is constructed, it will appear obvious that this map is indeed an equivalence.

Even if ξ can be pictured geometrically (e.g., if ξ is TM), there is seldom a geometric picture for ξ^*. Rather, ξ^* operates on ξ: If s is a section of ξ and σ is a section of ξ^*, then we can define a function from B to $\mathbb{R}$ by

$$p \mapsto \sigma(p)(s(p)) \qquad \begin{aligned} s(p) &\in \pi^{-1}(p) \\ \sigma(p) &\in \pi'^{-1}(p) = \pi^{-1}(p)^*. \end{aligned}$$

This function will be denoted simply by $\sigma(s)$.

When this construction is applied to the tangent bundle TM of M, the resulting bundle, denoted by T^*M, is called the **cotangent bundle** of M; the fibre of T^*M over p is $(M_p)^*$. Like TM, the cotangent bundle T^*M is actually a C^∞ vector bundle: since two trivializations x_* and y_* of TM are C^∞-related, the same is clearly true for x_*' and y_*' (in fact, $y_*' \circ (x_*')^{-1} = y_* \circ (x_*)^{-1}$). We can thus define C^∞, as well as continuous, sections of T^*M. If ω is a C^∞ section of T^*M and X is a C^∞ vector field, then $\omega(X)$ is the C^∞ function $p \mapsto \omega(p)(X(p))$.

If $f : M \to \mathbb{R}$ is a C^∞ function, then a C^∞ section df of T^*M can be defined by

$$df(p)(X) = X(f) \quad \text{for } X \in M_p.$$

The section df is called the **differential of** f. Suppose, in particular, that X is $dc/dt|_{t_0}$, where $c(t_0) = p$. Recall that

$$\left.\frac{dc}{dt}\right|_{t_0} = c_* \left(\left.\frac{d}{dt}\right|_{t_0}\right).$$

This means that

$$\begin{aligned} df\left(\left.\frac{dc}{dt}\right|_{t_0}\right) &= c_*\left(\left.\frac{d}{dt}\right|_{t_0}\right)(f) \\ &= \left.\frac{d}{dt}\right|_{t_0}(f \circ c) \\ &= (f \circ c)'(t_0) \quad \text{or} \quad \left.\frac{d(f(c(t)))}{dt}\right|_{t_0}. \end{aligned}$$

Adopting the elliptical notations

$$\frac{dc}{dt} \quad \text{for} \quad \left.\frac{dc}{dt}\right|_t, \qquad \frac{dg(t)}{dt} \quad \text{for} \quad g'(t),$$

this equation takes the nice form

$$df\left(\frac{dc}{dt}\right) = \frac{d(f(c(t)))}{dt}.$$

If (x, U) is a coordinate system, then the dx^i are sections of T^*M over U. Applying the definition, we see that

$$dx^i(p)\left(\left.\frac{\partial}{\partial x^j}\right|_p\right) = \delta^i_j.$$

Thus $dx^1(p), \ldots, dx^n(p)$ is just the basis of $M_p{}^*$ dual to the basis $\partial/\partial x^1|_p, \ldots, \partial/\partial x^n|_p$ of M_p.

This means that every section ω can be expressed uniquely on U as

$$\omega(p) = \sum_{i=1}^n \omega_i(p)\, dx^i(p),$$

for certain functions ω_i on U. The section ω is continuous or C^∞ if and only if the functions ω_i are.

We can also write

$$\omega = \sum_{i=1}^n \omega_i\, dx^i,$$

if we define sums of sections and products of functions and sections in the obvious way ("pointwise" addition and multiplication).

The section df must have some such expression. In fact, we obtain a classical formula:

1. THEOREM. If (x, U) is a coordinate system and f is a C^∞ function, then on U we have

$$df = \sum_{i=1}^{n} \frac{\partial f}{\partial x^i} \, dx^i.$$

PROOF. If $X_p \in M_p$ is

$$X_p = \sum_{i=1}^{n} a^i \left. \frac{\partial}{\partial x^i} \right|_p,$$

then

$$a^i = X_p(x^i) = dx^i(p)(X_p).$$

Thus

$$df(p)(X_p) = X_p(f) = \sum_{i=1}^{n} a^i \frac{\partial f}{\partial x^i}(p)$$

$$= \sum_{i=1}^{n} \frac{\partial f}{\partial x^i}(p) \, dx^i(p)(X_p). \; \diamond$$

Classical differential geometers (and classical analysts) did not hesitate to talk about "infinitely small" changes dx^i of the coordinates x^i, just as Leibnitz had. No one wanted to admit that this was nonsense, because true results were obtained when these infinitely small quantities were divided into each other (provided one did it in the right way).

Eventually it was realized that the closest one can come to describing an infinitely small change is to describe a direction in which this change is supposed to occur, i.e., a tangent vector. Since df is supposed to be the infinitesimal change of f under an infinitesimal change of the point, df must be a function of this change, which means that df should be a function on tangent vectors. The dx^i themselves then metamorphosed into functions, and it became clear that they must be distinguished from the tangent vectors $\partial/\partial x^i$.

Once this realization came, it was only a matter of making new definitions, which preserved the *old* notation, and waiting for everybody to catch up. In short, all classical notions involving infinitely small quantities became functions on tangent vectors, like df, except for quotients of infinitely small quantities, which became tangent vectors, like dc/dt.

Looking back at the classical works from our modern vantage point, one can usually see that, no matter how obscurely expressed, this point of view was in some sense the one always taken by classical geometers. In fact, the differential *df* was usually introduced in the following way:

CLASSICAL FORMULATION	MODERN FORMULATION
Let f be a function of the $x^1, \ldots, x^n$, say $f = f(x^1, \ldots, x^n)$.	Let f be a function on M, and x a coordinate system (so that $f = \bar{f} \circ x$ for some function $\bar{f}$ on $\mathbb{R}^n$, namely $\bar{f} = f \circ x^{-1}$).
Let x^i be functions of t, say $x^i = x^i(t)$. Then f becomes a function of t, $f(t) = f(x^1(t), \ldots, x^n(t))$.	Let $c \colon \mathbb{R} \to M$ be a curve. Then $f \circ c \colon \mathbb{R} \to \mathbb{R}$, where $$f \circ c(t) = \bar{f}(x^1 \circ c(t), \ldots, x^n \circ c(t)).$$
We now have $$\frac{df}{dt} = \sum_{i=1}^{n} \frac{\partial f}{\partial x^i} \frac{dx^i}{dt}.$$ (The classical notation, which suppresses the curve c, is still used by physicists, as we shall point out once again in Chapter 7.)	We now have $$(f \circ c)'(t)$$ $$= \sum_{i=1}^{n} D_i \bar{f}(x(c(t))) \cdot (x^i \circ c)'(t)$$ $$= \sum_{i=1}^{n} \frac{\partial f}{\partial x^i}(c(t)) \cdot (x^i \circ c)'(t)$$ or $$\frac{d(f(c(t)))}{dt} = \sum_{i=1}^{n} \frac{\partial f}{\partial x^i}(c(t)) \cdot \frac{dx^i(c(t))}{dt}.$$
Multiplying by dt gives $$df = \sum_{i=1}^{n} \frac{\partial f}{\partial x^i} dx^i.$$ (This equation signifies that true results are obtained by dividing by dt again, *no matter what the functions* $x^i(t)$ are. It is the closest approach in classical analysis to the realization of df as a function on tangent vectors.)	Consequently, $$df\left(\frac{dc}{dt}\right) = \sum_{i=1}^{n} \frac{\partial f}{\partial x^i}(c(t)) \cdot dx^i\left(\frac{dc}{dt}\right).$$ Since every tangent vector at $c(t)$ is of the form dc/dt, we have $$df = \sum_{i=1}^{n} \frac{\partial f}{\partial x^i} dx^i.$$

In preparation for our reading of Gauss and Riemann, we will continually examine the classical way of expressing all concepts which we introduce. After a while, the "translation" of classical terminology becomes only a little more difficult than the translation of the German in which it was written.

Recall that if $f: M \to N$ is C^∞, then there is a map $f_*: TM \to TN$; for each $p \in M$, we have a map $f_{*p}: M_p \to N_{f(p)}$. Since f_{*p} is a linear transformation between two vector spaces, it gives rise to a map

$$N_{f(p)}{}^* \to M_p{}^*.$$

Strict notational propriety would dictate that this map be denoted by $(f_{*p})^*$, but everyone denotes it simply by

$$f_p^*: N_{f(p)}{}^* \to M_p{}^*.$$

Notice that we cannot put all f_p^* together to obtain a bundle map from T^*N to T^*M; in fact, the same $q \in N$ may be $f(p_i)$ for more than one $p_i \in M$, and there is no reason why f_{*p_1} should equal f_{*p_2}. On the other hand, we can do something with the cotangent bundle that we could not do with the tangent bundle. Suppose ω is a section of T^*N. Then we can define a section η of T^*M as follows:

$$\eta(p) = \omega(f(p)) \circ f_{*p},$$

i.e.,

$$\eta(p)(X_p) = \omega(f(p))(f_{*p}X_p) \quad \text{for } X_p \in M_p.$$

(The complex symbolism tends to hide the simple idea: to operate on a vector, we push it over to N by f_*, and then operate on it by ω.) This section η is denoted, naturally enough, by $f^*\omega$. There is no corresponding way of transferring a vector field X on M over to a vector field on N.

Despite these differences, we can say, roughly, that a map $f: M \to N$ produces a map f_* going in the same direction on the tangent bundle and a map f^* going in the opposite direction on the cotangent bundle. Nowadays such situations are always distinguished by calling the things which go in the same direction "covariant" and the things which go in the opposite direction "contravariant". Classical terminology used these same words, and it just happens to have reversed this: a vector field is called a **contravariant vector field**, while a section of T^*M is called a **covariant vector field**. And no one has had the gall or authority to reverse terminology so sanctified by years of usage. So it's very easy to remember which kind of vector field is covariant, and which contravariant—it's just the opposite of what it logically ought to be.

The rationale behind the classical terminology can be seen by considering coordinate systems x on $\mathbb{R}^n$ which are linear transformations. In this case, if $x(v_i) = e_i$, then

$$x(a^1 v_1 + \cdots + a^n v_n) = (a^1, \ldots, a^n),$$

so the x coordinate system is just an "oblique Cartesian coordinate system".

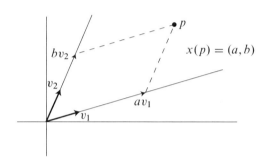

If x' is another such coordinate system, then $x'^j = \sum_{i=1}^n a_{ij} x^i$ for certain a_{ij}. Clearly $a_{ij} = \partial x'^j / \partial x^i$, so

$$(*) \qquad x'^j = \sum_{i=1}^n \frac{\partial x'^j}{\partial x^i} x^i;$$

this can be seen directly from the fact that the matrix $(\partial x'^j / \partial x^i)$ is the constant matrix $D(x' \circ x^{-1}) = x' \circ x^{-1}$. Comparing $(*)$ with

$$(**) \qquad dx'^j = \sum_{i=1}^n \frac{\partial x'^j}{\partial x^i} dx^i,$$

from Theorem 1, we see that the differentials dx^i "change in the same way" as the coordinates x^i, hence they are "covariant". Consequently, any combination

$$\omega = \sum_{i=1}^n \omega_i \, dx^i$$

is also called "covariant". Notice that if we also have

$$\omega = \sum_{i=1}^n \omega'_i \, dx'^i,$$

then we can express the ω'_i in terms of the ω_i. Substituting

$$dx^i = \sum_{j=1}^n \frac{\partial x^i}{\partial x'^j} \, dx'^j$$

into the first expression for ω and comparing coefficients with the second, we find that

$$\omega'_j = \sum_{i=1}^n \omega_i \frac{\partial x^i}{\partial x'^j}.$$

On the other hand, given two expressions

$$\sum_{i=1}^n a^i \frac{\partial}{\partial x^i} = \sum_{i=1}^n a'^i \frac{\partial}{\partial x'^i}$$

for a vector field, the functions a'^j must satisfy

$$a'^j = \sum_{i=1}^n a^i \frac{\partial x'^j}{\partial x^i}.$$

These expressions can always be remembered by noting that indices which are summed over always appear once "above" and once "below". (Coordinate functions $x^1, \ldots, x^n$ used to be denoted by $x_1, \ldots, x_n$. This suggested subscripts ω_i for covariant vector fields and superscripts a^i for contravariant vector fields. After this was firmly established, the indices on the x's were shifted upstairs again to make the summation convention work out.)

Covariant and contravariant vector fields, i.e., sections of T^*M and TM, respectively, are also called covariant and contravariant tensors (or tensor fields) of order 1, which is a warning that worse things are to come. We begin with some worse algebra.

If $V_1, \ldots, V_m$ are vector spaces, a function

$$T: V_1 \times \cdots \times V_m \to \mathbb{R}$$

is **multilinear** if

$$v \mapsto T(v_1, \ldots, v_{k-1}, v, v_{k+1}, \ldots, v_m)$$

is linear for each choice of $v_1, \ldots, v_{k-1}, v_{k+1}, \ldots, v_m$. The set of all such T is clearly a vector space. If $V_1, \ldots, V_m = V$, this vector space will be denoted by $\mathcal{T}^m(V)$. Notice that $\mathcal{T}^1(V) = V^*$. If $f : V \to W$ is a linear transformation, then there is a linear transformation $f^* : \mathcal{T}^m(W) \to \mathcal{T}^m(V)$, defined completely analogously to the case $m = 1$:

$$f^* T(v_1, \ldots, v_m) = T(f(v_1), \ldots, f(v_m)).$$

For $T \in \mathcal{T}^k(V)$, and $S \in \mathcal{T}^l(V)$ we can define the "tensor product" $T \otimes S \in \mathcal{T}^{k+l}(V)$ by

$$T \otimes S(v_1, \ldots, v_k, v_{k+1}, \ldots, v_{k+l}) = T(v_1, \ldots, v_k) \cdot S(v_{k+1}, \ldots, v_{k+l}).$$

Of course, $T \otimes S$ is not $S \otimes T$. On the other hand, $(S \otimes T) \otimes U = S \otimes (T \otimes U)$, so we can define n-fold tensor products unambiguously; this tensor product operation is itself multilinear, $(S_1 + S_2) \otimes T = S_1 \otimes T + S_2 \otimes T$, etc. In particular, if $v_1, \ldots, v_n$ is a basis for V and $v^*_1, \ldots, v^*_n$ is the dual basis for $V^* = \mathcal{T}^1(V)$, then the elements

$$v^*_{i_1} \otimes \cdots \otimes v^*_{i_k} \qquad 1 \le i_1, \ldots, i_k \le n$$

are easily seen to be a basis for $\mathcal{T}^k(V)$, which thus has dimension n^k.

We can use this new algebraic construction to obtain a new bundle from any vector bundle $\xi = \pi : E \to B$. We let

$$E' = \bigcup_{p \in B} \mathcal{T}^k(\pi^{-1}(p)),$$

and let

$$\pi' : E' \to B \quad \text{take} \quad \mathcal{T}^k(\pi^{-1}(p)) \text{ to } p.$$

If $U \subset B$ and

$$t : \pi^{-1}(U) \to U \times \mathbb{R}^n$$

is a trivialization, then the isomorphisms

$$t_p : \pi^{-1}(p) \to \{p\} \times \mathbb{R}^n$$

yield isomorphisms

$$(t_p^*)^{-1} : \mathcal{T}^k(\pi^{-1}(p)) \to \{p\} \times \mathcal{T}^k(\mathbb{R}^n).$$

If we choose an isomorphism $\mathcal{T}^k(\mathbb{R}^n) \to \mathbb{R}^{n^k}$ once and for all, these maps can be put together to give a map

$$t' : \pi'^{-1}(U) \to U \times \mathbb{R}^{n^k}.$$

We make $\pi' \colon E' \to B$ into a vector bundle $\mathcal{T}^k(\xi)$ by requiring that all such t' be local trivializations. The bundle ξ^* is the special case $k = 1$.

For the case of TM, the bundle $\mathcal{T}^k(TM)$ is called the **bundle of covariant tensors of order** k, and a section is called a **covariant tensor field of order** k. If (x, U) is a coordinate system, so that

$$dx^1(p), \ldots, dx^n(p)$$

is a basis for $(M_p)^*$, then the k-fold tensor products

$$dx^{i_1}(p) \otimes \cdots \otimes dx^{i_k}(p) \in \mathcal{T}^k(M_p) \quad 1 \le i_1, \ldots, i_k \le n$$

are a basis for $\mathcal{T}^k(M_p)$. Thus, on U every covariant tensor field A of order k can be written

$$A(p) = \sum_{i_1, \ldots, i_k} A_{i_1 \ldots i_k}(p) \, dx^{i_1}(p) \otimes \cdots \otimes dx^{i_k}(p),$$

or simply

$$A = \sum_{i_1, \ldots, i_k} A_{i_1 \ldots i_k} \, dx^{i_1} \otimes \cdots \otimes dx^{i_k},$$

where $dx^{i_1} \otimes \cdots \otimes dx^{i_k}$ now denotes a section of $\mathcal{T}^k(TM)$. If we also have

$$A = \sum_{i_1, \ldots, i_k} A'_{i_1 \ldots i_k} \, dx'^{i_1} \otimes \cdots \otimes dx'^{i_k},$$

then

$$A'_{\alpha_1 \ldots \alpha_k} = \sum_{i_1, \ldots, i_k} A_{i_1 \ldots i_k} \frac{\partial x^{i_1}}{\partial x'^{\alpha_1}} \cdots \frac{\partial x^{i_k}}{\partial x'^{\alpha_k}}$$

(the products are just ordinary products of functions). To derive this equation, we just use equation $(**)$ on page 114, and multilinearity of $\otimes$. The section A is continuous or C^∞ if and only if the functions $A_{i_1 \ldots i_k}$ are.

A covariant tensor field A of order k can just be thought of as an operation $\bar{A}$ on k vector fields $X_1, \ldots, X_k$ which yields a function:

$$\bar{A}(X_1, \ldots, X_k)(p) = A(p)(X_1(p), \ldots, X_k(p)).$$

Notice that $\bar{A}$ is multilinear on the set $\mathcal{V}$ of C^∞ vector fields:

$$\bar{A}(X_1, \ldots, X_i + X'_i, \ldots X_k) = \bar{A}(X_1, \ldots, X_i, \ldots, X_k) + \bar{A}(X_1, \ldots, X'_i, \ldots X_k)$$

$$\bar{A}(X_1, \ldots, a X_i, \ldots X_k) = a \bar{A}(X_1, \ldots, X_k).$$

Moreover, because $\bar{A}$ is defined "pointwise", it is actually linear *over the C^∞ functions $\mathcal{F}$*; i.e., if f is C^∞, then

$$\bar{A}(X_1, \ldots, f X_i, \ldots, X_k) = f \bar{A}(X_1, \ldots, X_i, \ldots, X_k),$$

for we have

$$\begin{aligned}
\bar{A}(X_1, \ldots, f X_i, \ldots, X_k)(p) &= A(p)(X_1(p), \ldots, f(p)X_i(p), \ldots, X_k(p)) \\
&= f(p)A(p)(X_1(p), \ldots, X_i(p), \ldots, X_k(p)) \\
&= f(p) \cdot \bar{A}(X_1, \ldots, X_i, \ldots, X_k)(p).
\end{aligned}$$

We are finally ready for another theorem, one that is used over and over.

2. THEOREM. If
$$\mathcal{A}: \underbrace{\mathcal{V} \times \cdots \times \mathcal{V}}_{k \text{ times}} \to \mathcal{F}$$

is linear over $\mathcal{F}$, then there is a unique tensor field A with $\mathcal{A} = \bar{A}$.

PROOF. Note first that if $v \in M_p$ is any tangent vector, then there is a vector field $X \in \mathcal{V}$ with $X(p) = v$. In fact, if (x, U) is a coordinate system and

$$v = \sum_{i=1}^n a^i \left. \frac{\partial}{\partial x^i} \right|_p,$$

then we can define

$$X = \begin{cases} f \sum_{i=1}^n a^i \dfrac{\partial}{\partial x^i} & \text{on } U \\ 0 & \text{outside } U, \end{cases}$$

where each a^i now denotes a constant function and f is a C^∞ function with $f(p) = 1$ and support $f \subset U$.

Now if $v_1, \ldots, v_k \in M_p$ are extended to vector fields $X_1, \ldots, X_k \in \mathcal{V}$ we clearly must define

$$A(p)(v_1, \ldots, v_k) = \mathcal{A}(X_1, \ldots, X_k)(p).$$

The problem is to prove that this is well-defined: If $X_i(p) = Y_i(p)$ for each i, we claim that

$$\mathcal{A}(X_1, \ldots, X_k)(p) = \mathcal{A}(Y_1, \ldots, Y_k)(p).$$

(The map $\mathcal{A}$ "lives at points", to use the in terminology.) For simplicity, take the case $k = 1$ (the general case is exactly analogous). The proof that $\mathcal{A}(X)(p) = \mathcal{A}(Y)(p)$ when $X(p) = Y(p)$ is in two steps.

(1) Suppose first that $X = Y$ in a neighborhood U of p. Let f be a C^∞ function with $f(p) = 1$ and support $f \subset U$. Then $fX = fY$, so

$$f\mathcal{A}(X) = \mathcal{A}(fX) = \mathcal{A}(fY) = f\mathcal{A}(Y);$$

evaluating at p gives

$$\mathcal{A}(X)(p) = \mathcal{A}(Y)(p).$$

(2) To prove the result, it obviously suffices to show that $\mathcal{A}(X)(p) = 0$ if $X(p) = 0$. Let (x, U) be a coordinate system around p, so that on U we can write

$$X = \sum_{i=1}^{n} b^i \frac{\partial}{\partial x^i} \quad \text{where all } b^i(p) = 0.$$

If g is 1 in a neighborhood V of p, and support $g \subset U$, then

$$Y = g \sum_{i=1}^{n} b^i \frac{\partial}{\partial x^i} = \sum_{i=1}^{n} b^i g \frac{\partial}{\partial x^i}$$

is a well-defined C^∞ vector field on all of M which equals X on V, so that

$$\mathcal{A}(X)(p) = \mathcal{A}(Y)(p), \quad \text{by (1)}.$$

Now

$$\mathcal{A}(Y)(p) = \sum_{i=1}^{n} b^i(p) \cdot \mathcal{A}\left(g\frac{\partial}{\partial x^i}\right)(p)$$

$$= 0, \quad \text{since } b^i(p) = 0. \; \clubsuit$$

Because of Theorem 2, we will never distinguish between the tensor field A and the operation $\bar{A}$, nor will we use the symbol $\bar{A}$ any longer. Note that Theorem 2 applies, in particular, to the case $k = 1$, where $\mathcal{T}^k(TM) = T^*M$, the cotangent bundle: a function from $\mathcal{V} \to \mathcal{F}$ which is linear over $\mathcal{F}$ comes from a covariant vector field ω. Just as with covariant vector fields, a C^∞ map $f: M \to N$ gives a map f^* taking covariant tensor fields A of order k on N to covariant tensor fields f^*A of order k on M:

$$f^*A(p)(X_{1_p}, \ldots X_{k_p}) = A(f(p))(f_*X_{1_p}, \ldots, f_*X_{k_p}).$$

Moreover, if A and B are covariant tensor fields of orders k and l, respectively, then we can define a new covariant tensor field $A \otimes B$ of order $k + l$:

$$(A \otimes B)(p) = A(p) \otimes B(p) \quad \text{(operating on } M_p \times \cdots \times M_p \quad k + l \text{ times)}.$$

Although covariant tensor fields will be our main concern, if only for the sake of completeness we should define contravariant tensor fields. Recall that a contravariant vector field is a section X of TM. So each $X_p \in M_p$. Now an element v of a vector space V can be thought of as a linear function $v: V^* \to \mathbb{R}$; we just define $v(\lambda)$ to be $\lambda(v)$. A contravariant tensor field of order k is just a section A of the bundle $\mathcal{T}^k(T^*M)$; thus, each $A(p)$ is a k-linear function on M_p^*. We could also use the notation $\mathcal{T}_k(TM)$, if we use $\mathcal{T}_k(V)$ to denote all k-linear functions on V^*. In local coordinates we can write

$$A(p) = \sum_{j_1, \ldots, j_k} A^{j_1 \ldots j_k}(p) \left. \frac{\partial}{\partial x^{j_1}} \right|_p \otimes \cdots \otimes \left. \frac{\partial}{\partial x^{j_k}} \right|_p$$

(remember that each $\partial/\partial x^j|_p$ operates on M_p^*), or simply

$$A = \sum_{j_1, \ldots, j_k} A^{j_1 \ldots j_k} \frac{\partial}{\partial x^{j_1}} \otimes \cdots \otimes \frac{\partial}{\partial x^{j_k}}.$$

If we have another such expression,

$$A = \sum_{j_1, \ldots, j_k} A'^{j_1 \ldots j_k} \frac{\partial}{\partial x'^{j_1}} \otimes \cdots \otimes \frac{\partial}{\partial x'^{j_k}},$$

then we easily compute that

$$A'^{\beta_1 \ldots \beta_k} = \sum_{j_1, \ldots, j_k} A^{j_1 \ldots j_k} \frac{\partial x'^{\beta_1}}{\partial x^{j_1}} \cdots \frac{\partial x'^{\beta_k}}{\partial x^{j_k}}.$$

A contravariant tensor field A of order k can be considered as an operator $\overline{A}$ taking k covariant vector fields $\omega_1, \ldots, \omega_k$ into a function:

$$\overline{A}(\omega_1, \ldots, \omega_k)(p) = A(p)(\omega_1(p), \ldots, \omega_k(p)).$$

Naturally, there is an analogue of Theorem 2, proved exactly the same way, that allows us to dispense with the notation $\overline{A}$, and to identify contravariant tensor fields of order k with operators on k covariant vector fields that are *linear over the C^∞ functions $\mathcal{F}$*.

Finally, we are ready to introduce "mixed" tensor fields. To make the introduction less painful, we consider a special case first. If V is a vector space, let $\mathcal{T}_1^1(V)$ denote all bilinear functions

$$T: V \times V^* \to \mathbb{R}.$$

A vector bundle $\xi = \pi: E \to B$ gives rise to a vector bundle $\mathcal{T}_1^1(\xi)$, obtained by replacing each fibre $\pi^{-1}(p)$ by $\mathcal{T}_1^1(\pi^{-1}(p))$. In particular, sections of $\mathcal{T}_1^1(TM)$ are called tensor fields, covariant of order 1 and contravariant of order 1.

There are all sorts of algebraic tricks one can play with $\mathcal{T}_1^1(V)$; although they should be kept to a minimum, certain ones are quite important. Let $End(V)$ denote the vector space of all linear transformations $T: V \to V$ ("endomorphisms" of V). Notice that each $S \in End(V)$ gives rise to a bilinear $\bar{S} \in \mathcal{T}_1^1(V)$,

$$\bar{S}: V \times V^* \to \mathbb{R},$$

by the formula

$$(*) \qquad\qquad \bar{S}(v, \lambda) = \lambda(S(v)).$$

Moreover, the correspondence $S \mapsto \bar{S}$ from $End(V)$ to $\mathcal{T}_1^1(V)$ is linear and one-one, for $\bar{S} = 0$ implies that $\lambda(S(v)) = 0$ for all λ, which implies that $S(v) = 0$, for all v. Since both $End(V)$ and $\mathcal{T}_1^1(V)$ have dimension n^2, this map is an isomorphism. The inverse, however, is not so easy to describe. Given $\bar{S}$, for each v the vector $S(v) \in V$ is merely determined by describing the action of a λ on it according to $(*)$. It is not hard to check that this isomorphism of $End(V)$ and $\mathcal{T}_1^1(V)$ makes the identity map $1: V \to V$ in $End(V)$ correspond to the "evaluation" map

$$e: V \times V^* \to \mathbb{R} \quad \text{in } \mathcal{T}_1^1(V)$$

given by

$$e(v, \lambda) = \lambda(v).$$

Generally speaking, our isomorphism can be used to transfer any operation from $End(V)$ to $\mathcal{T}_1^1(V)$. In particular, given a bilinear

$$T: V \times V^* \to \mathbb{R},$$

we can take the trace of the corresponding $S: V \to V$; this number is called the **contraction** of T. If $v_1, \ldots, v_n$ is a basis of V and

$$T = \sum_{i,j} T_i^j v^*_i \otimes v_j,$$

then we can find the matrix $A = (a_{ij})$ of S, defined by

$$S(v_i) = \sum_{j=1}^{n} a_{ji} v_j,$$

in terms of the T_i^j; in fact,

$$a_{ji} = v^*_j(S(v_i)) = T(v_i, v^*_j) = T_i^j.$$

Thus

$$\text{contraction of } T = \sum_{i=1}^{n} T_i^i.$$

(The term "contraction" comes from the fact that the number of indices is contracted from 2 to 0 by setting the upper and lower indices equal and summing.)

These identifications and operations can be carried out, fibre by fibre, in any fibre bundle $\mathcal{T}_1^1(\xi)$. Thus, a section A of $\mathcal{T}_1^1(\xi)$ can just as well be considered as a section of the bundle $End(\xi)$, obtained by replacing each fibre $\pi^{-1}(p)$ by $End(\pi^{-1}(p))$. In this case, each $A(p)$ is an endomorphism of $\pi^{-1}(p)$. Moreover, each section A gives rise to a *function*

$$(\text{contraction of } A) \colon B \to \mathbb{R}$$

defined by

$$p \mapsto \text{contraction of } A(p)$$
$$= \text{trace } A(p) \quad \text{if we consider } A(p) \in End(\pi^{-1}(p)).$$

In particular, given a tensor field A, covariant of order 1 and contravariant of order 1, which is a section of $\mathcal{T}_1^1(TM)$, we can consider each $A(p)$ as an endomorphism of M_p, and we obtain a function "contraction of A". If in a coordinate system

$$A = \sum_{i,j} A_i^j \, dx^i \otimes \frac{\partial}{\partial x^j},$$

then

$$(\text{contraction of } A) = \sum_{i=1}^{n} A_i^i.$$

The general notion of a mixed tensor field is a straightforward generalization. Define $\mathcal{T}_l^k(V)$ to be the set of all $(k+l)$-linear

$$T \colon \underbrace{V \otimes \cdots \otimes V}_{k \text{ times}} \times \underbrace{V^* \otimes \cdots \otimes V^*}_{l \text{ times}} \to \mathbb{R}.$$

Every bundle ξ gives rise to a bundle $\mathcal{T}_l^k(\xi)$. Sections of $\mathcal{T}_l^k(TM)$ are called tensor fields, covariant of order k and contravariant of order l, or simply of type $\binom{k}{l}$, an abbreviation that also saves everybody embarrassment about the use of the words "covariant" and "contravariant". Locally, a tensor field A of type $\binom{k}{l}$ can be expressed as

$$A = \sum_{\substack{i_1,\ldots,i_k \\ j_1,\ldots,j_l}} A^{j_1\ldots j_l}_{i_1\ldots i_k} \, dx^{i_1} \otimes \cdots \otimes dx^{i_k} \otimes \frac{\partial}{\partial x^{j_1}} \otimes \cdots \otimes \frac{\partial}{\partial x^{j_l}},$$

and if

$$A' = \sum_{\substack{i_1,\ldots,i_k \\ j_1,\ldots,j_l}} A'^{j_1\ldots j_l}_{i_1\ldots i_k} \, dx'^{i_1} \otimes \cdots \otimes dx'^{i_k} \otimes \frac{\partial}{\partial x'^{j_1}} \otimes \cdots \otimes \frac{\partial}{\partial x'^{j_l}},$$

then

$$(*) \qquad A'^{\beta_1\ldots\beta_l}_{\alpha_1\ldots\alpha_k} = \sum_{\substack{i_1,\ldots,i_k \\ j_1,\ldots,j_l}} A^{j_1\ldots j_l}_{i_1\ldots i_k} \frac{\partial x^{i_1}}{\partial x'^{\alpha_1}} \cdots \frac{\partial x^{i_k}}{\partial x'^{\alpha_k}} \frac{\partial x'^{\beta_1}}{\partial x^{j_1}} \cdots \frac{\partial x'^{\beta_l}}{\partial x^{j_l}}.$$

Classical differential geometry books are filled with monstrosities like this equation. In fact, the classical definition of a tensor field is: an assignment of n^{k+l} functions to every coordinate system so that $(*)$ holds between the n^{k+l} functions assigned to any two coordinate systems x and x'. (!) Or even, "a set of n^{k+l} functions which changes according to $(*)$". Consequently, in classical differential geometry, all important tensors are actually defined by defining the functions A_i^j, in terms of the coordinate system x, and then checking that $(*)$ holds.

Here is an important example. In every classical differential geometry book, one will find the following assertion: "The Kronecker delta δ_i^j is a tensor." In other words, it is asserted that if one chooses the same n^2 functions δ_i^j for each coordinate system, then $(*)$ holds, i.e.,

$$\delta_\alpha^\beta = \sum_{i,j} \delta_i^j \frac{\partial x^i}{\partial x'^\alpha} \frac{\partial x'^\beta}{\partial x^j};$$

this is certainly true, for

$$\sum_{i,j} \delta_i^j \frac{\partial x^i}{\partial x'^\alpha} \frac{\partial x'^\beta}{\partial x^j} = \sum_{i=1}^n \frac{\partial x^i}{\partial x'^\alpha} \frac{\partial x'^\beta}{\partial x^i} = \delta_\alpha^\beta.$$

From our point of view, what this equation shows is that

$$A = \sum_{i,j} \delta_i^j \, dx^i \otimes \frac{\partial}{\partial x^j}$$

is a certain tensor field, independent of the choice of the coordinate system x. To identify the mysterious map

$$A(p) \colon M_p \times M_p{}^* \to \mathbb{R},$$

we consider $v \in M_p$ and $\lambda \in M_p{}^*$ with the expressions

$$v = \sum_{\alpha=1}^{n} a^\alpha \left.\frac{\partial}{\partial x^\alpha}\right|_p, \quad \lambda = \sum_{\beta=1}^{n} b_\beta \, dx^\beta(p);$$

then

$$A(p)(v,\lambda) = \sum_{i,j} \delta_i^j \, dx^i(p) \otimes \left.\frac{\partial}{\partial x^j}\right|_p (v,\lambda)$$

$$= \sum_{i,j} \delta_i^j \, dx^i(p) \left(\sum_{\alpha=1}^{n} a^\alpha \left.\frac{\partial}{\partial x^\alpha}\right|_p\right) \cdot \left.\frac{\partial}{\partial x^j}\right|_p \left(\sum_{\beta=1}^{n} b_\beta \, dx^\beta(p)\right)$$

$$= \sum_{i,j} \delta_i^j \, a^i b_j$$

$$= \sum_{i=1}^{n} a^i b_i$$

$$= \lambda(v).$$

Thus $A(p)$ is just the evaluation map $M_p \times M_p{}^* \to \mathbb{R}$; considered as an endomorphism of M_p, it is just the identity map.

The contraction of a tensor is defined, classically, in a similar manner. Given a tensor, i.e., a collection of functions A_i^j, one for each coordinate system, satisfying

$$A'^\beta_\alpha = \sum_{i,j} A_i^j \frac{\partial x^i}{\partial x'^\alpha} \frac{\partial x'^\beta}{\partial x^j},$$

we note that

$$\sum_{\alpha=1}^{n} A'^{\alpha}_{\alpha} = \sum_{\alpha=1}^{n} \left(\sum_{i,j} A^{j}_{i} \frac{\partial x^{i}}{\partial x'^{\alpha}} \frac{\partial x'^{\alpha}}{\partial x^{j}} \right)$$

$$= \sum_{i,j} A^{j}_{i} \sum_{\alpha=1}^{n} \frac{\partial x^{i}}{\partial x'^{\alpha}} \frac{\partial x'^{\alpha}}{\partial x^{j}}$$

$$= \sum_{i,j} A^{j}_{i} \delta^{i}_{j}$$

$$= \sum_{i=1}^{n} A^{i}_{i},$$

so that this sum is a well-defined function. This calculation tends to obscure the one part which is really necessary—verification of the fact that the trace of a linear transformation, defined as the sum of the diagonal entries of its matrix, is independent of the basis with respect to which the matrix is written.

Incidentally, a tensor of type $\binom{k}{l}$ can be contracted with respect to any pair of upper and lower indices. For example, the functions

$$B^{\beta\gamma}_{\mu\nu} = \sum_{\alpha=1}^{n} A^{\alpha\beta\gamma}_{\mu\nu\alpha}$$

"transform correctly" if the $A^{\alpha\beta\gamma}_{\mu\nu\lambda}$ do. If we consider each $A(p) \in \mathcal{T}^{3}_{3}(M_{p})$, then we are taking $B(p) \in \mathcal{T}^{2}_{2}(M_{p})$ to be

$$B(p)(v_{1}, v_{2}, \lambda_{1}, \lambda_{2}) = \text{ contraction of: } (v, \lambda) \mapsto A(p)(v, v_{1}, v_{2}, \lambda_{1}, \lambda_{2}, \lambda).$$

While a contravariant vector field is classically a set of n functions which "transforms in a certain way", a vector at a single point p is classically just an assignment of n *numbers* $a^{1}, \ldots, a^{n}$ to each coordinate system x, such that the numbers $a'^{1}, \ldots, a'^{n}$ assigned to x' satisfy

$$a'^{j} = \sum_{i=1}^{n} a^{i} \frac{\partial x'^{j}}{\partial x^{i}}(p).$$

This is *precisely* the definition we adopted when we defined tangent vectors as equivalence classes $[x, a]_{p}$. The revolution in the modern approach is that the set of all vectors is made into a bundle, so that vector fields can be defined as sections, rather than as equivalence classes of sets of functions, and that all other types of tensors are constructed from this bundle. The tangent bundle itself was almost a victim of the excesses of revolutionary zeal. For a long time, the party line held that TM must be defined either as derivations, or as equivalence classes of curves; the return to the old definition was influenced by the "functorial" point of view of Theorems 3-1 and 3-4.

The modern revolt against the classical point of view has been so complete in certain quarters that some mathematicians will give a three page proof that avoids coordinates in preference to a three line proof that uses them. We won't go quite that far, but we will give an "invariant" definition (one that does not use a coordinate system) of any tensors that are defined. Unlike the "Kronecker delta" and contractions, such invariant definitions are usually not so easy to come by. As we shall see, invariant definitions of all the important tensors in differential geometry are made by means of Theorem 2. We seldom define $A(p)$ directly; instead we define a function $\mathcal{A}$ on vector fields, which miraculously turns out to be linear over the C^∞ functions $\mathcal{F}$, and hence must come from some A. At the appropriate time we will discuss whether or not this is all a big cheat.

PROBLEMS

1. Let $f: M^n \to N^m$, and suppose that (x, U) and (y, V) are coordinate systems around p and $f(p)$, respectively.

(a) If $g: N \to \mathbb{R}$, then

$$\frac{\partial(g \circ f)}{\partial x^i}(p) = \sum_{j=1}^{m} \frac{\partial g}{\partial y^j}(f(p)) \cdot \frac{\partial(y^j \circ f)}{\partial x^i}(p).$$

(Proposition 2-3 is the special case $f = $ identity.)

(b) Show that

$$f_* \left(\frac{\partial}{\partial x^i}\bigg|_p \right) = \sum_{j=1}^{m} \frac{\partial(y^j \circ f)}{\partial x^i}(p) \cdot \frac{\partial}{\partial y^j}\bigg|_{f(p)},$$

and, more generally, express $f_* \left(\sum_{i=1}^{n} a^i \partial/\partial x^i\big|_p \right)$ in terms of the $\partial/\partial y^j|_p$.

(c) Show that

$$(f^* dy^j)(p) = \sum_{i=1}^{n} \frac{\partial(y^j \circ f)}{\partial x^i}(p) \cdot dx^i(p).$$

(d) Express

$$f^* \left(\sum_{j_1, \dots, j_n} a_{j_1 \dots j_n} \, dy^{j_1} \otimes \cdots \otimes dy^{j_n} \right)$$

in terms of the dx^i.

2. If $f, g: M \to N$ are C^∞, show that

$$d(fg) = f \, dg + g \, df.$$

3. Let $f: M \to \mathbb{R}$ be C^∞. For $v \in M_p$, show that

$$f_*(v) = df(v)_{f(p)} \in \mathbb{R}_{f(p)}.$$

4. (a) Show that if the ordered bases $v_1, \dots, v_n$ and $w_1, \dots, w_n$ for V are equally oriented, then the same is true of the bases $v^*_1, \dots, v^*_n$ and $w^*_1, \dots, w^*_n$ for V^*.

(b) Show that a bundle ξ is orientable if and only if ξ^* is orientable.

5. The following statements and problems are all taken from Eisenhart's classical work *Riemannian Geometry*. In each case, check them, using the classical methods, and then translate the problem and solution into modern terms. An "invariant" is just a (well-defined) function. Remember that the summation convention is always used, so $\lambda^i \mu_i$ means $\sum_{i=1}^n \lambda^i \mu_i$. Hints and answers are given at the end, after (xiii).

(i) If the quantity $\lambda^i \mu_i$ is an invariant and either λ^i or μ_i are the components of an arbitrary [covariant or contravariant] vector field, the other sets are components of a vector field.

(ii) If $\lambda_{\alpha|}{}^i$ are the components of n vector fields [in an n-manifold], where i for $i = 1, \ldots, n$ indicates the component and α for $\alpha = 1, \ldots, n$ the vector, and these vectors are independent, that is, $\det(\lambda_{\alpha|}{}^i) \neq 0$, then any vector-field λ^i is expressible in the form

$$\lambda^i = a^\alpha \lambda_{\alpha|}{}^i,$$

where the a's are invariants.

(iii) If μ_i are the components of a given vector-field, any vector-field λ^i satisfying $\lambda^i \mu_i = 0$ is expressible linearly in terms of $n - 1$ independent vector fields $\lambda_{\alpha|}{}^i$ for $\alpha = 1, \ldots, n - 1$ which satisfy the equation.

(iv) If $a^{ij} = a^{ji}$ for the components of a tensor field in one coordinate system, then $a'^{ij} = a'^{ji}$ for the coordinates in any other coordinate system.

(v) If a^{ij} and b^{ij} are components of a tensor field, so are $a^{ij} + b^{ij}$. If a^{ij} and b_{kl} are components of a tensor field, so are $a^{ij} b_{kl}$.

(vi) If $a_{ij} \lambda^i \lambda^j$ is an invariant for λ^i an arbitrary vector, then $a_{ij} + a_{ji}$ are the components of a tensor; in particular, if $a_{ij} \lambda^i \lambda^j = 0$, then $a_{ij} + a_{ji} = 0$.

(vii) If $a_{ij} \lambda^i \lambda^j = 0$ for all vectors λ^i such that $\lambda^i \mu_i = 0$, where μ_i is a given covariant vector, if v^i is defined [c.f. (iii)] by $a_{ij} \lambda_{\alpha|}{}^i v^j = 0$, $\alpha = 1, \ldots, n-1$ and $\mu_i v^i \neq 0$, and by definition

$$a_{ij} v^i = \sigma_j \qquad v^i \mu_i = \tau,$$

then $(a_{ij} - \frac{1}{\tau} \mu_i \sigma_j) \xi^i \xi^j = 0$ is satisfied by every vector field ξ^i, and consequently

$$a_{ij} + a_{ji} = \frac{1}{\tau} (\mu_i \sigma_j + \mu_j \sigma_i).$$

(viii) If a_{rs} are the components of a tensor and b and c are invariants, show that if $b a_{rs} + c a_{sr} = 0$, then either $b = -c$ and a_{rs} is symmetric, or $b = c$ and a_{rs} is skew-symmetric.

(ix) By definition the *rank* of a tensor of the second order a_{ij} is the rank of the matrix (a_{ij}). Show that the rank is invariant under all transformations of coordinates.

(x) Show that the rank of the tensor of components $a_i b_j$, where a_i and b_j are the components of two vectors, is one; show that for the symmetric tensor $a_i b_j + a_j b_i$ the rank is two.

(xi) Show that the tensor equation $a^i_j \lambda_i = \alpha \lambda_j$, where α is an invariant, can be written in the form $(a^i_j - \alpha \delta^i_j) \lambda_i = 0$. Show also that $a^i_j = \delta^i_j \alpha$, if the equation is to hold for an arbitrary vector λ_i.

(xii) If $a^i_j \lambda_i = \alpha \lambda_j$ holds for all vectors λ_i such that $\mu^i \lambda_i = 0$, where μ^i is a given vector, then

$$a^i_j = \alpha \delta^i_j + \sigma_j \mu^i.$$

(xiii) If

$$\delta^{j_1 \ldots j_p}_{i_1 \ldots i_p} = \begin{cases} 0 & \text{if } j_\alpha = j_\beta \text{ for some } \alpha \neq \beta \text{ or } i_\alpha = i_\beta \text{ for some } \alpha \neq \beta \\ & \text{or if } \{j_1, \ldots, j_p\} \neq \{i_1, \ldots, i_p\} \\ 1 & \text{if } j_1, \ldots, j_p \text{ is an even permutation of } i_1, \ldots, i_p \\ -1 & \text{if } j_1, \ldots, j_p \text{ is an odd permutation of } i_1, \ldots, i_p \end{cases}$$

then $\delta^{j_1 \ldots j_p}_{i_1 \ldots i_p}$ are the components of a tensor in all coordinate systems.

HINTS AND ANSWERS.

(i) ω is determined if $\omega(X)$ is known for all X, and *vice versa*.

(iii) Given ω [with $\omega(p) \neq 0$ for all p], there are everywhere linearly independent vector fields $X_1, \ldots, X_{n-1}$ which span $\ker \omega$ at each point. (This is true only locally. For example, on $S^2 \times \mathbb{R}$ there is an ω such that $\ker \omega(p, t)$ consists of vectors tangent to $S^2 \times \{t\}$.)

(vi) For $T : V \times V \to \mathbb{R}$, let $T'(v, w) = T(w, v)$. Then $T + T'$ is determined by $S(v) = T(v, v)$. For, $T(v + w, v + w) = T(v, v) + T(v, w) + T(w, v) + T(w, w)$. Similarly, $T(v, v) = 0$ for all v implies that $T + T' = 0$.

(vii) Given ω [with $\omega(p) \neq 0$ for all p], choose Y complementary to $\ker \omega$ at all points. If $\sigma(Z) = T(Y, Z)$, then $T(Z, Z) = \omega(Z)\sigma(Z)/\omega(Y)$ for all vector fields Z.

(ix) $T : V \times V \to \mathbb{R}$ corresponds to $\bar{T} : V \to V^*$ [where $\bar{T}(v)(w) = T(v, w)$]. The rank of T may be defined as the rank of $\bar{T}$ (consider the matrix of $\bar{T}$ with respect to bases $v_1, \ldots, v_n$ and $v^*_1, \ldots, v^*_n$).

(xii) Let $V = M_p{}^*$. If $T : V \to V$ and $\mu \in V^*$ and $T(v) = \alpha v$ for all $v \in \ker \mu$, there is a y complementary to $\ker \mu$ such that

$$T(v) = \alpha v + \mu(v) y \quad \text{for all } v.$$

(Begin by choosing y_0 complementary to $\ker \mu$ and writing v uniquely as $v_0 + c y_0$ for $v_0 \in \ker \mu$.)

(xiii) Define

$$\delta: \underbrace{V \times \cdots \times V}_{p \text{ times}} \times \underbrace{V^* \times \cdots \times V^*}_{p \text{ times}} \to \mathbb{R}$$

by

$$\delta(v_1, \ldots, v_p, \lambda_1, \ldots, \lambda_p) = \det(\lambda_i(v_j)).$$

6. (a) Let $i_V : V \to V^{**}$ be the "natural isomorphism" $i_V(v)(\lambda) = \lambda(v)$. Show that for any linear transformation $f : V \to W$, the following diagram commutes:

$$\begin{array}{ccc} V & \xrightarrow{i_V} & V^{**} \\ f \downarrow & & \downarrow f^{**} \\ W & \xrightarrow{i_W} & W^{**} \end{array}$$

(b) Show that there do *not* exist isomorphisms $i_V : V \to V^*$ such that the following diagram always commutes.

$$\begin{array}{ccc} V & \xrightarrow{i_V} & V^* \\ f \downarrow & & \uparrow f^* \\ W & \xrightarrow{i_W} & W^* \end{array}$$

Hint: There does not even exist an isomorphism $i : \mathbb{R} \to \mathbb{R}^*$ which makes the diagram commute for all linear $f : \mathbb{R} \to \mathbb{R}$.

7. A *covariant functor* from (finite dimensional) vector spaces to vector spaces is a function F which assigns to every vector space V a vector space $F(V)$ and to every linear transformation $f : V \to W$ a linear transformation $F(f) : F(V) \to F(W)$, such that $F(1_V) = 1_{F(V)}$ and $F(g \circ f) = F(g) \circ F(v)$.

(a) The "identity functor", $F(V) = V$, $F(f) = f$ is a functor.
(b) The "double dual functor", $F(V) = V^{**}$, $F(f) = f^{**}$ is a functor.
(c) The "$\mathcal{T}_k$ functor", $F(V) = \mathcal{T}_k(V) = \mathcal{T}^k(V^*)$,

$$F(f)(T)(\lambda_1, \ldots, \lambda_k) = T(\lambda_1 \circ f, \ldots, \lambda_k \circ f)$$

is a functor.
(d) If F is any functor and $f : V \to W$ is an isomorphism, then $F(f)$ is an isomorphism.

A *contravariant functor* is defined similarly, except that $F(f) : F(W) \to F(V)$ and $F(g \circ f) = F(f) \circ F(g)$. Functors of more than one argument, covariant in some and contravariant in others, may also be defined.

(e) The "dual functor", $F(V) = V^*$, $F(f) = f^*$ is a contravariant functor.
(f) The "$\mathcal{T}^k$ functor", $F(V) = \mathcal{T}^k(V)$, $F(f) = f^*$ is a contravariant functor.

8. (a) Let $\mathrm{Hom}(V, W)$ denote all linear transformations from V to W. Choosing a basis for V and W, we can identify $\mathrm{Hom}(V, W)$ with the $m \times n$ matrices, and consequently give it the metric of $\mathbb{R}^{nm}$. Show that a different choice of bases leads to a homeomorphic metric on $\mathrm{Hom}(V, W)$.

(b) A functor F gives a map from $\mathrm{Hom}(V, W)$ to $\mathrm{Hom}(F(V), F(W))$. Call F *continuous* if this map is always continuous [using the metric in part (a)]. Show that if $\xi = \pi \colon E \to B$ is any vector bundle, and F is continuous, then there is a bundle $F(\xi) = \pi' \colon E' \to B$ for which $\pi'^{-1}(p) = F(\pi^{-1}(p))$, and such that to every trivialization

$$t \colon \pi^{-1}(U) \to U \times \mathbb{R}^n$$

corresponds a trivialization

$$t' \colon \pi'^{-1}(U) \to U \times F(\mathbb{R}^n).$$

(c) The functor $\mathcal{T}_1(V) = \mathcal{T}^1(V^*) = V^{**}$ is continuous. (The bundle $\mathcal{T}_1(TM)$ is just a case of the construction in (b).)

(d) Define a continuous contravariant functor F, and show how to construct a bundle $F(\xi)$.

(e) The functor $F(V) = V^*$ is continuous. (The bundle T^*M is a special case of the construction in (d).)

Generally, the same construction can be used when F is a functor of several arguments. The bundles $\mathcal{T}_l^k(M)$ are all special cases. See the next two problems for other examples, as well as an example of a functor which is not continuous.

9. (a) Let F be a functor from $\mathbf{V}^n$, the class of n-dimensional vector spaces, to $\mathbf{V}^k$. Given $A \in \mathrm{GL}(n, \mathbb{R})$ we can consider it as a map $A \colon \mathbb{R}^n \to \mathbb{R}^n$. Then $F(A) \colon F(\mathbb{R}^n) \to F(\mathbb{R}^n)$. Choose, once and for all, an isomorphism $F(\mathbb{R}^n) \to \mathbb{R}^k$. Then $F(A)$ can be considered as a map $h(A) \colon \mathbb{R}^k \to \mathbb{R}^k$. Show that $h \colon \mathrm{GL}(n, \mathbb{R}) \to \mathrm{GL}(k, \mathbb{R})$ is a homomorphism.

(b) How does the homomorphism h depend on the initial choice of the isomorphism $F(\mathbb{R}^n) \to \mathbb{R}^k$?

(c) Let $\mathbf{v} = (v_1, \ldots, v_n)$ and $\mathbf{w} = (w_1, \ldots, w_n)$ be ordered bases of V and let $\mathbf{e} = (e_1, \ldots, e_n)$ be the standard basis of $\mathbb{R}^n$. If $\mathbf{e} \to \mathbf{v}$ denotes the isomorphism taking e_i to v_i, show that the following diagram commutes

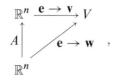

where $A = (a_{ij})$ is defined by

$$w_i = \sum_{j=1}^{n} a_{ji} v_j.$$

After identifying $F(\mathbb{R}^n)$ with $\mathbb{R}^k$, this means that

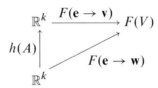

also commutes. This suggests a way of proving the following.

> THEOREM. If $h \colon \mathrm{GL}(n, \mathbb{R}) \to \mathrm{GL}(k, \mathbb{R})$ is any homomorphism, there is a functor $F_h \colon \mathbf{V}^n \to \mathbf{V}^k$ such that the homomorphism defined in part (a) is equal to h.

(d) For $q, q' \in \mathbb{R}^k$, define
$$(\mathbf{v}, q) \sim (\mathbf{w}, q')$$
if $q = h(A)q'$ where $w_i = \sum_{j=1}^{n} a_{ji} v_j$. Show that $\sim$ is an equivalence relation, and that every equivalence class contains exactly one element $(\mathbf{v}, q)$ for a given $\mathbf{v}$. We will denote the equivalence class of $(\mathbf{v}, q)$ by $[\mathbf{v}, q]$.

(e) Show that the operations
$$[\mathbf{v}, q_1] + [\mathbf{v}, q_2] = [\mathbf{v}, q_1 + q_2]$$
$$a \cdot [\mathbf{v}, q] = [\mathbf{v}, aq]$$

are well-defined operations making the set of all equivalence classes into a k-dimensional vector space $F_h(V)$.

(f) If $V, W \in \mathbf{V}^n$ and $f \colon V \to W$, choose ordered bases $\mathbf{v}, \mathbf{w}$, define A by $f(v_i) = \sum_{j=1}^{n} a_{ji} w_j$, and define

$$F_h(f)[\mathbf{v}, q] = [\mathbf{w}, h(A)(q)].$$

Show that this is a well-defined linear transformation, that F_h is a functor, and that $F_h(A) = h(A)$ when we identify $F_h(\mathbb{R}^n)$ with $\mathbb{R}^k$ by $[\mathbf{e}, q] \mapsto q$.

(g) Let $\alpha \colon \mathbb{R} \to \mathbb{R}$ be a non-continuous homomorphism (compare page 380), and let $h \colon \mathrm{GL}(n, \mathbb{R}) \to \mathrm{GL}(1, \mathbb{R}) = \mathbb{R}$ be $h(A) = \alpha(\det A)$. Then $F_h \colon \mathbf{V}^n \to \mathbf{V}^1$ is a non-continuous functor.

10. In classical tensor analysis there are, in addition to mixed tensor fields, other "quantities" which are defined as sets of functions which transform according to yet other rules. These new rules are of the form

$$A' = A \text{ operated on by } h\left(\frac{\partial x^\alpha}{\partial x'^\beta}\right).$$

For example, assignments of a single function a to each coordinate system x such that the function a' assigned to x' satisfies

$$a' = \det\left(\frac{\partial x^i}{\partial x'^j}\right) \cdot a$$

are called (**even**) **scalar densities**; assignments for which

$$a' = \left|\det\left(\frac{\partial x^i}{\partial x'^j}\right)\right| \cdot a$$

are called **odd scalar densities**. The Theorem in Problem 9 allows us to construct a bundle whose sections correspond to these classical entities (later we will have a more illuminating way):

(a) Let $h\colon GL(n, \mathbb{R}) \to GL(1, \mathbb{R})$ take A into multiplication by $\det A$. Let F_h be the functor given by the Theorem, and consider the 1-dimensional bundle $F_h(TM)$ obtained by replacing each fibre M_p with $F_h(M_p)$. If (x, U) is a coordinate system, then

$$\alpha_x(p) = \left[\left(\frac{\partial}{\partial x^1}\bigg|_p, \ldots, \frac{\partial}{\partial x^n}\bigg|_p\right), 1\right] \in F_h(M_p)$$

is non-zero, so every section on U can be expressed as $a \cdot \alpha_x$ for a unique function a. If x' is another coordinate system and $a \cdot \alpha_x = a' \cdot \alpha_{x'}$, show that

$$a' = \det\left(\frac{\partial x^i}{\partial x'^j}\right) \cdot a.$$

(b) If, instead, h takes A into multiplication by $|\det A|$, show that the corresponding equation is

$$a' = \left|\det\left(\frac{\partial x^i}{\partial x'^j}\right)\right| \cdot a.$$

(c) For this h, show that a non-zero element of $F_h(V)$ determines an orientation for V. Conclude that the bundle of odd scalar densities is *not* trivial if M is not orientable.

(d) We can identify $\mathcal{T}_l^k(\mathbb{R}^n)$ with $\mathbb{R}^{n^{k+l}}$ by taking

$$e^*{}_{i_1} \otimes \cdots \otimes e^*{}_{i_k} \otimes e_{j_1} \otimes \cdots \otimes e_{j_l} \mapsto (i_1, \ldots, i_k, j_1, \ldots, j_l)^{\text{th}} \text{ basis vector of } \mathbb{R}^{n^{k+l}}.$$

Recall that if $f: V \to V$, we define $\mathcal{T}_l^k(f): \mathcal{T}_l^k(V) \to \mathcal{T}_l^k(V)$ by

$$\mathcal{T}_l^k(f)(T)(v_1, \ldots, v_k, \lambda_1, \ldots, \lambda_l) = T(f(v_1), \ldots, f(v_k), \lambda_1 \circ f, \ldots, \lambda_l \circ f).$$

Given $A \in \text{GL}(n, \mathbb{R})$, we can consider it as a map $A: \mathbb{R}^n \to \mathbb{R}^n$. Then $\mathcal{T}_l^k(A): \mathcal{T}_l^k(\mathbb{R}^n) \to \mathcal{T}_l^k(\mathbb{R}^n)$ determines an element $\mathcal{T}_l^k(A)$ of $\text{GL}(n^{k+l}, \mathbb{R})$. Let $h: \text{GL}(n, \mathbb{R}) \to \text{GL}(n^{k+l}, \mathbb{R})$ be defined by

$$h(A) = (\det A)^w \mathcal{T}_l^k(A) \qquad w \text{ an integer.}$$

The bundle $F(TM)$ is called the bundle of (**even**) **relative tensors of type** $\binom{k}{l}$ **and weight** w. For $k = l = 0$ we obtain the bundle of (**even**) **relative scalars of weight** w [the (**even**) **scalar densities** are the (even) relative scalars of weight 1]. If $(\det A)^w$ is replaced by $|\det A|^w$ (w any real number), we obtain the bundle of **odd relative tensors of type** $\binom{k}{l}$ **and weight** w. Show that the transformation law for the components of sections of these bundles is

$$A'^{\beta_1 \ldots \beta_l}_{\alpha_1 \ldots \alpha_k} = \left[\det\left(\frac{\partial x^i}{\partial x'^j}\right)\right]^w \sum_{\substack{i_1, \ldots, i_k \\ j_1, \ldots, j_l}} A^{j_1 \ldots j_l}_{i_1 \ldots i_k} \frac{\partial x^{i_1}}{\partial x'^{\alpha_1}} \cdots \frac{\partial x^{i_k}}{\partial x'^{\alpha_k}} \frac{\partial x'^{\beta_1}}{\partial x^{j_1}} \cdots \frac{\partial x'^{\beta_l}}{\partial x^{j_l}}$$

(or the same formula with $\det(\partial x^i / \partial x'^j)$ replaced by $|\det(\partial x^i / \partial x'^j)|$).
(e) Define

$$\varepsilon_{i_1 \ldots i_n} = \begin{cases} +1 & \text{if } i_1, \ldots, i_n \text{ is an even permutation of } 1, \ldots, n \\ -1 & \text{if } i_1, \ldots, i_n \text{ is an odd permutation of } 1, \ldots, n \\ 0 & \text{if } i_\alpha = i_\beta \text{ for some } \alpha \neq \beta. \end{cases}$$

Show that there is a covariant relative tensor of weight -1 with these components in every coordinate system. Also show that $\varepsilon^{i_1 \ldots i_n} = \varepsilon_{i_1 \ldots i_n}$ are the components in every coordinate system of a certain contravariant relative tensor of weight 1. (See Problem 7-12 for a geometric interpretation of these relative tensors.)

CHAPTER 5

VECTOR FIELDS AND DIFFERENTIAL EQUATIONS

$\mathbf{W}$e return to a more detailed study of the tangent bundle TM, and its sections, i.e., vector fields. Let X be a vector field defined in a neighborhood of $p \in M$. We would like to know if there is a curve $\rho \colon (-\varepsilon, \varepsilon) \to M$ through p whose tangent vectors coincide with X, that is, a curve ρ with

$$\rho(0) = p$$

$$\rho_* \left(\frac{d}{dt}\bigg|_t \right) = \frac{d\rho}{dt}\bigg|_t = X(\rho(t)).$$

Since this a local question, we wish to introduce a coordinate system (x, U) around p and transfer the vector field X to $x(U) \subset \mathbb{R}^n$. Recall that, in general, $\alpha_* X$ does not make sense for C^∞ functions $\alpha \colon M \to N$. However, if α is a diffeomorphism, then we define

$$(\alpha_* X)_q = \alpha_* \big(X_{\alpha^{-1}(q)} \big) \qquad [\text{i.e.,} \ = \alpha_{*\alpha^{-1}(q)} \big(X_{\alpha^{-1}(q)} \big)].$$

It is not hard to check (Problem 1) that $\alpha_* X$ is C^∞ on $\alpha(M)$. In particular, we have a vector field $x_* X$ on $x(U) \subset \mathbb{R}^n$. There is a function $f \colon x(U) \to \mathbb{R}^n$ with

$$(x_* X)_q = f(q)_q \in \mathbb{R}^n{}_q,$$

i.e., $(x_* X)_q$ has "components" $f^1(q), \dots, f^n(q)$. Consider the curve $c = x \circ \rho$. The condition

$$\frac{d\rho}{dt} = X(\rho(t))$$

means that

$$\rho_* \left(\frac{d}{dt}\bigg|_t \right) = X(\rho(t));$$

135

hence

$$\frac{dc}{dt}\bigg|_t = x_* \rho_* \left(\frac{d}{dt}\bigg|_t\right) = x_*(X(\rho(t))) = (x_* X)_{x(\rho(t))}$$
$$= (x_* X)_{c(t)}.$$

If we use $c'(t)$ to denote the ordinary derivative of the $\mathbb{R}^n$-valued function c, then this equation finally becomes simply

$$c'(t) = f(c(t)).$$

This is a simple example of a differential equation for a function $c \colon \mathbb{R} \to \mathbb{R}^n$, which may also be considered as a system of n differential equations for the functions c^i,

$$c^{i\prime}(t) = f^i\big(c^1(t), \dots, c^n(t)\big) \qquad i = 1, \dots, n.$$

We also want the "initial conditions"

$$c^i(0) = x^i(p).$$

Solving a differential equation used to be described as "integrating" the equation (the process *is* integration when the equation has the special form $c'(t) = f(t)$ for $f \colon \mathbb{R} \to \mathbb{R}$, a form to which our particular equations never reduce); solutions were consequently called "integrals" of the equation. Part of this terminology is still preserved. A curve $\rho \colon (-\varepsilon, \varepsilon) \to M$ with

$$\rho(0) = p$$
$$\frac{d\rho}{dt} = X(\rho(t))$$

is called an **integral curve** for X with **initial condition** $\rho(0) = p$. Similar terminology is applied, of course, to the differential equations one obtains upon introducing a coordinate system. For quite some time, we will work entirely in Euclidean space, and for a while x, y, etc., will denote points of $\mathbb{R}^n$. If $U \subset \mathbb{R}^n$ is open and $f \colon U \to \mathbb{R}^n$, then a curve $c \colon (-\varepsilon, \varepsilon) \to M$ with

$$c(0) = x \qquad\qquad x \in U$$
$$c'(t) = f(c(t))$$

is called an **integral curve** for f with **initial condition** $c(0) = x$.

Before stating the main theorem about the existence and uniqueness of such integral curves, we consider some special cases.

The equation for a curve c with range $\mathbb{R}$,

$$c'(t) = -[c(t)]^2,$$

which would be written classically in terms of a function $y \colon \mathbb{R} \to \mathbb{R}$ as

$$\frac{dy}{dx} = -y^2,$$

is the special case $f(a) = -a^2$. The standard method of solving this equation is to write

$$\frac{dy}{-y^2} = dx$$

$$\int \frac{dy}{-y^2} = \int dx$$

$$\frac{1}{y} = x + C$$

$$y = \frac{1}{x + C}.$$

Thus the curves

$$c(t) = \frac{1}{t + C}$$

are supposed to be solutions. This can be checked directly if you don't believe the above manipulations. (They really do make sense; the equation in question asserts that $y' = f \circ y$, so

$$\left(\frac{1}{f} \circ y \right) \cdot y' = 1;$$

hence, if $F' = 1/f$, then

$$(F \circ y)' = 1$$
$$F(y(x)) = x + C$$

for some C.) To obtain the initial conditions $c(0) = a$, we must take

$$c(t) = \frac{1}{t + 1/a}.$$

This works in all cases except $a = 0$. In this case, the correct solution is

$$c(t) = 0 \qquad \text{for all } t$$

(which we missed by dividing by y). In terms of vector fields, the curves c are the integral curves of

$$X(a) = -a^2 \frac{d}{dt}.$$

Notice that no integral curve, except $c(t) = 0$, can be defined for all t, even though X is defined on all of $\mathbb{R}$. It might be thought that this somehow reflects the fact that $X(0) = 0$, but this has nothing to do with the case. For $a > 0$, the curve $c(t) = 1/(t + 1/a)$ is defined for all large t, and as $t \to \infty$ it approaches, but never reaches, 0. On the other hand, as $t \to -1/a$ the curve escapes to infinity because the vector field gets big too fast. This will continue to be true even if we modify the vector field near 0 so that it is never 0.

Another phenomenon is illustrated by the equation

$$c'(t) = c(t)^{2/3},$$

written classically as

$$\frac{dy}{dx} = y^{2/3}.$$

There are two different solutions with the initial condition $c(0) = 0$, namely

(1) $c(t) = 0$ for all t,

(2) $c(t) = \dfrac{1}{27} t^3$ for all t.

In this case, the function f, given by $f(a) = a^{2/3}$, is *not* differentiable. Uniqueness will always be insured when $f : U \to \mathbb{R}^n$ is C^1, but it can also be obtained with a rather less stringent condition. We say that the function f satisfies a **Lipschitz condition** on U if there is some K such that

$$|f(x) - f(y)| \leq K|x - y| \qquad \text{for all } x, y \in U.$$

Notice that $f(a) = a^{2/3}$ is not Lipschitz; in fact, there is no K with

$$|f(x) - f(0)| \leq K|x|$$

for x near 0, since

$$\frac{x^{2/3}}{x} = x^{-1/3} \to \pm\infty \quad \text{as} \quad x \to 0^{\pm}.$$

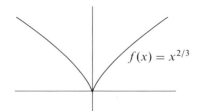

$f(x) = x^{2/3}$

A Lipschitz function is clearly continuous, but not necessarily differentiable (for example, $f(x) = |x|$). On the other hand, a C^1 function is locally Lipschitz, that is, it satisfies a Lipschitz condition in a neighborhood of each point—this follows from Lemma 2-5. A Lipschitz function is also clearly bounded on any bounded set.

The basic existence and uniqueness theorem for differential equations depends on a simple lemma about complete metric spaces.

1. THEOREM (THE CONTRACTION LEMMA). Let (M, ρ) be a non-empty complete metric space, and let $f: M \to M$ be a "contraction", that is, suppose there is some $C < 1$ such that

$$\rho(f(x), f(y)) \leq C\rho(x, y) \qquad \text{for all } x, y \in M.$$

Then there is a unique $x \in M$ such that $f(x) = x$ (the function f has a unique "fixed point").

PROOF. Notice that f is clearly continuous. Let $x_0 \in M$ and define a sequence $\{x_n\}$ inductively by

$$x_{n+1} = f(x_n),$$

i.e.,

$$x_{n+1} = f^n(x_0) = \underbrace{f \circ f \circ \cdots \circ f}_{n \text{ times}}(x_0).$$

Then an easy induction argument shows that

$$\rho(x_n, x_{n+1}) \leq C^n \rho(x_0, x_1).$$

Thus

$$\rho(x_n, x_{n+k}) \leq \rho(x_n, x_{n+1}) + \cdots + \rho(x_{n+k-1}, x_{n+k})$$
$$\leq (C^n + \cdots + C^{n+k-1})\rho(x_0, x_1).$$

Since $C < 1$, the sum $\sum_{n=0}^{\infty} C^n$ converges, so $C^n + \cdots + C^{n+k-1} \to 0$ as $n \to \infty$. Thus the sequence $\{x_n\}$ is Cauchy, so there is some x with

$$x = \lim_{n \to \infty} x_n.$$

Continuity of f then shows that

$$f(x) = \lim_{n \to \infty} f(x_n) = \lim_{n \to \infty} x_{n+1} = x. \quad \diamondsuit$$

We are going to apply the Contraction Lemma to certain spaces of functions. Recall that if (M, ρ) is a metric space and X is compact, then the set of all continuous functions $f : X \to M$ is a metric space if we define the metric σ by

$$\sigma(f, g) = \sup_{x \in X} \rho(f(x), g(x)).$$

If M is bounded, then we do not even need X to be compact. Moreover, if M is complete, then the new metric space is also complete; this is basically just the theorem that the uniform limit of continuous functions is continuous, plus the fact that each $\lim_{x \to \infty} f_n(x)$ exists since M is complete. In particular, if M is a compact subset of $\mathbb{R}^n$, then the set of all continuous functions $f : X \to M$ is complete with the metric

$$\sigma(f, g) = \|f - g\|, \qquad \text{where } \|f\| = \sup_{x \in X} |f(x)|.$$

Our basic strategy in solving differential equations will be to replace differentiable functions and derivatives by continuous functions and integrals. If $U \subset \mathbb{R}^n$ and $f : U \to \mathbb{R}^n$ is continuous, then a continuous function $\alpha : (-b, b) \to U$, defined on some interval around 0, clearly satisfies

(1)
$$\alpha'(t) = f(\alpha(t))$$
$$\alpha(0) = x$$

if it satisfies the integral equation

(2)
$$\alpha(t) = x + \int_0^t f(\alpha(u)) \, du,$$

where the integral of an $\mathbb{R}^n$-valued function is defined by integrating each component function separately. Conversely, if α satisfies (1), then α is differentiable, hence continuous; thus $\alpha' = f \circ \alpha$ is continuous, so

$$\alpha(t) - x = \alpha(t) - \alpha(0) = \int_0^t \alpha'(u) \, du = \int_0^t f(\alpha(u)) \, du.$$

For the proof of the basic theorem, we need only one simple estimate. If a continuous function $f : [a, b] \to \mathbb{R}^n$ satisfies $|f| \leq K$, then

$$\left| \int_a^b f(u) \, du \right| \leq K(b - a).$$

To prove this, we note that it is true for constant functions, hence for step functions, and thus for continuous functions, which are uniform limits on $[a, b]$ of step functions.

2. THEOREM. Let $f: U \to \mathbb{R}^n$ be any function, where $U \subset \mathbb{R}^n$ is open. Let $x_0 \in U$ and let $a > 0$ be a number such that the closed ball $\bar{B}_{2a}(x_0)$, of radius $2a$ and center x_0, is contained in U. Suppose that

(1) $|f| \leq L$ on $\bar{B}_{2a}(x_0)$

(2) $|f(x) - f(y)| \leq K|x - y|$ for $x, y \in \bar{B}_{2a}(x_0)$.

Choose $b > 0$ so that

(3) $b \leq a/L$

(4) $b < 1/K$.

Then for each $x \in \bar{B}_a(x_0)$ there is a unique $\alpha_x: (-b, b) \to U$ such that
$$\alpha_x{}'(t) = f(\alpha_x(t))$$
$$\alpha_x(0) = x.$$

PROOF. Choose $x \in \bar{B}_a(x_0)$, which will be fixed for the remainder of the proof. Let
$$M = \{\text{continuous } \alpha: (-b, b) \to \bar{B}_{2a}(x_0)\}.$$

Then M is a complete metric space. For each $\alpha \in M$, define a curve $S\alpha$ on $(-b, b)$ by
$$S\alpha(t) = x + \int_0^t f(\alpha(u)) \, du$$

(the integral exists since f is continuous on $\bar{B}_{2a}(x_0)$). The curve $S\alpha$ is clearly continuous. Moreover, for any $t \in (-b, b)$ we have
$$|S\alpha(t) - x| = \left| \int_0^t f(\alpha(u)) \, du \right|$$
$$< bL \qquad \text{by (1)}$$
$$\leq a \qquad \text{by (3)}.$$

Since $|x - x_0| \leq a$, it follows that $|S\alpha(t) - x_0| < 2a$, for all $t \in (-b, b)$, so

(∗) $\qquad S\alpha(t) \in B_{2a}(x_0) \subset \bar{B}_{2a}(x_0) \qquad$ for $t \in (-b, b)$.

Thus $S: M \to M$.

Now suppose $\alpha, \beta \in M$. Then
$$\|S\alpha - S\beta\| = \sup_t \left| \int_0^t f(\alpha(u)) - f(\beta(u)) \, du \right|$$
$$< bK \sup_{-b < u < b} |\alpha(u) - \beta(u)| \qquad \text{by (2)}$$
$$= bK \|\alpha - \beta\|.$$

Since we chose $bK < 1$ (by (4)), this shows that $S : M \to M$ is a contraction. Hence S has a unique fixed point:

> There is a unique $\alpha : (-b, b) \to \bar{B}_{2a}(x_0)$ with
> $$\alpha(t) = x + \int_0^t f(\alpha(u))\, du.$$

This, alas, is not quite what the theorem states. Having used the elegant Contraction Lemma, we pay for it by finishing off with a finicky detail:

> The map α is the unique $\beta : (-b, b) \to U$ satisfying
> $$\beta(t) = x + \int_0^t f(\beta(u))\, du.$$

Reason: We claim that any such β actually lies in $\bar{B}_{2a}(x_0)$, in fact, in $B_{2a}(x_0)$. Consider first numbers $t > 0$. We have already seen (statement (∗)) that for each t with $0 \le t < b$,

(∗∗) $\beta(t) = x + \displaystyle\int_0^t f(\beta(u))\, du$ is in $B_{2a}(x_0)$ [the *open* ball]

provided that

> $\beta(u) \in \bar{B}_{2a}(x_0)$ for all u with $0 \le u < t$,

so certainly if

> $\beta(u) \in B_{2a}(x_0)$ for all u with $0 \le u \le t$.

We can now use a simple least upper bound argument. Let

$$A = \{t : 0 \le t < b \text{ and } \beta(u) \in B_{2a}(x_0) \text{ for } 0 \le u < t\}.$$

Let $\alpha = \sup A$. Suppose $\alpha < b$. We clearly have $\beta(u) \in B_{2a}(x_0)$ for $0 \le u < \alpha$. So $\beta(\alpha) \in B_{2a}(x_0)$, by (∗∗). This clearly implies that $\beta(\alpha + s) \in B_{2a}(x_0)$ for sufficiently small $s > 0$, which contradicts the fact that $\alpha = \sup A$. So it must be that $\sup A = b$. A similar argument works for $-b < t \le 0$.

To sum up, the unique fixed point α_x of the map S is the unique curve with the desired properties. ❖

Notice that solutions of the differential equation

$$\alpha'(t) = f(\alpha(t))$$

remain solutions under additive changes of parameter; that is, if

$$\beta(t) = \alpha(t_0 + t),$$

then

$$\beta'(t) = \alpha'(t_0 + t) = f(\alpha(t_0 + t)) = f(\beta(t)).$$

This remark allows us to extend the uniqueness part of Theorem 2.

3. THEOREM. Suppose $f : U \to \mathbb{R}^n$ is locally Lipschitz, that is, around each point there is a ball on which f satisfies condition (2) of Theorem 2 for some K (and hence also condition (1) for some L). Let $x \in U$ and let α_1, α_2 be two maps on some open interval I with $\alpha_1(I), \alpha_2(I) \subset U$ and

$$\begin{aligned} \alpha_i'(t) &= f(\alpha_i(t)) \\ \alpha_i(0) &= x \end{aligned} \qquad i = 1, 2.$$

Then $\alpha_1 = \alpha_2$ on I.

PROOF. Suppose $\alpha_1(t_0) = \alpha_2(t_0)$ for some $t_0 \in I$. If we define

$$\beta_i(t) = \alpha_i(t_0 + t),$$

then the functions β_i satisfy the same differential equation, $\beta_i'(t) = f(\beta_i(t))$, and have the same initial condition $\beta_i(0) = \alpha_1(t_0) = \alpha_2(t_0) \in U$. Hence $\beta_1(t) = \beta_2(t)$ for sufficiently small t, by Theorem 1. Thus the set

$$\{t \in I : \alpha_1(t) = \alpha_2(t)\}$$

is open. It is clearly also closed and non-empty, so it equals I. ❖

We now revert to the situation in Theorem 2. We will write $\alpha_x(t)$ as $\alpha(t, x)$, so that we have a map

$$\alpha : (-b, b) \times B_a(x_0) \to U$$

satisfying

$$\alpha(0, x) = x$$

$$\frac{d}{dt}\alpha(t, x) = f(\alpha(t, x))$$

[i.e., $D_1\alpha(t, x) = f(\alpha(t, x))$], but we will frequently use $\partial/\partial t$ or d/dt in this discussion]. This map α is called a **local flow** for f in $(-b, b) \times B_a(x_0)$. To picture this map α, the best we can do is to draw the images of the integral

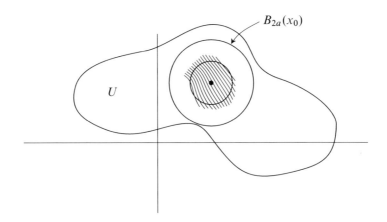

curves α_x. If $y = \alpha_x(t_0)$, then the integral curve α_x with the initial condition $\alpha_x(0) = x$ differs from the integral curve α_y with initial condition $\alpha_y(0) = y$ only by a change of parameter, so the two images overlap. For each fixed x, the map $t \mapsto \alpha(t, x)$ for $-b < t < b$ lies along *part* of the curve through x. On the other hand, if we fix t, then the map

$$x \mapsto \alpha(t, x)$$

gives the result of pushing each x along the integral curve through it, for a time interval of t. To focus attention on this map, we denote it by ϕ_t:

$$\phi_t(x) = \alpha(t, x) \quad [\, = \alpha_x(t)].$$

This map ϕ_t is always continuous. In fact, the whole flow α is continuous (as a function of both t and x):

4. THEOREM. If $f : U \to \mathbb{R}^n$ is locally Lipschitz, then the flow

$$\alpha : (-b, b) \times B_a(x_0) \to U$$

given by Theorem 2 is continuous.

PROOF. Let us denote the map S defined in the proof of Theorem 2 by S_x, to indicate explicitly the role of x. Then

$$\|\alpha_x - S_y \alpha_x\| = \|S_x \alpha_x - S_y \alpha_x\| = |x - y|.$$

Recall that

$$\|S\alpha - S\beta\| \le bK\|\alpha - \beta\|.$$

If S_y^n denotes the n-fold iterate of S_y, then

$$\|\alpha_x - S_y^n\alpha_x\| \le \|\alpha_x - S_y\alpha_x\| + \|S_y\alpha_x - S_y^2\alpha_x\| + \cdots + \|S_y^{n-1}\alpha_x - S_y^n\alpha_x\|$$

$$\le (1 + bK + \cdots + (bK)^{n-1})|x - y| \le \frac{1}{1 - bK}|x - y|.$$

Recall also that in Theorem 1 the fixed point α_y of S_y is the limit of $S_y^n\alpha$ for any α. Hence $\alpha_y = \lim\limits_{n\to\infty} S_y^n\alpha_x$, so we obtain

$$\|\alpha_x - \alpha_y\| \le \frac{1}{1 - bK}|x - y|.$$

Since $\|\alpha_x - \alpha_y\| = \sup\limits_t |\alpha(t, x) - \alpha(t, y)|$, this certainly proves continuity of α. ❖

If additional conditions are placed upon the map f, then further smoothness conditions can be proved for α. In fact,

If $f: U \to \mathbb{R}^n$ is C^k, then the flow $\alpha: (-b, b) \times B_a(x_0) \to U$ is also C^k.

Unfortunately, this is a very hard theorem. A clean exposition of the classical proof is given in Lang's *Introduction to Differentiable Manifolds* (2nd ed.), and a recently discovered proof can be found in Lang, *Real and Functional Analysis* (3rd ed.), pp. 371–379. In order to read this high-powered proof, you must first learn the elements of Banach spaces, including the Hahn-Banach theorem, and then read about differential calculus in Banach spaces, including the inverse and implicit function theorems (*Real and Functional Analysis*, pp. 360–365), but this is probably easier than reading the classical proof (and, besides, when you're finished you'll also know about Banach spaces, and differential calculus in Banach spaces).

We will just accept this fact. Notice that the maps ϕ_t are consequently C^∞ if f is C^∞.

Since the map

$$\alpha: (-b, b) \times B_a(x_0) \to U$$

satisfies $\alpha(0, x) = x$, we have

$$\alpha: \{0\} \times \bar{B}_{a/2}(x_0) \to \bar{B}_{a/2}(x_0) \subset B_a(x_0).$$

Continuity of α and compactness of $\{0\} \times \bar{B}_{a/2}(x_0)$ imply that there is some $\varepsilon > 0$ such that

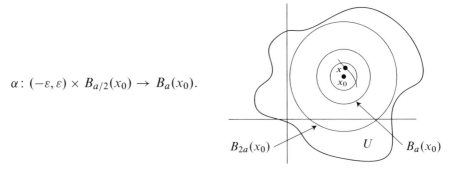

$$\alpha \colon (-\varepsilon, \varepsilon) \times B_{a/2}(x_0) \to B_a(x_0).$$

$B_{2a}(x_0)$ U $B_a(x_0)$

[If $x \in B_{a/2}(x_0)$, then the integral curve with initial condition x stays in $B_a(x_0)$ for $|t| < \varepsilon$.]

So if $|s| < \varepsilon$, and $x \in B_{a/2}(x_0)$, then the point $\alpha(s, x) \in B_a(x_0)$, so we can also define

$$\gamma(t) = \alpha(t, \alpha(s, x)) \qquad |t| < \varepsilon.$$

This satisfies

$$\gamma'(t) = f(\gamma(t))$$
$$\gamma(0) = \alpha(s, x).$$

We have also noted that

$$\beta(t) = \alpha(s + t, x), \qquad \text{defined for } |s + t| < \varepsilon,$$

satisfies

$$\beta'(t) = f(\beta(t))$$
$$\beta(0) = \alpha(s, x).$$

Consequently, $\beta(t) = \alpha(t, \alpha(s, x))$ for $|t| < \varepsilon$. In other words,

$$\text{if } \quad |s|, |t|, |s + t| < \varepsilon, \quad \text{then} \quad \alpha(t, \alpha(s, x)) = \alpha(s + t, x).$$

If we now let $\phi_t \colon B_{a/2}(x_0) \to \mathbb{R}^n$ be $\phi_t(x) = \alpha(t, x)$ for $x \in B_{a/2}(x_0)$, we can say:

$$\text{if } |s|, |t|, |s + t| < \varepsilon \text{ and } x, \phi_t(x) \in B_{a/2}(x_0), \text{ then}$$
$$\phi_s(\phi_t(x)) = \phi_{s+t}(x).$$

Roughly speaking, $\phi_{t+s} = \phi_t \circ \phi_s = \phi_s \circ \phi_t$. This shows, in particular, that for $|s| < \varepsilon$ each ϕ_s is a diffeomorphism, with inverse $\phi_s^{-1} = \phi_{-s}$. Everything we have said, since it is local, can be resaid, without requiring any more proof, on a manifold.

5. THEOREM. Let X be a C^∞ vector field on M, and let $p \in M$. Then there is an open set V containing p and an $\varepsilon > 0$, such that there is a unique collection of diffeomorphisms $\phi_t \colon V \to \phi_t(V) \subset M$ for $|t| < \varepsilon$ with the following properties:

(1) $\phi \colon (-\varepsilon, \varepsilon) \times V \to M$, defined by $\phi(t, p) = \phi_t(p)$, is C^∞.

(2) If $|s|, |t|, |s + t| < \varepsilon$, and $q, \phi_t(q) \in V$, then

$$\phi_{s+t}(q) = \phi_s \circ \phi_t(q).$$

(3) If $q \in V$, then X_q is the tangent vector at $t = 0$ of the curve $t \mapsto \phi_t(q)$.

The examples given previously show that we cannot expect ϕ_t to be defined for all t, or on all of M. In one case however, this can be attained. The **support** of a vector field X is just the closure of $\{p \in M \colon X_p \neq 0\}$.

6. THEOREM. If X has compact support (in particular, if M is compact), then there are diffeomorphisms $\phi_t \colon M \to M$ for all $t \in \mathbb{R}$ with properties (1), (2), (3).

PROOF. Cover support X by a finite number of open sets $V_1, \ldots, V_n$ given by Theorem 5 with corresponding $\varepsilon_1, \ldots, \varepsilon_n$ and diffeomorphisms ϕ_t^i. Let $\varepsilon = \min(\varepsilon_1, \ldots, \varepsilon_n)$. Notice that by uniqueness, $\phi_t^i(q) = \phi_t^j(q)$ for $q \in V_i \cap V_j$. So we can define

$$\phi_t(q) = \begin{cases} \phi_t^i(q) & \text{if } q \in V_i \\ q & \text{if } q \notin \text{support } X. \end{cases}$$

Clearly $\phi \colon (-\varepsilon, \varepsilon) \times M \to M$ is C^∞, and $\phi_{t+s} = \phi_t \circ \phi_s$ if $|t|, |s|, |t + s| < \varepsilon$, and each ϕ_t is a diffeomorphism.

To define ϕ_t for $|t| \geq \varepsilon$, write

$$t = k(\varepsilon/2) + r \qquad \text{with } k \text{ an integer, and } |r| < \varepsilon/2.$$

Let

$$\phi_t = \begin{cases} \phi_{\varepsilon/2} \circ \cdots \circ \phi_{\varepsilon/2} \circ \phi_r & [\phi_{\varepsilon/2} \text{ iterated } k \text{ times}] & \text{for } k \geq 0 \\ \phi_{-\varepsilon/2} \circ \cdots \circ \phi_{-\varepsilon/2} \circ \phi_r & [\phi_{-\varepsilon/2} \text{ iterated } -k \text{ times}] & \text{for } k < 0. \end{cases}$$

It is easy to check that this is the desired $\{\phi_t\}$. ❖

The unique collection $\{\phi_t\}$ given by Theorem 6, or more precisely, the map $t \mapsto \phi_t$ from $\mathbb{R}$ to the group of all diffeomorphisms of M, is called a 1-**parameter group of diffeomorphisms**, and is said to be **generated** by X. In the local case of Theorem 5, we obtain a "local 1-parameter group of local diffeomorphisms". The vector field X is sometimes called the "infinitesimal generator" of $\{\phi_t\}$ (vector fields used to be called "infinitesimal transformations").

Condition (3) in Theorem 5 can be rephrased in terms of the action of X_q on a C^∞ function $f : M \to \mathbb{R}$. Recall that

$$\frac{dc}{dt}(f) = \frac{df(c(t))}{dt} = (f \circ c)'(t).$$

Thus, to say that X_q is the tangent vector at $t = 0$ of the curve $t \mapsto \phi_t(q)$ amounts to saying that

$$(Xf)(q) = X_q f = \lim_{h \to 0} \frac{f(\phi_h(q)) - f(q)}{h}.$$

This equation will be used very frequently. The first use is to derive a corollary of Theorem 5 which allows us to simplify many calculations involving vector fields, and which also has important theoretical uses.

7. THEOREM. Let X be a C^∞ vector field on M with $X(p) \neq 0$. Then there is a coordinate system (x, U) around p such that

$$X = \frac{\partial}{\partial x^1} \quad \text{on} \quad U.$$

PROOF. It is easy to see that we can assume $M = \mathbb{R}^n$ (with the standard coordinate system $t^1, \ldots, t^n$, say), and $p = 0 \in \mathbb{R}^n$. Moreover, we can assume that $X(0) = \partial/\partial t^1\big|_0$. The idea of the proof is that in a neighborhood of 0 there is a unique integral curve through each point $(0, a^2, \ldots, a^n)$; if q lies on the integral

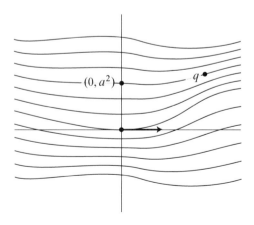

curve through this point, we will use $a^2, \ldots, a^n$ as the last $n-1$ coordinates of q and the time interval it takes the curve to get to q as the first coordinate. To do this, let X generate ϕ_t and consider the map χ defined on a neighborhood of 0 in $\mathbb{R}^n$ by

$$\chi(a^1, \ldots, a^n) = \phi_{a^1}(0, a^2, \ldots, a^n).$$

We compute that for $a = (a^1, \ldots, a^n)$,

$$\chi_* \left(\left. \frac{\partial}{\partial t^1} \right|_a \right)(f) = \left. \frac{\partial}{\partial t^1} \right|_a (f \circ \chi)$$

$$= \lim_{h \to 0} \frac{1}{h} [f(\chi(a^1 + h, a^2, \ldots, a^n)) - f(\chi(a))]$$

$$= \lim_{h \to 0} \frac{1}{h} [f(\phi_{a^1 + h}(0, a^2, \ldots, a^n)) - f(\chi(a))]$$

$$= \lim_{h \to 0} \frac{1}{h} [f(\phi_h(\chi(a))) - f(\chi(a))]$$

$$= (Xf)(\chi(a)).$$

Moreover, for $i > 1$ we can at least compute

$$\chi_* \left(\left. \frac{\partial}{\partial t^i} \right|_0 \right)(f) = \left. \frac{\partial}{\partial t^i} \right|_0 (f \circ \chi)$$

$$= \lim_{h \to 0} \frac{1}{h} [f(\chi(0, \ldots, h, \ldots, 0)) - f(0)]$$

$$= \lim_{h \to 0} \frac{1}{h} [f(0, \ldots, h, \ldots, 0) - f(0)]$$

$$= \left. \frac{\partial f}{\partial t^i} \right|_0.$$

Since $X(0) = \left. \partial/\partial t^1 \right|_0$ by assumption, this shows that $\chi_{*0} = I$ is non-singular. Hence $x = \chi^{-1}$ may be used as a coordinate system in a neighborhood of 0. This is the desired coordinate system, for it is easy to see that the equation $\chi_*(\partial/\partial t^1) = X \circ \chi$, which we have just proved, is equivalent to $X = \partial/\partial x^1$. ❖

The second use of the equation

$$(Xf)(p) = \lim_{h \to 0} \frac{1}{h} [f(\phi_h(p)) - f(p)]$$

is more comprehensive. The fact that Xf can be defined totally in terms of the diffeomorphisms ϕ_h suggests that an action of X on other objects can be obtained in a similar way. To emphasize the fundamental similarity of these notions, we first introduce the notation

$$L_X f \quad \text{for} \quad Xf.$$

We call $L_X f$ the (**Lie**) **derivative of** f **with respect to** X; it is another function, whose value at p is denoted variously by $(L_X f)(p) = L_X f(p) = (Xf)(p) = X_p(f)$. Now if ω is a C^∞ covariant vector field, we define a new covariant vector field, the **Lie derivative of** ω **with respect to** X, by

$$(L_X \omega)(p) = \lim_{h \to 0} \frac{1}{h}[(\phi_h{}^*\omega)(p) - \omega(p)].$$

This is the limit of certain members of $M_p{}^*$. Recall that if $X_p \in M_p$, then

$$(\phi_h{}^*\omega)(p)(X_p) = \omega(\phi_h(p))(\phi_{h*}X_p).$$

A fairly easy direct argument (Problem 8) shows that this limit always exists, and that the newly defined covariant vector field $L_X \omega$ is C^∞, but we will soon compute this vector field explicitly in a coordinate system, and these facts will then be obvious.

If Y is another vector field, we can define the **Lie derivative of** Y **with respect to** X,

$$(L_X Y)(p) = \lim_{h \to 0} \frac{1}{h}[Y_p - (\phi_{h*}Y)_p].$$

The vector field $\phi_{h*}Y$ appearing here is a special case of the vector field α_*Y defined at the beginning of the chapter, for $\alpha: M \to N$ a diffeomorphism and Y a vector field on M. Thus $(\phi_{h*}Y)_p = \phi_{h*}(Y_{\phi_{-h}(p)})$ is obtained by evaluating Y at $\phi_h{}^{-1}(p) = \phi_{-h}(p)$, and then moving it back to p by ϕ_{h*}.

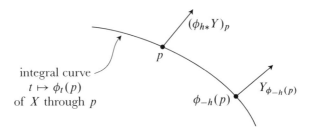

The definition of $L_X Y$ can be made to look more closely analogous to $L_X f$ and $L_X \omega$ in the following way. If $\alpha: M \to N$ is a diffeomorphism and Y is a vector field on the range N, then a vector field α^*Y on M can be defined by

$$(\alpha^*Y)_p = (\alpha^{-1})_*(Y_{\alpha(p)}).$$

Of course, $\alpha^*(Y)$ is just $(\alpha^{-1})_*Y$. Now notice that

$$\lim_{h\to 0}\frac{1}{h}[(\phi_h{}^*Y)_p - Y_p] = \lim_{h\to 0}\frac{Y_p - (\phi_h{}^*Y)_p}{-h} = \lim_{k\to 0}\frac{1}{k}[Y_p - (\phi_{-k}{}^*Y)_p]$$

$$= \lim_{k\to 0}\frac{1}{k}[Y_p - (\phi_{k*}Y)_p] = (L_XY)(p).$$

Nevertheless, we will stick to the original (equivalent) definition.

We now wish to compute $L_X\omega$ and L_XY in a coordinate system. The calculation is made a lot easier by first observing

8. **PROPOSITION.** If L_XY_i and $L_X\omega_i$ exist for $i = 1, 2$, then

(1) $L_X(Y_1 + Y_2) = L_XY_1 + L_XY_2$,

(2) $L_X(\omega_1 + \omega_2) = L_X\omega_1 + L_X\omega_2$.

If L_XY and $L_X\omega$ exist, then

(3) $L_X fY = Xf \cdot Y + f \cdot L_XY$,

(4) $L_X f \cdot \omega = Xf \cdot \omega + f \cdot L_X\omega$.

Finally, if $\omega(Y)$ denotes the function $p \mapsto \omega(p)(Y_p)$ and $L_X\omega$ and L_XY exist, then

(5) $L_X(\omega(Y)) = (L_X\omega)(Y) + \omega(L_XY)$.

PROOF. (1) and (2) are trivial. The remaining equations are all proved by the same trick, the one used in finding $(fg)'(x)$. We will do number (3) here.

$$(L_X fY)_p = \lim_{h\to 0}\frac{1}{h}[(fY)_p - (\phi_{h*}fY)_p]$$

$$= \lim_{h\to 0}\frac{1}{h}[f(p)Y_p - \phi_{h*}(fY)_{\phi_{-h}(p)}]$$

$$= \lim_{h\to 0}\frac{1}{h}[f(p)Y_p - f(\phi_{-h}(p))\phi_{h*}Y_{\phi_{-h}(p)}]$$

$$= \lim_{h\to 0} f(p)\frac{1}{h}[Y_p - \phi_{h*}Y_{\phi_{-h}(p)}]$$

$$+ \lim_{h\to 0}\left[\frac{f(p) - f(\phi_{-h}(p))}{h}\right]\phi_{h*}Y_{\phi_{-h}(p)}.$$

The first limit is clearly $f(p) \cdot L_X Y(p)$. In the second limit, the term in brackets approaches

$$\lim_{k \to 0} \frac{f(p) - f(\phi_k(p))}{-k} = Xf(p),$$

while an easy argument shows that $\phi_{h*} Y_{\phi_{-h}(p)} \to Y_p$. ❖

We are now ready to compute L_X in terms of a coordinate system (x, U) on M. Suppose $X = \sum_{i=1}^{n} a^i \partial/\partial x^i$. We first compute $L_X(dx^i)$. Recall (Problem 4-1) that if $f : M \to N$ and y is a coordinate system on N, then

$$f^*(dy^i) = \sum_{j=1}^{n} \frac{\partial(y^i \circ f)}{\partial x^j} \, dx^j.$$

We can apply this to ϕ_h^*, where y is x. Then

$$L_X(dx^i)(p) = \lim_{h \to 0} \frac{1}{h} [(\phi_h^*) dx^i(p) - dx^i(p)]$$

$$= \lim_{h \to 0} \frac{1}{h} \left[\sum_{j=1}^{n} \frac{\partial(x^i \circ \phi_h)}{\partial x^j}(p) \, dx^j(p) - dx^i(p) \right].$$

Now the coefficient of $dx^j(p)$ is

$$\lim_{h \to 0} \frac{1}{h} \left[\frac{\partial(x^i \circ \phi_h)}{\partial x^j} - \delta_j^i \right] = \lim_{h \to 0} \frac{1}{h} \left[\frac{\partial(x^i \circ \phi_h)}{\partial x^j}(p) - \frac{\partial(x^i \circ \phi_0)}{\partial x^j}(p) \right]$$

$$(*) \quad = \frac{\partial}{\partial x^j} \bigg|_p \lim_{h \to 0} \frac{1}{h} [(x^i \circ \phi_h) - (x^i \circ \phi_0)]$$

$$\{\text{this step will be justified in a moment}\}$$

$$= \frac{\partial}{\partial x^j} \bigg|_p X(x^i) = \frac{\partial a^i}{\partial x^j}(p).$$

To justify $(*)$ we note that the map $A(h, q) = x^i(\phi_h(q))$ is C^∞ from $\mathbb{R} \times M$ to $\mathbb{R}$; thus $\partial^2 A/\partial h \partial x^j = \partial^2 A/\partial x^j \partial h$, which is what the interchange of limits amounts to.

It now follows that

$$L_X \, dx^i = \sum_{j=1}^{n} \frac{\partial a^i}{\partial x^j} \, dx^j.$$

We could now use (2) and (4) of Proposition 8 to compute $L_X \omega$ in general, but we are really interested in computing $L_X Y$. To compute $L_X(\partial/\partial x^i)$ we could imitate the calculations of $L_X\, dx^i$; but there would be a complication, because ϕ_{h*} on vector fields involves one more composition than $\phi_h{}^*$ on covariant vector fields. The trick needed to deal with this complication has already been used to prove (3), (4), and (5) of Proposition 8, and we can now use (5) to get the answer immediately:

$$0 = L_X\, \delta^i_j = L_X\left[dx^i\left(\frac{\partial}{\partial x^j}\right)\right] = (L_X\, dx^i)\left(\frac{\partial}{\partial x^j}\right) + dx^i\left(L_X\frac{\partial}{\partial x^j}\right),$$

so

$$dx^i\left(L_X\frac{\partial}{\partial x^j}\right) = -\frac{\partial a^i}{\partial x^j};$$

thus,

$$L_X\frac{\partial}{\partial x^j} = -\sum_{i=1}^n \frac{\partial a^i}{\partial x^j}\frac{\partial}{\partial x^i}.$$

Using (3) we obtain

$$L_X\left(b^j\frac{\partial}{\partial x^j}\right) = L_X b^j \cdot \frac{\partial}{\partial x^j} + b^j L_X\left(\frac{\partial}{\partial x^j}\right)$$

$$= \sum_{i=1}^n a^i\frac{\partial b^j}{\partial x^i}\frac{\partial}{\partial x^j} - \sum_{i=1}^n b^j\frac{\partial a^i}{\partial x^j}\frac{\partial}{\partial x^i}.$$

Summing over j and then interchanging i and j in the second double sum we obtain

$$\boxed{L_X Y = \sum_{j=1}^n\left(\sum_{i=1}^n a^i\frac{\partial b^j}{\partial x^i} - b^i\frac{\partial a^j}{\partial x^i}\right)\frac{\partial}{\partial x^j}, \quad X = \sum_{i=1}^n a^i\frac{\partial}{\partial x^i}, \quad Y = \sum_{i=1}^n b^i\frac{\partial}{\partial x^i}.}$$

This somewhat complicated expression immediately leads to a much simpler coordinate-free expression for $L_X Y$. If $f\colon M \to N$ is a C^∞ function, then Yf is a function, so $XYf = X(Yf)$ makes sense. Clearly

$$X(Yf) = \sum_{i=1}^n a^i\frac{\partial}{\partial x^i}\left(\sum_{j=1}^n b^j\frac{\partial f}{\partial x^j}\right) = \sum_{i,j} a^i\frac{\partial b^j}{\partial x^i}\frac{\partial f}{\partial x^j} + a^i b^j\frac{\partial^2 f}{\partial x^j\partial x^i}.$$

The second partial derivatives which arise here cancel those in the expression for $Y(Xf)$, and we find that

$$L_X Y = XY - YX, \qquad \text{also denoted by } [X, Y].$$

Often, $[X, Y]$ (which is called the "bracket" of X and Y) is just defined as $XY - YX$; note that this means

$$[X, Y]_p(f) = X_p(Yf) - Y_p(Xf).$$

A straightforward verification shows that

$$[X, Y]_p(fg) = f(p)[X, Y]_p(g) + g(p)[X, Y]_p(f),$$

so that $[X, Y]_p$ is a derivation at p, and can therefore be considered as a member of M_p.

We are now in a very strange situation. Two vector fields $L_X Y$ and $[X, Y]$ have both been defined independently of any coordinate system, but they have been proved equal using a coordinate system. This sort of thing irks some people to no end. Fortunately, in this case the coordinate-free proof is short, though hardly obvious.

In Chapter 3 we proved a lemma which for the special case of $\mathbb{R}$ says that a C^∞ function $f : (-\varepsilon, \varepsilon) \to \mathbb{R}$ with $f(0) = 0$ can be written

$$f(t) = tg(t)$$

for a C^∞ function $g : (-\varepsilon, \varepsilon) \to \mathbb{R}$ with $g(0) = f'(0)$, namely

$$g(t) = \int_0^1 f'(st)\, ds.$$

This has an immediate generalization.

9. LEMMA. If $f : (-\varepsilon, \varepsilon) \times M \to \mathbb{R}$ is C^∞ and $f(0, p) = 0$ for all $p \in M$, then there is a C^∞ function $g : (-\varepsilon, \varepsilon) \times M \to \mathbb{R}$ with

$$f(t, p) = tg(t, p)$$
$$\frac{\partial f}{\partial t}(0, p) = g(0, p).$$

PROOF. Define

$$g(t, p) = \int_0^1 \frac{\partial}{\partial s} f(st, p)\, ds. \ \ \text{❖}$$

10. THEOREM. If X and Y are C^∞ vector fields, then

$$L_X Y = [X, Y].$$

PROOF. Let $f: M \to \mathbb{R}$ be C^∞. Let X generate ϕ_t, $|t| < \varepsilon$. By Lemma 9 there is a family of C^∞ functions g_t on M such that

$$f \circ \phi_t = f + t g_t$$
$$g_0 = Xf.$$

Then

$$(\phi_{h*}Y)_p(f) = \phi_{h*}(Y_{\phi_{-h}(p)})(f) = Y_{\phi_{-h}(p)}(f \circ \phi_h)$$
$$= Y_{\phi_{-h}(p)}(f + hg_h),$$

so

$$\lim_{h \to 0} \frac{1}{h}[Y_p - (\phi_{h*}Y)_p](f) = \lim_{h \to 0} \frac{1}{h}[(Yf)(p) - (Yf)(\phi_{-h}(p))]$$
$$- \lim_{h \to 0}(Yg_h)(\phi_{-h}(p))$$
$$= (L_X Yf)(p) - (Yg_0)(p)$$
$$= X_p(Yf) - Y_p(Xf). ❖$$

The equality $L_X Y = [X, Y] = XY - YX$ reveals certain facts about $L_X Y$ which are by no means obvious from the definition. Clearly

$$[X, Y] = -[Y, X], \quad \text{so} \quad [X, X] = 0.$$

Consequently,

$$L_X Y = -L_Y X, \quad \text{so} \quad L_X X = 0.$$

Since we obviously have $L_X(aY_1 + bY_2) = aL_X Y_1 + bL_X Y_2$, it follows immediately that L is also linear with respect to X:

$$L_{aX_1 + bX_2} Y = aL_{X_1} Y + bL_{X_2} Y.$$

Finally, a straightforward calculation proves the "Jacobi identity":

$$[X, [Y, Z]] + [Z, [X, Y]] + [Y, [Z, X]] = 0.$$

This equation is capable of two interpretations in terms of Lie derivatives:

(a) $L_X[Y, Z] = [L_X Y, Z] + [Y, L_X Z]$,

(b) as operators on C^∞ functions, we have
$L_{[X,Y]} = L_X \circ L_Y - L_Y \circ L_X$ (which might be written as $[L_X, L_Y]$).

Finally, note that $L_X Y$ is linear over constants only, *not* over the C^∞ functions $\mathcal{F}$. In fact, Proposition 8, or a simple calculation using the definition of $[X, Y]$, shows that

$$[fX, gY] = fg[X, Y] + f(Xg)Y - g(Yf)X.$$

Thus, the bracket operation $[\ ,\ ]$ is *not* a tensor—that is, $[X, Y]_p$ does not depend only on X_p and Y_p (which is not surprising—what can one do to two vectors in a vector space except take linear combinations of them?), but on the vector fields X and Y. In particular, even if $X_p = 0$, it does not necessarily follow that $[X, Y]_p = 0$—in the formula

$$[X, Y]_p(f) = X_p(Yf) - Y_p(Xf)$$

the first term $X_p(Yf)$ is zero, but the second may not be, for Xf may have a non-zero derivative in the Y_p direction even though $(Xf)(p) = 0$.

The bracket $[X, Y]$, although not a tensor, pops up in the definition of practically all other tensors, for reasons that will become more and more apparent. Before proceeding to examine its geometric interpretation, we will endeavor to become more at ease with the Lie derivative by taking time out to prove directly from the definition of $L_X Y$ two facts which are obvious from the definition of $[X, Y]$.

(1) $L_X X = 0$.

If X generates ϕ_t, it certainly suffices to show that $(\phi_{h*} X)_p = X_p$ for all h. Recall that $(\phi_{h*} X)_p = \phi_{h*} X_{\phi_{-h}(p)}$. Now $X_{\phi_{-h}(p)}$ is just the tangent vector at

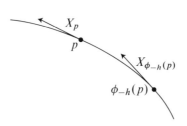

time $t = -h$ to the curve $t \mapsto \phi_t(p)$, and thus the tangent vector, at time $t = 0$, to the curve

$$\gamma(t) = \phi_{t-h}(p).$$

Thus $\phi_{h*} X_{\phi_{-h}(p)}$ is the tangent vector, at time $t = 0$, to the curve

$$\phi_h \circ \gamma(t) = \phi_h(\phi_{t-h}(p)) = \phi_t(p).$$

But this tangent vector is just X_p.

(2) If X_p and Y_p are both 0, then $L_X Y(p) = 0$.

Since $X_p = 0$, the unique integral curve c with $c(0) = p$ and $dc/dt = X(c(t))$ is simply $c(t) = p$ (an integral curve starting at p can never get away; conversely, of course, an integral curve starting at some other point can never get to p). Then $Y_p = 0$ and

$$(\phi_{h*}Y)_p = \phi_{h*}Y_{\phi_{-h}(p)} = \phi_{h*}Y_p = \phi_{h*}0 = 0,$$

so $L_X Y(p) = 0$.

To develop an interpretation of $[X, Y]$ we first prove two lemmas.

11. **LEMMA.** Let $\alpha \colon M \to N$ be a diffeomorphism and X a vector field on M which generates $\{\phi_t\}$. Then $\alpha_* X$ generates $\{\alpha \circ \phi_t \circ \alpha^{-1}\}$.

PROOF. We have

$$(\alpha_* X)_q(f) = [\alpha_* X_{\alpha^{-1}(q)}](f)$$
$$= X_{\alpha^{-1}(q)}(f \circ \alpha)$$
$$= \lim_{h \to 0} \frac{1}{h}[(f \circ \alpha)(\phi_h(\alpha^{-1}(q))) - (f \circ \alpha)(\alpha^{-1}(q))]$$
$$= \lim_{h \to 0} \frac{1}{h}[f(\alpha \circ \phi_h \circ \alpha^{-1}(q)) - f(q)]. \; ❖$$

12. **COROLLARY.** If $\alpha \colon M \to M$, then $\alpha_* X = X$ if and only if $\phi_t \circ \alpha = \alpha \circ \phi_t$ for all t.

13. **LEMMA.** Let X generate $\{\phi_t\}$ and Y generate $\{\psi_t\}$. Then $[X, Y] = 0$ if and only if $\phi_t \circ \psi_s = \psi_s \circ \phi_t$ for all s, t.

PROOF. If $\phi_t \circ \psi_s = \psi_s \circ \phi_t$ for all s, then $\phi_{t*}Y = Y$ by Corollary 12. If this is true for all t, then clearly $L_X Y = 0$.

Conversely, suppose that $[X, Y] = 0$, so that

$$(*) \qquad\qquad 0 = \lim_{h \to 0} \frac{1}{h}[Y_q - (\phi_{h*}Y)_q] \qquad \text{for all } q.$$

Given $p \in M$, consider the curve $c \colon (-\varepsilon, \varepsilon) \to M_p$ given by

$$c(t) = (\phi_{t*}Y)_p.$$

For the derivative, $c'(t)$, of this map into the vector space M_p we have

$$c'(t) = \lim_{h \to 0} \frac{1}{h}[c(t+h) - c(t)]$$

$$= \lim_{h \to 0} \frac{1}{h}[(\phi_{[t+h]*}Y)_p - (\phi_{t*}Y)_p]$$

$$= \lim_{h \to 0} \frac{1}{h}[\phi_{t*}(\phi_{h*}Y)_{\phi_{-t}(p)} - \phi_{t*}Y_{\phi_{-t}(p)}]$$

$$= \phi_{t*}\left\{\lim_{h \to 0} \frac{1}{h}[(\phi_{h*}Y)_{\phi_{-t}(p)} - Y_{\phi_{-t}(p)}]\right\}$$

$$= \phi_{t*}(0) \quad \text{using } (*) \text{ with } q = \phi_{-t}(p)$$

$$= 0.$$

Consequently $c(t) = c(0)$, so $\phi_{t*}Y = Y$. By Corollary 12, $\phi_t \circ \psi_s = \psi_s \circ \phi_t$ for all s, t. ❖

We have already shown that if $X(p) \neq 0$, then there is a coordinate system x with $X = \partial/\partial x^1$. If Y is another vector field, everywhere linearly independent of X, then we might expect to find a coordinate system with

$$(*) \qquad\qquad X = \frac{\partial}{\partial x^1}, \quad Y = \frac{\partial}{\partial x^2}.$$

However, a short calculation immediately gives the result

$$\left[\frac{\partial}{\partial x^1}, \frac{\partial}{\partial x^2}\right] = 0,$$

so there is no hope of finding a coordinate system satisfying $(*)$ unless $[X, Y] = 0$. The remarkable fact is that the condition $[X, Y] = 0$ is *sufficient*, as well as necessary, for the existence of the desired coordinate system.

14. THEOREM. If $X_1, \ldots, X_k$ are linearly independent C^∞ vector fields in a neighborhood of p, and $[X_\alpha, X_\beta] = 0$ for $1 \leq \alpha, \beta \leq k$, then there is a coordinate system (x, U) around p such that

$$X_\alpha = \frac{\partial}{\partial x^\alpha} \quad \text{on } U, \qquad \alpha = 1, \ldots, k.$$

PROOF. As in the proof of Theorem 7, we can assume that $M = \mathbb{R}^n$, that $p = 0$, and, by a linear change of coordinates, that

$$X_\alpha(0) = \left.\frac{\partial}{\partial t^\alpha}\right|_0 \qquad \alpha = 1, \ldots, k.$$

If X_α generates $\{\phi_t^\alpha\}$, define χ by

$$\chi(a^1,\ldots,a^n) = \phi_{a^1}^1(\phi_{a^2}^2(\ldots(\phi_{a^k}^k(0,\ldots,0,a^{k+1},\ldots,a^n))\ldots)).$$

As in the proof of Theorem 7, we can compute that

$$\chi_*\left(\frac{\partial}{\partial t^\alpha}\bigg|_0\right) = \begin{cases} X_\alpha(0) = \dfrac{\partial}{\partial t^\alpha}\bigg|_0 & \alpha = 1,\ldots,k \\[2ex] \dfrac{\partial}{\partial t^\alpha}\bigg|_0 & \alpha = k+1,\ldots,n. \end{cases}$$

Thus $x = \chi^{-1}$ can be used as a coordinate system in a neighborhood of $p = 0$. Moreover, just as before we see that

$$X_1 = \frac{\partial}{\partial x^1}.$$

Nothing said so far uses the hypothesis $[X_\alpha, X_\beta] = 0$. To make use of it, we appeal to Lemma 13; it shows that for each α between 1 and k, the map χ can also be written

$$\chi(a^1,\ldots,a^n) = \phi_{a^\alpha}^\alpha(\phi_{a^1}^1(\ldots(0,\ldots,0,a^{k+1},\ldots,a^n)\ldots)),$$

and our previous argument then shows that

$$X_\alpha = \frac{\partial}{\partial x^\alpha}. \quad \diamondsuit$$

We thus see that the bracket $[X,Y]$ measures, in some sense, the extent to which the integral curves of X and Y can be used to form the "coordinate lines" of a coordinate system. There is a more complicated, more difficult to prove, and less important result, which makes this assertion much more precise. If X and Y are two vector fields in a neighborhood of p, then for sufficiently small h we can

(1) follow the integral curve of X through p for time h;

(2) starting from that point, follow the integral curve of Y for time h;

(3) then follow the integral curve of X backwards for time h;

(4) then follow the integral curve of Y backwards for time h.

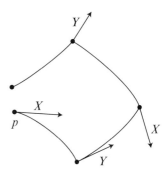

If there happens to be a coordinate system x with $x(p) = 0$ and

$$X = \frac{\partial}{\partial x^1}, \quad Y = \frac{\partial}{\partial x^2},$$

then these steps take us to points with coordinates

(1) $(h, 0, 0, \ldots, 0)$

(2) $(h, h, 0, \ldots, 0)$

(3) $(0, h, 0, \ldots, 0)$

(4) $(0, 0, 0, \ldots, 0)$,

so that this "parallelogram" is always closed. Even when X and Y are (linearly independent) vector fields with $[X, Y] \neq 0$, the parallelogram is "closed up to first order". The meaning of this phrase [an extension of the terminology "$c = \gamma$ up to first order at 0", which means that $c'(0) = \gamma'(0)$] is the following. Let $c(h)$ be the point which step (4) ends up at,

$$c(h) = \psi_{-h}(\phi_{-h}(\psi_h(\phi_h(0)))).$$

Then the curve c is the constant curve p up to first order, that is,

15. PROPOSITION. $c'(0) = 0$.

PROOF. If we define

$$\alpha_1(t, h) = \psi_t(\phi_h(p))$$
$$\alpha_2(t, h) = \phi_{-t}(\psi_h(\phi_h(p)))$$
$$\alpha_3(t, h) = \psi_{-t}(\phi_{-h}(\psi_h(\phi_h(p)))),$$

then

$$c(t) = \alpha_3(t, t).$$

Moreover,

(a) $\alpha_2(0, t) = \alpha_1(t, t)$

(b) $\alpha_3(0, t) = \alpha_2(t, t)$

and for any C^∞ function $f : M \to \mathbb{R}$,

(c)
$$\frac{\partial(f \circ \alpha_1)}{\partial t} = Yf \circ \alpha_1$$

(d)
$$\frac{\partial(f \circ \alpha_2)}{\partial t} = -Xf \circ \alpha_2$$

(e)
$$\frac{\partial(f \circ \alpha_3)}{\partial t} = -Yf \circ \alpha_3$$

while

(f)
$$\frac{\partial(f \circ \alpha_1)}{\partial h}(0, h) = Xf(\alpha_1(0, h)).$$

Consequently, repeated use of the chain rule gives

$$
\begin{aligned}
(f \circ c)'(0) = {}& D_1(f \circ \alpha_3)(0,0) + D_2(f \circ \alpha_3)(0,0) \\
= {}& D_1(f \circ \alpha_3)(0,0) \\
& + [D_1(f \circ \alpha_2)(0,0) + D_2(f \circ \alpha_2)(0,0)] \qquad \text{using (b)} \\
= {}& D_1(f \circ \alpha_3)(0,0) + D_1(f \circ \alpha_2)(0,0) \\
& + [D_1(f \circ \alpha_1)(0,0) + D_2(f \circ \alpha_1)(0,0)] \qquad \text{using (a).}
\end{aligned}
$$

Thus, (c), (d), (e), and (f) give

$$(f \circ c)'(0) = -Yf(p) - Xf(p) + Yf(p) + Xf(p) = 0. \; \diamondsuit$$

Whenever we have a curve $c : (-\varepsilon, \varepsilon) \to M$ with $c(0) = p$ and $c'(0) = 0 \in M_p$, we can define a new vector $c''(0)$ or $d^2c/dt^2\big|_0$ by

$$c''(0)(f) = (f \circ c)''(0).$$

A simple calculation shows, *using the assumption* $c'(0) = 0$, that this operator $c''(0)$ is a derivation, $c''(0) \in M_p$. (A more general construction is presented in Problem 17.) It turns out that for the curve c defined previously, the bracket $[X, Y]_p$ is related to this "second order" derivation. Until we get to Lie groups it will not be clear how anyone ever thought of the next theorem. The proof, which ends the chapter, but can easily be skipped, is an horrendous, but clever, calculation. It is followed by an addendum containing some additional important points about differential equations which are used later, and a second addendum concerning linearly independent vector fields in dimension 2.

16. THEOREM. $c''(0) = 2[X, Y]_p$.

PROOF. Using the notation of the previous proof, since $(f \circ c)(t) = (f \circ \alpha_3)(t, t)$ we have

$$(*) \quad (f \circ c)''(0) = D_{1,1}(f \circ \alpha_3)(0, 0) + 2D_{2,1}(f \circ \alpha_3)(0, 0) + D_{2,2}(f \circ \alpha_3)(0, 0).$$

Now

$$(1) \qquad\qquad D_{1,1}(f \circ \alpha_3)(0, 0) = D_1(-Yf \circ \alpha_3)(0, 0) \qquad \text{by (e)}$$
$$= YYf(p) \qquad\qquad\qquad \text{by (e).}$$

We also have

$$(2) \; 2D_{2,1}(f \circ \alpha_3)(0, 0)$$
$$= 2D_1(-Yf \circ \alpha_3) \qquad\qquad\qquad \text{by (e)}$$
$$= 2[D_1(Yf \circ \alpha_2)(0, 0)$$
$$\quad + D_2(Yf \circ \alpha_2)(0, 0)] \qquad\qquad \text{by (b) and the chain rule}$$
$$= 2XYf(p) - 2D_2(Yf \circ \alpha_2)(0, 0) \qquad \text{by (d)}$$
$$= 2XYf(p) - 2[D_1(Yf \circ \alpha_1)(0, 0)$$
$$\quad + D_2(Yf \circ \alpha_1)(0, 0)] \qquad\qquad \text{by (a) and the chain rule}$$
$$= 2XYf(p) - 2YYf(p) - 2XYf(p) \qquad \text{by (c) and (f).}$$

Since (b) gives

$$D_2(f \circ \alpha_3)(0, s) = D_1(f \circ \alpha_2)(s, s) + D_2(f \circ \alpha_2)(s, s),$$

we have

$$(3) \; D_{2,2}(f \circ \alpha_3)(0, 0) = D_{1,1}(f \circ \alpha_2)(0, 0) + 2D_{2,1}(f \circ \alpha_2)(0, 0)$$
$$\quad + D_{2,2}(f \circ \alpha_2)(0, 0)$$
$$= D_1(-Xf \circ \alpha_2)(0, 0) + 2D_2(-Xf \circ \alpha_2)(0, 0)$$
$$\quad + D_{2,2}(f \circ \alpha_2)(0, 0) \qquad \text{by (d)}$$
$$= XXf(p) - 2[D_1(Xf \circ \alpha_1)(0, 0) + D_2(Xf \circ \alpha_1)(0, 0)]$$
$$\quad + D_{2,2}(f \circ \alpha_2)(0, 0) \qquad \text{by (d) and the chain rule}$$
$$= XXf(p) - 2YXf(p) - 2XXf(p)$$
$$\quad + D_{2,2}(f \circ \alpha_2)(0, 0) \qquad \text{by (c) and (f).}$$

Finally, from

$$D_2(f \circ \alpha_2)(0, s) = D_1(f \circ \alpha_1)(s, s) + D_2(f \circ \alpha_2)(s, s) \quad \text{[from (a)]}$$

we have

$$
\begin{aligned}
(4) \qquad D_{2,2}(f \circ \alpha_2)(0, 0) &= D_{1,1}(f \circ \alpha_1)(0, 0) \\
&\quad + 2 D_{2,1}(f \circ \alpha_1)(0, 0) + D_{2,2}(f \circ \alpha_1)(0, 0) \\
&= YYf(p) + 2XYf(p) + XXf(p) \\
&\quad \text{by (c) and (f).}
\end{aligned}
$$

Substituting (1)–(4) in (∗) yields the theorem. ❖

ADDENDUM 1
DIFFERENTIAL EQUATIONS

Although we have always solved differential equations

$$\frac{\partial}{\partial t}\alpha(t,x) = f(\alpha(t,x))$$

with the initial condition

$$\alpha(0,x) = x,$$

we could just as well have required, for some t_0, that

$$\alpha(t_0,x) = x.$$

To prove this, one can replace 0 by t_0 everywhere in the proof of Theorem 2, or else just replace α by $t \mapsto \alpha(t - t_0, x)$.

Another omission in our treatment of differential equations is more glaring: the differential equations $\alpha'(t) = f(\alpha(t))$ do not even include simple equations of the form $\alpha'(t) = g(t)$, let alone equations like $\alpha'(t) = t\alpha(t)$. In general, we would like to solve equations

$$\frac{\partial}{\partial t}\alpha(t,x) = f(t,\alpha(t,x))$$
$$\alpha(0,x) = x,$$

where $f\colon (-c,c) \times U \to \mathbb{R}^n$. One way to do this is to replace $f(\alpha(t,x))$ by $f(t,\alpha(t,x))$ wherever it occurs in the proof. There is also a clever trick. Define

$$\bar{f}\colon (-c,c) \times U \to \mathbb{R}^{n+1}$$

by

$$\bar{f}(s,x) = (1, f(s,x)).$$

Then there is a flow $(\bar\alpha^1, \bar\alpha^2) = \bar\alpha\colon (-b,b) \times W \to \mathbb{R} \times \mathbb{R}^n$ with

$$\frac{\partial}{\partial t}\bar\alpha(t,s,x) = \bar{f}(\bar\alpha(t,s,x))$$
$$\bar\alpha(0,s,x) = (s,x).$$

For the first component function $\bar\alpha^1$ this means that

$$\frac{\partial}{\partial t}\bar\alpha^1(t,s,x) = 1$$
$$\bar\alpha^1(0,s,x) = s;$$

thus

$$\bar{\alpha}^1(t, s, x) = s + t.$$

For the second component $\bar{\alpha}^2$ we have

$$\frac{\partial}{\partial t}\bar{\alpha}^2(t, s, x) = f(\bar{\alpha}(t, s, x))$$
$$= f(\bar{\alpha}^1(t, s, x), \bar{\alpha}^2(t, s, x))$$
$$= f(s + t, \bar{\alpha}^2(t, s, x)).$$

Then

$$\beta(t, x) = \bar{\alpha}^2(t, 0, x)$$

is the desired flow with

$$\frac{\partial}{\partial t}\beta(t, x) = f(t, \beta(t, x))$$
$$\beta(0, x) = x.$$

Of course, we could also have arranged for $\beta(t_0, x) = x$ (by first finding $\bar{\alpha}$ with $\bar{\alpha}(t_0, s, x) = (s, x)$, *not* by considering the curve $t \mapsto \beta(t - t_0, x)$).

Finally, consider the special case of a *linear* differential equation

$$\alpha'(t) = g(t) \cdot \alpha(t),$$

where g is an $n \times n$ matrix-valued function on (a, b). In this case

$$f(t, x) = g(t) \cdot x.$$

If c is any $n \times n$ (constant) matrix, then

$$(c \cdot \alpha)'(t) = c \cdot \alpha'(t) = g(t) \cdot c \cdot \alpha(t)$$

so $c \cdot \alpha$ is also a solution of the same differential equation. This remark allows us to prove an important property of linear differential equations, distinguishing them from general differential equations $\alpha'(t) = f(t, \alpha(t))$, which may have solutions defined only on a small time interval, even if $f: (a, b) \times \mathbb{R}^n \to \mathbb{R}^n$ is C^∞.

17. PROPOSITION. If g is a continuous $n \times n$ matrix-valued function on (a, b), then the solutions of the equation

$$\alpha'(t) = g(t) \cdot \alpha(t)$$

can all be defined on (a, b).

PROOF. Notice that continuity of g implies that $f(t, x) = g(t) \cdot x$ is locally Lipschitz. So for any $t_0 \in (a, b)$ we can solve the equation, with any given initial condition, in a neighborhood of t_0. Extend it as far as possible. If the extended solution α is not defined for all t with $t_0 \leq t < b$, let t_1 be the least upper bound of the set of t's for which it is defined. Pick β with

$$\beta'(t) = g(t) \cdot \beta(t) \quad \text{for } t \text{ near } t_1$$
$$\beta(t_1) \neq 0.$$

Then $\beta(t^*) \neq 0$ for $t^* < t_1$ close enough to t_1. Hence there is c with

$$(c \cdot \beta)(t^*) = \alpha(t^*).$$

By uniqueness, $c \cdot \beta$ coincides with α on the interval where they are defined. Thus α may be extended past t_1 as $c \cdot \beta$, a contradiction. Similarly, α must be defined for all t with $a < t \leq t_0$. ❖

ADDENDUM 2
PARAMETER CURVES IN TWO DIMENSIONS

If $f: U \to M$ is an immersion from an open set $U \subset \mathbb{R}^n$ into an n-dimensional manifold M, the curve $t \mapsto f(a_1, \ldots, a_{i-1}, t, a_{i+1}, \ldots, a_n)$ is called a **parameter curve** in the ith direction. Given n vector fields $X_1, \ldots, X_n$ defined in a neighborhood of $p \in M$ and linearly independent at p, we know that there is usually no immersion $f: U \to M$ with $p \in f(U)$, whose parameter curves in the ith direction are the integral curves of the X_i—for we might not have $[X_i, X_j] = 0$. However, we might hope to find an immersion f for which the parameter curves in the ith direction lie along the integral curves of the X_i, but have different parameterizations. A simple example (Problem 20) shows that even this modest hope cannot be fulfilled in dimension 3.

On the other hand, in the special case of dimension 2, such an imbedding can be found:

18. PROPOSITION. Let X_1, X_2 be linearly independent vector fields in a neighborhood of a point p in a 2-dimensional manifold M. Then there is an imbedding $f: U \to M$, where $U \subset \mathbb{R}^2$ is open and $p \in f(U)$, whose i^{th} parameter lines lie along the integral curves of X_i.

PROOF. We can assume that $p = 0 \in \mathbb{R}^2$, and that $X_i(0) = (e_i)_0$. Every point q in a sufficiently small neighborhood of 0 is on a unique integral curve of X_1 through a point $(0, x^2(q))$—we proved precisely this fact in Theorem 7. Similarly, q is on a unique integral curve of X_2 through a point $(x^1(q), 0)$.

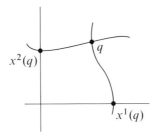

The map $q \mapsto (x^1(q), x^2(q))$ is C^∞, with Jacobian equal to I at 0 (these facts also follow from the proof of Theorem 7). Its inverse, in a sufficiently small neighborhood of 0, is the required diffeomorphism. ❖

We can always compose f with a map of the form $(x, y) \mapsto (\alpha(x), \beta(y))$ for diffeomorphisms α and β of $\mathbb{R}$, which gives us considerable flexibility. If, for example, $C \subset \mathbb{R}^2$ is the graph of a monotone function g, then the map

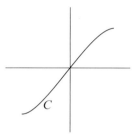

$(x, y) \mapsto (x, g(y))$ takes the diagonal $\{(x, x)\}$ to C. Moreover, for any particular parameterization $c = (c_1, c_2) \colon \mathbb{R} \to \mathbb{R}^2$ of C, we can further arrange that $c(t)$ maps to $(c(t), c(t))$, by composing with $(x, y) \mapsto (c_1^{-1}(x), y)$. Consequently, we can state

19. PROPOSITION. Let X_1, X_2 be linearly independent vector fields in a neighborhood of a point p in a 2-dimensional manifold M, and let c be a curve in M with $c(0) = p$ and $c'(t)$ never a multiple of X_1 or X_2. Then there is an imbedding $f \colon U \to M$, where $U \subset \mathbb{R}^2$ is open and $p \in f(U)$, whose i^{th} parameter lines lie along the integral curves of X_i, and for which $f(t, t) = c(t)$.

PROBLEMS

1. (a) If $\alpha: M \to N$ is C^∞, then $\alpha_*: TM \to TN$ is C^∞.
(b) If $\alpha: M \to N$ is a diffeomorphism, and X is a C^∞ vector field on M, then $\alpha_* X$ is a C^∞ vector field on N.
(c) If $\alpha: \mathbb{R} \to \mathbb{R}$ is $\alpha(t) = t^3$, then there is a C^∞ vector field X on $\mathbb{R}$ such that $\alpha_* X$ is not a C^∞ vector field.

2. Find a nowhere 0 vector field on $\mathbb{R}$ such that all integral curves can be defined only on some interval around 0.

3. Find an example of a complete metric space (M, ρ) and a function $f: M \to M$ such that $\rho(f(x), f(y)) < \rho(x, y)$ for all $x, y \in M$, but f has no fixed point.

4. Let $f: (-c, c) \times U \times V \to \mathbb{R}^n$ be C^∞, where $U, V \subset \mathbb{R}^n$ are open, and let $(x_0, y_0) \in U \times V$. Prove that there is a neighborhood W of (x_0, y_0) and a number $b > 0$ such that for each $(x, y) \in W$ there is a unique $\alpha = \alpha_{(x,y)}: (-b, b) \to U$ with $\alpha'(t) \in V$ for $t \in (-b, b)$ and

$$\begin{cases} \alpha''(t) = f(t, \alpha(t), \alpha'(t)) \\ \alpha(0) = x \\ \alpha'(0) = y. \end{cases}$$

Moreover, if we write $\alpha_{(x,y)}(t) = \alpha(t, x, y)$, then $\alpha: (-b, b) \times W \to U$ is C^∞. *Hint:* Consider the system of equations

$$\alpha'(t) = \beta(t)$$
$$\beta'(t) = f(t, \alpha(t), \beta(t)).$$

5. We sometimes have to solve equations "depending on parameters",

$$(*) \qquad \frac{\partial}{\partial t} \alpha(t, y, x) = f(t, y, \alpha(t, y, x))$$
$$\alpha(0, y, x) = x,$$

where $f: (-c, c) \times V \times U \to \mathbb{R}^n$, for open $U \subset \mathbb{R}^n$ and $V \subset \mathbb{R}^m$, and we are solving for $\alpha_{(y,x)}: (-b, b) \to U$ for each initial condition x and "parameter" y. For example, the equation

$$\alpha'(t) = y\alpha(t)$$
$$\alpha(0) = x,$$

with solution
$$\alpha(t) = xe^{yt},$$

is such a case.

(a) Define
$$\bar{f} : (-c, c) \times V \times U \to \mathbb{R}^m \times \mathbb{R}^n$$

by
$$\bar{f}(t, y, x) = (0, f(t, y, x)).$$

If $(\bar{\alpha}^1, \bar{\alpha}^2) = \bar{\alpha} : (-b, b) \times W \to \mathbb{R}^m \times \mathbb{R}^n$ is a flow for $\bar{f}$ in a neighborhood of (y_0, x_0), so that

$$\frac{\partial}{\partial t} \bar{\alpha}(t, y, x) = \bar{f}(t, \bar{\alpha}(t, y, x))$$
$$\bar{\alpha}(0, y, x) = (y, x),$$

show that we can write

$$\bar{\alpha}(t, y, x) = (y, \alpha(t, y, x))$$

for some α, and conclude that α satisfies (∗).

(b) Show that equations of the form

(∗∗)
$$\frac{\partial}{\partial t} \alpha(t, x) = f(t, x, \alpha(t, x))$$
$$\alpha(0, x) = x$$

can be reduced to equations of the form (∗) (and thus to equations

$$\frac{\partial}{\partial t} \alpha(t, x) = f(\alpha(t, x)),$$

ultimately). [When one proves that a C^k function $f : U \to \mathbb{R}^n$ has a C^k flow $\alpha : (-b, b) \times W \to U$, the hard part is to prove that if f is C^1, then α is differentiable with respect to the arguments in W, and that if the derivative with respect to these arguments is denoted by $D_2\alpha$, then

(∗∗∗) $D_1 D_2\alpha(t, x) = D_2 f(\alpha(t, x)) \cdot D_2\alpha(t, x)$

(a result which follows directly from the original equation

$$D_1\alpha(t, x) = f(\alpha(t, x))$$

if f is C^2, since $D_1 D_2 = D_2 D_1$). Since (∗∗∗) is an equation for $D_2\alpha$ of the form (∗∗), it follows that $D_2\alpha$ is differentiable if $D_2 f$ is C^1, i.e., if f is C^2. Differentiability of class C^k is then proved similarly, by induction.]

6. (a) Consider a linear differential equation

$$\alpha'(t) = g(t)\alpha(t),$$

where $g \colon \mathbb{R} \to \mathbb{R}$, so that we are solving for a real-valued function α. Show that all solutions are multiples of

$$\alpha(t) = e^{\int g(t)\,dt},$$

where $\int g(t)\,dt$ denotes some function G with $G'(t) = g$ (one can obtain all *positive* multiples simply by changing G). The remainder of this problem investigates the extent to which similar results hold for a system of linear differential equations.

(b) Let $A = (a_{ij})$ be an $n \times n$ matrix, and let $|A|$ denote the maximum of all $|a_{ij}|$. Show that

$$|A + B| \le |A| + |B|$$
$$|AB| \le n|A| \cdot |B|.$$

(c) Conclude that the infinite series of $n \times n$ matrices

$$\exp A = e^A = I + A + \frac{A^2}{2!} + \frac{A^3}{3!} + \frac{A^4}{4!} + \cdots$$

converges absolutely [in the sense that the $(i, j)^{\text{th}}$ entry of the partial sums converge absolutely for each (i, j)] and uniformly in any bounded set.

(d) Show that

$$\exp(TAT^{-1}) = T(\exp A)T^{-1}.$$

(e) If $AB = BA$, then

$$\exp(A + B) = (\exp A)(\exp B).$$

Hint: Write

$$\sum_{p=0}^{2N} \frac{(A+B)^p}{p!} = \left(\sum_{p=0}^{N} \frac{A^p}{p!}\right)\left(\sum_{p=0}^{N} \frac{B^p}{p!}\right) + R_N$$

and show that $|R_N| \to 0$ as $N \to \infty$.

(f) $(\exp A)(\exp -A) = I$, so $\exp A$ is always invertible.

(g) The map exp, considered as a map $\exp \colon \mathbb{R}^{n^2} \to \mathbb{R}^{n^2}$, is clearly differentiable (it is even analytic). Show that

$$\exp'(0)(B) = B \quad (= \exp(0) \cdot B).$$

(Notice that for $|A|$, the usual norm of $A \in \mathbb{R}^{n^2}$, we have $|A| \le |A| \le n|A|$.)

(h) Use the limit established in part (g) to show that $\exp'(A)(B) = \exp(A) \cdot B$ if $AB = BA$.

(i) Let $A: \mathbb{R} \to \mathbb{R}^{n^2}$ be differentiable, and let

$$B(t) = \exp(A(t)).$$

If $B'(t)$ denotes the matrix whose entries are the derivatives of the entries of B, show that

$$B'(t) = A'(t) \cdot \exp(A(t)),$$

provided that $A(t)A'(t) = A'(t)A(t)$. (This is clearly true if $A(s)A(t) = A(t)A(s)$ for all s, t.)

(j) Show that the linear differential equation

$$\alpha'(t) = g(t) \cdot \alpha(t)$$

has the solution

$$\alpha(t) = \exp\left(\int_0^t g(s)\, ds\right)$$

provided that $g(s)g(t) = g(t)g(s)$ for all s, t. (This certainly happens when $g(t)$ is a constant matrix A, so every system of linear equations with constant coefficients can be solved explicitly—the exponential of $\int_0^t g(s)\, ds = tA$ can be found by putting A in Jordan canonical form.)

7. Check that if the coordinate system x is $x = \chi^{-1}$, for $\chi: \mathbb{R}^n \to M$, then $X = \partial/\partial x^1$ is equivalent to $\chi_*(\partial/\partial t^1) = X \circ \chi$.

8. (a) Let M and N be C^∞ manifolds. For a C^∞ function $f: M \times N \to \mathbb{R}$ and $q \in N$, let $f(\cdot, q)$ denote the function from M to $\mathbb{R}$ defined by

$$p \mapsto f(p, q).$$

If (x, U) is a coordinate system on M, show that the function $\partial f/\partial x^i$, defined by

$$\frac{\partial f}{\partial x^i}(p, q) = \frac{\partial(f(\cdot, q))}{\partial x^i}(p),$$

is a C^∞ function on $M \times N$.

(b) If $\phi: (-\varepsilon, \varepsilon) \times M \to M$ is a 1-parameter group of diffeomorphisms, show that for every C^∞ function $f: M \to \mathbb{R}$, the limit

$$\lim_{h \to 0} \frac{1}{h}[f(\phi_h(p)) - f(p)]$$

exists, and defines a C^∞ function on M.

(c) If $\phi_*: (-\varepsilon, \varepsilon) \times TM \to TM$ is defined by

$$\phi_*(t, v) = \phi_{t*}(v),$$

show that ϕ_* is C^∞, and conclude that for every C^∞ vector field X and covariant vector field ω on M, the limit

$$\lim_{h \to 0} \frac{1}{h}[(\phi_h{}^*\omega)(X_p) - \omega(X_p)]$$

exists and defines a C^∞ function on M.

(d) Treat $L_X Y$ similarly.

9. Give the argument to show that $\phi_{h*}Y_{\phi_{-h}(p)} \to Y_p$ in the proof of Proposition 8.

10. (a) Prove that

$$L_X(f \cdot \omega) = Xf \cdot \omega + f \cdot L_X \omega$$
$$L_X[\omega(Y)] = (L_X \omega)(Y) + \omega(L_X Y).$$

(b) How would Proposition 8 have to be changed if we had defined $(L_X Y)(p)$ as

$$\lim_{h \to 0} \frac{1}{h}[(\phi_{h*}Y)_p - Y_p]?$$

11. (a) Show that

$$\phi^*(df)(Y) = Y(f \circ \phi).$$

(b) Using (a), show directly from the definition of L_X that for $Y \in M_p$,

$$[L_X \, df(p)](Y_p) = Y_p(L_X f),$$

and conclude that

$$L_X \, df = d(L_X f).$$

The formula for $L_X \, dx^i$, derived in the text, is just a special case derived in an unnecessarily clumsy way. In the next part we get a much simpler proof that $L_X Y = [X, Y]$, using the technique which appeared in the proof of Proposition 15.

(c) Let X and Y be vector fields on M, and $f: M \to \mathbb{R}$ a C^∞ function. If X generates $\{\phi_t\}$, define

$$\alpha(t, h) = Y_{\phi_{-t}(p)}(f \circ \phi_h).$$

Show that

$$D_1\alpha(0,0) = -X_p(Yf)$$
$$D_2\alpha(0,0) = Y_p(Xf).$$

Conclude that for $c(h) = \alpha(h,h)$ we have

$$-c'(0) = L_X Y(p)(f) = [X,Y]_p(f).$$

12. Check the Jacobi identity.

13. On $\mathbb{R}^3$ let X, Y, Z be the vector fields

$$X = z\frac{\partial}{\partial y} - y\frac{\partial}{\partial z}$$

$$Y = -z\frac{\partial}{\partial x} + x\frac{\partial}{\partial z}$$

$$Z = y\frac{\partial}{\partial x} - x\frac{\partial}{\partial y}.$$

(a) Show that the map

$$aX + bY + cZ \mapsto (a,b,c) \in \mathbb{R}^3$$

is an isomorphism (from a certain set of vector fields to $\mathbb{R}^3$) and that $[U,V] \mapsto$ the cross-product of the images of U and V.
(b) Show that the flow of $aX + bY + cZ$ is a rotation of $\mathbb{R}^3$ about some axis through 0.

14. If A is a tensor field of type $\binom{k}{l}$ on N and $\phi\colon M \to N$ is a diffeomorphism, we define $\phi^* A$ on M as follows. If $v_1,\ldots,v_k \in M_p$, and $\lambda_1,\ldots,\lambda_l \in M_p{}^*$, then

$$[\phi^* A(p)](v_1,\ldots,v_k,\lambda_1,\ldots,\lambda_l)$$
$$= A(\phi(p))(\phi_* v_1,\ldots,\phi_* v_k,(\phi^{-1})^*\lambda_1,\ldots,(\phi^{-1})^*\lambda_l).$$

(a) Check that under the identification of a vector field [or covariant vector field] with a tensor field of type $\binom{0}{1}$ [or type $\binom{1}{0}$] this agrees with our old $\phi^* Y$.
(b) If the vector field X on M generates $\{\phi_t\}$, and A is a tensor field of type $\binom{k}{l}$ on M, we define

$$(L_X A)(p) = \lim_{h \to 0} \frac{1}{h}[(\phi_h{}^* A)(p) - A(p)].$$

Show that

$$L_X(A + B) = L_X A + L_X B$$
$$L_X(A \otimes B) = (L_X A) \otimes B + A \otimes L_X B$$

(so that

$$L_X(fA) = X(f)A + f L_X A),$$

in particular).

(c) Show that

$$L_{X_1 + X_2} A = L_{X_1} A + L_{X_2} A.$$

Hint: We already know that it is true for A of type $\binom{0}{0}$, $\binom{0}{1}$, $\binom{1}{0}$.

(d) Let

$$C : \mathcal{T}_l^k(V) \to \mathcal{T}_{l-1}^{k-1}(V)$$

be any contraction

$$(CT)(v_1, \ldots, v_{k-1}, \lambda_1, \ldots, \lambda_{l-1})$$

$$= \text{contraction of}$$

$$(v, \lambda) \mapsto T(v_1, \ldots, v_{\alpha-1}, v, v_{\alpha+1}, \ldots, v_{k-1}, \lambda_1, \ldots, \lambda_{\beta-1}, \lambda, \lambda_{\beta+1}, \ldots, \lambda_{l-1}).$$

Show that

$$L_X(CA) = C(L_X A).$$

(e) Noting that $A(X_1, \ldots, X_k, \omega_1, \ldots, \omega_l)$ can be obtained by applying contractions repeatedly to $A \otimes X_1 \otimes \cdots \otimes X_k \otimes \omega_1 \otimes \cdots \otimes \omega_l$, use (d) to show that

$$L_X\big(A(X_1, \ldots, X_k, \omega_1, \ldots, \omega_l)\big)$$
$$= (L_X A)(X_1, \ldots, X_k, \omega_1, \ldots, \omega_l)$$
$$+ \sum_{i=1}^{k} A(X_1, \ldots, L_X X_i, \ldots, X_k, \omega_1, \ldots, \omega_l)$$
$$+ \sum_{i=1}^{l} A(X_1, \ldots, X_k, \omega_1, \ldots, L_X \omega_i, \ldots, \omega_l).$$

(f) If A has components $A_{i_1 \ldots i_k}^{j_1 \ldots j_l}$ in a coordinate system x and $X = \sum_{i=1}^{n} a^i \partial/\partial x^i$, show that the coordinates of $L_X A$ are given by

$$(L_X A)_{i_1 \ldots i_k}^{j_1 \ldots j_l} = \sum_{i=1}^{n} a^i \frac{\partial A_{i_1 \ldots i_k}^{j_1 \ldots j_l}}{\partial x^i} - \sum_{\alpha=1}^{k} \sum_{j=1}^{n} A_{i_1 \ldots i_k}^{j_1 \ldots j_{\alpha-1} j j_{\alpha+1} \ldots j_l} \frac{\partial a^{j_\alpha}}{\partial x^j}$$

$$+ \sum_{\alpha=1}^{l} \sum_{i=1}^{n} A_{i_1 \ldots i_{\alpha-1} i i_{\alpha+1} \ldots i_k}^{j_1 \ldots j_l} \frac{\partial a^i}{\partial x^{i_\alpha}}.$$

15. Let D be an operator taking the C^∞ functions $\mathcal{F}$ to $\mathcal{F}$, and the C^∞ vector fields $\mathcal{V}$ to $\mathcal{V}$, such that $D\colon \mathcal{F} \to \mathcal{F}$ and $D\colon \mathcal{V} \to \mathcal{V}$ are linear over $\mathbb{R}$ and

$$D(fY) = f \cdot DY + Df \cdot Y.$$

(a) Show that D has a unique extension to an operator taking tensor fields of type $\binom{k}{l}$ to themselves, such that

   (1) D is linear over $\mathbb{R}$

   (2) $D(A \otimes B) = DA \otimes B + A \otimes DB$

   (3) for any contraction C, $DC = CD$.

If we take $Df = Xf$ and $DY = L_X Y$, then this unique extension is L_X.

(b) Let A be a tensor field of type $\binom{1}{1}$, so that we can consider $A(p) \in End(M_p)$; then $A(X)$ is a vector field for each vector field X. Show that if we define $D_A f = 0$, $D_A X = A(X)$, then D_A has a unique extension satisfying (1), (2), and (3).

(c) Show that

$$(D_A \omega)(p) = -A(p)^*(\omega(p)).$$

(d) Show that

$$L_{fX} = f L_X - D_{X \otimes df}.$$

Hint: Check this for functions and vector fields first.

(e) If T is of type $\binom{2}{1}$, show that

$$(D_A T)^{ij}_k = \sum_{\alpha=1}^{n} T^{\alpha j}_k A^i_\alpha + \sum_{\alpha=1}^{n} T^{i\alpha}_k A^j_\alpha - \sum_{\alpha=1}^{n} T^{ij}_\alpha A^\alpha_k.$$

Generalize to tensors of type $\binom{k}{l}$.

16. (a) Let $f\colon \mathbb{R} \to \mathbb{R}$ satisfy $f'(0) = 0$. Define $g(t) = f(\sqrt{t})$ for $t \geq 0$. Show that the right-hand derivative

$$g'_+(0) = \lim_{h \to 0^+} \frac{g(h) - g(0)}{h} = \frac{f''(0)}{2}.$$

(Use Taylor's Theorem.)

(b) Given $c\colon \mathbb{R} \to M$ with $c'(0) = 0 \in M_p$, define $\gamma(t) = c(\sqrt{t})$ for $t \geq 0$. Show that the tangent vector $c''(0)$ defined by $c''(0)(f) = (f \circ c)''(0)$ can also be described by $c''(0) = 2\gamma'(0)$.

17. (a) Let $f: M \to \mathbb{R}$ have p as a critical point, so that $f_{*p} = 0$. Given vectors $X_p, Y_p \in M_p$, choose vector fields $\tilde{X}, \tilde{Y}$ with $\tilde{X}_p = X_p$ and $\tilde{Y}_p = Y_p$. Define

$$f_{**}(X_p, Y_p) = \tilde{X}_p(\tilde{Y}f).$$

Using the fact that $[X, Y]_p(f) = 0$, show that $f_{**}(X_p, Y_p)$ is symmetric, and conclude that it is well-defined.

(b) Show that

$$f_{**}\left(\sum_{i=1}^{n} a^i \left.\frac{\partial}{\partial x^i}\right|_p, \sum_{j=1}^{n} b^j \left.\frac{\partial}{\partial x^j}\right|_p\right) = \sum_{i,j=1}^{n} a^i b^j \frac{\partial^2 f}{\partial x^i \partial x^j}(p).$$

(c) The rank of $(\partial^2 f / \partial x^i \partial x^j(p))$ is independent of the coordinate system.

(d) Let $f: M \to N$ have p as a critical point. For $X_p, Y_p \in M$ and $g: N \to \mathbb{R}$ define

$$f_{**}(X, Y)(g) = \tilde{X}_p(\tilde{Y}(g \circ f)).$$

Show that

$$f_{**}: M_p \times M_p \to N_{f(p)}$$

is a well-defined bilinear map.

(e) If $c: \mathbb{R} \to M$ has 0 as a critical point, show that

$$c_{**}(0): \mathbb{R}_0 \times \mathbb{R}_0 \to M_{c(0)}$$

takes $(1_0, 1_0)$ to the tangent vector $c''(0)$ defined by $c''(0)(f) = (f \circ c)''(0)$.

18. Let c be the curve of Theorems 15 and 16. If x is a coordinate system around p with $x(p) = 0$, and

$$[X, Y]_p = \sum_{i=1}^{n} a^i \left.\frac{\partial}{\partial x^i}\right|_p,$$

show that

$$x^i(c(t)) = a^i t^2 + o(t^2),$$

where $o(t^2)$ denotes a function such that

$$\lim_{t \to 0} o(t^2)/t^2 = 0.$$

19. (a) If M is compact and 0 is a regular value of $f: M \to \mathbb{R}$, then there is a neighborhood U of $0 \in \mathbb{R}$ such that $f^{-1}(U)$ is diffeomorphic to $f^{-1}(0) \times U$,

by a diffeomorphism $\phi\colon f^{-1}(0) \times U \to f^{-1}(U)$ with $f(\phi(p,t)) = t$.   *Hint:*
Use Theorem 7 and a partition of unity to construct a vector field X on a
neighborhood of $f^{-1}(0)$ such that $f_* X = d/dt$.
(b) More generally, if M is compact and $q \in N$ is a regular value of $f\colon M \to
N$, then there is a neighborhood U of q and a diffeomorphism $\phi\colon f^{-1}(q) \times U \to
f^{-1}(U)$ with $f(\phi(p,q')) = q'$.
(c) It follows from (b) that if all points of N are regular values, then $f^{-1}(q_1)$
and $f^{-1}(q_2)$ are diffeomorphic for q_1, q_2 sufficiently close. If f is onto N, does
it follow that M is diffeomorphic to $f^{-1}(q) \times N$?

20. In $\mathbb{R}^3$, let Y and Z be unit vector fields always pointing along the y- and
z-axes, respectively, and let X will be a vector field one of whose integral curves
is the x-axis, while certain other integral curves are parabolas in the planes
$y = $ constant, as shown in the first part of the figure below. Using the second
part of the figure, show that Proposition 18 does not hold in dimension 3.

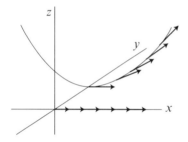 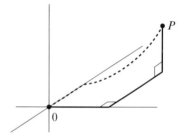

CHAPTER 6

INTEGRAL MANIFOLDS

PROLOGUE

In the previous chapter, we have seen that the integral curves of a vector field on a manifold M may be definable only for some small time interval, even though the vector field is C^∞ on all of M. We will now vary our question a little, so that global results can be obtained. Instead of a vector field, suppose that for each $p \in M$ we have a 1-dimensional subspace $\Delta_p \subset M_p$. The function Δ is called a 1-**dimensional distribution** (this kind of distribution has nothing whatsoever to do with the distributions of analysis, which include such things as the "δ-function"). Then Δ is spanned by a vector field *locally*; that is, we can choose (in many possible ways) a vector field X such that $0 \neq X_q \in \Delta_q$ for all q in some open set around p. We call Δ a C^∞ distribution if such a vector field X can be chosen to be C^∞ in a neighborhood of each point.

For a 1-dimensional distribution the notion of an integral curve makes no sense, but we define a (1-dimensional) submanifold N of M to be an **integral manifold** of Δ if for every $p \in N$ we have

$$i_*(N_p) = \Delta_p \quad \text{where} \quad i : N \to M \quad \text{is the inclusion map.}$$

For a given $p \in M$, we can always find an integral manifold N of a C^∞ distribution Δ with $p \in N$; we just choose a vector field X with $0 \neq X_q \in \Delta_q$ for q in a neighborhood of p, find an integral curve c of X with initial condition $c(0) = p$, and then forget about the parameterization of c, by defining N to be $\{c(t)\}$. This argument actually shows that for every $p \in M$ there is a coordinate system (x, U) such that for each fixed set of numbers $a^2, \ldots, a^n$, the set

$$\{q \in U : x^2(q) = a^2, \ldots, x^n(q) = a^n\}$$

is an integral manifold of Δ on U, and that these are the only integral manifolds in U.

This is still a local result, but because we are dealing with submanifolds, rather than curves with a particular parameterization, we can join overlapping integral submanifolds together. The entire manifold M can be written as a disjoint union of connected integral submanifolds of Δ, which locally look like

(rather than like

or something even more complicated). For example, there is a distribution on the torus whose integral manifolds all look like the dense 1-dimensional

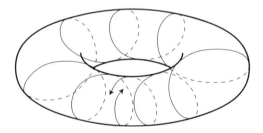

submanifold pictured in Chapter 2. On the other hand, there is a distribution on the torus which has one compact connected integral manifold, and all other

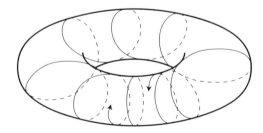

integral manifolds non-compact. It happens that the integral manifolds of these two distributions are also the integral curves for certain vector fields, but on the

Möbius strip there is a distribution which is spanned by a vector field only locally.

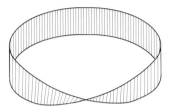

We are leaving out the details involved in fitting together these local integral manifolds because we will eventually do this over again in the higher dimensional case. For the moment we will investigate higher dimensional cases only locally.

A **k-dimensional distribution** on M is a function $p \mapsto \Delta_p$, where $\Delta_p \subset M_p$ is a k-dimensional subspace of M_p. For any $p \in M$ there is a neighborhood U and k vector fields $X_1, \ldots, X_k$ such that $X_1(q), \ldots, X_k(q)$ are a basis for Δ_q, for each $q \in U$. We call Δ a C^∞ distribution if it is possible to choose C^∞ vector fields $X_1, \ldots, X_q$ with this property, in a neighborhood of each point p. A (k-dimensional) submanifold N of M is called an **integral manifold** of Δ if for every $p \in N$ we have

$$i_*(N_p) = \Delta_p \quad \text{where} \quad i : N \to M \quad \text{is the inclusion map.}$$

Although the definitions given so far all look the same as the 1-dimensional case, the results will look very different. In general, integral manifolds *do not exist*, even locally.

As the simplest example, consider the 2-dimensional distribution Δ in $\mathbb{R}^3$ for which $\Delta_p = \Delta_{(a,b,c)}$ is spanned by

$$\left.\frac{\partial}{\partial x}\right|_p + b\left.\frac{\partial}{\partial z}\right|_p \quad \text{and} \quad \left.\frac{\partial}{\partial y}\right|_p.$$

Thus

$$\Delta_p = \left\{ r\left.\frac{\partial}{\partial x}\right|_p + s\left.\frac{\partial}{\partial y}\right|_p + br\left.\frac{\partial}{\partial z}\right|_p : r, s \in \mathbb{R} \right\}.$$

If we identify $T\mathbb{R}^3$ with $\mathbb{R}^3 \times \mathbb{R}^3$, then Δ_p consists of all $(r, s, br)_p$. Thus Δ_p may be pictured as the plane with the equation

$$z - c = b(x - a).$$

The figure below shows Δ_p for points $p = (a, b, 0)$. The plane $\Delta_{(a,b,c)}$ through (a, b, c) is just parallel to the one through $(a, b, 0)$.

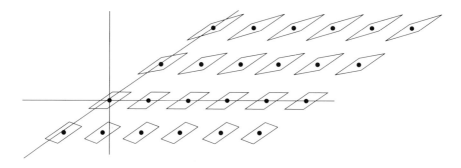

If you can picture this distribution, you can probably see that it has no integral manifolds; a proof can be given as follows. Suppose there were an integral manifold N of Δ with $0 \in N$. The intersection of N and $\{(0, y, z)\}$ would be a curve γ in the (y, z)-plane through 0 whose tangent vectors would have to lie in the intersection of $\Delta_{(0,y,z)}$ and the (y, z)-plane. The only such vectors have third component 0, so γ must be the y-axis. Now consider, for each fixed y_0, the intersection $N \cap \{(x, y_0, z)\}$. This will be a curve in the plane $\{(x, y_0, z)\}$ through $(0, y_0, 0)$, with all tangent vectors having slope y_0, so it must be the line $\{(x, y_0, y_0 x)\}$. Our integral manifold would have to look like the following picture. But this submanifold does not work. For example, its tangent space at $(1, 0, 0)$ contains vectors with third component non-zero.

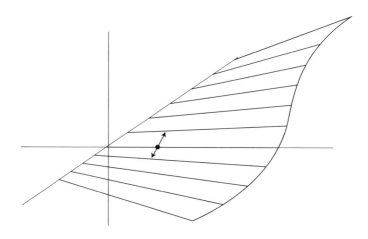

To see in greater detail what is happening here, consider the somewhat more general case where $\Delta_{(a,b,c)} = \Delta_p$ is

$$\Delta_p = \left\{ r \frac{\partial}{\partial x}\bigg|_p + s \frac{\partial}{\partial y}\bigg|_p + [rf(a,b) + sg(a,b)] \frac{\partial}{\partial z}\bigg|_p : r, s \in \mathbb{R} \right\};$$

geometrically, Δ_p is the plane with the equation

$$z - c = f(a,b)(x - a) + g(a,b)(y - b).$$

As in the first example, the plane $\Delta_{(a,b,c)}$ through (a,b,c) will be parallel to the one through $(a,b,0)$, since f and g depend only on a and b.

We now ask when the distribution Δ has an integral manifold N through each point. Since Δ_p is never perpendicular to the (x,y)-plane, the submanifold is given locally as the graph of a function:

$$N = \{(x,y,z) : z = \alpha(x,y)\}.$$

Now the tangent space at $p = (a,b,\alpha(a,b))$ is spanned by

$$\frac{\partial}{\partial x}\bigg|_p + \frac{\partial\alpha}{\partial x}(a,b) \frac{\partial}{\partial z}\bigg|_p,$$

$$\frac{\partial}{\partial y}\bigg|_p + \frac{\partial\alpha}{\partial y}(a,b) \frac{\partial}{\partial z}\bigg|_p.$$

These tangent vectors are in Δ_p if and only if

$$f(a,b) = \frac{\partial\alpha}{\partial x}(a,b),$$

$$g(a,b) = \frac{\partial\alpha}{\partial y}(a,b).$$

So we need to find a function $\alpha : \mathbb{R}^2 \to \mathbb{R}$ with

$$(*) \qquad \frac{\partial\alpha}{\partial x} = f, \qquad \frac{\partial\alpha}{\partial y} = g.$$

It is well-known that this is not always possible. By using the equality of mixed partial derivatives, we find a necessary condition on f and g:

(**)
$$\frac{\partial f}{\partial y} = \frac{\partial g}{\partial x}.$$

In our previous example,

$$f(a,b) = b, \qquad \frac{\partial f}{\partial y} = 1,$$

$$g(a,b) = 0, \qquad \frac{\partial g}{\partial x} = 0,$$

so this necessary condition is not satisfied. It is also well-known that the necessary condition (**) is *sufficient* for the existence of the function α satisfying (*) in a neighborhood of any point.

0. PROPOSITION. If $f, g \colon \mathbb{R}^2 \to \mathbb{R}$ satisfy

(**)
$$\frac{\partial f}{\partial y} = \frac{\partial g}{\partial x}$$

in a neighborhood of 0, and $z_0 \in \mathbb{R}$, then there is a function α, defined in a neighborhood of $0 \in \mathbb{R}^2$, such that

(*)
$$\alpha(0,0) = z_0$$
$$\frac{\partial \alpha}{\partial x} = f$$
$$\frac{\partial \alpha}{\partial y} = g.$$

PROOF. We first define $\alpha(x,0)$ so that $\alpha(0,0) = z_0$ and

(1)
$$\frac{\partial \alpha}{\partial x}(x,0) = f(x,0);$$

namely, we define

$$\alpha(x,0) = z_0 + \int_0^x f(t,0)\, dt.$$

Then, for each x, we define $\alpha(x, y)$ so that

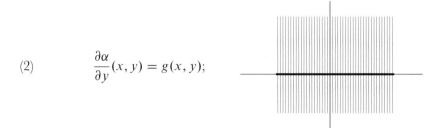

(2) $$\frac{\partial \alpha}{\partial y}(x, y) = g(x, y);$$

namely, we define

$$\alpha(x, y) = \alpha(x, 0) + \int_0^y g(x, t)\, dt$$

$$= z_0 + \int_0^x f(t, 0)\, dt + \int_0^y g(x, t)\, dt.$$

This construction does not use (∗∗), and always provide us with an α satisfying (2), $\partial \alpha / \partial y = g$. We claim that if (∗∗) holds, then also $\partial \alpha / \partial x = f$. To prove this, consider, for each fixed x, the function

$$y \mapsto \frac{\partial \alpha}{\partial x}(x, y) - f(x, y).$$

This is 0 for $y = 0$ by (1). To prove that it equals 0 for all y, we just have to show that its derivative is 0. But its derivative at y is

$$\frac{\partial^2 \alpha}{\partial y \partial x}(x, y) - \frac{\partial f}{\partial y}(x, y) = \frac{\partial}{\partial x}\left(\frac{\partial \alpha}{\partial y}\right)(x, y) - \frac{\partial f}{\partial y}(x, y)$$

$$= \frac{\partial g}{\partial x}(x, y) - \frac{\partial f}{\partial y}(x, y) \quad \text{by (2)}$$

$$= 0 \quad \text{by (∗∗).} \ \ \text{❖}$$

We are now ready to look at essentially the most general case of a 2-dimensional distribution in $\mathbb{R}^3$:

$$\Delta_p = \left\{ r \left.\frac{\partial}{\partial x}\right|_p + s \left.\frac{\partial}{\partial y}\right|_p + [r f(p) + s g(p)] \left.\frac{\partial}{\partial z}\right|_p : r, s \in \mathbb{R} \right\},$$

where $f, g \colon \mathbb{R}^3 \to \mathbb{R}$. Suppose that

$$N = \{(x, y, z) : z = \alpha(x, y)\}$$

is an integral manifold of Δ. The tangent space of N at $p = (a, b, \alpha(a, b))$ is spanned, once again, by

$$\left.\frac{\partial}{\partial x}\right|_p + \frac{\partial \alpha}{\partial x}(a, b) \left.\frac{\partial}{\partial z}\right|_p,$$

$$\left.\frac{\partial}{\partial y}\right|_p + \frac{\partial \alpha}{\partial y}(a, b) \left.\frac{\partial}{\partial z}\right|_p.$$

These tangent vectors are in Δ_p if and only if

(*)
$$f(a, b, \alpha(a, b)) = \frac{\partial \alpha}{\partial x}(a, b),$$

$$g(a, b, \alpha(a, b)) = \frac{\partial \alpha}{\partial y}(a, b).$$

In order to obtain necessary conditions for the existence of such a function α, we again use the equality of mixed partial derivatives. Thus (*) and the chain rule imply that

$$\frac{\partial^2 \alpha}{\partial y \partial x}(a, b) = \frac{\partial f}{\partial y}(a, b, \alpha(a, b)) + \frac{\partial f}{\partial z}(a, b, \alpha(a, b)) \cdot \frac{\partial \alpha}{\partial y}(a, b)$$

$$\|$$

$$\frac{\partial^2 \alpha}{\partial x \partial y}(a, b) = \frac{\partial g}{\partial x}(a, b, \alpha(a, b)) + \frac{\partial g}{\partial z}(a, b, \alpha(a, b)) \cdot \frac{\partial \alpha}{\partial y}(a, b).$$

This condition is not very useful, since it still involves the unknown function α, but we can substitute from (*) to obtain

$$\frac{\partial f}{\partial y}(a, b, \alpha(a, b)) + \frac{\partial f}{\partial z}(a, b, \alpha(a, b)) \cdot g(a, b, \alpha(a, b))$$

$$= \frac{\partial g}{\partial x}(a, b, \alpha(a, b)) + \frac{\partial g}{\partial z}(a, b, \alpha(a, b)) \cdot f(a, b, \alpha(a, b)).$$

Now we are looking for conditions which will be satisfied by f and g when there is an integral manifold of Δ *through every point*, which means that for each pair (a, b) these equations must hold no matter what $\alpha(a, b)$ is. Thus we obtain finally the necessary condition

(**)
$$\frac{\partial f}{\partial y} + \frac{\partial f}{\partial z} \cdot g = \frac{\partial g}{\partial x} + \frac{\partial g}{\partial z} \cdot f.$$

In this more general case, the necessary condition again turns out to be sufficient. In fact, there is no need to restrict ourselves to equations for a single function defined on $\mathbb{R}^2$; we can treat a system of partial differential equations for n functions on $\mathbb{R}^m$ (i.e., a partial differential equation for a function from $\mathbb{R}^m$ to $\mathbb{R}^n$). In the following theorem, we will use t to denote points in $\mathbb{R}^m$ and x for points in $\mathbb{R}^n$; so for a function $f: \mathbb{R}^m \times \mathbb{R}^n \to \mathbb{R}^k$ we use

$$\frac{\partial f}{\partial t^i} \quad \text{for} \quad D_i f,$$

$$\frac{\partial f}{\partial x^i} \quad \text{for} \quad D_{m+i} f.$$

1. THEOREM. Let $U \times V \subset \mathbb{R}^m \times \mathbb{R}^n$ be open, where U is a neighborhood of $0 \in \mathbb{R}^m$, and let $f_i: U \times V \to \mathbb{R}^n$ be C^∞ functions, for $i = 1, \ldots, m$. Then for every $x \in V$, there is at most one function

$$\alpha: W \to V,$$

defined in a neighborhood W of 0 in $\mathbb{R}^m$, satisfying

$$\alpha(0) = x$$
(*)
$$\frac{\partial \alpha}{\partial t^j}(t) = f_j(t, \alpha(t)) \quad \text{for all } t \in W.$$

(More precisely, any two such functions α_1 and α_2, defined on W_1 and W_2, agree on the component of $W_1 \cap W_2$ which contains 0.) Moreover, such a function exists (and is automatically C^∞) in some neighborhood W if and only if there is a neighborhood of $(0, x) \in U \times V$ on which

$$(**) \quad \frac{\partial f_j}{\partial t^i} - \frac{\partial f_i}{\partial t^j} + \sum_{k=1}^n \frac{\partial f_j}{\partial x^k} f_i^k - \sum_{k=1}^n \frac{\partial f_i}{\partial x^k} f_j^k = 0 \qquad i, j = 1, \ldots, m.$$

PROOF. Uniqueness will be obvious from the proof of existence. Necessity of the conditions $(**)$ is left to the reader as a simple exercise, and we will concern ourselves with proving existence if these conditions do hold. The proof will be like that of Proposition 0, with a different twist at the end.

We first want to define $\alpha(t, 0, \ldots, 0)$ so that

$$\alpha(0, 0, \ldots, 0) = x$$
(1)
$$\frac{\partial \alpha}{\partial t^1}(t, 0, \ldots, 0) = f_1(t, 0, \ldots, 0, \alpha(t, 0, \ldots, 0)).$$

To do this, we consider the ordinary differential equation

$$\beta_1(0) = x$$
$$\beta_1{}'(t) = f_1(t, 0, \ldots, 0, \beta_1(t)).$$

This equation has a unique solution, defined for $|t| < \varepsilon_1$. Define

$$\alpha(t, 0, \ldots, 0) = \beta_1(t) \qquad |t| < \varepsilon_1.$$

Then (1) holds for $|t| < \varepsilon_1$.

Now for each fixed t^1 with $|t^1| < \varepsilon_1$, consider the equation

$$\beta_2(0) = \alpha(t^1, 0, \ldots, 0)$$
$$\beta_2{}'(t) = f_2(t^1, t, 0, \ldots, 0, \beta_2(t)).$$

This has a unique solution for sufficiently small t. At this point the reader must refer back to Theorem 5-2, and verify the following assertion: If we choose ε_1 sufficiently small, then for $|t^1| < \varepsilon_1$ the solutions of the equations for β_2 with the initial conditions $\beta_2(0) = \alpha(t^1, 0, \ldots, 0)$ will each be defined for $|t| < \varepsilon_2$ for some $\varepsilon_2 > 0$. We then define

$$\alpha(t^1, t, 0, \ldots, 0) = \beta_2(t) \qquad |t^1| < \varepsilon_1, |t| < \varepsilon_2.$$

Then

$$\alpha(0, 0, 0, \ldots, 0) = x$$

(2)
$$\frac{\partial \alpha}{\partial t^2}(t^1, t, 0, \ldots, 0) = f_2(t^1, t, 0, \ldots, 0, \alpha(t^1, t, 0, \ldots, 0))$$

$$|t^1| < \varepsilon_1, |t| < \varepsilon_2.$$

We claim that for each fixed t^1 with $|t^1| < \varepsilon_1$ we also have, for all t with $|t| < \varepsilon_2$,

(3) $$0 = g(t) = \frac{\partial \alpha}{\partial t^1}(t^1, t, 0, \ldots, 0) - f_1(t^1, t, 0, \ldots, 0, \alpha(t^1, t, 0, \ldots, 0)).$$

Note first that

(4) $$g(0) = 0 \quad \text{by (1)}.$$

We now derive an equation for $g'(t)$. In the following, all expressions involving α are to be evaluated at $(t^1, t, 0, \ldots, 0)$ and all expressions involving f_i are to be evaluated at $(t^1, t, 0, \ldots, 0, \alpha(t^1, t, 0, \ldots, 0))$. We have

$$g'(t) = \frac{\partial^2 \alpha}{\partial t^2 \partial t^1} - \frac{\partial f_1}{\partial t^2} - \sum_{k=1}^{n} \frac{\partial f_1}{\partial x^k} \frac{\partial \alpha^k}{\partial t^2},$$

and thus

$$(5) \quad g'(t) = \frac{\partial}{\partial t^1}\left(\frac{\partial \alpha}{\partial t^2}\right) - \frac{\partial f_1}{\partial t^2} - \sum_{k=1}^{n} \frac{\partial f_1}{\partial x^k} f_2{}^k \qquad \text{by (2)}$$

$$= \frac{\partial f_2}{\partial t^1} + \sum_{k=1}^{n} \frac{\partial f_2}{\partial x^k} \frac{\partial \alpha^k}{\partial t^1} - \frac{\partial f_1}{\partial t^2} - \sum_{k=1}^{n} \frac{\partial f_1}{\partial x^k} f_2{}^k \qquad \text{by (2) again}$$

$$= \frac{\partial f_2}{\partial t^1} + \sum_{k=1}^{n} \frac{\partial f_2}{\partial x^k}\left[g^k(t) + f_1{}^k\right]$$

$$\quad - \frac{\partial f_1}{\partial t^2} - \sum_{k=1}^{n} \frac{\partial f_1}{\partial x^k} f_2{}^k \qquad \text{by definition, (3)}$$

$$= \sum_{k=1}^{n} \frac{\partial f_2}{\partial x^k} g^k(t) \qquad \text{by (**).}$$

Now equation (5) is a differential equation with a unique solution for each initial condition. The solution with initial condition $g(0) = 0$, given by (4), is clearly $g(t) = 0$ for all t. So (3) is true.

It is a simple exercise to continue the definition of α until it is eventually defined on $(-\varepsilon_1, \varepsilon_1) \times \cdots \times (-\varepsilon_n, \varepsilon_n)$ and satisfies (*). ❖

Theorem 1 essentially solves for us the problem of deciding which distributions have integral manifolds. Our investigation of the problem so far illustrates one basic fact about theorems in differential geometry:

> Many of the fundamental theorems of differential geometry fall into one of two classes. The first kind of theorem says that if one has a certain nice situation (e.g., a distribution with integral submanifolds through every point) then certain other conditions hold; these conditions are obtained by setting mixed partials equal, and are called "integrability conditions". The second kind of theorem justifies this terminology, by showing that the "integrability conditions" are sufficient for recovering the nice situation.

The remaining parts of our investigation, in which we will essentially begin anew, illustrates an even more important fact about the theorems of differential geometry:

> There are always incredibly concise and elegant ways to state the integrability conditions, and prove their sufficiency, without ever even mentioning partial derivatives.

LOCAL THEORY

If $f: M \to N$ is a C^∞ function, and X and Y are C^∞ vector fields on M and N, respectively, we say that X and Y are f-**related** if $f_{*p}(X_p) = Y_{f(p)}$ for each $p \in M$. If $g: N \to \mathbb{R}$ is a C^∞ function, then

$$Y_{f(p)}(g) = f_{*p} X_p(g)$$
$$= X_p(g \circ f),$$

so

$$(Yg) \circ f = X(f \circ g).$$

Conversely, if this is true for all C^∞ functions $g: N \to \mathbb{R}$, then X and Y are f-related.

Of course, a given vector field X may not be f-related to any vector field Y, nor must a given vector field Y be f-related to any vector field on M. In one case, the latter condition is fulfilled:

2. PROPOSITION. Let $f: M \to N$ be a C^∞ function such that f is an immersion. If Y is a C^∞ vector field on N with

$$Y_{f(p)} \in f_{p*}(M_p),$$

then there is a unique C^∞ vector field X on M which is f-related to Y.

PROOF. Clearly we must define X_p to be the unique element of M_p with $Y_{f(p)} = f_{p*} X_p$. To prove that X is C^∞, we use Theorem 2-10(2): there are coordinate systems (x, U) around $p \in M$ and (y, V) around $f(p) \in N$ such that

$$y \circ f \circ x^{-1}(a^1, \ldots, a^n) = (a^1, \ldots, a^n, 0, \ldots, 0).$$

This is easily seen to imply that

$$f_{p*}\left(\left. \frac{\partial}{\partial x^i} \right|_p \right) = \left. \frac{\partial}{\partial y^i} \right|_{f(p)}.$$

Thus if

$$Y = \sum_{i=1}^{n} \alpha^i \frac{\partial}{\partial y^i},$$

where α^i are C^∞ functions, then

$$X = \sum_{i=1}^{n} \beta^i \frac{\partial}{\partial x^i},$$

where $\alpha^i \circ f = \beta^i$. This implies that the functions β^i are C^∞ (Problem 3). ❖

The most important property of f-relatedness for us is the following:

3. PROPOSITION. If X_i and Y_i are f-related, for $i = 1, 2$, then $[X_1, X_2]$ and $[Y_1, Y_2]$ are f-related.

PROOF. If $g: N \to \mathbb{R}$ is C^∞, then

(1) $$(Y_i g) \circ f = X_i (g \circ f) \qquad i = 1, 2.$$

So

$$
\begin{aligned}
\{[Y_1, Y_2]g\} \circ f &= \{Y_1(Y_2 g)\} \circ f - \{Y_2(Y_1 g)\} \circ f \\
&= X_1([Y_2 g] \circ f) - X_2([Y_1 g] \circ f) \\
&\qquad \text{by (1), with } g \text{ replaced by } Y_2 g \text{ and } Y_1 g, \text{ respectively} \\
&= X_1(X_2(g \circ f)) - X_2(X_1(g \circ f)) \quad \text{by (1)} \\
&= [X_1, X_2](g \circ f). \; \clubsuit
\end{aligned}
$$

Now consider a k-dimensional distribution Δ. We will say that a vector field X **belongs to** Δ if $X_p \in \Delta_p$ for all p. Suppose that N is an integral manifold of Δ, and $i: N \to M$ is the inclusion map. If X and Y are two vector fields which belong to Δ, then for all $p \in N$ there are unique $\bar{X}_p, \bar{Y}_p \in N_p$ such that

$$X_p = i_* \bar{X}_p, \quad Y_p = i_* \bar{Y}_p.$$

In other words, X and $\bar{X}$ are i-related, and Y and $\bar{Y}$ are i-related. Proposition 2 shows that $\bar{X}$ and $\bar{Y}$ are C^∞ vector fields on N, and Proposition 3 then shows that $[\bar{X}, \bar{Y}]$ and $[X, Y]$ are i-related. Thus

$$i_*[\bar{X}, \bar{Y}]_p = [X, Y]_p.$$

Here $[\bar{X}, \bar{Y}]_p \in N_p$; this therefore shows that $[X, Y]_p \in \Delta_p$. Consequently, if there is an integral manifold of Δ through every point p, then $[X, Y]$ *also belongs to* Δ.

For a moment look back at the distribution Δ in $\mathbb{R}^3$ given by

$$
\Delta_p = \left\{ r \left. \frac{\partial}{\partial x} \right|_p + s \left. \frac{\partial}{\partial y} \right|_p + [rf(p) + sg(p)] \left. \frac{\partial}{\partial z} \right|_p : r, s \in \mathbb{R} \right\}.
$$

The vector fields

$$X = \frac{\partial}{\partial x} + f \frac{\partial}{\partial z}$$

$$Y = \frac{\partial}{\partial y} + g \frac{\partial}{\partial z}$$

belong to Δ. Using the formula on page 156, we see that

$$
[X, Y] = \left(\frac{\partial g}{\partial x} - \frac{\partial f}{\partial y} + f \frac{\partial g}{\partial z} - g \frac{\partial f}{\partial z} \right) \cdot \frac{\partial}{\partial z}.
$$

This belongs to Δ only when the expression in parentheses is 0, which is precisely the condition for Δ to have an integral manifold through every point.

In general, Δ is called **integrable** if $[X, Y]$ belongs to Δ whenever X and Y belong to Δ. This condition can be checked fairly easily:

4. PROPOSITION. If $X_1, \ldots, X_k$ span Δ in a neighborhood U of p, then Δ is integrable on U if and only if each $[X_i, X_j]$ is a linear combination

$$[X_i, X_j] = \sum_{\alpha=1}^{k} C_{ij}^{\alpha} X_\alpha$$

for C^∞ functions C_{ij}^{α}.

PROOF. Such functions clearly exist if Δ is integrable, since $[X_i, X_j]_q \in \Delta_q$, which is spanned by the $X_\alpha(q)$. Conversely, suppose such functions exist. If X and Y belong to Δ we can clearly write

$$X = \sum_{i=1}^{k} f_i X_i$$

$$Y = \sum_{i=1}^{k} g_i X_i.$$

To prove $[X, Y]$ belongs to Δ, it obviously suffices to treat each $[f_i X_i, g_j X_j]$ separately. Since we have

$$[fX, gY] = fg[X, Y] + f(Xg)Y - g(Yf)X,$$

clearly $[fX, gY]$ belongs to Δ if X, Y and $[X, Y]$ do. ❖

We are now ready for the main theorem. It is equivalent to Theorem 1; in fact, Theorem 1 can be derived from it (Problem 7). But the proof is quite different.

5. THEOREM (THE FROBENIUS INTEGRABILITY THEOREM; FIRST VERSION). Let Δ be a C^∞ integrable k-dimensional distribution on M. For every $p \in M$ there is a coordinate system (x, U) with

$$x(p) = 0$$

$$x(U) = (-\varepsilon, \varepsilon) \times \cdots \times (-\varepsilon, \varepsilon),$$

such that for each $a^{k+1}, \ldots, a^n$ with all $|a^i| < \varepsilon$, the set

$$\{q \in U : x^{k+1}(q) = a^{k+1}, \ldots, x^n(q) = a^n\}$$

is an integral manifold of Δ.

Any connected integral manifold of Δ restricted to U is contained in one of these sets.

PROOF. We can clearly assume that we are in $\mathbb{R}^n$, with $p = 0$. Moreover, we can assume that $\Delta_0 \subset \mathbb{R}^n{}_0$ is spanned by

$$\left.\frac{\partial}{\partial t^1}\right|_0, \ldots, \left.\frac{\partial}{\partial t^k}\right|_0.$$

Let $\pi: \mathbb{R}^n \to \mathbb{R}^k$ be projection onto the first k factors. Then $\pi_*: \Delta_0 \to \mathbb{R}^k{}_0$ is an isomorphism. By continuity, π_* is one-one on Δ_q for q near 0. So near 0, we can choose unique

$$X_1(q), \ldots, X_k(q) \in \Delta_q$$

so that

$$\pi_* X_i(q) = \left.\frac{\partial}{\partial t^i}\right|_{\pi(q)} \qquad i = 1, \ldots, k.$$

Then the vector fields X_i (on a neighborhood of $0 \in \mathbb{R}^n$) and $\partial/\partial t^i$ (on $\mathbb{R}^k$) are π-related. By Proposition 3,

$$\pi_*[X_i, X_j]_q = \left[\frac{\partial}{\partial t^i}, \frac{\partial}{\partial t^j}\right]_{\pi(q)}$$

$$= 0.$$

But, $[X_i, X_j]_q \in \Delta_q$ by assumption, and π_* is one-one on Δ_q. So $[X_i, X_j] = 0$. By Theorem 5-14, there is a coordinate system x such that

$$X_i = \frac{\partial}{\partial x^i} \qquad i = 1, \ldots, k.$$

The sets $\{q \in U : x^{k+1}(q) = a^{k+1}, \ldots, x^n(q) = a^n\}$ are clearly integral manifolds of Δ, since their tangent spaces are spanned by the $\partial/\partial x^i = X_i$ for $i = 1, \ldots, k$.

If N is a connected integral manifold of Δ restricted to U, with inclusion map $i: N \to U$, consider $d(x^m \circ i)$ for $k + 1 \leq m \leq n$. For any tangent vector X_q of N_q we have

$$d(x^m \circ i)(X_q) = X_q(x^m \circ i) = i_* X_q(x^m)$$

$$= 0,$$

since $i_* X_q \in \Delta_q$, which is spanned by the $\partial/\partial x^j|_q$ for $j = 1, \ldots, k$. Thus $d(x^m \circ i) = 0$, which implies that $x^m \circ i$ is constant on the connected manifold N. ❖

GLOBAL THEORY

In order to express the global results succinctly, we introduce the following terminology.

If M is a C^∞ manifold, a (usually disconnected) k-dimensional submanifold N of M is called a **foliation** of M if every point of M is in (some component of) N, and if around every point $p \in M$ there is a coordinate system (x, U), with

$$x(U) = (-\varepsilon, \varepsilon) \times \cdots \times (-\varepsilon, \varepsilon),$$

such that the components of $N \cap U$ are the sets of the form

$$\{q \in U : x^{k+1}(q) = a^{k+1}, \ldots, x^n(q) = a^n\} \qquad |a^i| < \varepsilon.$$

Each component of N is called a **folium** or **leaf** of the foliation N. Notice that two distinct components of $N \cap U$ might belong to the same leaf of the foliation.

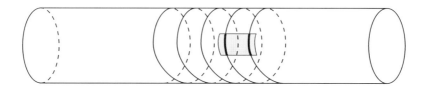

6. THEOREM. Let Δ be a C^∞ k-dimensional integrable distribution on M. Then M is foliated by an integral manifold of Δ (each component is called a **maximal integral manifold** of Δ).

PROOF. Using Theorem 1-2, we see that we can cover M by a sequence of coordinate systems (x_i, U_i) satisfying the conditions of Theorem 5. For such a coordinate system (x, U), let us call each set

$$\{q \in U : x^{k+1}(q) = a^{k+1}, \ldots, x^n(q) = a^n\}$$

a *slice* of U.

It is possible for a single slice S of U_i to intersect U_j in more than one slice of U_j, as shown below. But $S \cap U_j$ has at most countably many components, and each component is contained in a single slice of U_j by Theorem 5, so $S \cap U_j$ is contained in at most countably many slices of U_j.

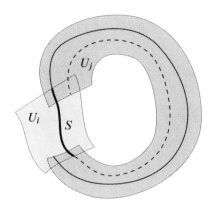

Given $p \in M$, choose a coordinate system (x_0, U_0) with $p \in U_0$, and let S_0 be the slice of U_0 containing p. A slice S of some U_i will be called *joined* to p if there is a sequence

$$0 = i_0, i_1, \ldots, i_l = i$$

and corresponding slices

$$S_0 = S_{i_0}, S_{i_1}, \ldots, S_{i_l} = S$$

with

$$S_{i_\alpha} \cap S_{i_{\alpha+1}} \neq \emptyset \qquad \alpha = 0, \ldots, l - 1.$$

Since there are at most countably many such sequences of slices for each sequence $i_0, \ldots, i_l$, and only countably many such sequences, there are at most countably many slices joined to p. Using Problem 3-1, we see that the union of all such slices is a submanifold of M. For $q \neq p$, the corresponding union is either equal to, or totally disjoint from, the first union. Consequently, M is foliated by the disjoint union of all such submanifolds; this disjoint union is clearly an integral manifold of Δ. ❖

[If we are allowing non-metrizable manifolds, the proof is even easier, since we do not have to find a countable number of coordinate systems for each leaf, and can merely describe the topology of the foliation as the smallest one which makes each slice an open set. In this case, however, the discussion to follow will not be valid—in fact, Appendix A describes a non-paracompact manifold which is foliated by a lower-dimensional *connected* submanifold.]

Notice that if (x, U) is a coordinate system of the sort considered in the proof of the theorem, then infinitely many slices of U may belong to the same folium.

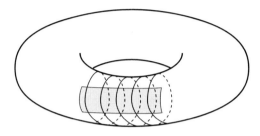

However, *at most countably many* slices can belong to the same folium; otherwise

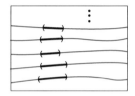

this folium would contain an uncountable disjoint family of open sets. This allows us to apply a proposition from Chapter 2.

7. THEOREM. Let M be a C^∞ manifold, and M_1 a folium of the foliation determined by some distribution Δ. Let P be another C^∞ manifold and $f : P \to M$ a C^∞ function with $f(P) \subset M_1$. Then f is C^∞ considered as a map into M_1.

PROOF. According to Proposition 2-11, it suffices to show that f is continuous as a map into M_1. Given $p \in P$, choose a coordinate system (x, U) around $f(p)$ such that the slices

$$\{q \in U : x^{k+1}(q) = a^{k+1}, \ldots, x^n(q) = a^n\}$$

are integral manifolds of Δ. Now f is continuous as a map into M, so f takes some neighborhood W of p into U; we can choose W to be connected. For $k+1 \leq i \leq n$, if we had $x^i(f(p')) \neq a^i$ for any $p' \in W$, then $x^i \circ f$ would take on all values between a^i and $x^i(f(p'))$, by continuity. This would mean that $f(W)$ contained points of uncountably many slices, contradicting the fact that $f(W) \subset M_1$.

Consequently, $x^i(f(p')) = a^i$ for all $p' \in W$. In other words, $f(W)$ is contained in the single slice of U which contains p. This makes it clear that f is continuous as a map into M_1. ❖

PROBLEMS

1. (a) Let $\xi = \pi\colon E \to B$ be an n-plane bundle, and $\xi' = \pi'\colon E' \to B$ a k-plane bundle such that $E' \subset E$. If $i\colon E' \to E$ is the inclusion map, and $1_B\colon B \to B$ the identity map, we say that ξ' is a **subbundle** of ξ if $(i, 1_B)$ is a bundle map. Show that a k-dimensional distribution on M is just a subbundle of TM.

(b) For the case of C^∞ bundles ξ and ξ' over a C^∞ manifold M, define a C^∞ subbundle, and show that a k-dimensional distribution is C^∞ if and only if it is a C^∞ subbundle.

2. (a) In the proof of Theorem 1, check the assertion about choosing ε_1 sufficiently small.

(b) Supply the proof of the uniqueness part of the theorem.

3. (a) In the proof of Proposition 2, show that

$$f_* \left(\left. \frac{\partial}{\partial x^i} \right|_p \right) = \left. \frac{\partial}{\partial y^i} \right|_{f(p)} .$$

(b) Complete the proof of Proposition 2 by showing that if

$$Y = \sum_{i=1}^n \alpha^i \frac{\partial}{\partial y^i},$$

so that

$$X = \sum_{i=1}^n \beta^i \frac{\partial}{\partial x^i},$$

with $\alpha^i \circ f = \beta^i$, then the functions β^i are C^∞.

4. In the proof of Proposition 4, show that the functions C_{ij}^α actually are C^∞.

5. Let $\Delta_1, \ldots, \Delta_h$ be integrable distributions on M, of dimensions $d_1, \ldots, d_h$. Suppose that for each $p \in M$,

$$M_p = (\Delta_1)_p \oplus \cdots \oplus (\Delta_h)_p.$$

Show that there is a coordinate system (x, U) around each point, such that Δ_1 is spanned by $\partial/\partial x^1, \ldots, \partial/\partial x^{d_1}$, etc.

6. Prove Theorem 1 from Theorem 5, by considering the distribution Δ in $\mathbb{R}^m \times \mathbb{R}^n$ (with coordinates t, x), defined by

$$\Delta_p = \left\{ \sum_{i=1}^m r^i \left.\frac{\partial}{\partial t^i}\right|_p + \sum_{k=1}^n \left(\sum_{i=1}^m r^i f_i^k(p) \right) \left.\frac{\partial}{\partial x^k}\right|_p : r \in \mathbb{R}^m \right\}.$$

Notice that even when the f_j do not depend on x, so that the equations are of the form

$$\frac{\partial \alpha}{\partial t^j}(t) = f_j(t),$$

with the integrability conditions

$$\frac{\partial f_j}{\partial t^i} = \frac{\partial f_i}{\partial t^j},$$

we nevertheless work in $\mathbb{R}^m \times \mathbb{R}^n$, rather than $\mathbb{R}^m$. This is connected with the classical technique of "introducing new independent variables".

7. This problem outlines another method of proving Theorem 1, by reducing the partial differential equations to ordinary equations along lines through the origin. A similar technique will be very important in Chapter II.7.

(a) If we want $\alpha(ut) = \beta(u, t)$ for some function $\beta \colon [0, \varepsilon) \times W \to V$, show that β must satisfy the equation

$$\frac{\partial \beta}{\partial u}(u, t) = \sum_{j=1}^m t^j \cdot f_j(ut, \beta(u, t))$$

$$\beta(0, t) = x.$$

We know that we can solve such equations (we need Problem 5-5, since the equation depends on the "parameter" $t \in \mathbb{R}^m$). One has to check that one ε can be picked which works for all $t \in W$.

(b) Show that

$$\beta(u, vt) = \beta(uv, t).$$

(Show that both functions satisfy the same differential equation as functions of u, with the same initial condition.) By shrinking W, we can consequently assume that $\varepsilon = 1$.

(c) Conclude that

$$\frac{\partial \beta}{\partial t^j}(v, t) = v \cdot \frac{\partial \beta}{\partial t^j}(1, vt).$$

(d) Use the integrability condition on f to show that

$$\frac{\partial \beta}{\partial t^j}(v,t) \quad \text{and} \quad v \cdot f_j(vt, \beta(v,t))$$

satisfy the same differential equation, as functions of v. Use (c) to conclude that the two functions are equal.

(e) Define $\alpha(t) = \beta(1,t)$. Noting that $\alpha(vt) = \beta(v,t)$, show that α satisfies the desired equation.

8. This problem is for those who know something about complex analysis. Let $f : \mathbb{C} \times \mathbb{C} \to \mathbb{C}$ be complex analytic. If we denote the coordinate functions in $\mathbb{C} \times \mathbb{C}$ by $z_1, z_2 = x_1, y_1, x_2, y_2$, then $f = u + iv$ satisfies the Cauchy-Riemann equations

$$\frac{\partial u}{\partial x_i} = \frac{\partial v}{\partial y_i}$$
$$\frac{\partial u}{\partial y_i} = -\frac{\partial v}{\partial x_i}$$

$$i = 1, 2.$$

Use Theorem 1 to prove that we can solve the equations

$$\frac{\partial \alpha^1}{\partial x} = u(x, y, \alpha^1(x, y), \alpha^2(x, y)) = \frac{\partial \alpha^2}{\partial y}$$

$$\frac{\partial \alpha^2}{\partial x} = v(x, y, \alpha^1(x, y), \alpha^2(x, y)) = -\frac{\partial \alpha^1}{\partial y}$$

in a neighborhood of $0 \in \mathbb{C}$ (or of any point $z_0 \in \mathbb{C}$), and conclude that the differential equation

$$\phi'(z) = f(z, \phi(z))$$

(in which $'$ denotes the complex derivative) has a solution in a neighborhood of z_0, with any given initial condition $\phi(z_0) = w_0$.

CHAPTER 7

DIFFERENTIAL FORMS

We turn our attention once more to tensor fields, but we will be concerned with a special kind of tensor field, the discussion of which requires some more algebraic preliminaries.

Let V be an n-dimensional vector space over $\mathbb{R}$. An element $T \in \mathcal{T}^k(V)$ is called **alternating** if

$$T(v_1, \ldots, v_i, \ldots, v_j, \ldots, v_k) = 0 \qquad \text{if} \quad v_i = v_j \ (i \neq j).$$

If T is alternating, then for any $v_1, \ldots, v_k$, we have

$$
\begin{aligned}
0 &= T(v_1, \ldots, v_i + v_j, \ldots, v_i + v_j, \ldots, v_k) \\
&= T(v_1, \ldots, v_i, \ldots, v_i, \ldots, v_k) + T(v_1, \ldots, v_i, \ldots, v_j, \ldots, v_k) \\
&\quad + T(v_1, \ldots, v_j, \ldots, v_i, \ldots, v_k) + T(v_1, \ldots, v_j, \ldots, v_j, \ldots, v_k) \\
&= 0 + T(v_1, \ldots, v_i, \ldots, v_j, \ldots, v_k) + T(v_1, \ldots, v_j, \ldots, v_i, \ldots, v_k) + 0.
\end{aligned}
$$

Therefore, T is **skew-symmetric**:

$$T(v_1, \ldots, v_i, \ldots, v_j, \ldots, v_k) = -T(v_1, \ldots, v_j, \ldots, v_i, \ldots, v_k).$$

Of course, if T is skew-symmetric, then T is also alternating. [This is not true in the special case of a vector space over a field where $1 + 1 = 0$; in this case, skew-symmetry is the same as symmetry, and the condition of being alternating is the stronger one.]

We will denote by $\Omega^k(V)$ the set of all alternating $T \in \mathcal{T}^k(V)$. It is clear that $\Omega^k(V) \subset \mathcal{T}^k(V)$ is a subspace of $\mathcal{T}^k(V)$. Moreover, if $f \colon V \to W$ is a linear transformation, then $f^* \colon \mathcal{T}^k(W) \to \mathcal{T}^k(V)$ preserves these subspaces— $f^* \colon \Omega^k(W) \to \Omega^k(V)$. Notice that $\Omega^1(V) = \mathcal{T}^1(V) = V^*$, so $\Omega^1(V)$ has dimension n. It is also convenient to set $\Omega^0(V) = \mathcal{T}^0(V) = \mathbb{R}$. At the moment it is not clear what the dimension of $\Omega^k(V)$ equals for $k > 1$, but one case is well-known. The most familiar example of an alternating T is the determinant function $\det \in \mathcal{T}^n(\mathbb{R}^n)$, considered as a function of the n rows of a matrix— we shall soon see that this function is, in a certain sense, the most general alternating function. Most discussions of the determinant begin by showing that of any two alternating n-linear functions on $\mathbb{R}^n$, one is a multiple of the other; in other words, $\dim \Omega^n(\mathbb{R}^n) \leq 1$. Then one proves $\dim \Omega^n(\mathbb{R}^n) = 1$ by actually constructing the non-zero function det (it follows, of course, that $\dim \Omega^n(V) = 1$ if V is any n-dimensional vector space). The construction of det is usually by a messy, explicit formula, which is a special case of the definition to follow.

Let S_k denote the set of all permutations of $\{1, \ldots, k\}$; an element $\sigma \in S_k$ is a function $i \mapsto \sigma(i)$. If $(v_1, \ldots, v_k)$ is a k-tuple (of any objects) we set

$$\sigma \cdot (v_1, \ldots, v_k) = (v_{\sigma(1)}, \ldots, v_{\sigma(k)}).$$

This definition has a built-in confusion. On the right side, the first element, for example, is the $\sigma(1)^{\text{st}}$ of the v's on the left side; if these v's have indices running in some order *other* than $1, \ldots, k$, then the first element on the right is *not* necessarily that v whose index is $\sigma(1)$. The simplest way to figure out something like $\sigma \cdot (v_3, v_2, v_1, \ldots)$ is to rename things: $v_3 = w_1, v_2 = w_2, v_1 = w_3, \ldots$. Thus warned, we compute

$$\sigma \cdot (\rho \cdot (v_1, \ldots, v_k)) = \sigma \cdot (v_{\rho(1)}, \ldots, v_{\rho(k)})$$

by setting

$$v_{\rho(1)} = w_1, \ldots, v_{\rho(k)} = w_k,$$

so that

$$\begin{aligned}
\sigma \cdot (\rho \cdot (v_1, \ldots, v_k)) &= \sigma \cdot (w_1, \ldots, w_k) \\
&= (w_{\sigma(1)}, \ldots, w_{\sigma(k)}) \\
&= (v_{\rho(\sigma(1))}, \ldots, v_{\rho(\sigma(k))}) \quad \text{since } w_\alpha = v_{\rho(\alpha)}.
\end{aligned}$$

Thus

> (*) $\qquad\qquad \sigma \cdot (\rho \cdot (v_1, \ldots, v_k)) = (\rho\sigma) \cdot (v_1, \ldots, v_k).$

Now for any $T \in \mathcal{T}^k(V)$ we define the "alternation of T"

$$\text{Alt } T = \frac{1}{k!} \sum_{\sigma \in S_k} \text{sgn } \sigma \cdot T \circ \sigma,$$

i.e.,

$$\text{Alt } T(v_1, \ldots, v_k) = \frac{1}{k!} \sum_{\sigma \in S_k} \text{sgn } \sigma \cdot T(v_{\sigma(1)}, \ldots, v_{\sigma(k)}),$$

where $\text{sgn } \sigma$ is $+1$ if σ is an even permutation and -1 if σ is odd.

1. PROPOSITION.

(1) If $T \in \mathcal{T}^k(V)$, then $\mathrm{Alt}(T) \in \Omega^k(V)$.

(2) If $\omega \in \Omega^k(V)$, then $\mathrm{Alt}\,\omega = \omega$.

(3) If $T \in \mathcal{T}^k(V)$, then $\mathrm{Alt}(\mathrm{Alt}(T)) = \mathrm{Alt}(T)$.

PROOF. Left to the reader (or see pp. 78–79 of *Calculus on Manifolds*). ❖

We now define, for $\omega \in \Omega^k(V)$ and $\eta \in \Omega^l(V)$, an element $\omega \wedge \eta \in \Omega^{k+l}(V)$, the **wedge product** of ω and η, by

$$\omega \wedge \eta = \frac{(k+l)!}{k!\,l!}\,\mathrm{Alt}(\omega \otimes \eta).$$

The funny coefficient is not essential, but it makes some things work out more nicely, as we shall soon see. It is clear that

(1) $\wedge$ is bilinear:
$$(\omega_1 + \omega_2) \wedge \eta = \omega_1 \wedge \eta + \omega_2 \wedge \eta$$
$$\omega \wedge (\eta_1 + \eta_2) = \omega \wedge \eta_1 + \omega \wedge \eta_2$$
$$a\omega \wedge \eta = \omega \wedge a\eta = a(\omega \wedge \eta)$$

(2) $f^*(\omega \wedge \eta) = f^*\omega \wedge f^*\eta.$

Moreover, it is easy to see that

(3) $\wedge$ is "anti-commutative": $\omega \wedge \eta = (-1)^{kl} \eta \wedge \omega.$

In particular, if k is odd then

$$\omega \wedge \omega = 0.$$

Finally, associativity of $\wedge$ is proved in the following way.

2. THEOREM.

(1) If $S \in \mathcal{T}^k(V)$ and $T \in \mathcal{T}^l(V)$ and $\mathrm{Alt}(S) = 0$, then
$$\mathrm{Alt}(S \otimes T) = \mathrm{Alt}(T \otimes S) = 0.$$

(2) $\mathrm{Alt}(\mathrm{Alt}(\omega \otimes \eta) \otimes \theta) = \mathrm{Alt}(\omega \otimes \eta \otimes \theta) = \mathrm{Alt}(\omega \otimes \mathrm{Alt}(\eta \otimes \theta)).$

(3) If $\omega \in \Omega^k(V)$, $\eta \in \Omega^l(V)$, $\theta \in \Omega^m(V)$, then
$$(\omega \wedge \eta) \wedge \theta = \omega \wedge (\eta \wedge \theta) = \frac{(k+l+m)!}{k!\,l!\,m!}\,\mathrm{Alt}(\omega \otimes \eta \otimes \theta).$$

PROOF. (1) We have

$$(k+l)! \operatorname{Alt}(S \otimes T)(v_1, \ldots, v_{k+l})$$

$$= \sum_{\sigma \in S_{k+l}} \operatorname{sgn} \sigma \cdot (S \otimes T) \cdot (\sigma \cdot (v_1, \ldots, v_{k+l}))$$

$$= \sum_{\sigma \in S_{k+l}} \operatorname{sgn} \sigma \cdot S(v_{\sigma(1)}, \ldots, v_{\sigma(k)}) \cdot T(v_{\sigma(k+1)}, \ldots, v_{\sigma(k+l)}).$$

Now let $G \subset S_{k+l}$ consist of all σ which leave $k+1, \ldots, k+l$ fixed. Then

$$\sum_{\sigma \in G} \operatorname{sgn} \sigma \cdot S(v_{\sigma(1)}, \ldots, v_{\sigma(k)}) \cdot T(v_{\sigma(k+1)}, \ldots, v_{\sigma(k+l)})$$

$$= \left[\sum_{\sigma' \in S_k} \operatorname{sgn} \sigma' \cdot S(v_{\sigma'(1)}, \ldots, v_{\sigma'(k)}) \right] \cdot T(v_{k+1}, \ldots, v_{k+l})$$

$$= 0.$$

Suppose now that $\sigma_0 \notin G$. Let $\sigma_0 G = \{\sigma_0 \sigma' : \sigma' \in G\}$. Then

$$\sum_{\sigma \in \sigma_0 G} \operatorname{sgn} \sigma \cdot (S \otimes T)(\sigma \cdot (v_1, \ldots, v_{k+l}))$$

$$= \operatorname{sgn} \sigma_0 \cdot \sum_{\sigma' \in G} \operatorname{sgn} \sigma' \cdot (S \otimes T)(\sigma' \cdot (\sigma_0 \cdot (v_1, \ldots, v_{k+l}))) \quad \text{by } (*).$$

We have just shown that this is 0 (since $\sigma_0 \cdot (v_1, \ldots, v_{k+l})$ is just some other $(k+l)$-tuple of vectors). Notice that $G \cap \sigma_0 G = \emptyset$, for if $\sigma \in G \cap \sigma_0 G$, then $\sigma = \sigma_0 \sigma'$ for some $\sigma' \in G$, so $\sigma_0 = \sigma(\sigma')^{-1} \in G$, a contradiction. We can then continue in this way, breaking S_{k+l} up into disjoint subsets, the sum over each being 0. The relation $\operatorname{Alt}(T \otimes S) = 0$ is proved similarly.

(2) Clearly

$$\operatorname{Alt}(\operatorname{Alt}(\eta \otimes \theta) - \eta \otimes \theta) = \operatorname{Alt}(\eta \otimes \theta) - \operatorname{Alt}(\eta \otimes \theta) = 0,$$

so (1) implies that

$$0 = \operatorname{Alt}(\omega \otimes [\operatorname{Alt}(\eta \otimes \theta) - \eta \otimes \theta])$$

$$= \operatorname{Alt}(\omega \otimes \operatorname{Alt}(\eta \otimes \theta)) - \operatorname{Alt}(\omega \otimes \eta \otimes \theta);$$

the other equality is proved similarly.

(3) We have

$$(\omega \wedge \eta) \wedge \theta = \frac{(k+l+m)!}{(k+l)! \, m!} \operatorname{Alt}((\omega \wedge \eta) \otimes \theta)$$

$$= \frac{(k+l+m)!}{(k+l)! \, m!} \frac{(k+l)!}{k! \, l!} \operatorname{Alt}(\omega \otimes \eta \otimes \theta).$$

The other equality is proved similarly. ❖

Notice that (2) just states that $\wedge$ is associative even if we had omitted the factor $(k+l)!/k!l!$ in the definition. On the other hand, the factor $1/k!$ in the definition of Alt is essential—without it, we would not have $\text{Alt}(\text{Alt } T) = \text{Alt } T$, and the first equation in the proof of (2) would fail. [If we had defined $\overline{\text{Alt}}$ just like Alt, but without the factor $1/k!$, then $\wedge$ could be defined by

$$\omega \wedge \eta = \frac{1}{k!l!}\overline{\text{Alt}}(\omega \otimes \eta).$$

This makes sense, *even over a field of finite characteristic*, because each term in the sum $\text{Alt}(\omega \otimes \eta)(v_1, \ldots, v_{k+l})$ occurs $k!l!$ times (since ω and η are alternating), and $1/k!l!$ can be interpreted as meaning that these $k!l!$ terms are replaced by just one.] The factor $(k+l)!/k!l!$ has been inserted into the definition of $\wedge$ for the following reason. If $v_1, \ldots, v_n$ is a basis of V, and $\phi_1, \ldots, \phi_n$ is the dual basis, then

$$\phi_1 \wedge \cdots \wedge \phi_n = \frac{(1 + \cdots + 1)!}{1! \cdots 1!} \text{Alt}(\phi_1 \otimes \cdots \otimes \phi_n)$$

$$= \sum_{\sigma \in S_n} \text{sgn}\,\sigma \cdot (\phi_1 \otimes \cdots \otimes \phi_n) \circ \sigma.$$

In particular,

$$(\phi_1 \wedge \cdots \wedge \phi_n)(v_1, \ldots, v_n) = 1.$$

(So if $v_1, \ldots, v_n$ is the standard basis for $\mathbb{R}^n$, then $\phi_1 \wedge \cdots \wedge \phi_n = \det$.) A basis for $\Omega^k(V)$ can now be described.

3. THEOREM. The set of all

$$\phi_{i_1} \wedge \cdots \wedge \phi_{i_k} \qquad 1 \leq i_1 < \cdots < i_k \leq n$$

is a basis for $\Omega^k(V)$, which therefore has dimension

$$\binom{n}{k} = \frac{n!}{k!\,(n-k)!}.$$

(In particular, $\Omega^k(V) = \{0\}$ for $k > n$.)

PROOF. If $\omega \in \Omega^k(V) \subset \mathcal{T}^k(V)$, we can write

$$\omega = \sum_{i_1, \ldots, i_k} a_{i_1 \ldots i_k}\, \phi_{i_1} \otimes \cdots \otimes \phi_{i_k}.$$

So

$$\omega = \text{Alt}(\omega) = \sum_{i_1,\dots,i_k} a_{i_1\dots i_k} \text{Alt}(\phi_{i_1} \otimes \cdots \otimes \phi_{i_k}).$$

Each $\text{Alt}(\phi_{i_1} \otimes \cdots \otimes \phi_{i_k})$ is either 0 or $= \pm(1/k!)\,\phi_{j_1} \wedge \cdots \wedge \phi_{j_k}$ for some $j_1 < \cdots < j_k$, so the elements $\phi_{j_1} \wedge \cdots \wedge \phi_{j_k}$ for $j_1 < \cdots < j_k$ span $\Omega^k(V)$. If

$$0 = \sum_{i_1 < \cdots < i_k} a_{i_1\dots i_k}\, \phi_{i_1} \wedge \cdots \wedge \phi_{i_k},$$

then applying both sides to $(v_{i_1},\dots,v_{i_k})$ gives $a_{i_1\dots i_k} = 0$. ❖

4. COROLLARY. If $\omega_1,\dots,\omega_k \in \Omega^1(V)$, then $\omega_1,\dots,\omega_k$ are linearly independent if and only if

$$\omega_1 \wedge \cdots \wedge \omega_k \neq 0.$$

PROOF. If $\omega_1,\dots,\omega_k$ are linearly independent, there is a basis $v_1,\dots,v_k,\dots,v_n$ of V such that the dual basis vectors $\phi_1,\dots,\phi_k,\dots,\phi_n$ satisfy $\phi_i = \omega_i$ for $1 \le i \le k$. Then $\omega_1 \wedge \cdots \wedge \omega_k$ is a basis element of $\Omega^k(V)$, so it is not 0.

  On the other hand, if

$$\omega_1 = a_2\omega_2 + \cdots + a_k\omega_k,$$

then

$$\omega_1 \wedge \omega_2 \wedge \cdots \wedge \omega_k = (a_2\omega_2 + \cdots + a_k\omega_k) \wedge \omega_2 \wedge \cdots \wedge \omega_k = 0. ❖$$

  To abbreviate formulas, it is convenient to let I denote a typical "multi-index" $(i_1,\dots,i_k)$, and let ϕ_I denote $\phi_{i_1} \wedge \cdots \wedge \phi_{i_k}$. Then every element of $\Omega^k(V)$ is uniquely expressible as

$$\sum_I a_I\, \phi_I.$$

Notice that Theorem 3 implies that every $\omega \in \Omega^k(\mathbb{R}^n)$ is a linear combination of the functions

$$(v_1,\dots,v_k) \mapsto \text{ determinant of a } k \times k \text{ minor of } \begin{pmatrix} v_1 \\ \vdots \\ v_k \end{pmatrix}.$$

  One more simple theorem is in order, before we proceed to apply our construction to manifolds.

5. THEOREM. Let $v_1, \ldots, v_n$ be a basis for V, let $\omega \in \Omega^n(V)$, and let

$$w_i = \sum_{j=1}^{n} \alpha_{ji} v_j \qquad i = 1, \ldots, n.$$

Then

$$\omega(w_1, \ldots, w_n) = \det(\alpha_{ij}) \cdot \omega(v_1, \ldots, v_n).$$

PROOF. Define $\eta \in \mathcal{T}^n(\mathbb{R}^n)$ by

$$\eta\big((a_{11}, \ldots, a_{n1}), \ldots, (a_{n1}, \ldots, a_{nn})\big) = \omega\left(\sum_{j=1}^{n} a_{j1} v_j, \ldots, \sum_{j=1}^{n} a_{jn} v_j\right).$$

Then clearly $\eta \in \Omega^n(\mathbb{R}^n)$, so $\eta = c \cdot \det$ for some $c \in \mathbb{R}$, and

$$c = \eta(e_1, \ldots, e_n) = \omega(v_1, \ldots, v_n). \; ❖$$

6. COROLLARY. If V is n-dimensional and $0 \neq \omega \in \Omega^n(V)$, then there is a unique orientation μ for V such that

$$[v_1, \ldots, v_n] = \mu \quad \text{if and only if} \quad \omega(v_1, \ldots, v_n) > 0.$$

With our new algebraic construction at hand, we are ready to apply it to vector bundles. If $\xi = \pi \colon E \to B$ is a vector bundle, we obtain a new bundle $\Omega^k(\xi)$ by replacing each fibre $\pi^{-1}(p)$ with $\Omega^k(\pi^{-1}(p))$. A section ω of $\Omega^k(\xi)$ is a function with $\omega(p) \in \Omega^k(\pi^{-1}(p))$ for each $p \in B$. If η is a section of $\Omega^l(\xi)$, then we can define a section $\omega \wedge \eta$ of $\Omega^{k+l}(\xi)$ by $(\omega \wedge \eta)(p) = \omega(p) \wedge \eta(p) \in \Omega^{k+l}(\pi^{-1}(p))$.

In particular, sections of $\Omega^k(TM)$, which are just alternating covariant tensor fields of order k, are called k-**forms** on M. A 1-form is just a covariant vector field. Since $\Omega^k(TM)$ can obviously be made into a C^∞ vector bundle, we can speak of C^∞ forms; all forms will be understood to be C^∞ forms unless the contrary is explicitly stated. Remember that covariant tensors actually map contravariantly: If $f \colon M \to N$ is C^∞, and ω is a k-form on N, then $f^*\omega$ is a k-form on M. We can also define $\omega_1 + \omega_2$ and $\omega \wedge \eta$. The following properties of k-forms are obvious from the corresponding properties for $\Omega^k(V)$:

$$(\omega_1 + \omega_2) \wedge \eta = \omega_1 \wedge \eta + \omega_2 \wedge \eta$$
$$\omega \wedge (\eta_1 + \eta_2) = \omega \wedge \eta_1 + \omega \wedge \eta_2$$
$$f\omega \wedge \eta = \omega \wedge f\eta = f(\omega \wedge \eta)$$
$$\omega \wedge \eta = (-1)^{kl} \eta \wedge \omega$$
$$f^*(\omega \wedge \eta) = f^*\omega \wedge f^*\eta.$$

If (x, U) is a coordinate system, then the $dx^i(p)$ are a basis for $M_p{}^*$, so the $dx^{i_1}(p) \wedge \cdots \wedge dx^{i_k}(p)$ $(i_1 < \cdots < i_k)$ are a basis for $\Omega^k(p)$. Thus every k-form ω can be written uniquely as

$$\omega = \sum_{i_1 < \cdots < i_k} \omega_{i_1 \ldots i_k} \, dx^{i_1} \wedge \cdots \wedge dx^{i_k}$$

or, if we denote $dx^{i_1} \wedge \cdots \wedge dx^{i_k}$ by dx^I for the multi-index $I = (i_1, \ldots, i_k)$,

$$\omega = \sum_I \omega_I \, dx^I.$$

The problem of finding the relationship between the ω_I and the functions ω'_I when

$$\omega = \sum_I \omega_I \, dx^I = \sum_I \omega'_I \, dy^I$$

is left to the reader (Problem 16), but we will do one special case here.

7. THEOREM. If $f : M \to N$ is a C^∞ function between n-manifolds, (x, U) is a coordinate system around $p \in M$, and (y, V) a coordinate system around $q = f(p) \in N$, then

$$f^*(g \, dy^1 \wedge \cdots \wedge dy^n) = (g \circ f) \cdot \det \left(\frac{\partial (y^i \circ f)}{\partial x^j} \right) dx^1 \wedge \cdots \wedge dx^n.$$

PROOF. It suffices to show that

$$f^*(dy^1 \wedge \cdots \wedge dy^n) = \det \left(\frac{\partial (y^i \circ f)}{\partial x^j} \right) dx^1 \wedge \cdots \wedge dx^n.$$

Now, by Problem 4-1,

$$f^*(dy^1 \wedge \cdots \wedge dy^n)(p) \left(\left. \frac{\partial}{\partial x^1} \right|_p , \ldots, \left. \frac{\partial}{\partial x^n} \right|_p \right)$$

$$= dy^1(q) \wedge \cdots \wedge dy^n(q) \left(f_* \left. \frac{\partial}{\partial x^1} \right|_p , \ldots, f_* \left. \frac{\partial}{\partial x^n} \right|_p \right)$$

$$= dy^1(q) \wedge \cdots \wedge dy^n(q) \left(\sum_{i=1}^n \frac{\partial (y^i \circ f)}{\partial x^1}(p) \left. \frac{\partial}{\partial y^i} \right|_q , \right.$$

$$\left. \ldots, \sum_{i=1}^n \frac{\partial (y^i \circ f)}{\partial x^n}(p) \left. \frac{\partial}{\partial y^i} \right|_q \right)$$

$$= \det \left(\frac{\partial (y^i \circ f)}{\partial x^j}(p) \right), \qquad \text{by Theorem 5.} \ \clubsuit$$

8. COROLLARY. If (x, U) and (y, V) are two coordinate systems on M and

$$g \, dy^1 \wedge \cdots \wedge dy^n = h \, dx^1 \wedge \cdots \wedge dx^n,$$

then

$$h = g \cdot \det \left(\frac{\partial y^i}{\partial x^j} \right).$$

PROOF. Apply the theorem with $f = $ identity map. ❖

[This corollary shows that n-forms are the geometric objects corresponding to the "even scalar densities" defined in Problem 4-10.]

If $\xi = \pi \colon E \to B$ is an n-plane bundle, then a *nowhere zero* section ω of $\Omega^n(\xi)$ has a special significance: For each $p \in B$, the non-zero $\omega(p) \in \Omega^n(\pi^{-1}(p))$ determines an orientation μ_p of $\pi^{-1}(p)$ by Corollary 6. It is easy to see that the collection of orientations $\{\mu_p\}$ satisfy the "compatability condition" set forth in Chapter 3, so that $\mu = \{\mu_p\}$ is an orientation of ξ. In particular, if there is a nowhere zero n-form ω on an n-manifold M, then M is orientable (i.e., the bundle TM is orientable). The converse also holds:

9. THEOREM. If a C^∞ manifold M is orientable, then there is an n-form ω on M which is nowhere 0.

PROOF. By Theorem 2-13 and 2-15, we can choose a cover $\mathcal{O}$ of M by a collection of coordinate systems $\{(x, U)\}$, and a partition of unity $\{\phi_U\}$ subordinate to $\mathcal{O}$. Let μ be an orientation of M. For each (x, U) choose an n-form ω_U on U such that for $v_1, \ldots, v_n \in M_p$, $p \in U$ we have

$$\omega_U(v_1, \ldots, v_n) > 0 \quad \text{if and only if} \quad [v_1, \ldots, v_n] = \mu_p.$$

Now let

$$\omega = \sum_{U \in \mathcal{O}} \phi_U \, \omega_U.$$

Then ω is a C^∞ n-form. Moreover, for every p, if $v_1, \ldots, v_n \in M_p$ satisfy $[v_1, \ldots, v_n] = \mu_p$, then *each*

$$(\phi_U \, \omega_U)(p)(v_1, \ldots, v_n) \geq 0,$$

and strict inequality holds for at least one U. Thus $\omega(p) \neq 0$. ❖

Notice that the bundle $\Omega^n(TM)$ is 1-dimensional. We have shown that if M is orientable, then $\Omega^n(TM)$ has a nowhere 0 section, which implies that it is trivial. Conversely, of course, if the bundle $\Omega^n(TM)$ is trivial, then it certainly has a nowhere 0 section, so M is orientable. [Generally, if ξ is a k-plane bundle, then $\Omega^k(\xi)$ is trivial if and only if ξ is orientable, provided that the base space B is "paracompact" (every open cover has a locally-finite refinement).]

Just as $\Omega^0(V)$ has been introduced as another name for $\mathbb{R}$, a 0-**form** on M will just mean a function f on M (and $f \wedge \omega$ will just mean $f \cdot \omega$). For every 0-form f we have the 1-form df (recall that $df(X) = X(f)$), which in a coordinate system (x, U) is given by

$$df = \sum_{j=1}^{n} \frac{\partial f}{\partial x^j} \, dx^j.$$

If ω is a k-form

$$\omega = \sum_I \omega_I \, dx^I,$$

then each $d\omega_I$ is a 1-form, and we can define a $(k+1)$-form $d\omega$, the **differential** of ω, by

$$d\omega = \sum_I d\omega_I \, dx^I$$

$$= \sum_I \sum_{\alpha=1}^{n} \frac{\partial \omega_I}{\partial x^\alpha} \, dx^\alpha \wedge dx^I.$$

It turns out that this definition does not depend on the coordinate system. This can be proved in several ways. The first way is to use a brute-force computation, comparing the coefficients ω'_I in the expression

$$\omega = \sum_I \omega'_I \, dx^I$$

with the ω_I.

The second method is a lot sneakier. We begin by finding some properties of $d\omega$ (still defined with respect to this particular coordinate system).

10. PROPOSITION.

(1) $d(\omega_1 + \omega_2) = d\omega_1 + d\omega_2.$

(2) If ω_1 is a k-form, then

$$d(\omega_1 \wedge \omega_2) = d\omega_1 \wedge \omega_2 + (-1)^k \omega_1 \wedge d\omega_2.$$

(3) $d(d\omega) = 0.$ Briefly, $d^2 = 0.$

PROOF. (1) is clear. To prove (2) we first note that because of (1) it suffices to consider only

$$\omega_1 = f \, dx^I$$
$$\omega_2 = g \, dx^J.$$

Then $\omega_1 \wedge \omega_2 = fg \, dx^I \wedge dx^J$ and

$$
\begin{aligned}
d(\omega_1 \wedge \omega_2) &= d(fg) \wedge dx^I \wedge dx^J \\
&= g \, df \wedge dx^I \wedge dx^J + f \, dg \wedge dx^I \wedge dx^J \\
&= d\omega_1 \wedge \omega_2 + (-1)^k f \, dx^I \wedge dg \wedge dx^J \\
&= d\omega_1 \wedge \omega_2 + (-1)^k \omega_1 \wedge d\omega_2.
\end{aligned}
$$

(3) It clearly suffices to consider only k-forms of the form

$$\omega = f \, dx^I.$$

Then

$$dw = \sum_{\alpha=1}^{n} \frac{\partial f}{\partial x^\alpha} \, dx^\alpha \wedge dx^I$$

so

$$d(dw) = \sum_{\alpha=1}^{n} \left(\sum_{\beta=1}^{n} \frac{\partial^2 f}{\partial x^\beta \partial x^\alpha} \, dx^\beta \wedge dx^\alpha \wedge dx^I \right).$$

In this sum, the terms

$$\frac{\partial^2 f}{\partial x^\beta \partial x^\alpha} \, dx^\beta \wedge dx^\alpha \wedge dx^I$$

and

$$\frac{\partial^2 f}{\partial x^\alpha \partial x^\beta} \, dx^\alpha \wedge dx^\beta \wedge dx^I$$

cancel in pairs. ❖

We next note that these properties characterize d on U.

11. PROPOSITION. Suppose d' takes k-forms on U to $(k+1)$-forms on U, for all k, and satisfies

(1) $d'(\omega_1 + \omega_2) = d'\omega_1 + d'\omega_2$.
(2) $d'(\omega_1 \wedge \omega_2) = d'\omega_1 \wedge \omega_2 + (-1)^k \omega_1 \wedge d'\omega_2$.
(3) $d'(d'f) = 0$.
(4) $d'f = $ (the old) df.

Then $d' = d$ on U.

PROOF. It is clearly enough to show that $d'\omega = d\omega$ when $\omega = f\,dx^I$. Now by (2),

$$d'(f\,dx^I) = d'f \wedge dx^I + f \wedge d'(dx^I)$$
$$= df \wedge dx^I + f \wedge d'(dx^I) \quad \text{by (4)}.$$

So it suffices to show that $d'(dx^I) = 0$, where

$$dx^I = dx^{i_1} \wedge \cdots \wedge dx^{i_k}$$
$$= d'x^{i_1} \wedge \cdots \wedge d'x^{i_k} \quad \text{by (4)}.$$

We will use induction on k. Assuming it for $k-1$ we have

$$d'(dx^I) = d'(d'x^{i_1} \wedge \cdots \wedge d'x^{i_k})$$
$$= d'(d'x^{i_1}) \wedge d'x^{i_2} \wedge \cdots \wedge d'x^{i_k}$$
$$\quad - d'x^{i_1} \wedge d'(d'x^{i_1} \wedge \cdots \wedge d'x^{i_k}) \qquad \text{by (2)}$$
$$= 0 - 0, \qquad \text{by (3) and the inductive hypothesis.} \;\diamond$$

12. COROLLARY. There is a unique operator d from the k-forms on M to the $(k+1)$-forms on M, for all k, satisfying

$$d(\omega_1 + \omega_2) = d\omega_1 + d\omega_2$$
$$d(\omega_1 \wedge \omega_2) = d\omega_1 \wedge \omega_2 + (-1)^k \omega_1 \wedge d\omega_2$$
$$d^2 = 0,$$

and agreeing with the old d on functions.

PROOF. For each coordinate system (x, U) we have a unique d_U defined. Given the form ω, and $p \in M$, pick any U with $p \in U$ and define

$$d\omega(p) = d_U(\omega|U)(p). \;\diamond$$

The third way of proving that the definition of d does not depend on the coordinate system is to give an invariant definition.

13. THEOREM. If ω is a k-form on M, then there is a unique $(k+1)$-form $d\omega$ on M such that for every set of vector fields $X_1, \ldots, X_{k+1}$ we have

$$(*) \quad d\omega(X_1, \ldots, X_{k+1})$$

$$= \sum_{i=1}^{k+1} (-1)^{i+1} X_i(\omega(X_1, \ldots, \widehat{X_i}, \ldots, X_{k+1}))$$

$$+ \sum_{1 \leq i < j \leq k+1} (-1)^{i+j} \omega([X_i, X_j], X_1, \ldots, \widehat{X_i}, \ldots, \widehat{X_j}, \ldots, X_{k+1})$$

$$(= \Sigma_1 + \Sigma_2, \text{ say})$$

where $\widehat{}$ over X_i indicates that it is omitted. This $(k+1)$-form agrees with $d\omega$ as defined previously.

PROOF. The operator which takes $(X_1, \ldots, X_{k+1})$ to $\Sigma_1 + \Sigma_2$ is clearly linear over $\mathbb{R}$. Moreover, it is actually linear *over the C^∞ functions $\mathcal{F}$*. In fact, if X_{i_0} is replaced by $f X_{i_0}$, then Σ_1 becomes

$$f\Sigma_1 + \sum_{i \neq i_0} (-1)^{i+1} (X_i f) \omega(X_1, \ldots, \widehat{X_i}, \ldots, X_{k+1}),$$

and using the formulas

$$[fX, Y] = f[X, Y] - Yf \cdot X$$
$$[X, fY] = f[X, Y] + Xf \cdot Y,$$

it is easily seen that Σ_2 becomes

$$f\Sigma_2 + \sum_{i < i_0} (-1)^{i+i_0} (X_i f) \omega(X_{i_0}, X_1, \ldots, \widehat{X_i}, \ldots, \widehat{X_{i_0}}, \ldots, X_{k+1})$$

$$- \sum_{i_0 < j} (-1)^{i_0+j} (X_j f) \omega(X_{i_0}, X_1, \ldots, \widehat{X_{i_0}}, \ldots, \widehat{X_j}, \ldots, X_{k+1});$$

a brief inspection then shows that $\Sigma_1 + \Sigma_2$ becomes $f\Sigma_1 + f\Sigma_2$.

Theorem 4-2 shows that there is a unique covariant tensor field $d\omega$ satisfying $(*)$. It is easy to check that $d\omega$ is alternating, so that it is a $(k+1)$-form.

To compute $d\omega$ in a coordinate system (x, U) it clearly suffices to compute $d(f\, dx^I)$. Moreover, by renumbering, we might as well assume

$$\omega = f\, dx^1 \wedge \cdots \wedge dx^k.$$

For $d\omega$, as for any form, we have

$$d\omega = \sum_{\alpha_1 < \cdots < \alpha_{k+1}} d\omega(\partial/\partial x^{\alpha_1}, \ldots, \partial/\partial x^{\alpha_{k+1}}) \, dx^{\alpha_1} \wedge \cdots \wedge dx^{\alpha_{k+1}}.$$

It is clear from $(*)$ that $d\omega(\partial/\partial x^{\alpha_1}, \ldots, \partial/\partial x^{\alpha_{k+1}}) = 0$

unless some $(\alpha_1, \ldots, \widehat{\alpha_i}, \ldots, \alpha_{k+1})$ is a permutation of $(1, \ldots, k)$.

Since the α's are increasing, this happens only if

$$(\alpha_1, \ldots, \alpha_{k+1}) = (1, \ldots, k, j) \qquad j > k,$$

in which case

$$d\omega(\partial/\partial x^{\alpha_1}, \ldots, \partial/\partial x^{\alpha_k}, \partial/\partial x^j) = (-1)^k \frac{\partial f}{\partial x^j},$$

so

$$\begin{aligned} d\omega &= \sum_{j > k} (-1)^k \frac{\partial f}{\partial x^k} \, dx^1 \wedge \cdots \wedge dx^k \wedge dx^j \\ &= \sum_{j > k} \frac{\partial f}{\partial x^j} \, dx^j \wedge dx^1 \wedge \cdots \wedge dx^k \\ &= \sum_{j=1}^{n} \frac{\partial f}{\partial x^j} \, dx^j \wedge dx^1 \wedge \cdots \wedge dx^k, \end{aligned}$$

which is just the old definition. ❖

This is our first real example of an invariant definition of an important tensor, and our first use of Theorem 4-2. We do not find $d\omega(p)(v_1, \ldots, v_{k+1})$ directly, but first find $d\omega(X_1, \ldots, X_{k+1})$, where X_i are vector fields extending v_i, and then evaluate this function at p. By some sort of magic, this turns out to be independent of the extensions $X_1, \ldots, X_{k+1}$. This may not seem to be much of an improvement over using a coordinate system and checking that the definition is independent of the coordinate system. But we can hardly hope for anything better. After all, although $d\omega(X_1, \ldots, X_{k+1})(p)$ does not depend on the values of X_i except at p, it *does* depend on the values of ω at points other than p—this must enter into our formula somehow. One other feature of our definition is common to most invariant definitions of tensors—the presence of a term involving brackets of various vector fields. This term is what makes the operator linear over the C^∞ functions, but it disappears in computations in a coordinate system.

In the particular case where ω is a 1-form, Theorem 13 gives the following formula.

$$dw(X, Y) = X(\omega(Y)) - Y(\omega(X)) - \omega([X, Y])$$

This enables us to state a second version of Theorem 6-5 (The Frobenius Integrability Theorem) in terms of differential forms. Define the ring $\Omega(M)$ to be the direct sum of the rings of l-forms on M, for all l. If Δ is a k-dimensional distribution on M, then $\mathcal{I}(\Delta) \subset \Omega(M)$ will denote the subring generated by the set of all forms ω with the property that (if ω has degree l)

$$\omega(X_1, \ldots, X_l) = 0 \qquad \text{whenever } X_1, \ldots, X_l \text{ belong to } \Delta.$$

It is clear that $\omega_1 + \omega_2 \in \mathcal{I}(\Delta)$ if $\omega_1, \omega_2 \in \mathcal{I}(\Delta)$, and that $\eta \wedge \omega \in \mathcal{I}(\Delta)$ if $\omega \in \mathcal{I}(\Delta)$ [thus, $\mathcal{I}(\Delta)$ is an ideal in the ring $\Omega(M)$]. Locally, the ideal $\mathcal{I}(\Delta)$ is generated by $n - k$ independent 1-forms $\omega^{k+1}, \ldots, \omega^n$. In fact, around any point $p \in M$ we can choose a coordinate system (x, U) so that

$$\left.\frac{\partial}{\partial x^1}\right|_p, \ldots, \left.\frac{\partial}{\partial x^k}\right|_p \quad \text{span } \Delta_p.$$

Then

$$dx^1(p) \wedge \cdots \wedge dx^k(p) \quad \text{is non-zero on } \Delta_p.$$

By continuity, the same is true for q sufficiently close to p, which by Corollary 4 implies that $dx^1(q), \ldots, dx^k(q)$ are linearly independent *in* Δ_q. Therefore, there are C^∞ functions f_β^α such that

$$dx^\alpha(q) = \sum_{\beta=1}^{k} f_\beta^\alpha(q)\, dx^\beta(q) \quad \text{restricted to } \Delta_q \qquad \alpha = k + 1, \ldots, n.$$

We can therefore let

$$\omega^\alpha = dx^\alpha - \sum_{\beta=1}^{k} f_\beta^\alpha\, dx^\beta.$$

14. PROPOSITION (THE FROBENIUS INTEGRABILITY THEOREM; SECOND VERSION). A distribution Δ on M is integrable if and only if $d(\mathcal{I}(\Delta)) \subset \mathcal{I}(\Delta)$.

PROOF. Locally we can choose 1-forms $\omega^1, \ldots, \omega^n$ which span $M_q{}^*$ for each q such that $\omega^{k+1}, \ldots, \omega^n$ generate $\mathcal{I}(\Delta)$. Let $X_1, \ldots, X_n$ be the vector fields with

$$\omega^i(X_j) = \delta^i_j.$$

Then $X_1, \ldots, X_k$ span Δ. So Δ is integrable if and only if there are functions C^β_{ij} with

$$[X_i, X_j] = \sum_{\beta=1}^{k} C^\beta_{ij} X_\beta \qquad i, j = 1, \ldots, k.$$

Now

$$d\omega^\alpha(X_i, X_j) = X_i(\omega^\alpha(X_j)) - X_j(\omega^\alpha(X_i)) - \omega^\alpha([X_i, X_j]).$$

For $1 \leq i, j \leq k$ and $\alpha > k$, the first two terms on the right vanish. So $d\omega^\alpha(X_i, X_j) = 0$ if and only if $\omega^\alpha([X_i, X_j]) = 0$. But each $\omega^\alpha([X_i, X_j]) = 0$ if and only if each $[X_i, X_j]$ belongs to Δ (i.e., if Δ is integrable), while each $d\omega^\alpha(X_i, X_j) = 0$ if and only if $d\omega^\alpha \in \mathcal{I}(\Delta)$. ❖

Notice that since the $\omega^i \wedge \omega^j$ $(i < j)$ span $\Omega^2(M_q)$ for each q, we can always write

$$d\omega^\alpha = \sum_{i<j} c^\alpha_{ij} \omega^i \wedge \omega^j$$

$$= \sum_j \theta^\alpha_j \wedge \omega^j \quad \text{for certain forms } \theta^\alpha_j.$$

If $\alpha > k$, and $i_0, j_0 \leq k$ are distinct, we have

$$0 = d\omega^\alpha(X_{i_0}, X_{j_0}) = \sum_j (\theta^\alpha_j \wedge \omega^j)(X_{i_0}, X_{j_0})$$

$$= \theta^\alpha_{j_0}(X_{i_0}),$$

so we can write the condition $d(\mathcal{I}(\Delta)) \subset \mathcal{I}(\Delta)$ as

$$d\omega^\alpha = \sum_{\beta > k} \theta^\alpha_\beta \wedge \omega^\beta.$$

Once we have introduced a coordinate system (x, U) such that the slices

$$\{q \in U : x^{k+1}(q) = a^{k+1}, \ldots, x^n(q) = a^n\}$$

are integral submanifolds of Δ, the forms $dx^{k+1}, \ldots, dx^n$ are a basis for $\mathcal{I}(\Delta)$, so $\omega^{k+1}, \ldots, \omega^n$ must be linear combinations of them. We therefore have the following.

15. COROLLARY. If $\omega^{k+1}, \ldots, \omega^n$ are linearly independent 1-forms in a neighborhood of $p \in M$, then there are 1-forms θ_β^α $(\alpha, \beta > k)$ with

$$d\omega^\alpha = \sum_\beta \theta_\beta^\alpha \wedge \omega^\beta$$

if and only if there are functions f_β^α, g^β $(\alpha, \beta > k)$ with

$$\omega^\alpha = \sum_\beta f_\beta^\alpha \, dg^\beta.$$

Although Theorem 13 warms the heart of many an invariant lover, the cases $k > 1$ will hardly ever be used (a very significant exception occurs in the last chapter of Volume V). Problem 18 gives another invariant definition of $d\omega$, using induction on the degree of ω, which is much simpler. The reader may reflect on the difficulties which would be involved in using the definition of Theorem 13 to prove the following important property of d:

16. PROPOSITION. If $f: M \to N$ is C^∞ and ω is a k-form on N, then

$$f^*(d\omega) = d(f^*\omega).$$

PROOF. For $p \in M$, let (x, U) be a coordinate system around $f(p)$. We can assume

$$\omega = g \, dx^{i_1} \wedge \cdots \wedge dx^{i_k}.$$

We will use induction on k. For $k = 0$ we have, tracing through some definitions,

$$f^*(dg)(X) = dg(f_*X) = [f_*X](g) = X(g \circ f)$$
$$= d(g \circ f)(X)$$

(and, of course, f^*g is to be interpreted as $g \circ f$). Assuming the formula for $k - 1$, we have

$$d(f^*\omega) = d\left((f^*g \, dx^{i_1} \wedge \cdots \wedge dx^{i_{k-1}}) \wedge f^*dx^{i_k}\right)$$
$$= d\left(f^*(g \, dx^{i_1} \wedge \cdots \wedge dx^{i_{k-1}})\right) \wedge f^*dx^{i_k} + 0$$
$$\text{since } df^*dx^{i_k} = dd(x^{i_k} \circ f) = 0$$
$$= f^*\left(d(g \, dx^{i_1} \wedge \cdots \wedge dx^{i_{k-1}})\right) \wedge f^*dx^{i_k}$$
$$\text{by the inductive hyposthesis}$$
$$= f^*(dg \wedge dx^{i_1} \wedge \cdots \wedge dx^{i_{k-1}}) \wedge f^*dx^{i_k}$$
$$= f^*(dg \wedge dx^{i_1} \wedge \cdots \wedge dx^{i_{k-1}} \wedge dx^{i_k})$$
$$= f^*(d\omega). \; ❖$$

One property of d qualifies, by the criterion of the previous chapter, as a basic theorem of differential geometry. The relation $d^2 = 0$ is just an elegant way of stating that mixed partial derivatives are equal. There is another set of terminology for stating the same thing. A form ω is called **closed** if $d\omega = 0$ and **exact** if $\omega = d\eta$ for some form η. (The terminology "exact" is classical— differential forms used to be called simply "differentials"; a differential was then called "exact" if it actually was the differential of something. The term "closed" is based on an analogy with chains, which will be discussed in the next chapter.) Since $d^2 = 0$, every exact form is closed. In other words, $d\omega = 0$ is a necessary condition for solving $\omega = d\eta$. If ω is a 1-form

$$\omega = \sum_{i=1}^{n} \omega_i \, dx^i,$$

then the condition $d\omega = 0$, i.e.,

$$\frac{\partial \omega_i}{\partial x^j} = \frac{\partial \omega_j}{\partial x^i}$$

is necessary for solving $\omega = df$, i.e.,

$$\frac{\partial f}{\partial x^i} = \omega_i.$$

Now we know from Theorem 6-1 that these conditions are also sufficient. For 2-forms the situation is more complicated, however. If ω is a 2-form on $\mathbb{R}^3$,

$$\omega = A \, dy \wedge dz - B \, dx \wedge dz + C \, dx \wedge dy,$$

then

$$\omega = d(P \, dx + Q \, dy + R \, dz)$$

if and only if

$$\frac{\partial R}{\partial y} - \frac{\partial Q}{\partial z} = A$$

$$\frac{\partial P}{\partial z} - \frac{\partial R}{\partial x} = B$$

$$\frac{\partial Q}{\partial x} - \frac{\partial P}{\partial y} = C.$$

The necessary condition, $d\omega = 0$, is

$$\frac{\partial A}{\partial x} + \frac{\partial B}{\partial y} + \frac{\partial C}{\partial z} = 0.$$

In general, we are dealing with a rather strange collection of partial differential equations (carefully selected so that we can get integrability conditions). It turns out that these necessary conditions are also sufficient: if ω is closed, then it is exact. Like our results about solutions to differential equations, this result is true only locally. The reasons for restricting ourselves to local results are now somewhat different, however. Consider the case of a closed 1-form ω on $\mathbb{R}^2$:

$$\omega = f\,dx + g\,dy, \qquad \text{with} \quad \frac{\partial f}{\partial y} = \frac{\partial g}{\partial x}.$$

We know how to find a function α on *all* of $\mathbb{R}^2$ with $\omega = d\alpha$, namely

$$\alpha(x, y) = \int_{x_0}^{x} f(t, y_0)\,dt + \int_{y_0}^{y} g(x, t)\,dt.$$

On the other hand, the situation is very different if ω is defined only on $\mathbb{R}^2 - \{0\}$. Recall that if $L \subset \mathbb{R}^2$ is $[0, \infty) \times \{0\}$, then

$$\theta \colon \mathbb{R}^2 - L \to \mathbb{R},$$

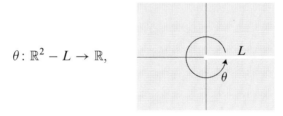

defined in Chapter 2, is C^∞; in fact,

$$(r, \theta) \colon \mathbb{R}^2 - L \to \{r : r > 0\} \times (0, 2\pi)$$

is the inverse of the map

$$(a, b) \mapsto (a \cos b, a \sin b),$$

whose derivative at (a, b) has determinant equal to $a \neq 0$. By deleting a different ray L_1 we can define a different function θ_1. Then $\theta_1 = \theta$ in the region A_1 and $\theta_1 = \theta + 2\pi$ in the region A_2. Consequently $d\theta$ and $d\theta_1$ agree on their common

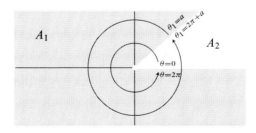

domain, so that together they define a 1-form ω on $\mathbb{R}^2 - \{0\}$. A computation (Problem 20) shows that

$$\omega = \frac{-y}{x^2 + y^2} \, dx + \frac{x}{x^2 + y^2} \, dy.$$

The 1-form ω is usually denoted by $d\theta$, but this is an abuse of notation, since $\omega = d\theta$ only on $\mathbb{R}^2 - L$. In fact, ω is not df for *any* C^1 function $f : \mathbb{R}^2 - \{0\} \to \mathbb{R}$. Indeed, if $\omega = df$, then

$$df = d\theta \quad \text{on} \quad \mathbb{R}^2 - L,$$

so $d(f - \theta) = 0$ on $\mathbb{R}^2 - L$, which implies that $\partial f/\partial x = \partial\theta/\partial x$ and $\partial f/\partial y = \partial\theta/\partial y$ and hence $f = \theta +$ constant on $\mathbb{R}^2 - L$, which is impossible. Nevertheless, $d\omega = 0$ [the two relations

$$d(d\theta) = 0 \quad \text{on} \quad \mathbb{R}^2 - L$$
$$d(d\theta_1) = 0 \quad \text{on} \quad \mathbb{R}^2 - L_1$$

clearly imply that this is so]. So ω is closed, but not exact. (It is still exact in a neighborhood of any point of $\mathbb{R}^2 - \{0\}$.)

Clearly ω is also not exact in any small region containing 0. This example shows that it is the shape of the region, rather than its size, that determines whether or not a closed form is necessarily exact.

A manifold M is called (**smoothly**) **contractible** to a point $p_0 \in M$ if there is a C^∞ function

$$H : M \times [0, 1] \to M$$

such that

$$H(p, 1) = p \qquad \text{for } p \in M.$$
$$H(p, 0) = p_0$$

For example, $\mathbb{R}^n$ is smoothly contractible to $0 \in \mathbb{R}^n$; we can define

$$H : \mathbb{R}^n \times [0, 1] \to \mathbb{R}^n$$

by

$$H(p, t) = tp.$$

More generally, $U \subset \mathbb{R}^n$ is contractible to $p_0 \in U$ if U has the property that

$p \in U$ implies $p_0 + t(p - p_0) \in U$ for $0 \leq t \leq 1$ (such a region U is called **star-shaped** with respect to p_0).

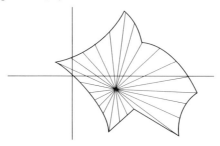

Of course, many other regions are also contractible to a point. If we think

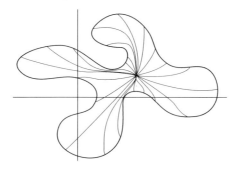

of $[0, 1]$ as representing time, then for each time t we have a map $p \mapsto H(p, t)$ of M into itself; at time 1 this is just the identity map, and at time 0 it is the constant map.

We will show that if M is smoothly contractible to a point, then every closed form on M is exact. (By the way, this result and our investigation of the form $d\theta$ prove the intuitively obvious fact that $\mathbb{R}^2 - \{0\}$ is *not* contractible to a point; the same result holds for $\mathbb{R}^n - \{0\}$, but we will not be in a position to prove this until the next chapter.) The trick in proving our result is to analyze $M \times [0, 1]$ (for any manifold M), and pay hardly any attention at all to H.

For $t \in [0, 1]$ we define

$$i_t : M \to M \times [0, 1]$$

by

$$i_t(p) = (p, t).$$

We claim that if ω is a form on $M \times [0, 1]$ with $d\omega = 0$, then

$$i_1{}^*\omega - i_0{}^*\omega \quad \text{is exact;}$$

we will see later (and you may try to convince yourself right now) that the theorem follows trivially from this.

Consider first a 1-form ω on $M \times [0, 1]$. We will begin by working in a coordinate system on $M \times [0, 1]$. There is an obvious function t on $M \times [0, 1]$ (namely, the projection π on the second coordinate), and if (x, U) is a coordinate system on M, while π_M is the projection on M, then

$$(x^1 \circ \pi_M, \ldots, x^n \circ \pi_M, t)$$

is a coordinate system on $U \times [0, 1]$. We will denote $x^i \circ \pi_M$ by $\bar{x}^i$, for convenience. It is easy to check (or should be) that

$$i_\alpha^* \left(\sum_{i=1}^n \omega_i \, d\bar{x}^i + f \, dt \right) = \sum_{i=1}^n \omega_i(\cdot, \alpha) \, dx^i,$$

where

$$\omega_i(\cdot, \alpha) \quad \text{denotes the function} \quad p \mapsto \omega_i(p, \alpha).$$

Now for $\omega = \sum_{i=1}^n \omega_i \, d\bar{x}^i + f \, dt$ we have

$$d\omega = [\text{terms not involving } dt] - \sum_{i=1}^n \frac{\partial \omega_i}{\partial t} \, d\bar{x}^i \wedge dt + \sum_{i=1}^n \frac{\partial f}{\partial \bar{x}^i} \, d\bar{x}^i \wedge dt.$$

So $d\omega = 0$ implies that

$$\frac{\partial \omega_i}{\partial t} = \frac{\partial f}{\partial \bar{x}^i}.$$

Consequently,

$$\omega_i(p, 1) - \omega_i(p, 0) = \int_0^1 \frac{\partial \omega_i}{\partial t}(p, t) \, dt$$

$$= \int_0^1 \frac{\partial f}{\partial \bar{x}^i}(p, t) \, dt,$$

so

$$(1) \qquad \sum_{i=1}^n \omega_i(p, 1) \, dx^i - \sum_{i=1}^n \omega_i(p, 0) \, dx^i = \sum_{i=1}^n \left(\int_0^1 \frac{\partial f}{\partial \bar{x}^i}(p, t) \, dt \right) dx^i.$$

If we define $g \colon M \to \mathbb{R}$ by

$$g(p) = \int_0^1 f(p, t) \, dt,$$

then

(2)
$$\frac{\partial g}{\partial x^i}(p) = \int_0^1 \frac{\partial f}{\partial \bar{x}^i}(p,t)\, dt.$$

Equations (1) and (2) show that

$$i_1{}^*\omega - i_0{}^*\omega = dg.$$

Now although we seem to be using a coordinate system, the function f, and hence g also, is really independent of the coordinate system. Notice that for the tangent space of $M \times [0,1]$ we have

(∗)
$$(M \times [0,1])_{(p,t)} = \ker \pi_* \oplus \ker \pi_{M*}.$$

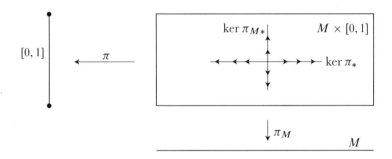

If a vector space V is a direct sum $V = V_1 \oplus V_2$ of two subspaces, then any $\omega \in \Omega^1(V)$ can be written

$$\omega = \omega_1 + \omega_2$$

where

$$\omega_1(v_1 + v_2) = \omega(v_1)$$
$$\omega_2(v_1 + v_2) = \omega(v_2).$$

Applying this to the decomposition (∗), we write the 1-form ω on $M \times [0,1]$ as $\omega_1 + \omega_2$; there is then a unique f with $\omega_2 = f\, dt$.

In general, for a k-form ω, it is easy to see (Problem 22) that we can write ω uniquely as

$$\omega = \omega_1 + (dt \wedge \eta)$$

where $\omega_1(v_1, \ldots, v_k) = 0$ if some $v_i \in \ker \pi_{M*}$, and η is a $(k-1)$-form with the analogous property. Define a $(k-1)$-form $I\omega$ on M as follows:

$$I\omega(p)(v_1,\ldots,v_{k-1}) = \int_0^1 \eta(p,t)(i_{t*}v_1,\ldots,i_{t*}v_{k-1})\,dt.$$

We claim that $d\omega = 0$ implies that $i_1^*\omega - i_0^*\omega = d(I\omega)$. Actually, it is easier to find a formula for $i_1^*\omega - i_0^*\omega$ that holds even when $d\omega \neq 0$.

17. THEOREM. For any k-form ω on $M \times [0,1]$ we have
$$i_1^*\omega - i_0^*\omega = d(I\omega) + I(d\omega).$$

(Consequently, $i_1^*\omega - i_0^*\omega = d(I\omega)$ if $d\omega = 0$.)

PROOF. Since $I\omega$ is already invariantly defined, we can just as well work in a coordinate system $(\bar{x}^1,\ldots,\bar{x}^n,t)$. The operator I is clearly linear, so we just have to consider two cases.

(1) $\omega = f\,d\bar{x}^{i_1} \wedge \cdots \wedge d\bar{x}^{i_k} = f\,d\bar{x}^I$. Then
$$d\omega = \underline{\quad\quad} + \frac{\partial f}{\partial t}\,dt \wedge d\bar{x}^I;$$

it is easy to see that

$$I(d\omega)(p) = \left(\int_0^1 \frac{\partial f}{\partial t}(p,t)\,dt\right) dx^I(p)$$
$$= [f(p,1) - f(p,0)]\,dx^I(p)$$
$$= i_1^*\omega(p) - i_0^*\omega(p).$$

Since $I\omega = 0$, this proves the result in this case.

(2) $\omega = f\,dt \wedge d\bar{x}^{i_1} \wedge \cdots \wedge d\bar{x}^{i_{k-1}} = f\,dt \wedge d\bar{x}^I$. Then $i_1^*\omega = i_0^*\omega = 0$. Now

$$I(d\omega)(p) = I\left(-\sum_{\alpha=1}^n \frac{\partial f}{\partial \bar{x}^\alpha}\,dt \wedge d\bar{x}^\alpha \wedge d\bar{x}^I\right)(p)$$
$$= -\sum_{\alpha=1}^n \left(\int_0^1 \frac{\partial f}{\partial \bar{x}^\alpha}(p,t)\,dt\right) dx^\alpha \wedge dx^I$$

and

$$d(I\omega) = d\left(\int_0^1 f(p,t)\,dt\right) dx^I$$
$$= \sum_{\alpha=1}^n \frac{\partial}{\partial x^\alpha}\left(\int_0^1 f(p,t)\,dt\right) dx^\alpha \wedge dx^I.$$

Clearly $I(d\omega) + d(I\omega) = 0$. ❖

18. COROLLARY. If M is smoothly contractible to a point $p_0 \in M$, then every closed form ω on M is exact.

PROOF. We are given $H \colon M \times [0, 1] \to M$ with

$$\begin{aligned} H(p, 1) &= p \\ H(p, 0) &= p_0 \end{aligned} \qquad \text{for all } p \in M.$$

Thus

$$\begin{aligned} H \circ i_1 &\colon M \to M \quad \text{is the identity} \\ H \circ i_0 &\colon M \to M \quad \text{is the constant map } p_0. \end{aligned}$$

So

$$\begin{aligned} \omega &= (H \circ i_1)^*(\omega) = i_1{}^*(H^*\omega) \\ 0 &= (H \circ i_0)^*(\omega) = i_0{}^*(H^*\omega). \end{aligned}$$

But

$$d(H^*\omega) = H^*(d\omega) = 0,$$

so

$$\begin{aligned} \omega - 0 &= i_1{}^*(H^*\omega) - i_0{}^*(H^*\omega) \\ &= d(I(H^*\omega)) \qquad \text{by the Theorem.} \; \diamond \end{aligned}$$

Corollary 18 is called the Poincaré Lemma by most geometers, while $d^2 = 0$ is called the Poincaré Lemma by some (I don't even know whether Poincaré had anything to do with it.) In the case of a star-shaped open subset U of $\mathbb{R}^n$, where we have an explicit formula for H, we can find (Problem 23) an explicit formula for $I(H^*\omega)$, for every form ω on U. Since the new form is given by an integral, we can solve the system of partial differential equations $\omega = d\eta$ explicitly in terms of integrals. There are classical theorems about vector fields in $\mathbb{R}^3$ which can be derived from the Poincaré Lemma and its converse (Problem 27), and originally d was introduced in order to obtain a uniform generalization of all these results. Even though the Poincaré Lemma and its converse fit very nicely into our pattern for basic theorems about differential geometry, it has always been something of a mystery to me just why d turns out to be so important. An answer to this question is provided by a theorem of Palais, *Natural Operations on Differential Forms*, Trans. Amer. Math. Soc. **92** (1959), 125–141. Suppose we have any operator D from k-forms to l-forms, such that the following diagram commutes for every C^∞ map $f: M \to N$ [it actually suffices to assume that the diagram commutes only for diffeomorphisms f].

$$
\begin{array}{ccc}
k\text{-forms on } M & \xleftarrow{\;f^*\;} & k\text{-forms on } N \\
\Big\downarrow{\scriptstyle D} & & \Big\downarrow{\scriptstyle D} \\
l\text{-forms on } M & \xleftarrow{\;f^*\;} & l\text{-forms on } M
\end{array}
$$

Palais' theorem says that, with few exceptions, $D = 0$. Roughly, these exceptional cases are the following. If $k = l$, then D can be a multiple of the identity map, but nothing else. If $l = k + 1$, then D can only be some multiple of d. (As a corollary, $d^2 = 0$, since d^2 makes the above diagram commute!) There is only one other case where a non-zero D exists—when k is the dimension of M and $l = 0$. In this case, D can be a multiple of "integration", which we discuss in the next chapter.

PROBLEMS

1. Show that if we define

$$\sigma \bullet (v_1, \ldots, v_k) = (v_{\sigma^{-1}(1)}, \ldots, v_{\sigma^{-1}(k)}),$$

then

$$\sigma \bullet \rho \bullet (v_1, \ldots, v_k) = \sigma\rho \bullet (v_1, \ldots, v_k).$$

2. Let $\overline{\text{Alt}}$ be Alt without the factor $1/k!$, and define $\omega \overline{\wedge} \eta = \overline{\text{Alt}}(\omega \otimes \eta)$. Show that $\overline{\wedge}$ is *not* associative. (Try $\omega, \eta \in \Omega^1(V)$ and $\theta \in \Omega^2(V)$.)

3. Let $S' \subset S_{k+l}$ be the subgroup of all σ which leave both sets $\{1, \ldots, k\}$ and $\{k+1, \ldots, k+l\}$ invariant. A *cross section* of S' is a subset $K \subset S_{k+l}$ containing exactly one element from each left coset of S'.

(a) Show that for any cross section K we have

$$\omega \wedge \eta(v_1, \ldots, v_{k+l}) = \sum_{\sigma \in K} \text{sgn}\, \sigma \cdot \omega \otimes \eta(v_{\sigma(1)}, \ldots, v_{\sigma(k+l)}).$$

This definition may be used even in a field of finite characteristic.
(b) Show from this definition that $\omega \wedge \eta$ is alternating, and $\omega \wedge \eta = (-1)^{kl} \eta \wedge \omega$. (Proving associativity is quite messy.)
(c) A permutation $\sigma \in S_{k+l}$ is called a *shuffle permutation* if $\sigma(1) < \sigma(2) < \cdots < \sigma(k)$ and $\sigma(k+1) < \sigma(k+2) < \cdots < \sigma(k+l)$. Show that the set of all shuffle permutations is a cross section of S'.

4. For $v \in V$ and $\omega \in \Omega^k(V)$, we define the **contraction** $v \lrcorner \omega \in \Omega^{k-1}(V)$ by

$$(v \lrcorner \omega)(v_1, \ldots, v_{k-1}) = \omega(v, v_1, \ldots, v_{k-1}).$$

This is sometime also called the *inner product* and the notation $i_v\omega$ is also used.

(a) Show that

$$v \lrcorner (w \lrcorner \omega) = -w \lrcorner (v \lrcorner \omega).$$

(b) Show that if $v_1, \ldots, v_n$ is a basis of V with dual basis $\phi_1, \ldots, \phi_n$, then

$$v_j \lrcorner (\phi_{i_1} \wedge \cdots \wedge \phi_{i_k}) = \begin{cases} 0 & j \neq \text{any } i_\alpha \\ (-1)^{\alpha-1} \phi_{i_1} \wedge \cdots \wedge \widehat{\phi_{i_\alpha}} \wedge \cdots \wedge \phi_{i_k} & \text{if } j = i_\alpha. \end{cases}$$

(c) Show that for $\omega_1 \in \Omega^k(V)$ and $\omega_2 \in \Omega^l(V)$ we have

$$v \lrcorner (\omega_1 \wedge \omega_2) = (v \lrcorner \omega_1) \wedge \omega_2 + (-1)^k \omega_1 \wedge (v \lrcorner \omega_2).$$

(Use (b) and linearity of everything.)

(d) Formula (c) can be used to give a definition of $\omega_1 \wedge \omega_2$ by induction on $k + l$ (which works for vector spaces over any field): If $\wedge$ is defined for forms of degree adding up to $< k + l$, we define

$$\omega_1 \wedge \omega_2(v_1, \ldots, v_{k+l}) = [(v_1 \lrcorner \omega_1) \wedge \omega_2](v_2, \ldots, v_{k+l})$$
$$+ (-1)^k [\omega_1 \wedge (v_1 \lrcorner \omega_2)](v_2, \ldots, v_{k+l}).$$

Show that with this definition $\omega_1 \wedge \omega_2$ is skew-symmetric (it is only necessary to check that interchanging v_1 and v_2 changes the sign of the right side).

(e) Prove by induction that $\wedge$ is bilinear and that $\omega_1 \wedge \omega_2 = (-1)^{kl} \omega_2 \wedge \omega_1$.

(f) If X is a vector field on M and ω a k-form on M we define a $(k-1)$-form $X \lrcorner \omega$ by

$$(X \lrcorner \omega)(p) = X(p) \lrcorner \omega(p).$$

Show that if ω_1 is a k-form, then

$$X \lrcorner (\omega_1 \wedge \omega_2) = (X \lrcorner \omega_1) \wedge \omega_2 + (-1)^k \omega_1 \wedge (X \lrcorner \omega_2).$$

5. Show that n functions $f_1, \ldots, f_n \colon M \to \mathbb{R}$ form a coordinate system in a neighborhood of $p \in M$ if and only if $df_1 \wedge \cdots \wedge df_n(p) \neq 0$.

6. An element $\omega \in \Omega^k(V)$ is called **decomposable** if $\omega = \phi_1 \wedge \cdots \wedge \phi_k$ for some $\phi_i \in V^* = \Omega^1(V)$.

(a) If $\dim V \leq 3$, then every $\omega \in \Omega^2(V)$ is decomposable.

(b) If ϕ_i, $i = 1, \ldots, 4$ are independent, then $\omega = (\phi_1 \wedge \phi_2) + (\phi_3 \wedge \phi_4)$ is not decomposable. *Hint*: Look at $\omega \wedge \omega$.

7. For any $\omega \in \Omega^k(V)$, we define the **annihilator** of ω to be

$$Ann(\omega) = \{\phi \in V^* \colon \phi \wedge \omega = 0\}.$$

(a) Show that

$$\dim Ann(\omega) \leq k,$$

and that equality holds if and only if ω is decomposable.

(b) Every subspace of V^* is $Ann(\omega)$ for some decomposable ω, which is unique up to a multiplicative constant.

(c) If ω_1 and ω_2 are decomposable, then $Ann(\omega_1) \subset Ann(\omega_2)$ if and only if $\omega_2 = \omega_1 \wedge \eta$ for some η.

(d) If ω_i are decomposable, then $Ann(\omega_1) \cap Ann(\omega_2) = \{0\}$ if and only if $\omega_1 \wedge \omega_2 \neq 0$. In this case,

$$Ann(\omega_1) + Ann(\omega_2) = Ann(\omega_1 \wedge \omega_2).$$

(e) If V has dimension n, then any $\omega \in \Omega^{n-1}(V)$ is decomposable.

(f) Since $v_i \in V$ can be regarded as elements of V^{**}, we can consider $v_1 \wedge \cdots \wedge v_k \in \Omega^k(V^*)$. Reformulate parts (a)–(d) in terms of this $\wedge$ product.

8. (a) Let $\omega \in \Omega^2(V)$. Show that there is a basis $\phi_1, \ldots, \phi_n$ of V^* such that

$$\omega = (\phi_1 \wedge \phi_2) + \cdots + (\phi_{2r-1} \wedge \phi_{2r}).$$

Hint: If

$$\omega = \sum_{i<j} a_{ij} \psi_i \wedge \psi_j,$$

choose ϕ_1 involving $\psi_1, \psi_3, \ldots, \psi_n$ and ϕ_2 involving $\psi_2, \ldots, \psi_n$ so that

$$\omega = \phi_1 \wedge \phi_2 + \omega',$$

where ω' does not involve ψ_1 or ψ_2.

(b) Show that the r-fold wedge product $\omega \wedge \cdots \wedge \omega$ is non-zero and decomposable, and that the $(r+1)$-fold wedge product is 0. Thus r is well-determined; it is called the **rank** of ω.

(c) If $\omega = \sum_{i<j} a_{ij} \psi_i \wedge \psi_j$, show that the rank of ω is the rank of the matrix (a_{ij}).

9. If $v_1, \ldots, v_n$ is a basis for V and $w_i = \sum_{j=1}^n \alpha_{ji} v_j$, show that

$$\det(\alpha_{ij}) w^*_1 \wedge \cdots \wedge w^*_n = v^*_1 \wedge \cdots \wedge v^*_n.$$

10. Let $A = (a_{ij})$ be an $n \times n$ matrix. Let $1 \leq p \leq n$ be fixed, and let $q = n - p$. For $H = h_1 < \cdots < h_p$ and $K = k_1 < \cdots < k_q$, let

$$B^H = \det \begin{pmatrix} a_{1,h_1} & \cdots & a_{1,h_p} \\ \vdots & & \vdots \\ a_{p,h_1} & \cdots & a_{p,h_p} \end{pmatrix}, \qquad C^K = \det \begin{pmatrix} a_{p+1,k_1} & \cdots & a_{p+1,k_q} \\ \vdots & & \vdots \\ a_{n,k_1} & \cdots & a_{n,k_q} \end{pmatrix}.$$

(a) If $v_1, \ldots, v_n$ is a basis of V and

$$w_i = \sum_{j=1}^n a_{ji} v_j,$$

show that

$$w_1 \wedge \cdots \wedge w_p = \sum_H B^H v_H$$

$$w_{p+1} \wedge \cdots \wedge w_n = \sum_K C^K v_K.$$

(b) Let $H' = \{1, \ldots, n\} - H$ (arranged in increasing order). Show that

$$v_H \wedge v_K = \begin{cases} 0 & K \neq H' \\ e_{H,H'} \, v_1 \wedge \cdots \wedge v_n & K = H', \end{cases}$$

where $e_{H,H'}$ is the sign of the permutation

$$\begin{pmatrix} 1 & . & . & . & . & . & . & . & . & . & n \\ h_1, h_2, \ldots, h_p, k_1, \ldots, k_q \end{pmatrix}.$$

(c) Prove "Laplace's expansion"

$$\det A = \sum_H e_{H,H'} \, B^H C^{H'}.$$

11. (Cartan's Lemma) Let $\phi_1, \ldots, \phi_k \in V^*$ be independent and suppose that $\psi_1, \ldots, \psi_k \in V^*$ satisfy

$$(\phi_1 \wedge \psi_1) + \cdots + (\phi_k \wedge \psi_k) = 0.$$

Then

$$\psi_i = \sum_{j=1}^{k} a_{ji} \phi_j, \quad \text{where } a_{ji} = a_{ij}.$$

12. In addition to forms, we can consider sections of bundles constructed from TM using Ω and other operations. For example, if $\xi = \pi \colon E \to B$ is a vector bundle, we can consider $\Omega^k(\xi^*)$, the bundle whose fibre at p is $\Omega^k([\pi^{-1}(p)]^*)$. Since we can regard

$$\frac{\partial}{\partial x^i} \quad \text{as an element of} \quad (M_p)^{**},$$

any section of $\Omega^n(T^*M)$ can be written locally as

$$h \frac{\partial}{\partial x^1} \wedge \cdots \wedge \frac{\partial}{\partial x^n}.$$

(a) Show that if

$$g \frac{\partial}{\partial y^1} \wedge \cdots \wedge \frac{\partial}{\partial y^n} = h \frac{\partial}{\partial x^1} \wedge \cdots \wedge \frac{\partial}{\partial x^n},$$

then

$$h = g \cdot \left[\det \left(\frac{\partial y^i}{\partial x^j} \right) \right]^{-1}.$$

This shows that sections of $\Omega^n(T^*M)$ are the geometric objects corresponding to the (even) relative scalars of weight -1 in Problem 4-10.

(b) Let $\mathcal{T}_l^{k[m]}(V)$ denote the vector space of all multilinear functions

$$\underbrace{V \times \cdots \times V}_{k \text{ times}} \times \underbrace{V^* \times \cdots \times V^*}_{l \text{ times}} \to \Omega^m(V).$$

Show that sections of $\mathcal{T}_l^{k[n]}(TM)$ correspond to (even) relative tensors of type $\binom{k}{l}$ and weight 1. (Notice that if $v_1, \ldots, v_n$ is a basis for V, then elements of $\Omega^n(V)$ can be represented by real numbers [times the element $v^*_1 \wedge \cdots \wedge v^*_n$].)

(c) If $\mathcal{T}_{l[m]}^k(V)$ is defined similarly, except that $\Omega^m(V)$ is replaced by $\Omega^m(V^*)$, show that sections of $\mathcal{T}_{l[n]}^k(TM)$ correspond to (even) relative tensors of type $\binom{k}{l}$ and weight -1.

(d) Show that the covariant relative tensor of type $\binom{0}{n}$ and weight 1 defined in Problem 4-10, with components $\varepsilon^{i_1 \cdots i_n}$, corresponds to the map

$$\underbrace{V^* \times \cdots \times V^*}_{n \text{ times}} \to \Omega^n(V)$$

given by $(\phi_1, \ldots, \phi_n) \mapsto \phi_1 \wedge \cdots \wedge \phi_n$. Interpret the relative tensor with components $\varepsilon_{i_1 \ldots i_n}$ similarly.

(e) Suppose $\Omega^{n;w}(V)$ denotes all functions $\eta \colon V \times \cdots \times V \to \mathbb{R}$ which are of the form

$$\eta(v_1, \ldots, v_n) = [\omega(v_1, \ldots, v_n)]^w \qquad w \text{ an integer}$$

for some $\omega \in \Omega^n(V)$. Let $\mathcal{T}_l^{k[n;w]}(V)$ be defined like $\mathcal{T}_l^{k[n]}$, except that $\Omega^n(V)$ is replaced by $\Omega^{n;w}(V)$. Show that sections of $\mathcal{T}_l^{k[n;w]}(TM)$ correspond to (even) relative tensors of type $\binom{k}{l}$ and weight w. Similarly for $\mathcal{T}_{l[n;w]}^k$.

(f) For those who know about tensor products $V \otimes W$ and exterior algebras $\Lambda^k(V)$, these results can all be restated. We can identify $\mathcal{T}_l^k(V)$ with

$$\bigotimes^k V^* \otimes \bigotimes^l V = \underbrace{V^* \otimes \cdots \otimes V^*}_{k \text{ times}} \otimes \underbrace{V \otimes \cdots \otimes V}_{l \text{ times}}.$$

Since $\Omega^m(V) \approx \Lambda^m(V^*) \approx [\Lambda^m(V)]^*$, we can identify

$$\mathcal{T}_l^{k[m]}(V) \quad \text{with} \quad \bigotimes^k V^* \otimes \bigotimes^l V \otimes \Lambda^m(V)$$

$$\mathcal{T}_{l[m]}^k(V) \quad \text{with} \quad \bigotimes^k V^* \otimes \bigotimes^l V \otimes \Lambda^m(V^*).$$

Consider, more generally,

$$\mathcal{T}_l^{k[m;w]}(V) = \bigotimes{}^k V^* \otimes \bigotimes{}^l V \otimes \bigotimes{}^w \Lambda^m(V)$$

$$\mathcal{T}_{l[m;w]}^k(V) = \bigotimes{}^k V^* \otimes \bigotimes{}^l V \otimes \bigotimes{}^w \Lambda^m(V^*).$$

Noting that $\Lambda^n(V) \otimes \cdots \otimes \Lambda^n(V)$ is always 1-dimensional, show that sections of $\mathcal{T}_l^{k[n;w]}(TM)$ and $\mathcal{T}_{l[n;w]}^k(TM)$ correspond to (even) relative tensors of type $\binom{k}{l}$ and weight w and $-w$, respectively.

13. (a) If V has dimension n and $A: V \to V$ is a linear transformation, then the map $A^*: \Omega^n(V) \to \Omega^n(V)$ must be multiplication by some constant c. Show that $c = \det A$. (This may be used as a definition of $\det A$.)
(b) Conclude that $\det AB = (\det A)(\det B)$.

14. Recall that the characteristic polynomial of $A: V \to V$ is

$$\begin{aligned}
\chi(\lambda) &= \det(\lambda I - A) \\
&= \lambda^n - (\text{trace } A)\lambda^{n-1} + \cdots + (-1)^n \det A \\
&= \lambda^n - c_1 \lambda^{n-1} + c_2 \lambda^{n-2} + \cdots + (-1)^n c_n.
\end{aligned}$$

(a) Show that $c_k = \text{trace of } A^*: \Omega^k(V) \to \Omega^k(V)$.
(b) Conclude that $c_k(AB) = c_k(BA)$.
(c) Let $\delta_{i_1 \ldots i_k}^{j_1 \ldots j_k}$ be as defined in Problem 4-5(xiii). If $A: V \to V$ has a matrix (a_i^j) (with respect to some basis), show that

$$c_k(A) = \frac{1}{k!} \sum_{\substack{i_1,\ldots,i_k \\ j_1,\ldots,j_k}} a_{i_1}^{j_1} a_{i_2}^{j_2} \cdots a_{i_k}^{j_k} \delta_{j_1 \ldots j_k}^{i_1 \ldots i_k}.$$

Thus, if δ is as defined on page 130, and A is a tensor of type $\binom{1}{1}$, then the function $p \mapsto c_k(A(p))$ can be defined as a $(2k)$-fold contraction of

$$\underbrace{A \otimes \cdots \otimes A}_{k \text{ times}} \otimes \delta.$$

15. Let $P(X_{ij})$ be a polynomial in n^2 variables. For every $n \times n$ matrix $A = (a_{ij})$ we then have a number $P(a_{ij})$. Call P **invariant** if $P(A) = P(BAB^{-1})$ for all A and all invertible B. This problem outlines a proof that any invariant P is a polynomial in the polynomials $c_1, \ldots, c_n$ defined in Problem 14. We will need the algebraic result that any symmetric polynomial $Q(y_1, \ldots, y_n)$ in the n variables $y_1, \ldots, y_n$ can be written as a polynomial in $\sigma_1, \ldots, \sigma_n$, where σ_i is the ith elementary symmetric polynomial of $y_1, \ldots, y_n$. Recall that the σ_i can be defined by the equation

$$\prod_{i=1}^{n}(y - y_i) = y^n - \sigma_1 y^{n-1} + \cdots + (-1)^n \sigma_n.$$

Thus, they are the coefficients, up to sign, of the polynomial with roots $y_1, \ldots, y_n$. Since the eigenvalues $\lambda_1, \ldots, \lambda_n$ of a matrix A are, by definition, the roots of the polynomial $\chi(\lambda)$, it follows that

$$c_i(A) = \sigma_i(\lambda_1, \ldots, \lambda_n).$$

We will first consider matrices A over the complex numbers $\mathbb{C}$ (the coefficients of P may also be complex).

(a) Define $Q(y_1, \ldots, y_n)$ to be $P(A)$ where A is the diagonal matrix

$$\begin{pmatrix} y_1 & & 0 \\ & \ddots & \\ 0 & & y_n \end{pmatrix}.$$

Then there is a polynomial R such that

$$Q(y_1, \ldots, y_n) = R(\sigma_1(y_1, \ldots, y_n), \ldots, \sigma_n(y_1, \ldots, y_n)).$$

The polynomial R has real coefficients if P does.

(b) $P(A) = R(c_1(A), \ldots, c_n(A))$ for all diagonalizable A.

(c) The **discriminant** $D(A)$ is defined as $\prod_{i \neq j}(\lambda_i - \lambda_j)^2$, where λ_i are the eigenvalues of A. Show that $D(A)$ can be written as a polynomial in the entries of A.

(d) Show that $P(A) = R(c_1(A), \ldots, c_n(A))$ whenever $D(A) \neq 0$. Conclude, by continuity, that the equation holds for all matrices A over $\mathbb{C}$. (This last conclusion follows even if $\mathbb{C}$ is replaced by some other field, since the set where $D \neq 0$ is Zariski-dense; this is "the principal of irrelevance of algebraic inequalities", compare pg. V.375.)

Now suppose that the coefficients of P are real and that $P(A) = P(BAB^{-1})$ for all real A and real invertible B.

(e) The same equation holds for complex A and complex invertible B. (Regard the equation as n^2 polynomial equations in the a_{ij} and b_{ij}.)

16. (a) Let $v_1, \ldots, v_n$ be a basis for V, and let $w_1, \ldots, w_k \in V$ be given by

$$w_i = \sum_{j=1}^{n} \alpha_{ji} v_j.$$

For $\omega \in \Omega^k(V)$ show that

$$\omega(w_1, \ldots, w_k) = \sum_{I=i_1 < \cdots < i_k} \alpha_I \, \omega(v_{i_1}, \ldots, v_{i_k}),$$

where α_I is the determinant of the $k \times k$ submatrix of (α_{ij}) obtained by selecting rows $i_1, \ldots, i_k$.

(b) Generalize Theorem 7 and Corollary 8 to k-forms.

(c) Check directly from (b) that the definition of d does not depend on the coordinate system.

17. Show that $d(\sum_{i<j} \alpha_{ij} \, dx^i \wedge dx^j) = 0$ if and only if

$$\frac{\partial \alpha_{ij}}{\partial x^k} - \frac{\partial \alpha_{ik}}{\partial x^j} + \frac{\partial \alpha_{jk}}{\partial x^i} = 0 \quad \text{for all } i < j < k.$$

18. In Problem 5-14 we defined $L_X A$ for any tensor field A.

(a) Show that if ω is a k-form, then so is $L_X \omega$.

(b) Show that

$$L_X(\omega_1 \wedge \omega_2) = L_X \omega_1 \wedge \omega_2 + \omega_1 \wedge L_X \omega_2.$$

(c) Using 5-14(e), show that

$$X(\omega(X_1, \ldots, X_k)) = L_X(\omega(X_1, \ldots, X_k))$$
$$= L_X \omega(X_1, \ldots, X_k)$$
$$+ \sum_{i=1}^{k} (-1)^{i+1} \omega([X, X_i], X_1, \ldots, \widehat{X_i}, \ldots, X_k).$$

(d) Deduce the following two expressions:

$$d\omega(X_1, \ldots, X_{k+1})$$
$$= \sum_{i=1}^{k+1} (-1)^{i+1} L_{X_i} \omega(X_1, \ldots, \widehat{X_i}, \ldots, X_{k+1})$$
$$+ \sum_{i<j} (-1)^{i+j+1} \omega([X_i, X_j], X_1, \ldots, \widehat{X_i}, \ldots, \widehat{X_j}, \ldots, X_{k+1})$$

$$dω(X_1, \ldots, X_{k+1})$$
$$= \frac{1}{2} \sum_{i=1}^{k+1} (-1)^{i+1} \{ X_i(ω(X_1, \ldots, \widehat{X_i}, \ldots, X_{k+1}))$$
$$+ L_{X_i} ω(X_1, \ldots, \widehat{X_i}, \ldots, X_{k+1}) \}$$

(e) Show that
$$X \lrcorner \, dω = L_X ω - d(X \lrcorner \, ω),$$

i.e.,
$$dω(X_1, \ldots, X_{k+1}) = (L_{X_1} ω)(X_2, \ldots, X_{k+1}) - d(X_1 \lrcorner \, ω)(X_2, \ldots, X_{k+1}).$$

(This may be used to give an inductive definition of d.)
(f) Using (e), show that $d(L_X ω) = L_X(dω)$.

19. Let a_{ij} be n^2 functions on $\mathbb{R}^n$ with $a_{ij} = a_{ji}$. Show that in order for there to be functions $u_1, \ldots, u_n$ in a neighborhood of any point in $\mathbb{R}^n$ with

$$a_{ij} = \frac{1}{2} \left(\frac{\partial u_i}{\partial x^j} + \frac{\partial u_j}{\partial x^i} \right)$$

it is necessary and sufficient that

$$\frac{\partial^2 a_{ij}}{\partial x^k \partial x^l} - \frac{\partial^2 a_{ik}}{\partial x^j \partial x^l} = \frac{\partial^2 a_{lj}}{\partial x^k \partial x^i} - \frac{\partial^2 a_{lk}}{\partial x^j \partial x^i} \qquad \text{for all } i, j, k, l.$$

Hint: First set up partial differential equations for the functions $f_{jk} = \partial u_j / \partial x^k - \partial u_k / \partial x^j$, and use Theorem 6-1.

20. Compute that
$$\text{``}dθ\text{''} = \frac{x \, dy - y \, dx}{x^2 + y^2}.$$

(At most places $θ = \arctan y/x \; [+ \text{a constant}]$.)

21. (a) If $ω$ is a 1-form $f \, dx$ on $[0, 1]$ with $f(0) = f(1)$, show that there is a unique number $λ$ such that $ω - λ \, dx = dg$ for some function g with $g(0) = g(1)$. *Hint:* Integrate the equation $ω - λ \, dx = dg$ on $[0, 1]$ to find $λ$.
(b) Let $i: S^1 → \mathbb{R}^2 - \{0\}$ be the inclusion, and let $σ' = i^*(dθ)$. If $c: [0, 1] → S^1$ is

$$c(x) = (\cos 2π x, \sin 2π x),$$

show that
$$c^*(σ') = 2π \, dx.$$

(c) If $ω$ is a closed 1-form on S^1 show that there is a unique number $λ$ such that $ω - λσ'$ is exact.

22. (a) Show that every $\omega \in \Omega^k(V_1 \oplus V_2)$ can be written as a sum of forms $\omega_1 \wedge \omega_2$ where ω_1 has degree α and ω_2 has degree $\beta = k - \alpha$ and

$$\omega_1(v_1, \ldots, v_\alpha) = 0 \quad \text{if some } v_i \in V_2$$
$$\omega_2(v_1, \ldots, v_\beta) = 0 \quad \text{if some } v_i \in V_1.$$

(b) If dim $V_2 = 1$, and $0 \neq \lambda \in V_2{}^*$, then ω can be written uniquely as $\omega_1 + (\omega_2 \wedge \lambda)$, where ω_1 is a k-form and ω_2 is a $(k-1)$-form such that

$$\omega_1(v_1, \ldots, v_k) = 0 \quad \text{if some } v_i \in V_2$$
$$\omega_2(v_1, \ldots, v_{k-1}) = 0 \quad \text{if some } v_i \in V_2.$$

23. Let $U \subset \mathbb{R}^n$ be an open set star-shaped with respect to 0, and define $H: U \times [0,1] \to U$ by $H(p,t) = tp$. If

$$\omega = \sum_{i_1 < \cdots < i_k} \omega_{i_1 \ldots i_k} \, dx^{i_1} \wedge \cdots \wedge dx^{i_k}$$

on U, show that

$$I(H^*\omega)$$

$$= \sum_{i_1 < \cdots < i_k} \sum_{\alpha=1}^{k} (-1)^{\alpha-1} \left(\int_0^1 t^{k-1} \omega_{i_1 \ldots i_k}(tx) \, dt \right) x^{i_\alpha} \, dx^{i_1} \wedge \cdots \wedge \widehat{dx^{i_\alpha}} \wedge \cdots \wedge dx^{i_k}.$$

24. (a) Let $U \subset \mathbb{R}^2$ be a bounded open set such that $\mathbb{R}^2 - U$ is connected. Show that U is diffeomorphic to $\mathbb{R}^2$, and hence smoothly contractible to a point. (The converse is proved in Problem 8-9.) *Hint:* Obtain U as an increasing union of sets, the kth set being a finite union of squares containing the set of points in U whose distance from boundary U is $\leq 1/k$.

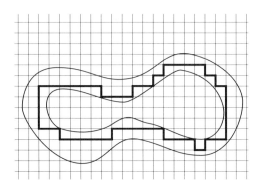

(b) Find a bounded open set $U \subset \mathbb{R}^3$ such that $\mathbb{R}^3 - U$ is connected, but U is not contractible to a point.

25. Let $U \subset \mathbb{R}^n$ be an open set star-shaped with respect to 0. Is U homeomorphic to $\mathbb{R}^n$? (It would certainly appear so, but the "obvious" proof does not work, since the length of rays from 0 to the boundary of the set could vary discontinuously.)

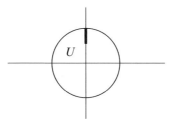

26. Let $\langle \ , \ \rangle$ be the usual inner product on $\mathbb{R}^n$,

$$\langle a, b \rangle = \sum_{i=1}^{n} a^i b^i.$$

(a) If $v_1, \ldots, v_{n-1} \in \mathbb{R}^n$, show that there is a unique vector $v_1 \times \cdots \times v_{n-1} \in \mathbb{R}^n$ with

$$\langle v_1 \times \cdots \times v_{n-1}, w \rangle = \det \begin{pmatrix} w \\ v_1 \\ \vdots \\ v_{n-1} \end{pmatrix} \quad \text{for all } w \in \mathbb{R}^n.$$

(b) Show that $\times \cdots \times \in \Omega^{n-1}(\mathbb{R}^n)$, and express it in terms of the e^*_i, using the expansion of a matrix by minors.

(c) For $\mathbb{R}^3$ show that

$$v \times w = (v^2 w^3 - v^3 w^2, \ v^3 w^1 - v^1 w^3, \ v^1 w^2 - v^2 w^1).$$

(First find all $e_i \times e_j$.)

27. (a) If $f \colon \mathbb{R}^n \to \mathbb{R}$, define a vector field grad f, the **gradient** of f, on $\mathbb{R}^n$ by

$$\text{grad } f = \sum_{i=1}^{n} \frac{\partial f}{\partial x^i} \cdot \frac{\partial}{\partial x^i} = \sum_{i=1}^{n} D_i f \cdot \frac{\partial}{\partial x^i}.$$

Introducing the formal symbolism

$$\nabla = \sum_{i=1}^{n} D_i \frac{\partial}{\partial x^i},$$

we can write grad $f = \nabla f$. If $(\text{grad } f)(p) = w_p$, show that

$$D_v f(p) = \langle v, w \rangle,$$

where $D_v f(p)$ denotes the directional derivative in the direction v at p (or simply $v_p(f)$, if we regard $v_p \in \mathbb{R}^n{}_p$). Conclude that $\nabla f(p)$ is the direction in which f is changing fastest at p.

(b) If $X = \sum_{i=1}^n a^i \partial/\partial x^i$ is a vector field on $\mathbb{R}^n$, we define the **divergence** of X as

$$\text{div } X = \sum_{i=1}^n \frac{\partial a^i}{\partial x^i}.$$

(Symbolically, we can write $\text{div } X = \langle \nabla, X \rangle$.) We also define, for $n = 3$,

$$\text{curl } X \ (= \nabla \times X)$$
$$= \left(\frac{\partial a^3}{\partial x^2} - \frac{\partial a^2}{\partial x^3} \right) \frac{\partial}{\partial x^1} + \left(\frac{\partial a^1}{\partial x^3} - \frac{\partial a^3}{\partial x^1} \right) \frac{\partial}{\partial x^2} + \left(\frac{\partial a^2}{\partial x^1} - \frac{\partial a^1}{\partial x^2} \right) \frac{\partial}{\partial x^3}.$$

Define forms

$$\omega_X = a^1 \, dx + a^2 \, dy + a^3 \, dz$$
$$\eta_X = a^1 \, dy \wedge dz + a^2 \, dz \wedge dx + a^3 \, dx \wedge dy.$$

Show that

$$df = \omega_{\text{grad } f}$$
$$d(\omega_X) = \eta_{\text{curl } X}$$
$$d(\eta_X) = (\text{div } X) \, dx \wedge dy \wedge dz.$$

(c) Conclude that

$$\text{curl grad } f = 0$$
$$\text{div curl } X = 0.$$

(d) If X is a vector field on a star-shaped open set $U \subset \mathbb{R}^n$ and $\text{curl } X = 0$, then $X = \text{grad } f$ for some function $f : U \to \mathbb{R}$. Similarly, if $\text{div } X = 0$, then $X = \text{curl } Y$ for some vector field Y on U.

CHAPTER 8

INTEGRATION

The basic concept of this chapter generalizes line and surface integrals, which first arose from very physical considerations. Suppose, for example, that $c \colon [0, 1] \to \mathbb{R}^2$ is a curve and $\omega = f\,dx + g\,dy$ is a 1-form on $\mathbb{R}^2$ (where $f, g \colon \mathbb{R}^2 \to \mathbb{R}$, and x and y denote the coordinate functions on $\mathbb{R}^2$). If we choose a partition $0 = t_0 < \cdots < t_n = 1$ of $[0, 1]$, then we can divide the curve c into n pieces, the ith piece going from $c(t_{i-1})$ to $c(t_i)$. When the differences $t_i - t_{i-1}$ are small, each such piece is approximately a straight segment, with

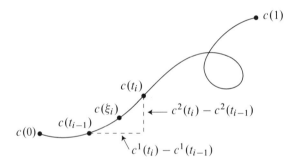

horizontal projection $c^1(t_i) - c^1(t_{i-1})$ and vertical projection $c^2(t_i) - c^2(t_{i-1})$. We can choose points $c(\xi_i)$ on each piece by choosing points $\xi_i \in [t_{i-1}, t_i]$. For each partition P and each such choice $\xi = (\xi_1, \ldots, \xi_n)$, consider the sum

$$S(P, \xi) = \sum_{i=1}^{n} f(c(\xi_i))[c^1(t_i) - c^1(t_{i-1})] + g(c(\xi_i))[c^2(t_i) - c^2(t_{i-1})].$$

If these sums approach a limit as the "mesh" $\|P\|$ of P approaches 0, that is, as the maximum of $t_i - t_{i-1}$ approaches 0, then the limit is denoted by

$$\int_c f\,dx + g\,dy.$$

(This is a complicated limit. To be precise, if $\|P\| = \max_i \{t_i - t_{i-1}\}$, then the equation

$$\lim_{\|P\| \to 0} S(P, \xi) = \int_c f\,dx + g\,dy$$

239

means: for all $\varepsilon > 0$, there is a $\delta > 0$ such that for all partitions P with $\|P\| < \delta$, we have

$$\left| S(P,\xi) - \int_c f\,dx + g\,dy \right| < \varepsilon$$

for all choices ξ for P.)

The limit which we have just defined is called a "line integral"; it has a natural physical interpretation. If we consider a "force field" on $\mathbb{R}^2$, described by the vector field

$$f\frac{\partial}{\partial x} + g\frac{\partial}{\partial y}$$

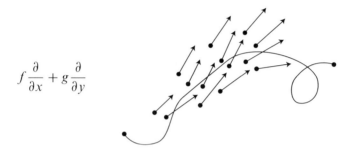

then $S(P,\xi)$ is the "work" involved in moving a unit mass along the curve c in the case where c is actually a straight line between t_{i-1} and t_i and f and g are constant along these straight line segments; the limit is the natural definition of the work done in the general case. (In classical terminology, the differential $f\,dx + g\,dy$ would be described as the work done by the force field on an "infinitely small" displacement with components dx, dy; the integral is the "sum" of these infinitely small displacements.)

Before worrying about how to compute this limit, consider the special case where

$$c(t) = (t, y_0).$$

In this case, $c^1(t_i) - c^1(t_{i-1}) = t_i - t_{i-1}$, while $c^2(t_i) - c^2(t_{i-1}) = 0$, so

$$S(P,\xi) = \sum_{i=1}^n f(\xi_i, y_0)(t_i - t_{i-1}).$$

These sums approach

$$\int_c f\,dx + g\,dy = \int_0^1 f(x, y_0)\,dx.$$

On the other hand, if

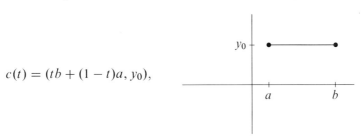

$$c(t) = (tb + (1 - t)a, y_0),$$

then $c^1(t_i) - c^1(t_{i-1}) = (b - a)(t_i - t_{i-1})$, so

$$S(P, \xi) = (b - a) \cdot \sum_{i=1}^n f(\xi_i b + (1 - \xi_i)a, y_0)(t_i - t_{i-1}).$$

These sums approach

$$(b - a) \int_0^1 f(xb + (1 - x)a, y_0)\,dx = \int_a^b f(x, y_0)\,dx.$$

In general, for any curve c, we have, by the mean value theorem,

$$
\begin{aligned}
c^1(t_i) - c^1(t_{i-1}) &= c^{1\prime}(\alpha_i)(t_i - t_{i-1}) & \alpha_i \in [t_{i-1}, t_i] \\
c^2(t_i) - c^2(t_{i-1}) &= c^{2\prime}(\beta_i)(t_i - t_{i-1}) & \beta_i \in [t_{i-1}, t_i].
\end{aligned}
$$

So

$$S(P, \xi) = \sum_{i=1}^n \left\{ f(c(\xi_i))c^{1\prime}(\alpha_i) + g(c(\xi_i))c^{2\prime}(\beta_i) \right\} (t_i - t_{i-1}).$$

A somewhat messy argument (Problem 1) shows that these sums approach what it looks like they should approach, namely

$$\int_0^1 [f(c(t))c^{1\prime}(t) + g(c(t))c^{2\prime}(t)]\,dt.$$

Physicists' notation (or abuse thereof) makes it easy to remember this result. The components c^1, c^2 of c are denoted simply by x and y [i.e., x denotes

$x \circ c$ and y denotes $y \circ c$; this is indicated classically by saying "let $x = x(t)$, $y = y(t)$"]. The above integral is then written

$$\int_c f \, dx + g \, dy = \int_0^1 \left[f(x, y) \frac{dx}{dt} + g(x, y) \frac{dy}{dt} \right] dt.$$

In preference to this physical interpretation of "line integrals", we can introduce a more geometrical interpretation. Recall that $dc/dt(\xi_i)$ denotes the

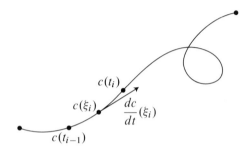

tangent vector of c at time ξ_i. Then the sums

$$(*) \qquad \sum_{i=1}^{n} \omega(c(\xi_i)) \left(\frac{dc}{dt}(\xi_i) \right) \cdot (t_i - t_{i-1})$$

$$= \sum_{i=1}^{n} [f(c(\xi_i))c^{1\prime}(\xi_i) + g(c(\xi_i))c^{2\prime}(\xi_i)] \cdot (t_i - t_{i-1})$$

clearly also approach

$$\int_0^1 [f(c(t))c^{1\prime}(t) + g(c(t))c^{2\prime}(t)] \, dt.$$

Consider the special case where c goes with constant velocity on each (t_{i-1}, t_i).

If we choose any $\xi_i \in (t_{i-1}, t_i)$, then

$$\text{length of } \frac{dc}{dt}(\xi_i) = \text{the constant speed on } (t_{i-1}, t_i)$$

$$= \frac{\text{length of the segment from } c(t_{i-1}) \text{ to } c(t_i)}{t_i - t_{i-1}},$$

so

$$\left[\text{length of } \frac{dc}{dt}(\xi_i)\right] \cdot (t_i - t_{i-1}) = \text{length of segment from } c(t_{i-1}) \text{ to } c(t_i).$$

In this case,

$$\sum_{i=1}^{n} \left[\text{length of } \frac{dc}{dt}(\xi_i)\right] \cdot (t_i - t_{i-1})$$

is the length of c, and the limit of such sums, for a general c, can be used as a definition of the length of c. The line integral

$$\int_c \omega = \text{limit of the sums } (*)$$

can be thought of as the "length" of c, when our ruler is changing continuously in a way specified by ω: Notice that the restriction of $\omega(c(t))$ to the 1-dimensional subspace of $\mathbb{R}^2_{c(t)}$ spanned by dc/dt is a constant times "signed length". The natural way to specify a continuously changing length along c is to specify a length on its tangent vectors; this is the modern counterpart of the classical conception, whereby the curve c is divided into infinitely small parts, the infinitely small piece at $c(t)$, with components dx, dy, having length $f(c(t)) \, dx + g(c(t)) \, dy$.

Before pushing this geometrical interpretation too far, we should note that there is no 1-form ω on $\mathbb{R}^2$ such that

$$\int_c \omega = \text{length of } c \qquad \text{for all curves } c.$$

It is true that for a given one-one curve c we can produce a form ω which works for c; we choose $\omega(c(t)) \in \Omega^1(\mathbb{R}^2_{c(t)})$ so that

$$\omega(c(t)) \left(\frac{dc}{dt}\right) = 1,$$

kernel $\omega(c(t))$

(choosing the kernel of ω arbitrarily), and then extend ω to $\mathbb{R}^2$. But if c is not one-one this may be impossible; for example, in the situation shown below,

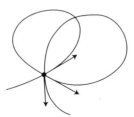

there is no element of $\Omega^1(\mathbb{R}^2{}_{c(t)})$ which has the value 1 on all three vectors. In general, given any ω on $\mathbb{R}^2$ which is everywhere non-zero, the subspaces $\Delta_p = \ker \omega(p)$ form a 1-dimensional distribution on $\mathbb{R}^2$; any curve contained in an integral submanifold of Δ will have "length" 0. Later we will see a way of circumventing this difficulty, if we are interested in obtaining the ordinary length of a curve. For the present, we note that the sums $(*)$, used to define this generalized "length", make sense even if c is a curve in a manifold M (where there is no notion of "length"), and ω is a 1-form on M, so we can define $\int_c \omega$ as the limit of these sums.

One property of line integrals should be mentioned now, because it is obvious with our original definition and merely true for our new definition. If $p \colon [0,1] \to [0,1]$ is a one-one increasing function from $[0,1]$ *onto* $[0,1]$, then the curve $c \circ p \colon [0,1] \to M$ is called a **reparameterization** of c—it has exactly the same image as c, but transverses it at a different rate. Every sum $S(P,\xi)$ for c is clearly equal to a sum $S(P',\xi')$ for $c \circ p$, and conversely, so it is clear from our first definition that for a curve $c \colon [0,1] \to \mathbb{R}^2$ we have

$$\int_c \omega = \int_{c \circ p} \omega$$

("the integral of ω over c is independent of the parameterization"). This is no longer so clear when we consider the sums $(*)$ for a curve $c \colon [0,1] \to M$, nor is it clear even for a curve $c \colon [0,1] \to \mathbb{R}^2$, but in this case we can proceed right to the integral these sums approach, namely

$$\int_0^1 [f(c(t))c^{1\prime}(t) + g(c(t))c^{2\prime}(t)]\,dt.$$

The result then follows from a calculation: the substitution $t = p(u)$ gives

$$\int_0^1 [f(c(t))c^{1\prime}(t) + g(c(t))c^{2\prime}(t)]\, dt$$

$$= \int_{p^{-1}(0)}^{p^{-1}(1)} [f(c(p(u)))c^{1\prime}(p(u)) + g(c(p(u)))c^{2\prime}(p(u))]p'(u)\, du$$

$$= \int_0^1 [f(c \circ p(u))(c \circ p)^{1\prime}(u) + g(c \circ p(u))(c \circ p)^{2\prime}(u)]\, du.$$

For a curve in $\mathbb{R}^n$, and a 1-form $\omega = \sum_{i=1}^n \omega_i\, dx^i$, there is a similar calculation; for a general manifold M, we can introduce a coordinate system for our calculations if $c([0, 1])$ lies in one coordinate system, or break c up into several pieces otherwise. We are being a bit sloppy about all this because we are about to introduce yet a third definition, which will eventually become our formal choice. Consider once again the case of a 1-form on $\mathbb{R}^2$, where

$$\int_c \omega = \int_0^1 [f(c(t))c^{1\prime}(t) + g(c(t))c^{2\prime}(t)]\, dt.$$

Notice that if t is the standard coordinate system on $\mathbb{R}$, then for the map $c \colon [0, 1] \to \mathbb{R}^2$ we have

$$c^*(f\, dx + g\, dy) = (f \circ c)c^*(dx) + (g \circ c)c^*(dy)$$
$$= (f \circ c)\, d(x \circ c) + (g \circ c)\, d(y \circ c)$$
$$= (f \circ c)c^{1\prime}\, dt + (g \circ c)c^{2\prime}\, dt,$$

so that formally we just integrate $c^*(f\, dx + g\, dy)$; to be precise, we write $c^*(f\, dx + g\, dy) = h\, dt$ (in the unique possible way), and take the integral of h on $[0, 1]$.

Everything we have said for curves $c \colon [0, 1] \to \mathbb{R}^n$ could be generalized to functions $c \colon [0, 1]^2 \to \mathbb{R}^n$. If x and y are the coordinate functions on $\mathbb{R}^2$, let

$$\frac{\partial c}{\partial x} = c_* \left(\frac{\partial}{\partial x} \right)$$

$$\frac{\partial c}{\partial y} = c_* \left(\frac{\partial}{\partial y} \right).$$

For a pair of partitions $s_0 < \cdots < s_m$ and $t_0 < \cdots < t_n$ of $[0, 1]$, if we choose

$\xi_{ij} \in [s_{i-1}, s_i] \times [t_{j-1}, t_j]$ and ω is a 2-form on $\mathbb{R}^n$, then

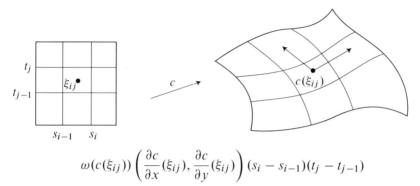

$$\omega(c(\xi_{ij})) \left(\frac{\partial c}{\partial x}(\xi_{ij}), \frac{\partial c}{\partial y}(\xi_{ij}) \right) (s_i - s_{i-1})(t_j - t_{j-1})$$

is a "generalized area" of the parallelogram spanned by

$$\frac{\partial c}{\partial x}(\xi_{ij}), \quad \frac{\partial c}{\partial y}(\xi_{ij}).$$

The limit of sums of these terms can be thought of as a "generalized area" of c. To make a long story short, we now proceed with the formal definitions.

A C^∞ function $c \colon [0,1]^k \to M$ is called a **singular k-cube** in M (the word "singular" indicates that c is not necessarily one-one). We will let $[0,1]^0 = \mathbb{R}^0 = 0 \in \mathbb{R}$, so that a singular 0-cube c is determined by the one point $c(0) \in M$. The inclusion map of $[0,1]^k$ in $\mathbb{R}^k$ will be denoted by $I^k \colon [0,1]^k \to \mathbb{R}^k$; it is called the **standard k-cube**.

If ω is a k-form on $[0,1]^k$, and $x^1, \ldots, x^k$ are the coordinate functions, then ω can be written uniquely as

$$\omega = f \, dx^1 \wedge \cdots \wedge dx^k.$$

We define

$$\int_{[0,1]^k} \omega \quad \text{to be} \quad \int_{[0,1]^k} f \quad \left(\begin{array}{l} = \int_{[0,1]^k} f(x^1, \ldots, x^k) \, dx^1 \ldots dx^k \\ \text{in classical notation, which modern} \\ \text{notation attempts to mimic as far} \\ \text{as logic permits} \end{array} \right).$$

If ω is a k-form on M, and c is a singular k-cube in M, we define

$$\int_c \omega = \int_{[0,1]^k} c^*\omega,$$

where the right hand side has just been defined. For $k = 0$, we have a special definition: a 0-form is a function f, and for a singular 0-cube c we define

$$\int_c f = f(c(0)).$$

1. PROPOSITION. Let $c \colon [0,1]^n \to \mathbb{R}^n$ be a one-one singular n-cube with $\det c' \geq 0$ on $[0,1]^n$. Let ω be the n-form

$$\omega = f\, dx^1 \wedge \cdots \wedge dx^n.$$

Then

$$\int_c \omega = \int_{c([0,1]^n)} f.$$

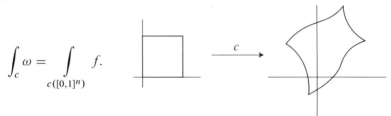

PROOF. By definition,

$$\int_c \omega = \int_{[0,1]^n} c^*(\omega)$$

$$= \int_{[0,1]^n} (f \circ c)(\det c')\, dx^1 \wedge \cdots \wedge dx^n \quad \text{by Theorem 7-7}$$

$$= \int_{[0,1]^n} (f \circ c)|\det c'|\, dx^1 \wedge \cdots \wedge dx^n \quad \text{by assumption}$$

$$= \int_{c([0,1]^n)} f \qquad \text{by the change of variable formula.} \ \clubsuit$$

2. COROLLARY. Let $p \colon [0,1]^k \to [0,1]^k$ be one-one onto with $\det p' \geq 0$, let c be a singular k-cube in M and let ω be a k-form on M. Then

$$\int_c \omega = \int_{c \circ p} \omega.$$

PROOF. We have

$$\int_{c \circ p} \omega = \int_{[0,1]^k} (c \circ p)^*\omega = \int_{[0,1]^k} p^*(c^*\omega)$$

$$= \int_{[0,1]^k} c^*(\omega) \quad \text{by the Proposition, since } p \text{ is onto}$$

$$= \int_c \omega. \ \clubsuit$$

The map $c \circ p \colon [0,1]^k \to M$ is called a **reparameterization** of c if $p \colon [0,1]^k \to [0,1]^k$ is a C^∞ one-one onto map with $\det p' \neq 0$ everywhere (so that p^{-1} is also C^∞); it is called **orientation preserving** or **orientation reversing** depending on whether $\det p' > 0$ or $\det p' < 0$ everywhere. The corollary thus shows independence of parameterization, provided it is orientation preserving; an orientation reversing reparameterization clearly changes the sign of the integral. Notice that there would be no such result if we tried to define the integral over c of a C^∞ function $f \colon M \to \mathbb{R}$ by the formula

$$\int_{[0,1]^k} f \circ c.$$

For example, if $c \colon [0,1] \to M$ then

$$\int_0^1 f(c(t))\, dt \quad \text{is generally} \quad \neq \int_0^1 f(c(p(t)))\, dt.$$

From a formal point of view, differential forms are the things we integrate because they transform correctly (i.e., in accordance with Theorem 7-7, so that the change of variable formula will pop up); functions on a manifold cannot be integrated (we can integrate a function f on the manifold $\mathbb{R}^k$ only because it gives us a form $f\, dx^1 \wedge \cdots \wedge dx^k$).

Our definition of the integral of a k-form ω over a singular k-cube c can immediately be generalized. A k-**chain** is simply a formal (finite) sum of singular k-cubes multiplied by integers, e.g.,

$$1c_1 - 2c_2 + 3c_3.$$

The k-chain $1c_1 = 1 \cdot c_1$ will also be denoted simply by c_1. We add k-chains, and multiply them by integers, purely formally, e.g.,

$$2(c_1 + 3c_4) + (-2)(c_1 + c_3 + c_2) = -2c_2 - 2c_3 + 6c_4.$$

Moreover, we define the integral of ω over a k-chain $c = \sum_i a_i c_i$ in the obvious way:

$$\int_{\sum_i a_i c_i} \omega = \sum_i a_i \int_{c_i} \omega.$$

The reason for introducing k-chains is that to every k-chain c (which may be just a singular k-cube) we wish to associate a $(k-1)$-chain ∂c, which is called the **boundary** of c, and which is supposed to be the sum of the various singular

$(k-1)$-cubes around the boundary of each singular k-cube in c. In practice, it

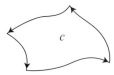

is convenient to modify this idea. The boundary of I^2, for example, will not be the sum of the four singular 1-cubes indicated below on the left, but the sum,

with the indicated coefficients, of the four singular 1-cubes shown on the right. (Notice that this will not change the integral of a 1-form over ∂I^2.) For each i with $1 \le i \le n$ we first define two singular $(n-1)$-cubes $I^n_{(i,0)}$ and $I^n_{(i,1)}$ (the $(i,0)$-face and $(i,1)$-face of I^n) as follows: If $x \in [0,1]^{n-1}$, then

$$I^n_{(i,0)}(x) = I^n(x^1,\ldots,x^{i-1},0,x^i,\ldots,x^{n-1})$$
$$= (x^1,\ldots,x^{i-1},0,x^i,\ldots,x^{n-1}),$$
$$I^n_{(i,1)}(x) = I^n(x^1,\ldots,x^{i-1},1,x^i,\ldots,x^{n-1})$$
$$= (x^1,\ldots,x^{i-1},1,x^i,\ldots,x^{n-1}).$$

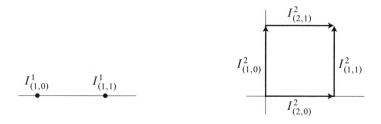

The (i, α)-face of a singular n-cube c is defined by

$$c_{(i,\alpha)} = c \circ (I^n_{(i,\alpha)}).$$

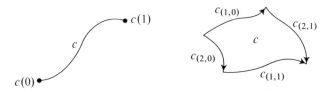

Now we define

$$\partial c = \sum_{i=1}^{n} \sum_{\alpha=0,1} (-1)^{i+\alpha} c_{(i,\alpha)}.$$

Finally, the boundary of an n-chain $\sum_i a_i c_i$ is defined by

$$\partial \left(\sum_i a_i c_i \right) = \sum_i a_i \partial(c_i).$$

These definitions all make sense only for $n \geq 1$. For the case of a 0-cube $c \colon [0, 1]^0 \to M$, which we will usually simply identify with the point $P = c(0)$, we define ∂c to be the number $1 \in \mathbb{R}$, and for a 0-chain $\sum_i a_i c_i$ we define

$$\partial \left(\sum_i a_i c_i \right) = \sum_i a_i \partial(c_i) = \sum_i a_i.$$

Notice that for a 1-cube $c \colon [0, 1] \to M$ we have

$$\partial c = c_{(1,1)} - c_{(1,0)},$$

so

$$\partial(\partial c) = 1 - 1 = 0.$$

We also have, for a singular 2-cube $c \colon [0, 1]^2 \to M$,

$$\partial c = c_{(1,1)} - c_{(2,1)} - c_{(1,0)} + c_{(2,0)},$$
$$\partial(\partial c) = (R - Q) - (R - S)$$
$$\qquad - (S - P) + (Q - P)$$
$$= 0.$$

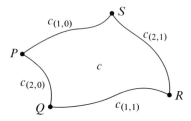

From a picture it can be checked that this also happens for a singular 3-cube, a good exercise because this involves figuring out just what the boundary of a 3-cube looks like. In general, we have:

3. PROPOSITION. If c is any n-chain in M, then $\partial(\partial c) = 0$. Briefly, $\partial^2 = 0$.

PROOF. Let $i \leq j \leq n - 1$, and consider $(I^n_{(i,\alpha)})_{(j,\beta)}$. For $x \in [0,1]^{n-2}$, we have, from the definition

$$(I^n_{(i,\alpha)})_{(j,\beta)}(x) = I^n_{(i,\alpha)}(I^{n-1}_{(j,\beta)}(x))$$
$$= I^n_{(i,\alpha)}(x^1,\ldots,x^{j-1},\beta,x^j,\ldots,x^{n-2})$$
$$= I^n(x^1,\ldots,x^{i-1},\alpha,x^i,\ldots,x^{j-1},\beta,x^j,\ldots,x^{n-2}).$$

Similarly,

$$(I^n_{(j+1,\beta)})_{(i,\alpha)} = I^n_{(j+1,\beta)}(I^{n-1}_{(i,\alpha)}(x))$$
$$= I^n_{(j+1,\beta)}(x^1,\ldots,x^{i-1},\alpha,x^i,\ldots,x^{n-2})$$
$$= I^n(x^1,\ldots,x^{i-1},\alpha,x^i,\ldots,x^{j-1},\beta,x^j,\ldots,x^{n-2}).$$

Thus $(I^n_{(i,\alpha)})_{(j,\beta)} = (I^n_{(j+1,\beta)})_{(i,\alpha)}$ for $i \leq j \leq n - 1$. It follows easily for any singular n-cube c that $(c_{(i,\alpha)})_{(j,\beta)} = (c_{(j+1,\beta)})_{(i,\alpha)}$ for $i \leq j \leq n - 1$. Now

$$\partial(\partial c) = \partial\left(\sum_{i=1}^{n} \sum_{\alpha=0,1} (-1)^{i+\alpha} c_{(i,\alpha)} \right)$$
$$= \sum_{i=1}^{n} \sum_{\alpha=0,1} \sum_{j=1}^{n-1} \sum_{\beta=0,1} (-1)^{i+\alpha+j+\beta} (c_{(i,\alpha)})_{(j,\beta)}.$$

In this sum, $(c_{(i,\alpha)})_{(j,\beta)}$ and $(c_{(j+1,\beta)})_{(i,\alpha)}$ occur with opposite signs. Therefore all terms cancel in pairs, and $\partial(\partial c) = 0$. Since the theorem is true for singular n-cubes, it is clearly also true for singular n-chains. ❖

Notice that for some n-chains c we have not only $\partial(\partial c) = 0$, but even $\partial c = 0$. For example, this is the case if $c = c_1 - c_2$, where c_1 and c_2 are two 1-cubes with $c_1(0) = c_2(0)$ and $c_1(1) = c_2(1)$. If c is just a singular 1-cube itself, then

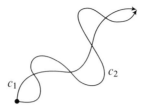

$\partial c = 0$ precisely when $c(0) = c(1)$, i.e., when c is a "closed" curve. In general,

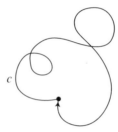

any k-chain c is called **closed** if $\partial c = 0$.

Recall that a differential form ω with $d\omega = 0$ is also called "closed"; this terminology has been purposely chosen to parallel the terminology for chains (on the other hand, a chain of the form ∂c is not described, reciprocally, by the classical term of "exact", but is simply called "a boundary"). This parallel terminology was not chosen merely because of the formal similarities between d and ∂, expressed by the relations $d^2 = 0$ and $\partial^2 = 0$. The connection between forms and chains goes much deeper than that. For example, we have seen that on $\mathbb{R}^2 - \{0\}$ there is a 1-form "$d\theta$" which is closed but not exact. There is also a 1-chain c which is closed but not a boundary, namely, a closed curve encircling

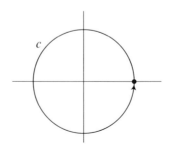

the point 0 once. Although it is intuitively clear that c is not the boundary of a 2-chain in $\mathbb{R}^2 - \{0\}$, the simplest proof uses the theorem which establishes the connection between forms, chains, d, and ∂.

4. THEOREM (STOKES' THEOREM). If ω is a $(k-1)$-form on M and c is a k-chain in M, then

$$\int_c d\omega = \int_{\partial c} \omega.$$

PROOF. Most of the proof involves the special case where ω is a $(k-1)$-form on $\mathbb{R}^k$ and $c = I^k$. In this case, ω is a sum of $(k-1)$-forms of the type

$$f \, dx^1 \wedge \cdots \wedge \widehat{dx^i} \wedge \cdots \wedge dx^k,$$

and it suffices to prove the theorem for each of these. We now compute. First, a little notation translation shows that

$$\int_{[0,1]^{k-1}} I_{(j,\alpha)}^k{}^*(f \, dx^1 \wedge \cdots \wedge \widehat{dx^i} \wedge \cdots \wedge dx^k)$$

$$= \begin{cases} 0 & \text{if } j \neq i \\ \displaystyle\int_{[0,1]^k} f(x^1,\ldots,\alpha,\ldots,x^k) \, dx^1 \ldots dx^k & \text{if } j = i. \end{cases}$$

Therefore

$$\int_{\partial I^k} f \, dx^1 \wedge \cdots \wedge \widehat{dx^i} \wedge \cdots \wedge dx^k$$

$$= \sum_{j=1}^{k} \sum_{\alpha=0,1} (-1)^{j+\alpha} \int_{[0,1]^{k-1}} I_{(j,\alpha)}^k{}^*(f \, dx^1 \wedge \ldots \wedge \widehat{dx^i} \wedge \cdots \wedge dx^k)$$

$$= (-1)^{i+1} \int_{[0,1]^k} f(x^1,\ldots,1,\ldots,x^k) \, dx^1 \ldots dx^k$$

$$+ (-1)^i \int_{[0,1]^k} f(x^1,\ldots,0,\ldots,x^k) \, dx^1 \ldots dx^k.$$

On the other hand,

$$\int_{I^k} d(f\, dx^1 \wedge \cdots \wedge \widehat{dx^i} \wedge \cdots \wedge dx^k)$$

$$= \int_{[0,1]^k} D_i f\, dx^i \wedge dx^1 \wedge \cdots \wedge \widehat{dx^i} \wedge \cdots \wedge dx^k$$

$$= (-1)^{i-1} \int_{[0,1]^k} D_i f.$$

By Fubini's theorem and the fundamental theorem of calculus we have

$$\int_{I^k} d(f\, dx^1 \wedge \cdots \wedge \widehat{dx^i} \wedge \cdots \wedge dx^k)$$

$$= (-1)^{i-1} \int_0^1 \cdots \left(\int_0^1 D_i f(x^1,\ldots,x^k)\, dx^i \right) dx^1 \ldots \widehat{dx^i} \ldots dx^k$$

$$= (-1)^{i-1} \int_0^1 \cdots \int_0^1 \Big[f(x^1,\ldots,1,\ldots,x^k)$$

$$- f(x^1,\ldots,0,\ldots,x^k) \Big]\, dx^1 \ldots \widehat{dx^i} \ldots dx^k$$

$$= (-1)^{i-1} \int_{[0,1]^k} f(x^1,\ldots,1,\ldots,x^k)\, dx^1 \ldots dx^k$$

$$+ (-1)^i \int_{[0,1]^k} f(x^1,\ldots,0,\ldots,x^k)\, dx^1 \ldots dx^k.$$

Thus

$$\int_{I^k} d\omega = \int_{\partial I^k} \omega.$$

For an arbitrary singular k-cube, chasing through the definitions shows that

$$\int_{\partial c} \omega = \int_{\partial I^k} c^* \omega.$$

Therefore

$$\int_c d\omega = \int_{I^k} c^*(d\omega) = \int_{I^k} d(c^*\omega) = \int_{\partial I^k} c^*\omega = \int_{\partial c} \omega.$$

The theorem clearly follows for k-chains also. ❖

Notice that Stokes' Theorem not only uses the fundamental theorem of calculus, but actually becomes that theorem when $c = I^1$ and $\omega = f$.

As an application of Stokes' Theorem, we show that the curve $c : [0, 1] \to \mathbb{R}^2 - \{0\}$ defined by

$$c(t) = (\cos 2\pi t, \sin 2\pi t),$$

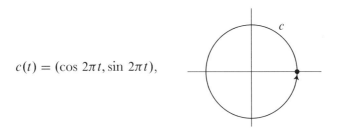

although closed, is not ∂c^2 for any 2-chain c^2. If we did have $c = \partial c^2$, then we would have

$$\int_c d\theta = \int_{\partial c^2} d\theta = \int_{c^2} d(d\theta) = \int_{c^2} 0 = 0.$$

But a straightforward computation (which will be good for the soul) shows that

$$\int_c d\theta = \int_c \frac{-y}{x^2 + y^2}\, dx + \frac{x}{x^2 + y^2}\, dy = 2\pi.$$

[There is also a non-computational argument, using the fact that "$d\theta$" really is $d\theta$ for $\theta : \mathbb{R}^2 - ([0, \infty) \times \{0\}) \to \mathbb{R}$: We have

$$\int_{c\,|\,[\varepsilon, 1 - \varepsilon]} d\theta = \theta(1 - \varepsilon) - \theta(\varepsilon),$$

and $\theta(1 - \varepsilon) - \theta(\varepsilon) \to 2\pi$ as $\varepsilon \to 0$.]

Although we used this calculation to show that c is not a boundary, we could just as well have used it to show that $\omega = $ "$d\theta$" is not exact. For, if we had $\omega = df$ for some C^∞ function $f : \mathbb{R}^2 - \{0\} \to \mathbb{R}$, then we would have

$$2\pi = \int_c \omega = \int_c df = \int_{\partial c} f = \int_0 f = 0.$$

We were previously able to give a simpler argument to show that "$d\theta$" is not exact, but Stokes' Theorem is the tool which will enable us to deal with forms on $\mathbb{R}^n - \{0\}$. For example, we will eventually obtain a 2-form ω on $\mathbb{R}^3 - \{0\}$,

$$\omega = \frac{x\, dy \wedge dz - y\, dx \wedge dz + z\, dx \wedge dy}{(x^2 + y^2 + z^2)^{3/2}}$$

which is closed but not exact. For the moment we are keeping the origin of ω a secret, but a straightforward calculation shows that $d\omega = 0$. To prove that ω is not exact we will want to integrate it over a 2-chain which "fills up" the 2-sphere $S^2 \subset \mathbb{R}^3 - \{0\}$. There are lots of ways of doing this, but they all turn out to give the same result. In fact, we first want to describe a way of integrating n-forms over n-manifolds. This is possible only when M is orientable; the reason will be clear from the next result, which is basic for our definition.

5. THEOREM. Let M be an n-manifold with an orientation μ, and let c_1, c_2: $[0, 1]^n \to M$ be two singular n-cubes which can be extended to be diffeomorphisms in a neighborhood of $[0, 1]^n$. Assume that c_1 and c_2 are both *orientation preserving* (with respect to the orientation μ on M, and the usual orientation on $\mathbb{R}^n$). If ω is an n-form on M such that

$$\text{support } \omega \subset c_1([0, 1]^n) \cap c_2([0, 1]^n),$$

then

$$\int_{c_1} \omega = \int_{c_2} \omega.$$

PROOF. We want to use Corollary 2, and write

$$\int_{c_2} \omega = \int_{c_2 \circ (c_2^{-1} \circ c_1)} \omega = \int_{c_1} \omega.$$

The only problem is that $c_2^{-1} \circ c_1$ is not defined on all of $[0, 1]^n$ (it does satisfy $\det(c_2^{-1} \circ c_1)' \geq 0$, since c_1 and c_2 are both orientation preserving). However, a glance at the proof of Corollary 2 will show that the result still follows, because of the fact that support $\omega \subset c_1([0, 1]^n) \cap c_2([0, 1]^n)$. ❖

The common number $\int_c \omega$, for singular n-cubes $c : [0, 1]^n \to M$ with support $\omega \subset c([0, 1]^n)$ and c orientation preserving, will be denoted by

$$\int_M \omega.$$

If ω is an arbitrary n-form on M, then there is a cover $\mathcal{O}$ of M by open sets U, each contained in some $c([0, 1]^n)$, where c is a singular n-cube of this sort; if Φ is a partition of unity subordinate to this cover, then

$$\int_M \phi \cdot \omega$$

is defined for each $\phi \in \Phi$. We wish to define

$$\int_M \omega = \sum_{\phi \in \Phi} \int_M \phi \cdot \omega.$$

We will adopt this definition only when ω has compact support, in which case the sum is actually finite, since support ω can intersect only finitely many of the sets $\{p : \phi(p) \neq 0\}$, which form a locally finite collection. If we have another partition of unity Ψ (subordinate to a cover $\mathcal{O}'$), then

$$\sum_{\phi \in \Phi} \int_M \phi \cdot \omega = \sum_{\phi \in \Phi} \int_M \sum_{\psi \in \Psi} \psi \cdot \phi \cdot \omega = \sum_{\phi \in \Phi} \sum_{\psi \in \Psi} \int_M \psi \cdot \phi \cdot \omega;$$

these sums are all finite, and the last sum can clearly also be written as

$$\sum_{\psi \in \Psi} \sum_{\phi \in \Phi} \int_M \phi \cdot \psi \cdot \omega = \sum_{\psi \in \Psi} \int_M \psi \cdot \omega,$$

so that our definition does not depend on the partition. (We really should denote this sum by

$$\int_{(M,\mu)} \omega;$$

for the orientation $-\mu$ of M we clearly have

$$\int_{(M,-\mu)} \omega = -\int_{(M,\mu)} \omega.$$

However, we usually omit explicit mention of μ.)

With minor modifications we can define $\int_M \omega$ even if M is an n-manifold-with-boundary. If $M \subset \mathbb{R}^n$ is an n-dimensional manifold-with-boundary and $f : M \to \mathbb{R}$ has compact support, then

$$\int_M f \, dx^1 \wedge \cdots \wedge dx^n = \int_M f,$$

where the right hand side denotes the ordinary integral. This is a simple consequence of Proposition 1. Likewise, if $f : M^n \to N^n$ is a diffeomorphism onto, and ω is an n-form with compact support on N, then

$$\int_M f^* \omega = \begin{cases} \displaystyle\int_N \omega & \text{if } f \text{ is orientation preserving} \\[4mm] \displaystyle-\int_N \omega & \text{if } f \text{ is orientation reversing.} \end{cases}$$

Although n-forms can be integrated only over orientable manifolds, there is a way of discussing integration on non-orientable manifolds. Suppose that ω is a function on M such that for each $p \in M$ we have

$$\omega(p) = |\eta_p| \qquad \text{for some} \quad \eta_p \in \Omega^n(M_p),$$

i.e., for any n vectors $v_1, \ldots, v_n \in M_p$ we have

$$\omega(p)(v_1, \ldots, v_n) = |\eta_p(v_1, \ldots, v_n)| \geq 0.$$

Such a function ω is called a **volume element**—on each vector space it determines a way of measuring n-dimensional volume (not signed volume). If (x, U) is a coordinate system, then on U we can write

$$\omega = f \, |dx^1 \wedge \cdots \wedge dx^n| \qquad \text{for } f \geq 0;$$

we call ω a C^∞ volume element if f is C^∞. One way of obtaining a volume element is to begin with an n-form η and then define $\omega(p) = |\eta(p)|$. However, not every volume element arises in this way—the form η_p may not vary continuously with p. For example, consider the Möbius strip M, imbedded in $\mathbb{R}^3$. Since M_p can be considered as a subspace of $\mathbb{R}^3{}_p$, we can define

$$\omega(p)(v_p, w_p) = \text{ area of parallelogram spanned by } v \text{ and } w.$$

It is not hard to see that ω is a volume element; locally, ω is of the form $\omega = |\eta|$ for an n-form η. But this cannot be true on all of M, since there is no n-form η on M which is everywhere non-zero.

Theorem 7-7 has an obvious modification for volume elements:

7-7′. THEOREM. If $f : M \to N$ is a C^∞ function between n-manifolds, (x, U) is a coordinate system around $p \in M$, and (y, V) a coordinate system around $q = f(p) \in N$, then for non-negative $g : V \to \mathbb{R}$ we have

$$f^*\big(g \, |dy^1 \wedge \cdots \wedge dy^n|\big) = (g \circ f) \cdot \left| \det \left(\frac{\partial(y^i \circ f)}{\partial x^j} \right) \right| \cdot |dx^1 \wedge \cdots \wedge dx^n|.$$

PROOF. Go through the proof of Theorem 7-7, putting in absolute value signs in the right place. ❖

7-8′. COROLLARY. If (x, U) and (y, V) are two coordinate systems on M and

$$g \, |dy^1 \wedge \cdots \wedge dy^n| = h \, |dx^1 \wedge \cdots \wedge dx^n| \qquad g, h \geq 0$$

then

$$h = g \cdot \left| \det \left(\frac{\partial y^i}{\partial x^j} \right) \right|.$$

[This corollary shows that volume elements are the geometric objects corresponding to the "odd scalar densities" defined in Problem 4-10.]

It is now an easy matter to integrate a volume element ω over any manifold. First we define

$$\int_{[0,1]^n} \omega = \int_{[0,1]^n} f \quad \text{for } \omega = f \, |dx^1 \wedge \cdots \wedge dx^n|, \quad f \geq 0.$$

Then for an n-chain $c \colon [0, 1]^n \to M$ we define

$$\int_c \omega = \int_{[0,1]^n} c^* \omega.$$

Theorem 7-7′ shows that Proposition 1 holds for a volume element $\omega = f \, |dx^1 \wedge \cdots \wedge dx^n|$ even if $\det c'$ is not ≥ 0. Thus Corollary 2 holds for volume elements even if $\det p'$ is not ≥ 0. From this we conclude that Theorem 5 holds for volume elements ω on any manifold M, without assuming c_1, c_2 orientation preserving (or even that M is orientable). Consequently we can define $\int_M \omega$ for any volume element ω with compact support.

Of course, when M is orientable these considerations are unnecessary. For, there is a nowhere zero n-form η on M, and consequently any volume element ω can be written

$$\omega = f |\eta|, \quad f \geq 0.$$

If we choose an orientation μ for M such that $\omega(v_1, \ldots, v_n) > 0$ for $v_1, \ldots, v_n$ positively oriented, then we can define

$$\int_M \omega = \int_{(M,\mu)} f \eta.$$

Volume elements will be important later, but for the remainder of this chapter we are concerned only with integrating forms over oriented manifolds. In fact, our main result about integrals of forms over manifolds, an analogue of Stokes' Theorem about the integral of forms over chains, does not work for volume elements.

Recall from Problem 3-16 that if M is a manifold-with-boundary, and $p \in \partial M$, then certain vectors $v \in M_p$ can be distinguished by the fact that for any coordinate system $x: U \to \mathbb{H}^n$ around p, the vector $x_*(v) \in \mathbb{H}^n{}_{f(p)}$ points "outwards". We call such vectors $v \in M_p$ "outward pointing". If M has an

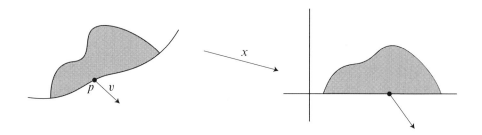

orientation μ, we define the **induced orientation** $\partial \mu$ for ∂M by the condition that $[v_1, \ldots, v_{n-1}] \in (\partial \mu)_p$ if and only if $[w, v_1, \ldots, v_{n-1}] \in \mu_p$ for every outward pointing $w \in M_p$. If μ is the usual orientation of $\mathbb{H}^n$, then for $p = (a, 0) \in \mathbb{H}^n$ we have

$$\mu_p = [(e_1)_p, \ldots, (e_n)_p] = (-1)^{n-1}[(e_n)_p, (e_1)_p, \ldots, (e_{n-1})_p]$$
$$= (-1)^n[(-e_n)_p, (e_1)_p, \ldots, (e_{n-1})_p].$$

Since $(-e_n)_p$ is an outward pointing vector, this shows that the induced orientation on $\mathbb{R}^{n-1} \times \{0\} = \partial \mathbb{H}^n$ is $(-1)^n$ times the usual one. The reason for this choice is the following. Let c be an orientation preserving singular n-cube in (M, μ) such that $\partial M \cap c([0, 1]^n) = c_{(n,0)}([0, 1]^{n-1})$. Then $c_{(n,0)}: [0, 1]^{n-1} \to$

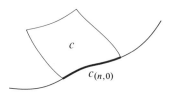

$(\partial M, \partial \mu)$ is orientation preserving for even n, and orientation reversing for odd n. If ω is an $(n-1)$-form on M whose support is contained in the interior of the image of c (this interior contains points in the image of $c_{(n,0)}$), it follows that

$$\int_{c_{(n,0)}} \omega = (-1)^n \int_{\partial M} \omega.$$

support ω

But $c_{(n,0)}$ appears with coefficient $(-1)^n$ in ∂c. So

$$(*) \qquad \int_{\partial c} \omega = \int_{(-1)^n c_{(n,0)}} \omega = (-1)^n \int_{c_{(n,0)}} \omega = \int_{\partial M} \omega.$$

If it were not for this choice of $\partial \mu$ we would have some unpleasant minus signs in the following theorem.

6. THEOREM (STOKES' THEOREM). If M is an oriented n-dimensional manifold-with-boundary, and ∂M is given the induced orientation, and ω is an $(n-1)$-form on M with compact support, then

$$\int_M d\omega = \int_{\partial M} \omega.$$

PROOF. Suppose first that there is an orientation preserving singular n-cube c in $M - \partial M$ such that support $\omega \subset$ interior of image c. Then

$$\int_M d\omega = \int_c d\omega = \int_{\partial c} \omega \qquad \text{by Theorem 4}$$
$$= 0 \qquad \text{since support } \omega \subset \text{ interior of image } c,$$

while we clearly have

$$\int_{\partial M} \omega = 0.$$

Suppose next that there is an orientation preserving singular n-cube c in M such that $\partial M \cap c([0,1]^n) = c_{(n,0)}([0,1]^{n-1})$, and support $\omega \subset$ interior of image c. Then once again

$$\int_M d\omega = \int_c d\omega = \int_{\partial c} \omega = \int_{\partial M} \omega \quad \text{by } (*).$$

In general, there is an open cover $\mathcal{O}$ of M and a partition of unity Φ subordinate to $\mathcal{O}$ such that for each $\phi \in \Phi$ the form $\phi \cdot \omega$ is one of the two sorts already considered. We have

$$0 = d(1) = d\left(\sum_{\phi \in \Phi} \phi \right) = \sum_{\phi \in \Phi} d\phi,$$

so

$$\sum_{\phi \in \Phi} d\phi \wedge \omega = 0.$$

Since ω has compact support, this is really a finite sum, and we conclude that

$$\sum_{\phi \in \Phi} \int_M d\phi \wedge \omega = 0.$$

Therefore

$$\int_M d\omega = \sum_{\phi \in \Phi} \int_M \phi \cdot d\omega = \sum_{\phi \in \Phi} \int_M d\phi \wedge \omega + \phi \cdot d\omega$$

$$= \sum_{\phi \in \Phi} \int_M d(\phi \cdot \omega) = \sum_{\phi \in \Phi} \int_{\partial M} \phi \cdot \omega = \int_{\partial M} \omega. \; \clubsuit$$

One of the simplest applications of Stokes' Theorem occurs when the oriented n-manifold (M, μ) is compact (so that every form has compact support) and $\partial M = \emptyset$. In this case, if η is any $(n-1)$-form, then

$$\int_M d\eta = \int_{\partial M} \eta = 0.$$

Therefore we can find an n-form ω on M which is *not* exact (even though it must be closed, because all $(n+1)$-forms on M are 0), simply by finding an ω with

$$\int_M \omega \neq 0.$$

Such a form ω always exists. Indeed we have seen that there is a form ω such that for $v_1, \ldots, v_n \in M_p$ we have

$(*)$ $\omega(v_1, \ldots, v_n) > 0 \quad \text{if } [v_1, \ldots, v_n] = \mu_p.$

If $c: [0,1]^n \to (M, \mu)$ is orientation preserving, then the form $c^* \omega$ on $[0,1]^n$ is clearly

$$g \, dx^1 \wedge \cdots \wedge dx^n \qquad \text{for some } g > 0 \text{ on } [0,1]^n,$$

so $\int_c \omega > 0$. It follows that $\int_M \omega > 0$. There is, moreover, no need to choose a form ω with $(*)$ holding everywhere—we can allow the $>$ sign to be replaced by $\geq$. Thus we can even obtain a non-exact n-form on M which has support contained in a coordinate neighborhood.

This seemingly minor result already proves a theorem: a compact oriented manifold is not smoothly contractible to a point. As we have already emphasized, it is the "shape" of M, rather than its "size", which determines whether or not every closed form on M is exact. Roughly speaking, we can obtain more information about the shape of M by analyzing more closely the extent to which closed forms are not necessarily exact. In particular, we would now like to ask just how many non-exact n-forms there are on a compact oriented n-manifold M. Naturally, if ω is not exact, then the same is true for $\omega + d\eta$ for any $(n-1)$-form η, so we really want to consider ω and $\omega + d\eta$ as equivalent. There is, of course, a standard way of doing this, by considering quotient spaces. We will apply this construction not only to n-forms, but to forms of any degree.

For each k, the collection $Z^k(M)$ of all closed k-forms on M is a vector space. The space $B^k(M)$ of all exact k-forms is a subspace (since $d^2 = 0$), so we can form the quotient vector space

$$H^k(M) = Z^k(M)/B^k(M);$$

this vector space $H^k(M)$ is called the k-**dimensional de Rham cohomology vector space** of M. [*de Rham's Theorem* states that this vector space is isomorphic to a certain vector space defined purely in terms of the topology of M (for any space M), called the "k-dimensional cohomology group of M with real coefficients"; the notation Z^k, B^k is chosen to correspond to the notation used in algebraic topology, where these groups are defined.]

An element of $H^k(M)$ is an equivalence class $[\omega]$ of a closed k-form ω, two closed k-forms ω_1 and ω_2 being equivalent if and only if their difference is exact. In terms of these vector spaces, the Poincaré Lemma says that $H^k(\mathbb{R}^n) = 0$ (the vector space containing only 0) if $k > 0$, or more generally, $H^k(M) = 0$ if M is contractible and $k > 0$.

To compute $H^0(M)$ we note first that $B^0(M) = 0$ (there are no non-zero exact 0-forms, since there are no non-zero (-1)-forms for them to be the differential of). So $H^0(M)$ is the same as the vector space of all C^∞ functions $f : M \to \mathbb{R}$ with $df = 0$. If M is connected, the condition $df = 0$ implies that f is constant, so $H^0(M) \approx \mathbb{R}$. (In general, the dimension of $H^0(M)$ is the number of components of M.)

Aside from these trivial remarks, we presently know only one other fact about $H^k(M)$—if M is compact and oriented, then $H^n(M)$ has dimension ≥ 1. The further study of $H^k(M)$ requires a careful look at spheres and Euclidean space.

On $S^{n-1} \subset \mathbb{R}^n - \{0\}$ there is a natural choice of an $(n-1)$-form σ' with $\int_{S^{n-1}} \sigma' > 0$: for $(v_1)_p, \ldots, (v_{n-1})_p \in S^{n-1}{}_p$, we define

$$\sigma'(p)\big((v_1)_p, \ldots, (v_{n-1})_p\big) = \det \begin{pmatrix} p \\ v_1 \\ \vdots \\ v_{n-1} \end{pmatrix}.$$

Clearly this is > 0 if $(v_1)_p, \ldots, (v_{n-1})_p$ is a positively oriented basis. In fact, we defined the orientation of S^{n-1} in precisely this way—this orientation is just the induced orientation when S^{n-1} is considered as the boundary of the unit ball $\{p \in \mathbb{R}^n : |p| \leq 1\}$ with the usual orientation. Using the expansion of a determinant by minors along the top row we see that σ' is the restriction to S^{n-1} of the form σ on $\mathbb{R}^n$ defined by

$$\sigma = \sum_{i=1}^n (-1)^{i-1} x^i \, dx^1 \wedge \cdots \wedge \widehat{dx^i} \wedge \cdots \wedge dx^n.$$

The form σ' on S^{n-1} will now be used to find an $(n-1)$-form on $\mathbb{R}^n - \{0\}$ which is closed but not exact (thus showing that $H^{n-1}(\mathbb{R}^n - \{0\}) \neq 0$). Consider the map $r \colon \mathbb{R}^n - \{0\} \to S^{n-1}$ defined by

$$r(p) = \frac{p}{|p|} = \frac{p}{v(p)}.$$

Clearly $r(p) = p$ if $p \in S^{n-1}$; otherwise said, if $i \colon S^{n-1} \to \mathbb{R}^n - \{0\}$ is the inclusion, then

$$r \circ i = \text{ identity of } S^{n-1}.$$

(In general, if $A \subset X$ and $r \colon X \to A$ satisfies $r(a) = a$ for $a \in A$, then r is called a **retraction** of X onto A.)

Clearly, $r^*\sigma'$ is closed:

$$d(r^*\sigma') = r^* d\sigma' = 0.$$

However, it is not exact, for if $r^*\sigma' = d\eta$, then

$$\sigma' = i^* r^* \sigma' = d i^* \eta;$$

but we know that σ' is not exact.

It is a worthwhile exercise to compute by brute force that

$$\text{for } n = 2, \quad r^*\sigma' = \frac{x\,dy - y\,dx}{x^2 + y^2} = \frac{x\,dy - y\,dx}{v^2} = d\theta$$

$$\text{for } n = 3, \quad r^*\sigma' = \frac{x\,dy \wedge dz - y\,dx \wedge dz + z\,dx \wedge dy}{(x^2 + y^2 + z^2)^{3/2}}$$

$$= \frac{1}{v^3}[x\,dy \wedge dz - y\,dx \wedge dz + z\,dx \wedge dy].$$

Since we will actually need to know $r^*\sigma'$ in general, we evaluate it in another way:

7. LEMMA. If σ is the form on $\mathbb{R}^n$ defined by

$$\sigma = \sum_{i=1}^{n}(-1)^{i-1}x^i\,dx^1 \wedge \cdots \wedge \widehat{dx^i} \wedge \cdots \wedge dx^n,$$

and σ' is the restriction $i^*\sigma$ of σ to S^{n-1}, then

(∗)
$$r^*\sigma'(p) = \frac{\sigma(p)}{|p|^n}.$$

So

$$r^*\sigma' = \frac{1}{v^n}\sum_{i=1}^{n}(-1)^{i-1}x^i\,dx^1 \wedge \cdots \wedge \widehat{dx^i} \wedge \cdots \wedge dx^n.$$

PROOF. At any point $p \in \mathbb{R}^n - \{0\}$, the tangent space $\mathbb{R}^n{}_p$ is spanned by p_p and the vectors v_p in the tangent space of the sphere $S^{n-1}(|p|)$ of radius $|p|$. So it suffices to check that both sides of (∗) give the same result when applied to $n - 1$ vectors each of which is one of these two sorts. Now p_p is the tangent vector of a curve γ lying along the straight line through 0 and p; this curve is taken to the single point $r(p)$ by r, so $r_*(p_p) = 0$. On the other hand,

$$\sigma(p)\big(p_p, (v_1)_p, \ldots, (v_{n-2})_p\big) = \det\begin{pmatrix} p \\ p \\ v_1 \\ \vdots \\ v_{n-2} \end{pmatrix} = 0.$$

So it suffices to apply both sides of (∗) to vectors in the tangent space of $S^{n-1}(|p|)$. Thus (Problem 15), it suffices to show that for such vectors v_p we have

$$r_*(v_p) = \frac{1}{|p|}v_{r(p)}.$$

But this is almost obvious, since the vector v_p is the tangent vector of a circle γ lying in $S^{n-1}(|p|)$, and the curve $r \circ \gamma$ lies in S^{n-1} and goes $1/|p|$ as far in the same time. ❖

8. COROLLARY (INTEGRATION IN "POLAR COORDINATES"). Let $f : B \to \mathbb{R}$, where
$$B = \{p \in \mathbb{R}^n : |p| \le 1\},$$
and define $g : S^{n-1} \to \mathbb{R}$ by
$$g(p) = \int_0^1 u^{n-1} f(u \cdot p) \, du.$$

Then
$$\int_B f = \int_B f \, dx^1 \wedge \cdots \wedge dx^n = \int_{S^{n-1}} g\sigma'.$$

PROOF. Consider $S^{n-1} \times [0, 1]$ and the two projections

$$\pi_1 : S^{n-1} \times [0, 1] \to S^{n-1}$$
$$\pi_2 : S^{n-1} \times [0, 1] \to [0, 1].$$

Let us use the abbreviation

$$\sigma' \wedge dt = \pi_1{}^*\sigma' \wedge \pi_2{}^* dt.$$

If (y, U) is a coordinate system on S^{n-1}, with a corresponding coordinate system $(\bar{y}, t) = (y \circ \pi_1, \pi_2)$ on $S^{n-1} \times [0, 1]$, and $\sigma' = \alpha \, dy^1 \wedge \cdots \wedge dy^{n-1}$, then clearly
$$\sigma' \wedge dt = \alpha \circ \pi_1 \, d\bar{y}^1 \wedge \cdots \wedge d\bar{y}^{n-1} \wedge dt.$$

From this it is easy to see that if we define $h : S^{n-1} \times [0, 1] \to \mathbb{R}$ by
$$h(p, u) = u^{n-1} f(u \cdot p),$$

then
$$\int_{S^{n-1}} g\sigma' = (-1)^{n-1} \int_{S^{n-1} \times [0,1]} h\sigma' \wedge dt.$$

Now we can define a diffeomorphism $\phi : B - \{0\} \to S^{n-1} \times (0, 1]$ by
$$\phi(p) = (r(p), v(p)) = (p/|p|, |p|).$$

Then

$$
\begin{aligned}
\phi^*(\sigma' \wedge dt) &= \phi^*(\pi_1{}^*\sigma' \wedge \pi_2{}^*dt) \\
&= \phi^*\pi_1{}^*\sigma' \wedge \phi^*\pi_2{}^*dt \\
&= (\pi_1 \circ \phi)^*\sigma' \wedge (\pi_2 \circ \phi)^*dt \\
&= r^*\sigma' \wedge v^*dt \\
&= \frac{1}{v^n}\left(\sum_{i=1}^{n}(-1)^{i-1}x^i\,dx^1 \wedge \cdots \wedge \widehat{dx^i} \wedge \cdots \wedge dx^n\right) \wedge \sum_{i=1}^{n}\frac{x^i}{v}\,dx^i \\
&= \frac{(-1)^{n-1}}{v^{n+1}}\sum_{i=1}^{n}(x^i)^2\,dx^1 \wedge \cdots \wedge dx^n \\
&= \frac{(-1)^{n-1}}{v^{n-1}}\,dx^1 \wedge \cdots \wedge dx^n.
\end{aligned}
$$

Hence

$$
\begin{aligned}
\phi^*(h\sigma' \wedge dt) &= (h \circ \phi)\phi^*(\sigma' \wedge dt) \\
&= v^{n-1}f \cdot \frac{(-1)^{n-1}}{v^{n-1}}\,dx^1 \wedge \cdots \wedge dx^n \\
&= (-1)^{n-1}f\,dx^1 \wedge \cdots \wedge dx^n.
\end{aligned}
$$

So,

$$
\begin{aligned}
\int_B f\,dx^1 \wedge \cdots \wedge dx^n &= (-1)^{n-1}\int_{B-\{0\}} \phi^*(h\sigma' \wedge dt) \\
&= (-1)^{n-1}\int_{S^{n-1}\times(0,1]} h\sigma' \wedge dt \\
&= \int_{S^{n-1}} g\sigma'.
\end{aligned}
$$

(This last step requires some justification, which should be supplied by the reader, since the forms involved do not have compact support on the manifolds $B - \{0\}$ and $S^{n-1} \times (0, 1]$ where they are defined.) ❖

We are about ready to compute $H^k(M)$ in a few more cases. We are going to reduce our calculations to calculations within coordinate neighborhoods, which are submanifolds of M, but not compact. It is therefore necessary to introduce another collection of vector spaces, which are interesting in their own right.

The **de Rham cohomology vector spaces with compact supports** $H_c^k(M)$ are defined as

$$H_c^k(M) = Z_c^k(M)/B_c^k(M),$$

where $Z_c^k(M)$ is the vector space of closed k-forms with compact support, and $B_c^k(M)$ is the vector space of all k-forms $d\eta$ where η is a $(k-1)$-form with compact support. Of course, if M is compact, then $H_c^k(M) = H^k(M)$. Notice that $B_c^k(M)$ is *not* the same as the set of all exact k-forms with compact support. For example, on $\mathbb{R}^n$, if $f \geq 0$ is a function with compact support, and $f > 0$ at some point, then

$$\omega = f\,dx^1 \wedge \cdots \wedge dx^n$$

is exact (every closed form on $\mathbb{R}^n$ is) and has compact support, but ω is not $d\eta$ for any form η with compact support. Indeed, if $\omega = d\eta$ where η has compact support, then by Stokes' Theorem

$$\int_{\mathbb{R}^n} \omega = \int_{\mathbb{R}^n} d\eta = \int_{\partial \mathbb{R}^n} \eta = 0.$$

This example shows that $H_c^n(\mathbb{R}^n) \neq 0$, and a similar argument shows that if M is any orientable manifold, then $H_c^n(M) \neq 0$. We are now going to show that for any connected orientable manifold M we actually have

$$H_c^n(M) \approx \mathbb{R}.$$

This means that if we choose a fixed ω with $\int_M \omega \neq 0$, then for any n-form ω' with compact support there is a real number a such that $\omega' - a\omega$ is exact. The number a can be described easily: if

$$\omega' - a\omega = d\eta,$$

then

$$\int_M \omega' - \int_M a\omega = \int_M d\eta = 0,$$

so

$$a = \int_M \omega' \bigg/ \int_M \omega;$$

the problem, of course, is showing that η exists. Notice that the assertion that $H_c^n(M) \approx \mathbb{R}$ is equivalent to the assertion that

$$[\omega] \mapsto \int_M \omega$$

is an isomorphism of $H_c^n(M)$ with $\mathbb{R}$, i.e., to the assertion that a closed form ω with compact support is the differential of another form with compact support if $\int_M \omega = 0$.

9. THEOREM. If M is a connected orientable n-manifold, then $H_c^n(M) \approx \mathbb{R}$.

PROOF. We will establish the theorem in three steps:

(1) The theorem is true for $M = \mathbb{R}$.

(2) If the theorem is true for $(n-1)$-manifolds, in particular for S^{n-1}, then it is true for $\mathbb{R}^n$.

(3) If the theorem is true for $\mathbb{R}^n$, then it is true for any connected oriented n-manifold.

Step 1. Let ω be a 1-form on $\mathbb{R}$ with compact support such that $\int_{\mathbb{R}} \omega = 0$. There is some function f (not necessarily with compact support) such that $\omega = df$. Since support ω is compact, $df = 0$ outside some interval $[-N, N]$, so f is a

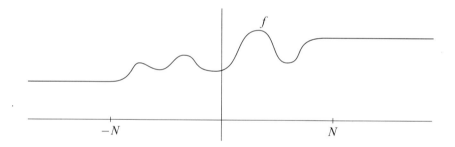

constant c_1 on $(-\infty, -N)$ and a constant c_2 on (N, ∞). Moreover,

$$0 = \int_{\mathbb{R}} \omega = \int_{\mathbb{R}} df = \int_{\mathbb{R}} f'(t)\, dt = c_2 - c_1.$$

Therefore $c_1 = c_2 = c$ and we have

$$\omega = d(f - c)$$

where $f - c$ has compact support.

Step 2. Let $\omega = f\, dx^1 \wedge \cdots \wedge dx^n$ be an n-form with compact support on $\mathbb{R}^n$ such that $\int_{\mathbb{R}^n} \omega = 0$. For simplicity assume that support $\omega \subset \{p \in \mathbb{R}^n : |p| < 1\}$. We know that there is an $(n-1)$-form η on $\mathbb{R}^n$ such that $\omega = d\eta$. In fact, from Problem 7-23, we have an explicit formula for η,

$$\eta(p) = \sum_{i=1}^{n} (-1)^{i-1} \left(\int_0^1 t^{n-1} f(t \cdot p)\, dt \right) x^i\, dx^1 \wedge \cdots \wedge \widehat{dx^i} \wedge \cdots \wedge dx^n.$$

Using the substitution $u = |p|t$ this becomes

$$\eta(p) = \left(\int_0^{|p|} u^{n-1} f\left(u \cdot \frac{p}{|p|} \right) du \right) \frac{1}{|p|^n}$$

$$\times \sum_{i=1}^n (-1)^{i-1} x^i \, dx^1 \wedge \cdots \wedge \widehat{dx^i} \wedge \cdots \wedge dx^n$$

$$= \left(\int_0^{|p|} u^{n-1} f\left(u \cdot \frac{p}{|p|} \right) du \right) \cdot r^* \sigma'(p) \qquad \text{by Lemma 7.}$$

Define $g \colon S^{n-1} \to \mathbb{R}$ by

$$g(p) = \int_0^1 u^{n-1} f(u \cdot p) \, du.$$

On the set $A = \{ p \in \mathbb{R}^n : |p| > 1 \}$ we have $f = 0$, so on A we have

$$\eta(p) = \left(\int_0^1 u^{n-1} f\left(u \cdot \frac{p}{|p|} \right) du \right) \cdot r^* \sigma'(p),$$

or

$$\eta = (g \circ r) \cdot r^* \sigma' = r^* (g\sigma').$$

Moreover, by Corollary 8 we have for the $(n-1)$-form $g\sigma'$ on S^{n-1},

$$\int_{S^{n-1}} g\sigma' = \int_B f \, dx^1 \wedge \cdots \wedge dx^n$$

$$= \int_{\mathbb{R}^n} \omega = 0.$$

Thus, by the hypothesis for *Step 2*,

$$g\sigma' = d\lambda \quad \text{for some } (n-2)\text{-form } \lambda \text{ on } S^{n-1}.$$

Hence

$$\eta = r^*(d\lambda) = d(r^*\lambda).$$

Let $h \colon \mathbb{R}^n \to [0, 1]$ be any C^∞ function with $h = 1$ on A and $h = 0$ in a neighborhood of 0. Then $hr^*\lambda$ is a C^∞ form on $\mathbb{R}^n$ and

$$\omega = d\eta = d(\eta - d(hr^*\lambda));$$

the form $\eta - d(hr^*\lambda)$ has compact support, since on A we have

$$\eta - d(hr^*\lambda) = \eta - d(r^*\lambda) = 0.$$

Step 3. Choose an n-form ω such that $\int_M \omega \neq 0$ and ω has compact support contained in an open set $U \subset M$, with U diffeomorphic to $\mathbb{R}^n$. If ω' is any other n-form with compact support, we want to show that there is a number c and a form η with compact support such that

$$\omega' = c\omega + d\eta.$$

Using a partition of unity, we can write

$$\omega' = \phi_1 \omega' + \cdots + \phi_k \omega'$$

where each $\phi_i \omega'$ has compact support contained in some open set $U_i \subset M$ with U_i diffeomorphic to $\mathbb{R}^n$. It obviously suffices to find c_i and η_i with $\phi_i \omega' = c_i \omega + d\eta_i$, for each i. In other words, we can assume ω' has support contained in some open $V \subset M$ which is diffeomorphic to $\mathbb{R}^n$.

Using the connectedness of M, it is easy to see that there is a sequence of open sets

$$U = V_1, \ldots, V_r = V$$

diffeomorphic to $\mathbb{R}^n$, with $V_i \cap V_{i+1} \neq \emptyset$. Choose forms ω_i with support $\omega_i \subset$

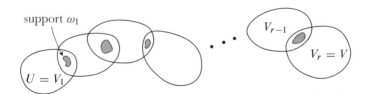

$V_i \cap V_{i+1}$ and $\int_{V_i} \omega_i \neq 0$. Since we are assuming the theorem for $\mathbb{R}^n$ we have

$$\omega_1 - c_1 \omega = d\eta_1$$
$$\omega_2 - c_2 \omega_1 = d\eta_2$$

$$\vdots$$

$$\omega' - c_r \omega_{r-1} = d\eta_r,$$

where all η_i have compact support ($\subset V_i$). From this we clearly obtain the desired result. ❖

The method used in the last step can be used to derive another result.

10. THEOREM. If M is any connected non-orientable n-manifold, then $H_c^n(M) = 0$.

PROOF. Choose an n-form ω with compact support contained in an open set U diffeomorphic to $\mathbb{R}^n$, such that $\int_U \omega \neq 0$ (this integral makes sense, since U is orientable). It obviously suffices to show that $\omega = d\eta$ for some form η with compact support. Consider a sequence

$$U = V_1, \ldots, V_r = V$$

of coordinate systems (V_i, x_i) where each $x_i \circ x_{i+1}^{-1}$ is orientation preserving. Choose the forms ω_i in *Step 3* so that, using the orientation of V_i which makes $x_i \colon V_i \to \mathbb{R}^n$ orientation preserving, we have $\int_{V_i} \omega_i > 0$; then also $\int_{V_{i+1}} \omega_i > 0$. Consequently, the numbers

$$c_i = \int_{V_i} \omega_i \bigg/ \int_{V_i} \omega_{i-1} \quad \text{are positive.}$$

It follows that

$$\omega_i = c\omega + d\eta \quad \text{where } c > 0.$$

Now if M is unorientable, there is such a sequence where $V_r = V_1$ but $x_r \circ x_1^{-1}$ is orientation *reversing*. Taking $\omega' = -\omega$, we have

$$-\omega = c\omega + d\eta \qquad \text{for } c > 0$$

so

$$(-c - 1)\omega = d\eta \qquad \text{for } -c - 1 \neq 0. \quad \diamond$$

We can also compute $H^n(M)$ for non-compact M.

11. THEOREM. If M is a connected non-compact n-manifold (orientable or not), then $H^n(M) = 0$.

PROOF. Consider first an n-form ω with support contained in a coordinate neighborhood U which is diffeomorphic to $\mathbb{R}^n$. Since M is not compact, there is an infinite sequence

$$U = U_1, U_2, U_3, U_4, \ldots$$

of such coordinate neighborhoods such that $U_i \cap U_{i+1} \neq \emptyset$, and such that the sequence is eventually in the complement of any compact set.

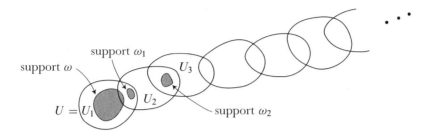

Now choose n-forms ω_i with compact support contained in $U_i \cap U_{i+1}$, such that $\int_{U_i} \omega_i \neq 0$. There are constants c_i and forms η_i with compact support $\subset U_i$ such that

$$\omega = c_1\omega_1 + d\eta_1$$
$$\omega_i = c_{i+1}\omega_{i+1} + d\eta_{i+1} \quad i \geq 1.$$

Then

$$\omega = d\eta_1 + c_1\omega_1$$
$$= d\eta_1 + c_1 d\eta_2 + c_1 c_2 \omega_2$$
$$= d\eta_1 + c_1 d\eta_2 + c_1 c_2 d\eta_3 + c_1 c_2 c_3 \omega_3$$
$$= \cdots.$$

Since any point $p \in M$ is eventually in the complement of the U_i's, we have

$$\omega = d\eta_1 + c_1 d\eta_2 + c_1 c_2 d\eta_3 + c_1 c_2 c_3 d\eta_4 + \cdots,$$

where the right side makes sense since the U_i are eventually outside of any compact set.

Now it can be shown (Problem 20) that there is actually such a sequence $U_1, U_2, U_3, \ldots$ whose union is all of M (repetitions are allowed, and U_i may intersect several U_j for $j < i$, but the sequence is still eventually outside of any compact set). The cover $\mathcal{O} = \{U\}$ is then locally finite. Let $\{\phi_U\}$ be a partition of unity subordinate to $\mathcal{O}$. If ω is an n-form on M, then for each U_i we have seen that

$$\phi_{U_i}\omega = d\eta_i \quad \text{where } \eta_i \text{ has support contained in } U_i \cup U_{i+1} \cup U_{i+2} \cup \cdots.$$

Hence

$$\omega = \sum_{i=1}^{\infty} \phi_{U_i}\omega = \sum_{i=1}^{\infty} d\eta_i = d\left(\sum_{i=1}^{\infty} \eta_i\right). \; \diamond$$

SUMMARY OF RESULTS

(1) For $\mathbb{R}^n$ we have

$$H^k(\mathbb{R}^n) \approx \begin{cases} \mathbb{R} & k = 0 \\ 0 & k > 0. \end{cases}$$

(2) If M is a connected n-manifold, then

$$H^0(M) \approx \mathbb{R}$$

$$H_c^n(M) \approx \begin{cases} \mathbb{R} & \text{if } M \text{ is orientable} \\ 0 & \text{if } M \text{ is non-orientable} \end{cases}$$

$$H^n(M) \approx \begin{cases} H_c^n & \text{if } M \text{ is compact} \\ 0 & \text{if } M \text{ is not compact.} \end{cases}$$

We also know that $H^{n-1}(\mathbb{R}^n - \{0\}) \neq 0$, but we have not listed this result, since we will eventually improve it. In order to proceed further with our computations we need to examine the behavior of the de Rham cohomology vector spaces under C^∞ maps $f \colon M \to N$. If ω is a closed k-form on N, then $f^*\omega$ is also closed ($df^*\omega = f^*d\omega = 0$), so f^* takes $Z^k(N)$ to $Z^k(M)$. On the other hand, f^* also takes $B^k(N)$ to $B^k(M)$, since $f^*(d\eta) = d(f^*\eta)$. This shows that f^* induces a map

$$Z^k(N)/B^k(N) \to Z^k(M)/B^k(M),$$

also denoted by f^*:

$$f^* \colon H^k(N) \to H^k(M).$$

For example, consider the case $k = 0$. If N is connected, then $H^0(N)$ is just the collection of constant functions $c \colon N \to \mathbb{R}$. Then $f^*(c) = c \circ f$ is also a constant function. If M is connected, then $f^* \colon H^0(N) \to H^0(M)$ is just the identity map under the natural identification of $H^0(N)$ and $H^0(M)$ with $\mathbb{R}$. If M is disconnected, with components M_α, $\alpha \in A$, then $H^0(M)$ is isomorphic to the direct sum

$$\bigoplus_{\alpha \in A} \mathbb{R}_\alpha, \quad \text{where each } \mathbb{R}_\alpha \approx \mathbb{R};$$

the map f^* takes $c \in \mathbb{R}$ into the element of $\bigoplus \mathbb{R}_\alpha$ with α^{th} component equal to c. If N is also disconnected, with components N_β, $\beta \in B$, then

$$f^* \colon \bigoplus_{\beta \in B} \mathbb{R}_\beta \to \bigoplus_{\alpha \in A} \mathbb{R}_\alpha$$

takes the element $\{c_\beta\}$ of $\bigoplus_{\beta \in B} \mathbb{R}_\beta$ to $\{c_\alpha'\}$, where $c_\alpha' = c_\beta$ when $f(M_\alpha) \subset N_\beta$.

A more interesting case, and the only one we are presently in a position to look at, is the map

$$f^* : H^n(N) \to H^n(M)$$

when M and N are both compact connected oriented n-manifolds. There is no natural way to make $H^n(M)$ isomorphic to $\mathbb{R}$, so we really want to compare

$$\int_M f^*\omega \quad \text{and} \quad \int_N \omega$$

for ω an n-form on N. Choose one ω_0 with $\int_N \omega_0 \neq 0$. Then there is some number a such that

$$\int_M f^*\omega_0 = a \cdot \int_N \omega_0.$$

Since $\omega \mapsto \int_M \omega$ is an isomorphism of $H^n(M)$ and $\mathbb{R}$ (and similarly for N) it follows that for *every* form ω we have

$$\int_M f^*\omega = a \cdot \int_N \omega.$$

The number $a = \deg f$, which depends only on f, is called the **degree** of f. If M and N are not compact, but f is proper (the inverse image of any compact set is compact), then we have a map

$$f^* : H^n_c(N) \to H^n_c(M)$$

and a number $\deg f$, such that

$$\int_M f^*\omega = (\deg f) \int_N \omega$$

for all forms ω on N with compact support. Until one sees the proof of the next theorem, it is almost unbelievable that this number is *always an integer*.

12. THEOREM. Let $f : M \to N$ be a proper map between two connected oriented n-manifolds (M, μ) and (N, ν). Let $q \in N$ be a regular value of f. For each $p \in f^{-1}(q)$, let

$$\text{sign}_p f = \begin{cases} 1 & \text{if } f_{*p} : M_p \to N_q \text{ is orientation preserving} \\ & \text{(using the orientations } \mu_p \text{ for } M_p \text{ and } \nu_q \text{ for } N_q) \\ -1 & \text{if } f_{*p} \text{ is orientation reversing.} \end{cases}$$

Then

$$\deg f = \sum_{p \in f^{-1}(q)} \text{sign}_p f \qquad (= 0 \text{ if } f^{-1}(p) = \emptyset).$$

PROOF. Notice first that regular values exist, by Sard's Theorem. Moreover, $f^{-1}(q)$ is finite, since it is compact and consists of isolated points, so the sum above is a finite sum.

Let $f^{-1}(q) = \{p_1, \ldots, p_k\}$. Choose coordinate systems (U_i, x_i) around p_i such that all points in U_i are regular values of f, and the U_i are disjoint. We want to choose a coordinate system (V, y) around q such that $f^{-1}(V) = U_1 \cup \cdots \cup U_k$. To do this, first choose a compact neighborhood W of q, and let

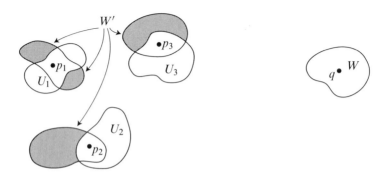

$W' \subset M$ be the compact set

$$W' = f^{-1}(W) - (U_1 \cup \cdots \cup U_k).$$

Then $f(W')$ is a closed set which does not contain q. We can therefore choose $V \subset W - f(W')$. This ensures that $f^{-1}(V) \subset U_1 \cup \cdots \cup U_k$. Finally, redefine U_i to be $U_i \cap f^{-1}(V)$.

Now choose ω on N to be $\omega = g \, dy^1 \wedge \cdots \wedge dy^n$ where $g \geq 0$ has compact support contained in V. Then support $f^*\omega \subset U_1 \cup \cdots \cup U_k$. So

$$\int_M f^*\omega = \sum_{i=1}^{k} \int_{U_i} f^*\omega.$$

Since f is a diffeomorphism from each U_i to V we have

$$\int_{U_i} f^*\omega = \int_V \omega \quad \text{if } f \text{ is orientation preserving}$$

$$= -\int_V \omega \quad \text{if } f \text{ is orientation reversing.}$$

Since f is orientation preserving [or reversing] precisely when $\text{sign}_p f = 1$ [or -1] this proves the theorem. ❖

As an immediate application of the theorem, we compute the degree of the "antipodal map" $A: S^n \to S^n$ defined by $A(p) = -p$. We have already seen that A is orientation preserving or reversing at all points, depending on whether n is odd or even. Since $A^{-1}(p)$ consists of just one point, we conclude that

$$\deg A = (-1)^{n-1}.$$

We can draw an interesting conclusion from this result, but we need to introduce another important concept first. Two functions $f, g: M \to N$ between two C^∞ manifolds are called (**smoothly**) **homotopic** if there is a smooth function

$$H: M \times [0,1] \to N$$

with

$$\begin{aligned} H(p,0) &= f(p) \\ H(p,1) &= g(p) \end{aligned} \qquad \text{for all } p \in M;$$

the map H is called a (**smooth**) **homotopy** between f and g. Notice that M is smoothly contractible to a point $p_0 \in M$ if and only if the identity map of M is homotopic to the constant map p_0. Recall that for every k-form ω on $M \times [0,1]$ we defined a $(k-1)$-form $I\omega$ on M such that

$$i_1{}^*\omega - i_0{}^*\omega = d(I\omega) + I(d\omega).$$

We used this fact to show that all closed forms on a smoothly contractible manifold are exact. We can now prove a more general result.

13. THEOREM. If $f, g: M \to N$ are smoothly homotopic, then the maps

$$f^*: H^k(N) \to H^k(M)$$
$$g^*: H^k(N) \to H^k(M)$$

are equal, $f^* = g^*$.

PROOF. By assumption, there is a smooth map $H: M \times [0,1] \to N$ with

$$\begin{aligned} f &= H \circ i_0 \\ g &= H \circ i_1. \end{aligned}$$

Any element of $H^k(N)$ is the equivalence class $[\omega]$ of some closed k-form ω on N. Then

$$\begin{aligned} g^*\omega - f^*\omega &= (H \circ i_1)^*\omega - (H \circ i_0)^*\omega \\ &= i_1{}^*(H^*\omega) - i_0{}^*(H^*\omega) \\ &= d(IH^*\omega) + I(dH^*\omega) \\ &= d(IH^*\omega) + 0. \end{aligned}$$

But this means that $g^*([\omega]) = f^*([\omega])$. ❖

14. COROLLARY. If M and N are compact oriented n-manifolds and the maps $f, g: M \to N$ are homotopic, then $\deg f = \deg g$.

15. COROLLARY. If n is even, then there does not exist a nowhere zero vector field on S^n.

PROOF. We have already seen that the degree of the antipodal map $A: S^n \to S^n$ is $(-1)^{n-1}$. Since the identity map has degree 1, A is not homotopic to the identity for n even. But if there is a nowhere zero vector field on S^n, then we can construct a homotopy between A and the identity map as follows. For each p, there is a unique great semi-circle γ_p from p to $A(p) = -p$ whose tangent vector at p is a multiple of $X(p)$. Define

$$H(p, t) = \gamma_p(t). \; \diamondsuit$$

For n odd we can explicitly construct a nowhere zero vector field on S^n. For $p = (x_1, \ldots, x_{n+1}) \in S^n$ we define

$$X(p) = (-x_1, x_0, -x_3, x_2, \ldots, -x_{n+1}, x_n);$$

this is perpendicular to $p = (x_1, x_2, \ldots, x_{n+1})$, and therefore in $S^n{}_p$. (On S^1 this gives the standard picture.) The vector field on S^n can then be used to give

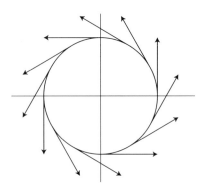

a homotopy between A and the identity map.

For another application of Theorem 13, consider the retraction

$$r: \mathbb{R}^n - \{0\} \to S^{n-1} \qquad r(p) = p/|p|.$$

If $i: S^{n-1} \to \mathbb{R}^n - \{0\}$ is the inclusion, then

$$r \circ i: S^{n-1} \to S^{n-1} \text{ is the identity 1 of } S^{n-1}.$$

The map

$$i \circ r \colon \mathbb{R}^n - \{0\} \to \mathbb{R}^n - \{0\} \qquad\qquad i \circ r(p) = p/|p|$$

is, of course, not the identity, but it *is* homotopic to the identity; we can define the homotopy H by

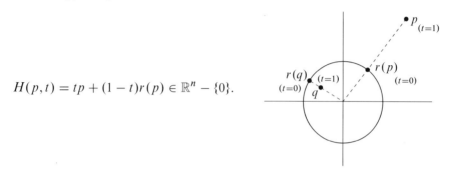

$$H(p,t) = tp + (1-t)r(p) \in \mathbb{R}^n - \{0\}.$$

A retraction with this property is called a **deformation retraction**. Whenever r is a deformation retraction, the maps $(r \circ i)^*$ and $(i \circ r)^*$ are the identity. Thus, for the case of $S^{n-1} \subset \mathbb{R}^n - \{0\}$, we have

$$H^k(S^{n-1}) \xrightarrow{\;r^*\;} H^k(\mathbb{R}^n - \{0\})$$

$$H^k(\mathbb{R}^n - \{0\}) \xrightarrow{\;i^*\;} H^k(S^{n-1})$$

and

$$r^* \circ i^* = (i \circ r)^* = \text{ identity of } H^k(\mathbb{R}^n - \{0\})$$

$$i^* \circ r^* = (r \circ i)^* = \text{ identity of } H^k(S^{n-1}).$$

So i^* and r^* are inverses of each other. Thus

$$H^k(S^{n-1}) \approx H^k(\mathbb{R}^n - \{0\}) \quad \text{for all } k.$$

In particular, we have $H^{n-1}(\mathbb{R}^n - \{0\}) \approx \mathbb{R}$. A generator of $H^{n-1}(\mathbb{R}^n - \{0\})$ is the closed form $r^*\sigma'$.

We are now going to compute $H^k(\mathbb{R}^n - \{0\})$ for all k. We need one further observation. The manifold

$$M \times \{0\} \subset M \times \mathbb{R}^l$$

is clearly a deformation retraction of $M \times \mathbb{R}^l$. So $H^k(M) \approx H^k(M \times \mathbb{R}^l)$ for all l.

16. THEOREM. For $0 < k < n - 1$ we have $H^k(\mathbb{R}^n - \{0\}) = H^k(S^{n-1}) = 0$.

PROOF. Induction on n. The first case where there is anything to prove is $n = 3$. We claim $H^1(\mathbb{R}^3 - \{0\}) = 0$.

Let ω be a closed 1-form on $\mathbb{R}^3$. Let A and B be the open sets

$$A = \mathbb{R}^3 - \{(0,0) \times (-\infty, 0]\}$$
$$B = \mathbb{R}^3 - \{(0,0) \times [0, \infty)\}.$$

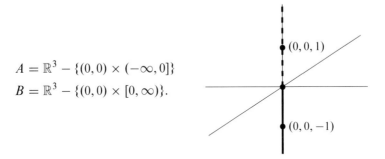

Since A and B are both star-shaped (with respect to the points $(0,0,1)$ and $(0,0,-1)$, respectively), there are 0-forms f_A and f_B on A and B with

$$\omega = df_A \quad \text{on } A$$
$$\omega = df_B \quad \text{on } B.$$

Now

$$d(f_A - f_B) = 0 \quad \text{on } A \cap B,$$

and

$$A \cap B = [\mathbb{R}^2 - \{0\}] \times \mathbb{R},$$

so clearly $f_A - f_B$ is a constant c on $A \cap B$. Thus ω is exact, for

$$\omega = d(f_A - c) \qquad \text{on } A$$
$$\omega = d(f_B) \qquad \text{on } B$$

and $f_A - c = f_B$ on $A \cap B$.

If ω is a closed 1-form on $\mathbb{R}^4$, there is a similar argument, using

$$A = \mathbb{R}^4 - \{(0,0,0) \times (-\infty, 0]\}$$
$$B = \mathbb{R}^4 - \{(0,0,0) \times [0, \infty)\}.$$

If ω is a closed 2-form on $\mathbb{R}^4$, then we obtain 1-forms η_A and η_B with

$$\omega = d\eta_A \quad \text{on } A$$
$$\omega = d\eta_B \quad \text{on } B.$$

Now

$$d(\eta_A - \eta_B) = 0 \quad \text{on } A \cap B$$

and

$$H^1(A \cap B) = H^1([\mathbb{R}^3 - \{0\}] \times \mathbb{R}) \approx H^1(\mathbb{R}^3 - \{0\}) = 0.$$

So $\eta_A - \eta_B = d\lambda$ for some 0-form λ on $A \cap B$. Unlike the previous case, we cannot simply consider $\eta_A - d\lambda$, since this is not defined on A. To circumvent this difficulty, note that there is a partition of unity $\{\phi_A, \phi_B\}$ for the cover $\{A, B\}$ of $\mathbb{R}^3 - \{0\}$:

$$\phi_A + \phi_B = 1$$
$$d\phi_A + d\phi_B = 0$$
$$\text{support } \phi_A \subset A$$
$$\text{support } \phi_B \subset B.$$

Now, if

$$\phi_B \lambda \quad \text{denotes} \quad \begin{cases} \phi_B \lambda & \text{on } A \cap B \\ 0 & \text{on } A - (A \cap B), \end{cases}$$

and similarly for $\phi_A \lambda$, then

$$\phi_B \lambda \quad \text{is a } C^\infty \text{ form on } A$$
$$\phi_A \lambda \quad \text{is a } C^\infty \text{ form on } B.$$

On $A \cap B$ we have

$$\begin{aligned} \eta_A - d(\phi_B \lambda) &= \eta_A - \phi_B\, d\lambda - d\phi_B \wedge \lambda \\ &= \eta_A + (\phi_A - 1)\, d\lambda + d\phi_A \wedge \lambda \\ &= \eta_A - d\lambda + d(\phi_A \lambda) \\ &= \eta_B + d(\phi_A \lambda). \end{aligned}$$

So we can define a C^∞ form on $\mathbb{R}^n - \{0\} = A \cup B$ by letting it be $\eta_A - d(\phi_B \lambda)$ on A, and $\eta_B + d(\phi_A \lambda)$ on B. Clearly,

$$\begin{aligned} \omega = d\eta_A &= d(\eta_A - d(\phi_B \lambda)) \quad \text{on } A \\ &= d\eta_B = d(\eta_B + d(\phi_A \lambda)) \quad \text{on } B, \end{aligned}$$

so ω is exact.

The general inductive step is similar. ❖

We end this chapter with one more calculation, which we will need in Chapter 11.

17. THEOREM. For $0 \leq k < n$ we have $H_c^k(\mathbb{R}^n) = 0$.

PROOF. The proof that $H_c^0(\mathbb{R}^n) = 0$ is left to the reader.

Let ω be a k-form on $\mathbb{R}^n$ with compact support, $0 < k < n$. We know that $\omega = d\eta$ for some $(k-1)$-form η on $\mathbb{R}^n$. Let B be a closed ball containing support ω. Then on $A = \mathbb{R}^n - B$ we have $d\eta = 0$. Since A is diffeomorphic to

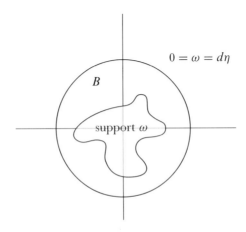

$\mathbb{R}^n - \{0\}$ and $k - 1 < n - 1$ we have from Theorem 16 that

$$\eta = d\lambda \quad \text{for some } (k-2)\text{-form } \lambda \text{ on } A.$$

Let $f \colon \mathbb{R}^n \to [0, 1]$ be a C^∞ function with $f = 0$ in a neighborhood of B and $f = 1$ on $\mathbb{R}^n - 2B$, where $2B$ denotes the ball of twice the radius of B. Then $d(f\lambda)$ makes sense on all of $\mathbb{R}^n$ and

$$\omega = d\eta = d(\eta - d(f\lambda));$$

the form $\eta - d(f\lambda)$ clearly has compact support contained in $2B$. ❖

PROBLEMS

1. *The Riemann integral versus the Darboux integral.* Let $f : [a, b] \to \mathbb{R}$ be bounded. For a partition $P = \{t_0 < \cdots < t_n\}$ of $[a, b]$, let $m_i = m_i(f)$ be the inf of f on $[t_{i-1}, t_i]$ and define $M_i = M_i(f)$ similarly. A *choice* for P is an n-tuple $\xi = (\xi_1, \ldots, \xi_n)$ with $\xi_i \in [t_{i-1}, t_i]$. We define the "lower sum", "upper sum", and "Riemann sum" for a partition P and choice ξ by

$$L(f, P) = \sum_{i=1}^{n} m_i(f) \cdot (t_i - t_{i-1})$$

$$U(f, P) = \sum_{i=1}^{n} M_i(f) \cdot (t_i - t_{i-1})$$

$$S(f, P, \xi) = \sum_{i=1}^{n} f(\xi_i)(t_i - t_{i-1}).$$

Clearly $L(f, P) \le S(f, P, \xi) \le U(f, P)$. We call f **Darboux integrable** if the sup of all $L(f, P)$ equals the inf of all $U(f, P)$; this sup or inf is called the **Darboux integral** of f on $[a, b]$. We call f **Riemann integrable** if

$$\lim_{\|P\| \to 0} S(f, P, \xi) \quad \text{exists};$$

the limit is called the **Riemann integral** of f on $[a, b]$.

(a) We can define $S(f, P, \xi)$ even if f is not bounded. Show however, that $\lim_{\|P\| \to 0} S(f, P, \xi)$ cannot exist if f is unbounded.

(b) If f is continuous on $[a, b]$, then f is Riemann and Darboux integrable on $[a, b]$, and the two integrals are equal. (Use uniform continuity of f on $[a, b]$.)

(c) If f is Riemann integrable on $[a, b]$, then f is Darboux integrable on $[a, b]$ and the two integrals are equal.

(d) Let $m \le f \le M$ on $[a, b]$. Let $P = \{s_0 < \cdots < s_m\}$ and $Q = \{t_0 < \cdots < t_n\}$ be two partitions of $[a, b]$. For each $i = 1, \ldots, n$, let

$$e_i = \text{length of } [t_{i-1}, t_i]$$
$$\quad - \text{sum of lengths of all } [s_{\alpha-1}, s_\alpha] \text{ which are contained in } [t_{i-1}, t_i].$$

$[s_{\alpha-1}, s_\alpha]$'s contained in $[t_{i-1}, t_i]$
shaded lengths ▬ add up to e_i

Show that, if M_i denotes the sup of f on $[t_{i-1}, t_i]$, then

$$U(f, P) \leq U(f, Q) + \sum_{i=1}^{n}(M - M_i)e_i$$

$$\leq U(f, Q) + (M - m)\sum_{i=1}^{n}e_i.$$

There is a similar result for lower sums.

(e) Show that $\sum_{i=1}^{n}e_i \to 0$ as $\|P\| \to 0$, and deduce Darboux's Theorem:

$$\lim_{\|P\|\to 0}U(f, P) = \inf\{U(f, Q) : Q \text{ a partition of } [a, b]\}$$

$$\lim_{\|P\|\to 0}L(f, P) = \sup\{L(f, Q) : Q \text{ a partition of } [a, b]\}.$$

(f) If f is Darboux integrable on $[a, b]$, then f is Riemann integrable on $[a, b]$.

(g) (Osgood's Theorem). Let f and g be integrable on $[a, b]$. Show that for choices ξ, ξ' for P,

$$\lim_{\|P\|\to 0}\sum_{i=1}^{n}f(\xi_i)g(\xi'_i)(t_i - t_{i-1}) = \int_{a}^{b}fg.$$

Hint: If $|g| \leq M$ on $[a, b]$, then $|f(\xi'_i)g(\xi'_i) - f(\xi_i)g(\xi'_i)| \leq M|f(\xi'_i) - f(\xi_i)|.$

(h) Show that $\int_c f\,dx + g\,dy$, defined as a limit of sums, equals

$$\int_{a}^{b}[f(c(t))c^{1\prime}(t) + g(c(t))c^{2\prime}(t)]\,dt.$$

2. Compute $\int_c d\theta = \int_{[0,1]}c^* d\theta$, where $c(t) = (\cos 2\pi t, \sin 2\pi t)$ on $[0, 1]$.

3. For n an integer, and $R > 0$, let $c_{R,n}\colon [0, 1] \to \mathbb{R}^2 - \{0\}$ be defined by

$$c_{R,n}(t) = (R\cos 2n\pi t, R\sin 2n\pi t).$$

(a) Show that there is a singular 2-cube $c\colon [0, 1]^2 \to \mathbb{R}^2 - \{0\}$ such that $c_{R_1,n} - c_{R_2,n} = \partial c$.

(b) If $c\colon [0, 1] \to \mathbb{R}^2 - \{0\}$ is any curve with $c(0) = c(1)$, show that there is some n such that $c - c_{1,n}$ is a boundary in $\mathbb{R}^2 - \{0\}$.

(c) Show that n is unique. It is called the **winding number** of c around 0.

4. Let $f: \mathbb{C} \to \mathbb{C}$ be a polynomial, $f(z) = z^n + a_1 z^{n-1} + \cdots + a_n$, where $n \geq 1$. Define $c_{R,f}: [0, 1] \to \mathbb{C}$ by $c_{R,f} = f \circ c_{R,1}$.

(a) Show that if R is large enough, then $c_{R,f} - c_{R,n}$ is the boundary of a chain in $\mathbb{C} - \{0\}$. *Hint:* Note that $c_{R^n,n}(t) = [c_{R,1}(t)]^n$, and write

$$f(z) = z^n \left(1 + \frac{a_1}{z} + \cdots + \frac{a_n}{z^n}\right).$$

(b) Show that $f(z) = 0$ for some $z \in \mathbb{C}$ ("Fundamental Theorem of Algebra"). *Hint:* If $f(z) \neq 0$ for all z with $|z| \leq R$, then $c_{R,f} - c_{0,f}$ is a boundary.

5. Some approaches to integration use singular simplexes instead of singular cubes. Although Stokes' Theorem becomes more complicated, there are some advantages in using singular simplexes, as indicated in the next Problem.

Let $\Delta_n \subset \mathbb{R}^n$ be the set of all $x \in \mathbb{R}^n$ such that

$$0 \leq x^i \leq 1, \qquad \sum_{i=1}^{n} x^i \leq 1.$$

A **singular n-simplex** in M is a C^∞ function $c: \Delta_n \to M$, and an **n-chain** is a formal sum of singular n-simplexes. As before, let $I^n: \Delta_n \to \mathbb{R}^n$ be the inclusion map. Define $\partial_i: \Delta_{n-1} \to \Delta_n$ by

$$\partial_0(x) = \left(\left[1 - \sum_{i=1}^{n-1} x^i\right], x^1, \ldots, x^{n-1}\right)$$
$$\partial_i(x) = (x^1, \ldots, x^{i-1}, 0, x^i, \ldots, x^{n-1}) \qquad 0 < i \leq n,$$

and for singular n-simplexes c, define $\partial_i c = c \circ \partial_i$. Then we define

$$\partial c = \sum_{i=0}^{n} (-1)^i \partial_i c.$$

(a) Describe geometrically the images $\partial_i(\Delta_{n-1})$ in Δ_n.
(b) Show that $\partial^2 = 0$.

(c) Show that if $\omega = f\,dx^1 \wedge \cdots \wedge \widehat{dx^i} \wedge \cdots \wedge dx^n$ is an $(n-1)$-form on $\mathbb{R}^n$, then

$$\int_{I^n} d\omega = \int_{\partial I^n} \omega.$$

(Imitate the proof for cubes.)

(d) Define $\int_c \omega$ for any k-chain c in M and k-form ω on M, and prove that

$$\int_c d\omega = \int_{\partial c} \omega$$

for any $(k-1)$-form ω.

6. Every $x \in \Delta_{k+1}$ can be written as tx', for $0 \le t \le 1$, and $x' \in \partial_0(\Delta_k)$.

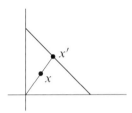

Moreover, x' is unique except when $t = 0$. For any singular k-simplex $c \colon \Delta_k \to \mathbb{R}^n$, define $\bar{c} \colon \Delta_{k+1} \to \mathbb{R}^n$ by

$$\bar{c}(x) = t \cdot c(x').$$

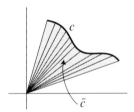

We then define $\bar{c}$ for chains c in the obvious way.

(a) Show that $\partial c = 0$ implies that $c = \partial \bar{c}$.

(b) Let $c \colon [0,1] \to \mathbb{R}^2$ be a closed curve. Show that c is *not* the boundary of any sum σ of singular 2-cubes. *Hint:* If $\partial \sigma = \sum_i a_i c_i$, what can be said about $\sum_i a_i$?

(c) Show that we do have $c = \partial \sigma + c'$ where c' is *degenerate*, that is, $c'([0,1])$ is a point.

(d) If $c_1(0) = c_2(0)$ and $c_1(1) = c_2(1)$, show that $c_1 - c_2$ is a boundary, using either simplexes or cubes.

7. Let ω be a 1-form on a manifold M. Suppose that $\int_c \omega = 0$ for every closed curve c in M. Show that ω is exact. *Hint*: If we do have $\omega = df$, then for any curve c we have

$$\int_c \omega = f(c(1)) - f(c(0)).$$

8. A manifold M is called **simply-connected** if M is connected and if every smooth map $f: S^1 \to M$ is smoothly contractible to a point. [Actually, any space M (not necessarily a manifold) is called simply-connected if it is connected and any continuous $f: S^1 \to M$ is (continuously) contractible to a point. It is not hard to show that for a manifold we may insert "smooth" at both places.]

(a) If M is smoothly contractible to a point, then M is simply-connected.

(b) S^1 is not simply-connected.

(c) S^n is simply-connected for $n > 1$. *Hint*: Show that a smooth $f: S^1 \to S^n$ is not onto.

(d) If M is simply-connected and $p \in M$, then any smooth map $f: S^1 \to M$ is smoothly contractible to p.

(e) If $M = U \cup V$ where U and V are simply-connected open subsets with $U \cap V$ connected, then M is simply-connected. (This gives another proof that S^n is simply-connected for $n > 1$.) *Hint*: Given $f: S^1 \to M$, partition S^1 into a finite number of intervals each of which is taken into either U or V.

(f) If M is simply-connected, then $H^1(M) = 0$. (See Problem 7.)

9. (a) Let $U \subset \mathbb{R}^2$ be a bounded open set such that $\mathbb{R}^2 - U$ is not connected. Show that U is not smoothly contractible to a point. (Converse of

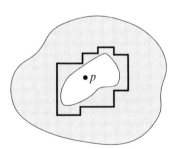

Problem 7-24.) *Hint*: If p is in a bounded component of $\mathbb{R}^2 - U$, show that there is a curve in U which "surrounds" p.

(b) A bounded connected open set $U \subset \mathbb{R}^2$ is smoothly contractible to a point if and only if it is simply-connected.

(c) This is false for open subsets of $\mathbb{R}^3$.

10. Let ω be an n-form on an oriented manifold M^n. Let Φ and Ψ be two partitions of unity by functions with compact support, and suppose that

$$\sum_{\phi \in \Phi} \int_M \phi \cdot |\omega| < \infty.$$

(a) This implies that $\sum_{\phi \in \Phi} \int_M \phi \cdot \omega$ converges absolutely.
(b) Show that

$$\sum_{\phi \in \Phi} \int_M \phi \cdot \omega = \sum_{\phi \in \Phi} \sum_{\psi \in \Psi} \int_M \psi \cdot \phi \cdot \omega,$$

and show the same result with ω replaced by $|\omega|$. (Note that for each ϕ, there are only finitely many ψ which are non-zero on support ϕ.)
(c) Show that $\sum_{\psi \in \Psi} \int_M \psi \cdot |\omega| < \infty$, and that

$$\sum_{\phi \in \Phi} \int_M \phi \cdot \omega = \sum_{\psi \in \Psi} \int_M \psi \cdot \omega.$$

We define this common sum to be $\int_M \omega$.
(d) Let $A_n \subset (n, n+1)$ be closed sets. Let $f : \mathbb{R} \to \mathbb{R}$ be a C^∞ function with $\int_{A_n} f = (-1)^n/n$ and support $f \subset \bigcup_n A_n$. Find two partitions of unity Φ and Ψ such that $\sum_{\phi \in \Phi} \int_{\mathbb{R}} \phi \cdot f\, dx$ and $\sum_{\psi \in \Psi} \int_{\mathbb{R}} \psi \cdot f\, dx$ converge absolutely to different values.

11. Following Problem 7-12, define geometric objects corresponding to odd relative tensors of type $\binom{k}{l}$ and weight w (w any real number).

12. (a) Let M be $\{(x, y) \in \mathbb{R}^2 : |(x, y)| < 1\}$, together with a proper portion

of its boundary, and let $\omega = x\, dy$. Show that

$$\int_M d\omega \neq \int_{\partial M} \omega,$$

even though both sides make sense, using Problem 10. (No computations needed—note that equality would hold if we had the entire boundary.)
(b) Similarly, find a counterexample to Stokes' Theorem when $M = (0, 1)$ and ω is a 0-form whose support is not compact.
(c) Examine a partition of unity for $(0, 1)$ by functions with compact support to see just why the proof of Stokes' Theorem breaks down in this case.

13. Suppose M is a compact orientable n-manifold (with no boundary), and θ is an $(n-1)$-form on M. Show that $d\theta$ is 0 at some point.

14. Let $M_1, M_2 \subset \mathbb{R}^n$ be compact n-dimensional manifolds-with-boundary with $M_2 \subset M_1 - \partial M_1$. Show that for any closed $(n-1)$-form ω on M_1,

$$\int_{\partial M_1} \omega = \int_{\partial M_2} \omega.$$

15. Account for the factor $1/|p|^n$ in Lemma 7 (we have $r_*(v_p) = (1/|p|)v_{r(p)}$, but this only accounts for a factor of $1/|p|^{n-1}$, since there are $n-1$ vectors $v_1, \ldots, v_{n-1}$).

16. Use the formula for $r^* dx^i$ (Problem 4-1) to compute $r^* \sigma'$. (Note that

$$r^* \sigma' = r^* i^* \sigma = (i \circ r)^* \sigma;$$

the map $i \circ r \colon \mathbb{R}^n - \{0\} \to \mathbb{R}^n - \{0\}$ is just r, considered as a map into $\mathbb{R}^n - \{0\}$.)

17. (a) Let M^n and N^m be oriented manifolds, and let ω and η be an n-form and an m-form with compact support, on M and N, respectively. We will orient $M \times N$ by agreeing that $v_1, \ldots, v_n, w_1, \ldots, w_m$ is positively oriented in $(M \times N)_{(p,q)} \approx M_p \oplus N_q$ if $v_1, \ldots, v_n$ and $w_1, \ldots, w_m$ are positively oriented in M_p and N_q, respectively. If $\pi_i \colon M \times N \to M$ or N is projection on the ith factor, show that

$$\int_{M \times N} \pi_1^* \omega \wedge \pi_2^* \eta = \int_M \omega \cdot \int_N \eta.$$

(b) If $h \colon M \times N \to \mathbb{R}$ is C^∞, then

$$\int_{M \times N} h\, \pi_1^* \omega \wedge \pi_2^* \eta = \int_M g\omega,$$

where

$$g(p) = \int_N h(p, \cdot)\eta, \qquad h(p, \cdot) = q \mapsto h(p,q).$$

(c) Every $(m+n)$-form on $M \times N$ is $h\, \pi_1^* \omega \wedge \pi_2^* \eta$ for some ω and η.

18. (a) Let $p \in \mathbb{R}^n - \{0\}$. Let $w_1, \ldots, w_{n-2} \in \mathbb{R}^n{}_p$ and let $v \in \mathbb{R}^n{}_p$ be $(\lambda p)_p$ for some $\lambda \in \mathbb{R}$. Show that

$$r^* \sigma'(v, w_1, \ldots, w_{n-2}) = 0.$$

(b) Let $M \subset \mathbb{R}^n - \{0\}$ be a compact $(n-1)$-manifold-with-boundary which is the union of segments of rays through 0. Show that $\int_M r^* \sigma' = 0$.

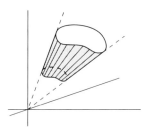

(c) Let $M \subset \mathbb{R}^n - \{0\}$ be a compact $(n-1)$-manifold-with-boundary which intersects every ray through 0 at most once, and let $C(M) = \{\lambda p : p \in M, \lambda \geq 0\}$.

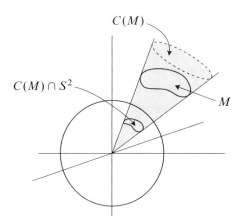

Show that

$$\int_M r^* \sigma' = \int_{C(M) \cap S^2} r^* \sigma'.$$

The latter integral is the measure of the solid angle subtended by M. For this reason we often denote $r^* \sigma'$ by $d\Theta_n$.

19. For all $(x, y, z) \in \mathbb{R}^3$ except those with $x = 0$, $y = 0$, $z \in (-\infty, 0]$, we define $\phi(x, y, z)$ to be the angle between the positive z-axis and the ray from 0 through (x, y, z).

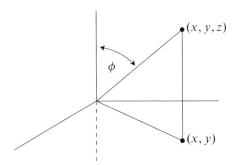

(a) $\phi(x, y, z) = \arctan\left(\sqrt{x^2 + y^2}/z\right)$ (with appropriate conventions).

(b) If $v(p) = |p|$, and θ is considered as a function on $\mathbb{R}^3$, $\theta(x, y, z) = \arctan y/x$, then (v, θ, ϕ) is a coordinate system on the set of all points (x, y, z) in $\mathbb{R}^3$ except those with $y = 0$, $x \in [0, \infty)$ or with $x = 0$, $y = 0$, $z \in (-\infty, 0]$.

(c) If v is a longitudinal unit tangent vector on the sphere $S^2(r)$ of radius r, then $d\phi(v) = 1$. If w points along a meridian through $p = (x, y, z) \in S^2(r)$,

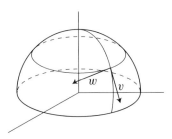

then

$$d\theta(w_p) = \frac{1}{\sqrt{x^2 + y^2}}.$$

(d) If θ and ϕ are taken to mean the restrictions of θ and ϕ to [certain portions of] S^2, then

$$\sigma' = h \, d\theta \wedge d\phi,$$

where $h \colon S^2 \to \mathbb{R}$ is

$$h(x, y, z) = -\sqrt{x^2 + y^2} \quad \text{(the minus sign comes from the orientation).}$$

(e) Conclude that

$$\sigma' = d(-\cos \phi \, d\theta).$$

(f) Let $r_2\colon \mathbb{R}^2 - \{0\} \to S^1$ be the retraction, so that $d\theta = r_2{}^*i^*\sigma$, for the form σ on $\mathbb{R}^2$. Show that

$$r_2{}^*d\theta = d\theta.$$

If $\pi\colon \mathbb{R}^3 \to \mathbb{R}^2$ is the projection, then the form $d\theta$ on [part of] $\mathbb{R}^3$ is just $\pi^*d\theta$, for the form $d\theta$ on [part of] $\mathbb{R}^2$. Use this to show that

$$r^*d\theta = d\theta.$$

(g) Also prove this directly by using the result in part (c), and the fact that $r_*(v_p) = v_{r(p)}/|p|$ for v tangent to $S^2(|p|)$.

(h) Conclude that

$$d\Theta_3 = r^*\sigma' = d(-\cos(\phi \circ r)\, d\theta)$$
$$= d(-\cos\phi\, d\theta).$$

(i) Similarly, express $d\Theta_n$ on $\mathbb{R}^n - \{0\}$ in terms of $d\Theta_{n-1}$ on $\mathbb{R}^{n-1} - \{0\}$.

20. Prove that a connected manifold is the union $U_1 \cup U_2 \cup U_3 \cup \cdots$, where the U_i are coordinate neighborhoods, with $U_i \cap U_j \neq \emptyset$, and the sequence is eventually outside of any compact set.

21. Let $f\colon M^n \to N^n$ be a proper map between oriented n-manifolds such that $f_*\colon M_p \to N_{f(p)}$ is orientation preserving whenever p is a regular point. Show that if N is connected, then either f is onto N, or else all points are critical points of f.

22. (a) Show that a polynomial map $f\colon \mathbb{C} \to \mathbb{C}$, given by $f(z) = z^n + a_1 z^{n-1} + \cdots + a_n$, is proper $(n \geq 1)$.

(b) Let $f'(z) = nz^{n-1} + (n-1)a_1 z^{n-2} + \cdots + a_{n-1}$. Show that we have $f'(z) = \lim_{w \to 0} [f(z+w) - f(z)]/w$, where w varies over complex numbers.

(c) Write $f(x+iy) = u(x,y) + iv(x,y)$ for real-valued functions u and v. Show that

$$f'(x+iy) = \frac{\partial u}{\partial x}(x,y) + i\frac{\partial v}{\partial x}(x,y)$$
$$= \frac{\partial v}{\partial y}(x,y) - i\frac{\partial u}{\partial y}(x,y).$$

Hint: Choose w to be a real h, and then to be ih.

(d) Conclude that

$$|f'(x+iy)|^2 = \det Df(x,y),$$

where f' is defined in part (b), while Df is the linear transformation defined for any differentiable $f : \mathbb{R}^2 \to \mathbb{R}^2$.

(e) Using Problem 21, give another proof of the Fundamental Theorem of Algebra.

(f) There is a still simpler argument, not using Problem 21 (which relies on many theorems of this chapter). Show directly that if $f : M \to N$ is proper, then the number of points in $f^{-1}(a)$ is a locally constant function on the set of regular values of f. Show that this set is connected for a polynomial $f : \mathbb{C} \to \mathbb{C}$, and conclude that f takes on all values.

23. Let $M^{n-1} \subset \mathbb{R}^n$ be a compact oriented manifold. For $p \in \mathbb{R}^n - M$, choose an $(n-1)$-sphere Σ around p such that all points inside Σ are in $\mathbb{R}^n - M$. Let $r_p : \mathbb{R}^n - \{p\} \to \Sigma$ be the obvious retraction. Define the **winding number** $w(p)$ of M around p to be the degree of $r_p | M$.

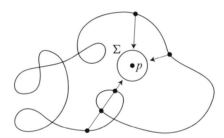

(a) Show that this definition agrees with that in Problem 3.

(b) Show that this definition does not depend on the choice of Σ.

(c) Show that w is constant in a neighborhood of p. Conclude that w is constant on each component of $\mathbb{R}^n - M$.

(d) Suppose M contains a portion A of an $(n-1)$-plane. Let p and q be points

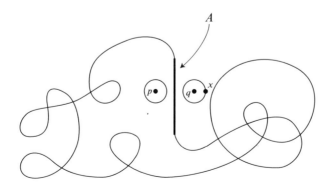

close to this plane, but on opposite sides. Show that $w(q) = w(p) \pm 1$. (Show that $r_q|M$ is homotopic to a map which equals $r_p|M$ on $M - A$ and which does not take any point of A onto the point x in the figure.)

(e) Show that, in general, if M is orientable, then $\mathbb{R}^n - M$ has at least 2 components. The next few Problems show how to prove the same result even if M is not orientable. More precise conclusions are drawn in Chapter 11.

24. Let M and N be compact n-manifolds, and let $f, g : M \to N$ be smoothly homotopic, by a smooth homotopy $H : M \times [0, 1] \to N$.

(a) Let $q \in N$ be a regular value of H. Let $\#f^{-1}(q)$ denote the (finite) number of points in $f^{-1}(q)$. Show that

$$\#f^{-1}(q) \equiv \#g^{-1}(q) \quad (\text{mod } 2).$$

Hint: $H^{-1}(q)$ is a compact 1-manifold-with-boundary. The number of points in its boundary is clearly even. (This is one place where we use the stronger form of Sard's Theorem.)

(b) Show, more generally, that this result holds so long as q is a regular value of both f and g.

25. For two maps $f, g : M \to N$ we will write $f \simeq g$ to indicate that f is smoothly homotopic to g.

(a) If $f \simeq g$, then there is a smooth homotopy $H' : M \times [0, 1] \to N$ such that

$$H'(p, t) = f(p) \quad \text{for } t \text{ in a neighborhood of } 0,$$
$$H'(p, t) = g(p) \quad \text{for } t \text{ in a neighborhood of } 1.$$

(b) $\simeq$ is an equivalence relation.

26. If f is smoothly homotopic to g by a smooth homotopy H such that $p \mapsto H(p, t)$ is a diffeomorphism for each t, we say that f is **smoothly isotopic** to g.

(a) Being smoothly isotopic is an equivalence relation.

(b) Let $\phi : \mathbb{R}^n \to \mathbb{R}$ be a C^∞ function which is positive on the interior of the unit ball, and 0 elsewhere. For $p \in S^{n-1}$, let $H : \mathbb{R} \times \mathbb{R}^n \to \mathbb{R}^n$ satisfy

$$\frac{\partial H(t, x)}{\partial t} = \phi(H(t, x)) \cdot p$$

$$H(0, x) = x.$$

(Each solution is defined for all t, by Theorem 5-6.) Show that each $x \mapsto H(t, x)$ is a diffeomorphism, which is smoothly isotopic to the identity, and leaves all points outside the unit ball fixed.

(c) Show that by choosing suitable p and t we can make $H(t,0)$ be any point in the interior of the unit ball.

(d) If M is connected and $p, q \in M$, then there is a diffeomorphism $f : M \to M$ such that $f(p) = q$ and f is smoothly isotopic to the identity.

(e) Use part (d) to give an alternate proof of *Step 3* of Theorem 9.

(f) If M and N are compact n-manifolds, and $f : M \to N$, then for regular values $q_1, q_2 \in N$ we have

$$\# f^{-1}(q_1) \equiv \# f^{-1}(q_2) \quad (\text{mod } 2)$$

(where $\# f^{-1}(q)$ is defined in Problem 24). This number is called the **mod 2 degree** of f.

(g) By replacing "degree" with "mod 2 degree" in Problem 23, show that if $M \subset \mathbb{R}^n$ is a compact $(n-1)$-manifold, then $\mathbb{R}^n - M$ has at least 2 components.

27. Let $\{X^t\}$ be a C^∞ family of C^∞ vector fields on a compact manifold M. (To be more precise, suppose X is a C^∞ vector field on $M \times [0, 1]$; then $X^t(p)$ will denote $\pi_{M*} X_{(p,t)}$.) From the addendum to Chapter 5, and the argument which was used in the proof of Theorem 5-6, it follows that there is a C^∞ family $\{\phi_t\}$ of diffeomorphisms of M [not necessarily a 1-parameter group], with $\phi_0 = $ identity, which is generated by $\{X^t\}$, i.e., for any C^∞ function $f : M \to \mathbb{R}$ we have

$$(X^t f)(p) = \lim_{h \to 0} \frac{f(\phi_{t+h}(p)) - f(\phi_t(p))}{h}.$$

For a family ω_t of k-forms on M we define the k-form

$$\dot{\omega}_t = \lim_{h \to 0} \frac{\omega_{t+h} - \omega_t}{h}.$$

(a) Show that for $\eta(t) = \phi_t{}^* \omega_t$ we have

$$\dot{\eta}_t = \phi_t{}^*(L_{X^t} \omega_t + \dot{\omega}_t).$$

(b) Let ω_0 and ω_1 be nowhere zero n-forms on a compact oriented n-manifold M, and define

$$\omega_t = (1 - t)\omega_0 + t\omega_1.$$

Show that the family ϕ_t of diffeomorphisms generated by $\{X^t\}$ satisfies

$$\phi_t{}^* \omega_t = \omega_0 \qquad \text{for all } t$$

if and only if

$$L_{X^t} \omega_t = \omega_0 - \omega_1.$$

(c) Using Problem 7-18, show that this holds if and only if

$$d(X^t \lrcorner \omega_t) = \omega_0 - \omega_1.$$

(d) Suppose that $\int_M \omega_0 = \int_M \omega_1$, so that $\omega_0 - \omega_1 = d\lambda$ for some λ. Show that there is a diffeomorphism $f_1 \colon M \to M$ such that $\omega_0 = f_1{}^*\omega_1$.

28. Let $f \colon M^k \to \mathbb{R}^n$ and $g \colon N^l \to \mathbb{R}^n$ be C^∞ maps, where M and N are compact oriented manifolds, $n = k + l + 1$, and $f(M) \cap g(N) = \emptyset$. Define

$$\alpha_{f,g} \colon M \times N \to S^{n-1} \subset \mathbb{R}^n - \{0\}$$

by

$$\alpha_{f,g}(p,q) = r(g(q) - f(p)) = \frac{g(q) - f(p)}{|g(q) - f(p)|}.$$

We define the **linking number** of f and g to be

$$\ell(f,g) = \deg \alpha_{f,g},$$

where $M \times N$ is oriented as in Problem 18.

(a) $\ell(f,g) = (-1)^{kl+1}\ell(g,f)$.
(b) Let $H \colon M \times [0,1] \to \mathbb{R}^n$ and $K \colon N \times [0,1] \to \mathbb{R}^n$ be smooth homotopies with

$$H(p,0) = f(p) \qquad K(q,0) = g(q)$$
$$H(p,1) = \bar{f}(p) \qquad K(q,1) = \bar{g}(q)$$

such that

$$\{H(p,t) : p \in M\} \cap \{K(q,t) : q \in N\} = \emptyset \quad \text{for every } t.$$

Show that
$$\ell(f,g) = \ell(\bar{f},\bar{g}).$$

(c) For $f,g \colon S^1 \to \mathbb{R}^3$ show that

$$\ell(f,g) = \frac{-1}{4\pi} \int_0^1 \int_0^1 \frac{A(u,v)}{[r(u,v)]^3} \, du \, dv,$$

where

$$r(u, v) = |g(v) - f(u)|$$

$$A(u, v) = \det \begin{pmatrix} (f^1)'(u) & (f^2)'(u) & (f^3)'(u) \\ (g^1)'(v) & (g^2)'(v) & (g^3)'(v) \\ g^1(v) - f^1(u) & g^2(v) - f^2(u) & g^3(v) - f^3(u) \end{pmatrix}$$

(the factor $1/4\pi$ comes from the fact that $\int_{S^2} \sigma' = 4\pi$ [Problem 9-14]).

(d) Show that $\ell(f, g) = 0$ if f and g both lie in the same plane (first do it for (x, y)-plane). The next problem shows how to determine $\ell(f, g)$ without calculating.

29. (a) For $(a, b, c) \in \mathbb{R}^3$ define

$$d\Theta_{(a,b,c)} = \frac{(x - a)\, dy \wedge dz - (y - b)\, dx \wedge dz + (z - c)\, dx \wedge dy}{[(x - a)^2 + (y - b)^2 + (z - c)^2]^{3/2}}.$$

For a compact oriented 2-manifold-with-boundary $M \subset \mathbb{R}^3$ and $(a, b, c) \notin M$, let

$$\Omega(a, b, c) = \int_M d\Theta_{(a,b,c)}.$$

Let (a, b, c) and (a', b', c') be points close to $p \in M$, on opposite sides of M. Suppose (a, b, c) is on the same side as a vector $w_p \in \mathbb{R}^3{}_p - M_p$ for which the

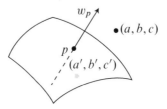

triple $w_p, (v_1)_p, (v_2)_p$ is positively oriented in $\mathbb{R}^3{}_p$ when $(v_1)_p, (v_2)_p$ is positively oriented in M_p. Show that

$$\lim_{\substack{(a,b,c) \to p \\ (a',b',c') \to p}} \Omega(a, b, c) - \Omega(a', b', c') = -4\pi.$$

Hint: First show that if $M = \partial N$, then $\Omega(a, b, c) = -4\pi$ for $(a, b, c) \in N - M$ and $\Omega(a, b, c) = 0$ for $(a, b, c) \notin N$.

(b) Let $f\colon S^1 \to \mathbb{R}^3$ be an imbedding such that $f(S^1) = \partial M$ for some compact oriented 2-manifold-with-boundary M. (An M with this property always exists. See Fort, *Topology of 3-Manifolds*, pg. 138.) Let $g\colon S^1 \to \mathbb{R}^3$ and suppose

The figure on the left shows a *non*-orientable surface whose boundary is the "trefoil" knot,

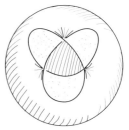

but the surface on the right—including the hemisphere behind the plane of the paper— *is* orientable.

that when $g(t) = p \in M$ we have $dg/dt \notin M_p$. Let n^+ be the number of intersections where dg/dt points in the same direction as the vector w_p of part (a), and n^- the number of other intersections. Show that

$$n = n^+ - n^- = \frac{-1}{4\pi} \int_{S^1} g^*(d\Omega).$$

(c) Show that

$$\frac{\partial \Omega}{\partial a}(a,b,c) = \int_{S^1} f^* \left(\frac{(y-b)\,dz - (z-c)\,dy}{|(x,y,z)|^3} \right)$$

$$\frac{\partial \Omega}{\partial b}(a,b,c) = \int_{S^1} f^* \left(\frac{(z-c)\,dx - (x-a)\,dz}{|(x,y,z)|^3} \right)$$

$$\frac{\partial \Omega}{\partial c}(a,b,c) = \int_{S^1} f^* \left(\frac{(x-a)\,dy - (y-b)\,dx}{|(x,y,z)|^3} \right).$$

(d) Show that $n = \ell(f,g)$. Compute $\ell(f,g)$ for the pairs shown below.

30. (a) Let $p, q \in \mathbb{R}^n$ be distinct. Choose open sets $A, B \subset \mathbb{R}^n - \{p, q\}$ so that A and B are diffeomorphic to $\mathbb{R}^n - \{0\}$, and $A \cap B$ is diffeomorphic to $\mathbb{R}^n$. Using an argument similar to that in the proof of Theorem 16, show that

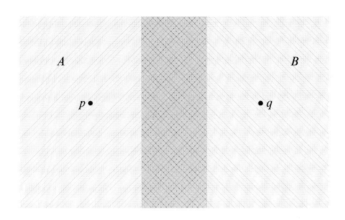

$H^k(\mathbb{R}^n - \{p, q\}) = 0$ for $0 < k < n - 1$, and that $H^{n-1}(\mathbb{R}^n - \{p, q\})$ has dimension 2.

(b) Find the de Rham cohomology vector spaces of $\mathbb{R}^n - F$ where $F \subset \mathbb{R}^n$ is a finite set.

31. We define the **cup product** $\cup : H^k(M) \times H^l(M) \to H^{k+l}(M)$ by

$$[\omega] \cup [\eta] = [\omega \wedge \eta].$$

(a) Show that $\cup$ is well-defined, i.e., $\omega \wedge \eta$ is exact if ω is exact and η is closed.
(b) Show that $\cup$ is bilinear.
(c) If $\alpha \in H^k(M)$ and $\beta \in H^l(M)$, then $\alpha \cup \beta = (-1)^{kl} \beta \cup \alpha$.
(d) If $f : M \to N$, and $\alpha \in H^k(N)$, $\beta \in H^l(N)$, then

$$f^*(\alpha \cup \beta) = f^*\alpha \cup f^*\beta.$$

(e) The **cross-product** $\times : H^k(M) \times H^l(N) \to H^{k+l}(M \times N)$ is defined by

$$[\omega] \times [\eta] = [\pi_M{}^*\omega \wedge \pi_N{}^*\eta].$$

Show that $\times$ is well-defined, and that

$$\alpha \times \beta = \pi_M{}^*\alpha \cup \pi_N{}^*\beta.$$

(f) If $\Delta: M \to M \times M$ is the "diagonal map", given by $\Delta(p) = (p, p)$, show that

$$\alpha \cup \beta = \Delta^*(\alpha \times \beta).$$

32. On the n-dimensional torus

$$T^n = \underbrace{S^1 \times \cdots \times S^1}_{n \text{ times}}$$

let $d\theta^i$ denote $\pi_i{}^* d\theta$, where $\pi_i : T^n \to S^1$ is projection on the ith factor.

(a) Show that all $d\theta^{i_1} \wedge \cdots \wedge d\theta^{i_k}$ represent different elements of $H^k(T^n)$, by finding submanifolds of T^n over which they have different integrals. Hence $\dim H^k(T^n) \geq \binom{n}{k}$. Equality is proved in the Problems for Chapter 11.

(b) Show that every map $f: S^n \to T^n$ has degree 0. *Hint*: Use Problem 25.

CHAPTER 9

RIEMANNIAN METRICS

In previous chapters we have exploited nearly every construction associated with vector spaces, and thus with bundles, but there has been one notable exception—we have never mentioned inner products. The time has now come to make use of this neglected tool.

An **inner product** on a vector space V over a field F is a bilinear function from $V \times V$ to F, denoted by $(v, w) \mapsto \langle v, w \rangle$, which is **symmetric**,

$$\langle v, w \rangle = \langle w, v \rangle,$$

and **non-degenerate**: if $v \neq 0$, then there is some $w \neq 0$ such that

$$\langle w, v \rangle \neq 0.$$

For us, the field F will always be $\mathbb{R}$.

For each r with $0 \leq r \leq n$, we can define an inner product $\langle \ , \ \rangle_r$ on $\mathbb{R}^n$ by

$$\langle a, b \rangle_r = \sum_{i=1}^{r} a^i b^i - \sum_{i=r+1}^{n} a^i b^i;$$

this is non-degenerate because if $a \neq 0$, then

$$\left\langle (a^1, \ldots, a^n), (a^1, \ldots, a^r, -a^{r+1}, \ldots, -a^n) \right\rangle_r = \sum_{i=1}^{n} (a^i)^2 > 0.$$

In particular, for $r = n$ we obtain the "usual inner product", $\langle \ , \ \rangle$ on $\mathbb{R}^n$,

$$\langle a, b \rangle = \sum_{i=1}^{n} a^i b^i.$$

For this inner product we have $\langle a, a \rangle > 0$ for any $a \neq 0$. In general, a symmetric bilinear function $\langle \ , \ \rangle$ is called **positive definite** if

$$\langle v, v \rangle > 0 \qquad \text{for all } v \neq 0.$$

A positive definite bilinear function $\langle \ , \ \rangle$ is clearly non-degenerate, and consequently an inner product.

301

Notice that an inner product $\langle \ , \ \rangle$ on V is an element of $\mathcal{T}^2(V)$, so if $f \colon W \to V$ is a linear transformation, then $f^*\langle \ , \ \rangle$ is a symmetric bilinear function on W. This symmetric bilinear function may be degenerate even if f is one-one, e.g., if $\langle \ , \ \rangle$ is defined on $\mathbb{R}^2$ by

$$\langle a, b \rangle = a^1 b^1 - a^2 b^2,$$

and $f \colon \mathbb{R} \to \mathbb{R}^2$ is

$$f(a) = (a, a).$$

However, $f^*\langle \ , \ \rangle$ is clearly non-degenerate if f is an isomorphism *onto* V. Also, if $\langle \ , \ \rangle$ is positive definite, then $f^*\langle \ , \ \rangle$ is positive definite if and only if f is one-one.

For any basis $v_1, \ldots, v_n$ of V, with corresponding dual basis $v^*{}_1, \ldots, v^*{}_n$, we can write

$$\langle \ , \ \rangle = \sum_{i,j=1}^{n} g_{ij} v^*{}_i \otimes v^*{}_j.$$

In this expression,

$$g_{ij} = \langle v_i, v_j \rangle,$$

so symmetry of $\langle \ , \ \rangle$ implies that the matrix (g_{ij}) is symmetric,

$$g_{ij} = g_{ji}.$$

The matrix (g_{ij}) has another important interpretation. Since an inner product $\langle \ , \ \rangle$ is linear in the second argument, we can define a linear functional $\phi_v \in V^*$, for each $v \in V$, by

$$\phi_v(w) = \langle v, w \rangle.$$

Since $\langle \ , \ \rangle$ is linear in the first argument, the map $v \mapsto \phi_v$ is a linear transformation from V to V^*. Non-degeneracy of $\langle \ , \ \rangle$ implies that $\phi_v \neq 0$ if $v \neq 0$. Thus, if V is finite dimensional, an inner product $\langle \ , \ \rangle$ gives us an isomorphism $\alpha \colon V \to V^*$, with

$$\langle v, w \rangle = \alpha(v)(w).$$

Clearly, the matrix (g_{ij}) is just the matrix of $\alpha \colon V \to V^*$ with respect to the bases $\{v_i\}$ for V and $\{v^*{}_i\}$ for V^*. Thus, non-degeneracy of $\langle \ , \ \rangle$ is equivalent to the condition that

$$(g_{ij}) \text{ is non-singular}, \qquad \det(g_{ij}) \neq 0.$$

Positive definiteness of $\langle \ , \ \rangle$ corresponds to the more complicated condition that the matrix (g_{ij}) be "positive definite", meaning that

$$\sum_{i=1}^{n} g_{ij} a^i a^j > 0 \qquad \text{for all } a_1, \ldots, a_n \text{ with at least one } a^i \neq 0.$$

Given any *positive definite* inner product $\langle\ ,\ \rangle$ on V we define the associated **norm** $\|\ \|$ by

$$\|v\| = \sqrt{\langle v, v \rangle} \qquad \text{(the positive square root is to be taken).}$$

In $\mathbb{R}^n$ we denote the norm corresponding to $\langle\ ,\ \rangle$ simply by

$$|a| = \sqrt{\langle a, a \rangle} = \sqrt{\sum_{i=1}^{n} (a^i)^2}\,.$$

The principal properties of $\|\ \|$ are the following.

1. **THEOREM.** For all $v, w \in V$ we have

    (1) $\|av\| = |a| \cdot \|v\|$.

    (2) $|\langle v, w \rangle| \leq \|v\| \cdot \|w\|$, with equality if and only if v and w are linearly dependent (Schwarz inequality).

    (3) $\|v + w\| \leq \|v\| + \|w\|$ (Triangle inequality).

PROOF. (1) is trivial.
(2) If v and w are linearly dependent, equality clearly holds. If not, then $0 \neq \lambda v - w$ for all $\lambda \in \mathbb{R}$, so

$$0 < \|\lambda v - w\|^2 = \langle \lambda v - w, \lambda v - w \rangle$$
$$= \lambda^2 \|v\|^2 - 2\lambda \langle v, w \rangle + \|w\|^2.$$

So the right side is a quadratic equation in λ with no real solution, and its discriminant must be negative. Thus

$$4\langle v, w \rangle^2 - 4\|v\|^2 \|w\|^2 < 0.$$

$$
\begin{aligned}
(3) \qquad \|v + w\|^2 &= \langle v + w, v + w \rangle \\
&= \|v\|^2 + \|w\|^2 + 2\langle v, w \rangle \\
&\leq \|v\|^2 + \|w\|^2 + 2\|v\| \cdot \|w\| \qquad \text{by (2)} \\
&= (\|v\| + \|w\|)^2. \; \clubsuit
\end{aligned}
$$

The function $\|\ \|$ has certain unpleasant properties—for example, the function $|\ |$ on $\mathbb{R}^n$ is not differentiable at $0 \in \mathbb{R}^n$—which do not arise for the function $\|\ \|^2$. This latter function is a "quadratic function" on V—in terms of a basis $\{v_i\}$ for V it can be written as a "homogeneous polynomial of degree 2" in the components,

$$\left\| \sum_{i=1}^{n} a^i v_i \right\|^2 = \sum_{i,j=1}^{n} g_{ij} a^i a^j\,.$$

More succinctly,

$$\| \ \|^2 = \sum_{i,j=1}^{n} g_{ij} v^*_i \cdot v^*_j.$$

An invariant definition of a quadratic function can be obtained (Problem 1) from the following observation.

2. THEOREM (POLARIZATION IDENTITY). If $\| \ \|$ is the norm associated to an inner product $\langle \ , \ \rangle$ on V, then

(1) $\langle v, w \rangle = \frac{1}{2}[\|v + w\|^2 - \|v\|^2 - \|w\|^2]$

(2) $\langle v, w \rangle = \frac{1}{4}[\|v + w\|^2 - \|v - w\|^2].$

PROOF. Compute. ❖

Theorem 2 shows that two inner products which induce the same norm are themselves equal. Similarly, if $f \colon V \to V$ is norm preserving, that is, $\|f(v)\| = \|v\|$ for all $v \in V$, then f is also inner product preserving, that is, $\langle f(v), f(w) \rangle = \langle v, w \rangle$ for all $v, w \in V$.

We will now see that, "up to isomorphism", there is only one positive definite inner product.

3. THEOREM. If $\langle \ , \ \rangle$ is a positive definite inner product on an n-dimensional vector space V, then there is a basis $v_1, \ldots, v_n$ for V such that $\langle v_i, v_j \rangle = \delta_{ij}$. (Such a basis is called **orthonormal** with respect to $\langle \ , \ \rangle$.) Consequently, there is an isomorphism $f \colon \mathbb{R}^n \to V$ such that

$$\langle\!\langle a, b \rangle\!\rangle = \langle f(a), f(b) \rangle, \qquad a, b \in \mathbb{R}^n.$$

In other words,

$$f^* \langle \ , \ \rangle = \langle\!\langle \ , \ \rangle\!\rangle.$$

PROOF. Let $w_1, \ldots, w_n$ be any basis for V. We obtain the desired basis by applying the "Gram-Schmidt orthonormalization process" to this basis:

Since $w_1 \neq 0$, we can define

$$v_1 = \frac{w_1}{\|w_1\|},$$

and clearly $\|v_1\| = 1$. Suppose that we have constructed $v_1, \ldots, v_k$ so that

$$\langle v_i, v_j \rangle = \delta_{ij} \qquad 1 \le i, j \le k$$

and

$$\text{span } v_1, \ldots, v_k = \text{span } w_1, \ldots, w_k.$$

Then w_{k+1} is linearly independent of $v_1, \ldots, v_k$. Let

$$w'_{k+1} = w_{k+1} - \langle v_1, v_{k+1} \rangle v_1 - \cdots - \langle v_k, v_{k+1} \rangle v_k \neq 0.$$

It is easy to see that

$$\langle w'_{k+1}, v_i \rangle = 0 \qquad i = 1, \ldots, k.$$

So we can define

$$v_{k+1} = \frac{w'_{k+1}}{\|w'_{k+1}\|},$$

and continue inductively. ❖

A positive definite inner product $\langle \ , \ \rangle$ on V is sometimes called a *Euclidean metric* on V. This is because we obtain a metric ρ on V by defining

$$\rho(v, w) = \|v - w\|.$$

The "triangle inequality" (Theorem 1(3)) shows that this is indeed a metric. We also call $\|v\|$ the **length** of v.

We have only one more algebraic trick to play. Recall that an inner product $\langle \ , \ \rangle$ on V provides an isomorphism $\alpha \colon V \to V^*$ with

$$\alpha(v)(w) = \langle v, w \rangle.$$

Using the natural isomorphism $i \colon V \to V^{**}$, defined by

$$i(v)(\lambda) = \lambda(v),$$

we obtain an isomorphism

$$\beta \colon V^* \xrightarrow{\ \alpha^{-1}\ } V \xrightarrow{\ i\ } (V^*)^*.$$

We can now use β to define a bilinear function $\langle \ , \ \rangle^*$ on V^* by

$$\langle \lambda, \mu \rangle^* = \beta(\lambda)(\mu) = i\alpha^{-1}(\lambda)(\mu) = \mu(\alpha^{-1}(\lambda)).$$

Now, the symmetry of $\langle \ , \ \rangle$ can be expressed by the equation

$$\alpha(v)(w) = \alpha(w)(v).$$

Letting

$$\alpha(v) = \lambda, \quad \alpha(w) = \mu,$$

this can be written

$$\lambda(\alpha^{-1}(\mu)) = \mu(\alpha^{-1}(\lambda)),$$

which shows that $\langle \ , \ \rangle^*$ is also symmetric,

$$\langle \mu, \lambda \rangle^* = \langle \lambda, \mu \rangle^*.$$

Consequently $\langle \ , \ \rangle^*$ is an inner product on the dual space V^* (in fact, the one which produces β).

To see what this all means, choose a basis $\{v_i\}$ for V, let $\{v^*_i\}$ be the dual basis for V^*, and let

$$\langle \ , \ \rangle = \sum_{i,j=1}^{n} g_{ij} v^*_i \otimes v^*_j.$$

Then

(g_{ij}) is the matrix of $\quad \alpha: V \to V^* \quad$ with respect to $\{v_i\}$ and $\{v^*_i\}$

so

$(g_{ij})^{-1}$ is the matrix of $\alpha^{-1}: V^* \to V \quad$ with respect to $\{v^*_i\}$ and $\{v_i\}$

so

$(g_{ij})^{-1}$ is the matrix of $\quad \beta: V^* \to V^{**}$ with respect to $\{v^*_i\}$ and $\{v^{**}_i\}$.

Thus, if we let g^{ij} be the entries of the inverse matrix, $(g^{ij}) = (g_{ij})^{-1}$, so that

$$\sum_{k=1}^{n} g^{ik} g_{kj} = \delta^i_j,$$

then

$$\langle \ , \ \rangle^* = \sum_{i,j=1}^{n} g^{ij} v^{**}_i \otimes v^{**}_j$$

$$= \sum_{i,j=1}^{n} g^{ij} v_i \otimes v_j, \quad \text{if we consider } v_i \in V^{**}.$$

One can check directly (Problem 9), without the invariant definition, that this equation defines $\langle \ , \ \rangle^*$ independently of the choice of basis.

Notice that if $\langle \ , \ \rangle$ is positive definite, so that

$$\alpha(v)(v) > 0 \qquad \text{for } v \neq 0,$$

then, letting $\alpha(v) = \lambda$, we have

$$\lambda(\alpha^{-1}(\lambda)) = \beta(\lambda)(\lambda) > 0 \qquad \text{for } \lambda \neq 0,$$

so $\langle \ , \ \rangle^*$ is also positive definite. This can also be checked directly from the definition in terms of a basis. In the positive definite case, the simplest way to describe $\langle \ , \ \rangle^*$ is as follows: The basis $v^*_1, \ldots, v^*_n$ of V^* is orthonormal with respect to $\langle \ , \ \rangle^*$ if and only if $v_1, \ldots, v_n$ is orthonormal with respect to $\langle \ , \ \rangle$.

Similar tricks can be used (Problem 4) to produce an inner product on all the vector spaces $\mathcal{T}^k(V)$, $\mathcal{T}_k(V) = \mathcal{T}^k(V^*)$, and $\Omega^k(V)$. However, we are interested in only one case, which we will not describe in a completely invariant way. The vector space $\Omega^n(V)$ is 1-dimensional, so to produce an inner product on it, we need only describe which two elements, ω and $-\omega$, will have length 1. Let $v_1, \ldots, v_n$ and $w_1, \ldots, w_n$ be two bases of V which are orthonormal with respect to $\langle \ , \ \rangle$. If we write

$$w_i = \sum_{j=1}^{n} \alpha_{ji} v_j,$$

then

$$\delta_{ij} = \langle w_i, w_j \rangle = \left\langle \sum_{k=1}^{n} \alpha_{ki} v_k, \sum_{l=1}^{n} \alpha_{lj} v_l \right\rangle = \sum_{k,l=1}^{n} \alpha_{ki} \alpha_{lj} \langle v_k, v_l \rangle$$

$$= \sum_{k=1}^{n} \alpha_{ki} \alpha_{kj}.$$

So the transpose matrix $A^{\mathbf{t}}$ of $A = (\alpha_{ij})$ satisfies $A \cdot A^{\mathbf{t}} = I$, which implies that $\det A = \pm 1$. It follows from Theorem 7-5 that for any $\omega \in \Omega^n(V)$ we have

$$\omega(v_1, \ldots, v_n) = \pm \omega(w_1, \ldots, w_n).$$

It clearly follows that

$$v^*_1 \wedge \cdots \wedge v^*_n = \pm w^*_1 \wedge \cdots \wedge w^*_n.$$

We have thus distinguished two elements of $\Omega^n(V)$; they are both of the form $v^*_1 \wedge \cdots \wedge v^*_n$ for $\{v_i\}$ an orthonormal basis of V. We will call these two elements the **elements of norm** 1 in $\Omega^n(V)$. If we also have an orientation μ, then we can further distinguish the one which is positive when applied to any $(v_1, \ldots, v_n)$ with $[v_1, \ldots, v_n] = \mu$; we will call it the **positive element of norm** 1 in $\Omega^n(V)$.

To express the elements of norm 1 in terms of an arbitrary basis $w_1, \ldots, w_n$, we choose an orthonormal basis $v_1, \ldots, v_n$ and write

$$w_i = \sum_{j=1}^{n} \alpha_{ji} v_j.$$

Problem 7-9 implies that

$$\det(\alpha_{ij}) \, w^*_1 \wedge \cdots \wedge w^*_n = v^*_1 \wedge \cdots \wedge v^*_n.$$

If we write

$$\langle \, , \, \rangle = \sum_{i,j=1}^{n} g_{ij} \, w^*_i \otimes w^*_j,$$

then

$$g_{ij} = \langle w_i, w_j \rangle = \left\langle \sum_{k=1}^{n} \alpha_{ki} v_k, \sum_{l=1}^{n} \alpha_{lj} v_l \right\rangle$$

$$= \sum_{k=1}^{n} \alpha_{ki} \alpha_{kj},$$

so if $A = (\alpha_{ij})$, then

$$\det(g_{ij}) = \det(A^{\mathbf{t}} \cdot A) = (\det A)^2.$$

In particular, $\det(g_{ij})$ *is always positive.* Consequently, the elements of norm 1 in $\Omega^n(V)$ are

$$\pm \sqrt{\det(g_{ij})} \, w^*_1 \wedge \cdots \wedge w^*_n \qquad\qquad g_{ij} = \langle w_i, w_j \rangle.$$

We now apply our new tool to vector bundles. If $\xi = \pi \colon E \to B$ is a vector bundle, we define a **Riemannian metric** on ξ to be a function $\langle \, , \, \rangle$ which assigns to each $p \in B$ a positive definite inner product $\langle \, , \, \rangle_p$ on $\pi^{-1}(p)$, and which is continuous in the sense that for any two continuous sections $s_1, s_2 \colon B \to E$, the function

$$\langle s_1, s_2 \rangle = p \mapsto \langle s_1(p), s_2(p) \rangle_p$$

is also continuous. If ξ is a C^∞ vector bundle over a C^∞ manifold we can also speak of C^∞ Riemannian metrics.

[Another approach to the definition can be given. Let $Euc(V)$ be the set of all positive definite inner products on V. If we replace each $\pi^{-1}(p)$ by $Euc(\pi^{-1}(p))$, and let

$$Euc(\xi) = \bigcup_{p \in B} Euc(\pi^{-1}(p)),$$

then a Riemannian metric on ξ can be defined to be a section of $Euc(\xi)$. The only problem is that $Euc(V)$ is not a vector space; the new object $Euc(\xi)$ that we obtain is not a vector bundle at all, but an instance of a more general structure, a fibre bundle.]

4. THEOREM. Let $\xi = \pi : E \to M$ be a $[C^\infty]$ k-plane bundle over a C^∞ manifold M. Then there is a $[C^\infty]$ Riemannian metric on ξ.

PROOF. There is an open locally finite cover $\mathcal{O}$ of M by sets U for which there exists $[C^\infty]$ trivializations

$$t_U : \pi^{-1}(U) \to U \times \mathbb{R}^k.$$

On $U \times \mathbb{R}^k$, there is an obvious Riemannian metric,

$$\langle (p, a), (p, b) \rangle_p = \langle a, b \rangle.$$

For $v, w \in \pi^{-1}(p)$, define

$$\langle v, w \rangle_p^U = \langle t_U(v), t_U(w) \rangle_p.$$

Then $\langle \ , \ \rangle^U$ is a $[C^\infty]$ Riemannian metric for $\xi | U$. Let $\{\phi_U\}$ be a partition of unity subordinate to $\mathcal{O}$. We define $\langle \ , \ \rangle$ by

$$\langle v, w \rangle_p = \sum_{U \in \mathcal{O}} \phi_U(p) \langle v, w \rangle_p^U \qquad v, w \in \pi^{-1}(p).$$

Then $\langle \ , \ \rangle$ is continuous $[C^\infty]$ and each $\langle \ , \ \rangle_p$ is a symmetric bilinear function on $\pi^{-1}(p)$. To show that it is positive definite, note that

$$\langle v, v \rangle_p = \sum_{U \in \mathcal{O}} \phi_U(p) \langle v, v \rangle_p^U;$$

each $\phi_U(p) \langle v, v \rangle_p^U \geq 0$, and for some U strict inequality holds. ❖

[The same argument shows that any vector bundle over a paracompact space has a Riemannian metric.]

Notice that the argument in the final step would not work if we had merely picked non-degenerate inner products $\langle \ , \ \rangle^U$. In fact (Problem 7), there is no $\langle \ , \ \rangle$ on TS^2 which gives a symmetric bilinear function on each S^2_p which is not positive definite or negative definite but is still non-degenerate.

As an application of Theorem 4, we settle some questions which have till now remained unanswered.

5. COROLLARY. If $\xi = \pi : E \to M$ is a k-plane bundle, then $\xi \simeq \xi^*$.

PROOF. Let $\langle \ , \ \rangle$ be a Riemannian metric for ξ. Then for each $p \in M$, we have an isomorphism

$$\alpha_p : \pi^{-1}(p) \to [\pi^{-1}(p)]^*$$

defined by

$$\alpha_p(v)(w) = \langle v, w \rangle_p \qquad v, w \in \pi^{-1}(p).$$

Continuity of $\langle \ , \ \rangle$ implies that the union of all α_p is a homeomorphism from E to $E' = \bigcup_{p \in M} [\pi^{-1}(p)]^*$. ❖

6. COROLLARY. If $\xi = \pi : E \to M$ is a 1-plane bundle, then ξ is trivial if and only if ξ is orientable.

PROOF. The "only if" part is trivial. If ξ has an orientation μ and $\langle \ , \ \rangle$ is a Riemannian metric on M then there is a unique

$$s(p) \in \pi^{-1}(p)$$

with

$$\langle s(p), s(p) \rangle_p = 1, \qquad [s(p)] = \mu_p.$$

Clearly s is a section; we then define an equivalence $f : E \to M \times \mathbb{R}$ by

$$f(\lambda s(p)) = (p, \lambda).$$

ALTERNATIVE PROOF. We know (see the discussion after Theorem 7-9) that if ξ is orientable, then there is a nowhere 0 section of

$$\Omega^1(\xi) = \xi^*,$$

so that ξ^* is trivial. But $\xi \simeq \xi^*$. ❖

All these considerations take on special significance when our bundle is the tangent bundle TM of a C^∞ manifold M. In this case, a C^∞ Riemannian metric $\langle \ , \ \rangle$ for TM, which gives a positive definite inner product $\langle \ , \ \rangle_p$ on each M_p, is called a **Riemannian metric on** M. If (x, U) is a coordinate system on M, then on U we can write our Riemannian metric $\langle \, , \, \rangle$ as

$$\langle \, , \, \rangle = \sum_{i,j=1}^{n} g_{ij} \, dx^i \otimes dx^j,$$

where the C^∞ functions g_{ij} satisfy $g_{ij} = g_{ji}$, since $\langle \, , \, \rangle$ is symmetric, and $\det(g_{ij}) > 0$ since $\langle \, , \, \rangle$ is positive definite. A Riemannian metric $\langle \, , \, \rangle$ on M is, of course, a covariant tensor of order 2. So for every C^∞ map $f \colon N \to M$ there is a covariant tensor $f^*\langle \, , \, \rangle$ on N, which is clearly symmetric; it is a Riemannian metric on N if and only if f is an immersion (f_{*p} is one-one for all $p \in N$).

The Riemannian metric $\langle \, , \, \rangle^*$, which $\langle \, , \, \rangle$ induces on the dual bundle T^*M, is a contravariant tensor of order 2, and we can write it as

$$\langle \, , \, \rangle^* = \sum_{i,j=1}^{n} g^{ij} \frac{\partial}{\partial x^i} \otimes \frac{\partial}{\partial x^j}.$$

Our discussion of inner products induced on V^* shows that for each p, the matrix $(g^{ij}(p))$ is the inverse of the matrix $(g_{ij}(p))$; thus

$$\sum_{k=1}^{n} g_{ik} g^{kj} = \delta_i^j.$$

Similarly, for each $p \in M$ the Riemannian metric $\langle \, , \, \rangle$ on M determines two elements of $\Omega^n(M_p)$, the elements of norm 1. We have seen that they can be written

$$\pm \sqrt{\det(g_{ij}(p))} \, dx^1(p) \wedge \cdots \wedge dx^n(p).$$

If M has an orientation μ, then μ_p allows us to pick out the positive element of norm 1, and we obtain an n-form on M; if $x \colon U \to \mathbb{R}^n$ is *orientation preserving*, then on U this form can be written

$$\sqrt{\det(g_{ij})} \, dx^1 \wedge \cdots \wedge dx^n.$$

Even if M is not orientable, we obtain a "volume element" on M, as defined in Chapter 8; in a coordinate system (x, U) it can be written as

$$\sqrt{\det(g_{ij})} \, |dx^1 \wedge \cdots \wedge dx^n|.$$

This volume element is denoted by dV, even though it is usually not d of anything (even when M is orientable and it can be considered to be an n-form), and is called the **volume element determined by** $\langle\ ,\ \rangle$. We can then define the **volume** of M as

$$\int_M dV.$$

This certainly makes sense if M is compact, and in the non-compact case (see Problem 8-10) it either converges to a definite number, or becomes arbitrarily large over compact subsets of M, in which case we say that M has "infinite volume".

If M is an n-dimensional manifold (-with-boundary) in $\mathbb{R}^n$, with the "usual Riemannian metric"

$$\langle\ ,\ \rangle = \sum_{i=1}^{n} dx^i \otimes dx^i,$$

then $g_{ij} = \delta_{ij}$, so

$$dV = |dx^1 \wedge \cdots \wedge dx^n|,$$

and "volume" becomes ordinary volume.

There is an even more important construction associated with a Riemannian metric on M, which will occupy us for the rest of the chapter. For every C^∞ curve $\gamma \colon [a,b] \to M$, we have tangent vectors

$$\gamma'(t) = \frac{d\gamma}{dt} \in M_{\gamma(t)},$$

and can therefore use $\langle\ ,\ \rangle$ to define their length

$$\left\| \frac{d\gamma}{dt} \right\| = \sqrt{\left\langle \frac{d\gamma}{dt}, \frac{d\gamma}{dt} \right\rangle} \quad \left(= \sqrt{\left\langle \frac{d\gamma}{dt}, \frac{d\gamma}{dt} \right\rangle_{\gamma(t)}}, \quad \text{to be precise} \right).$$

We can then define the **length** of γ from a to b,

$$L_a^b(\gamma) = \int_a^b \left\| \frac{d\gamma}{dt} \right\| dt \quad \left(= \int_a^b \| \gamma'(t) \| \, dt \right).$$

If γ is merely *piecewise smooth*, meaning that there is a partition $a = t_0 < \cdots < t_n = b$ of $[a,b]$ such that γ is smooth on each $[t_{i-1}, t_i]$ (with possibly different

left- and right-hand derivatives at $t_1, \ldots, t_{n-1}$), we can define the **length** of γ by

$$L_a^b(\gamma) = \sum_{i=1}^{n} L_{t_{i-1}}^{t_i}(\gamma \,|\, [t_{i-1}, t_i]).$$

Whenever there is no possibility of misunderstanding we will denote L_a^b simply by L. A little argument shows (Problem 15) that for piecewise smooth curves in $\mathbb{R}^n$, with the usual Riemannian metric

$$\sum_{i=1}^{n} dx^i \otimes dx^i,$$

this definition agrees with the definition of length as the least upper bound of the lengths of inscribed polygonal curves.

We can also define a function $s \colon [a, b] \to \mathbb{R}$, the "arclength function of γ" by

$$s(t) = L_a^t(\gamma) = \int_a^t \left\| \frac{d\gamma}{dt} \right\| \, dt.$$

Naturally,

$$(*) \qquad\qquad\qquad s'(t) = \left\| \frac{d\gamma}{dt} \right\|.$$

Consequently $d\gamma/dt$ has constant length 1 precisely when $s(t) = t + \text{constant}$, thus precisely when $s(t) = t - a$. Then

$$b - a = s(b) = L_a^b(\gamma).$$

We can reparameterize γ to be a curve on $[0, b-a]$ by defining

$$\bar{\gamma}(t) = \gamma(t - a).$$

For the new curve $\bar{\gamma}$ we have

$$\text{new } s(t) = L_0^t(\bar{\gamma}) = L_a^{t+a}(\gamma) = \text{old } s(t+a) - \text{old } s(a)$$
$$= t.$$

If γ satisfies $s(t) = t$ we say that γ is **parameterized by arclength** (and then often use s instead of t to denote the argument in the domain of γ).

Classically, the norm $\| \ \|$ on M was denoted by ds. (This makes some sort of sense even in modern notation; equation $(*)$ says that for each curve γ and corresponding $s\colon [a,b] \to \mathbb{R}$ we have

$$|ds| = \gamma^*(\| \ \|)$$

on $[a,b]$.) Consequently, in classical books one usually sees the equation

$$ds^2 = \sum_{i,j=1}^{n} g_{ij}\, dx^i dx^j.$$

Nowadays, this is sometimes interpreted as being the equivalent of the modern equation $\langle \ , \ \rangle = \sum_{i,j=1}^{n} g_{ij}\, dx^i \otimes dx^j$, but what it always actually meant was

$$\| \ \|^2 = \sum_{i,j=1}^{n} g_{ij}\, dx^i dx^j.$$

The symbol $dx^i dx^j$ appearing here is *not* a classical substitute for $dx^i \otimes dx^j$ — the value $(dx^i dx^j)(p)$ of $dx^i dx^j$ at p should not be interpreted as a bilinear function at all, but as the quadratic function

$$v \mapsto dx^i(p)(v) \cdot dx^j(p)(v) \qquad v \in M_p,$$

and we would use the same symbol today. The classical way of indicating $dx^i \otimes dx^j$ was very strange: one wrote

$$\sum_{i,j=1}^{n} g_{ij}\, dx^i \delta x^j \qquad \text{where } dx \text{ and } \delta x \text{ are independent infinitesimals.}$$

(Classically, the Riemannian metric was not a function on tangent vectors, but the inner product of two "infinitely small displacements" dx and δx.)

Consider now a Riemannian metric $\langle \ , \ \rangle$ on a *connected* manifold M. If $p, q \in M$ are any two points, then there is at least one piecewise smooth curve $\gamma\colon [a,b] \to M$ from p to q (there is even a smooth curve from p to q). Define

$$d(p,q) = \inf\{L(\gamma)\colon \gamma \text{ a piecewise smooth curve from } p \text{ to } q\}.$$

It is clear that $d(p,q) \geq 0$ and $d(p,p) = 0$. Moreover, if $r \in M$ is a third point, then for any $\varepsilon > 0$, we can choose piecewise smooth curves

$$\gamma_1\colon [a,b] \to M \quad \text{from } p \text{ to } q \text{ with } L(\gamma_1) - d(p,q) < \varepsilon$$
$$\gamma_2\colon [b,c] \to M \quad \text{from } q \text{ to } r \text{ with } L(\gamma_2) - d(q,r) < \varepsilon.$$

If we define $\gamma\colon [a,c] \to M$ to be γ_1 on $[a,b]$ and γ_2 on $[b,c]$, then γ is a piecewise smooth curve from p to r and

$$L(\gamma) = L(\gamma_1) + L(\gamma_2) < d(p,q) + d(q,r) + 2\varepsilon.$$

Since this is true for all $\varepsilon > 0$, it follows that

$$d(p,r) \leq d(p,q) + d(q,r).$$

[If we did not allow piecewise smooth curves, there would be difficulties in fitting together γ_1 and γ_2, but d would still turn out to be the same (Problem 17).] The function $d\colon M \times M \to \mathbb{R}$ has all properties for a metric, except that it is not so clear that $d(p,q) > 0$ for $p \neq q$. This is made clear in the following.

7. THEOREM. The function $d\colon M \times M \to \mathbb{R}$ is a metric on M, and if $\rho\colon M \times M \to \mathbb{R}$ is the original metric on M (which makes M a manifold), then (M,d) is homeomorphic to (M,ρ).

PROOF. Both parts of the theorem are obviously consequences of the following

7'. LEMMA. Let U be an open neighborhood of the closed ball $B = \{p \in \mathbb{R}^n : |p| \leq 1\}$, let $\langle\ ,\ \rangle_e$ be the "Euclidean" or usual Riemannian metric on U,

$$\langle\ ,\ \rangle_e = \sum_{i=1}^{n} dx^i \otimes dx^i,$$

and let $\langle\ ,\ \rangle$ be any other Riemannian metric. Let $|\ | = \|\ \|_e$ and $\|\ \|$ be the corresponding norms. Then there are numbers $m, M > 0$ such that

$$m \cdot |\ | \leq \|\ \| \leq M \cdot |\ | \qquad \text{on } B,$$

and consequently for any curve $\gamma\colon [a,b] \to B$ we have

$$mL_e(\gamma) \leq L(\gamma) \leq ML_e(\gamma).$$

PROOF. Define $G\colon B \times S^{n-1} \to \mathbb{R}$ by

$$G(p,a) = \|a_p\|_p.$$

Then G is continuous and positive. Since $B \times S^{n-1}$ is compact there are numbers $m, M > 0$ such that

$$m < G < M \qquad \text{on } B \times S^{n-1}.$$

Now if $p \in B$ and $0 \neq b_p \in \mathbb{R}^n{}_p$, let $a \in S^{n-1}$ be $a = b/|b|$. Then

$$m|b| < |b|G(p,a) < M|b|;$$

since

$$|b|G(p,a) = |b| \cdot \|a_p\|_p = \|(|b|a)_p\|_p = \|b\|_p,$$

this gives the desired inequality (which clearly also holds for $b = 0$). ❖

Notice that the distance $d(p,q)$ defined by our metric need not be $L(\gamma)$ for any piecewise smooth curve from p to q. For example, the manifold M might be $\mathbb{R}^2 - \{0\}$, and q might be $-p$. Of course, if $d(p,q) = L(\gamma)$ for some γ,

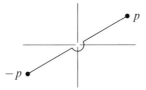

then γ is clearly a shortest piecewise smooth curve from p to q (there might be more than one shortest curve, e.g., the two semi-circles between the points p and $-p$ on S^1).

In order to investigate the question of shortest curves more thoroughly, we have to employ techniques from the "calculus of variations". As an introduction to such techniques, we consider first a simple problem of this sort. Suppose we are given a (suitably differentiable) function

$$F : \mathbb{R} \times \mathbb{R} \times \mathbb{R} \to \mathbb{R}.$$

We seek, among all functions $f : [a,b] \to \mathbb{R}$ with $f(a) = a'$ and $f(b) = b'$ one

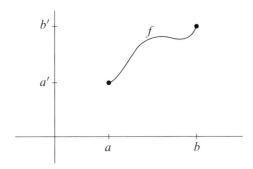

which will maximize (or minimize) the quantity

$$\int_a^b F(t, f(t), f'(t))\, dt.$$

For example, if

$$F(t, x, y) = \sqrt{1 + y^2},$$

then we are looking for a function f on $[a, b]$ which makes the curve $t \mapsto (t, f(t))$ between (a, a') and (b, b') of shortest length

$$\int_a^b \sqrt{1 + [f'(t)]^2} \, dt.$$

As a second example, if

$$F(t, x, y) = 2\pi x \sqrt{1 + y^2},$$

then we are trying to minimize the area of the surface obtained by revolving the graph of f around the x-axis, which is given (Problem 12) by

$$2\pi \int_a^b f(t) \sqrt{1 + [f'(t)]^2} \, dt.$$

To approach this sort of problem we recall first the methods used for solving the much simpler problem of determining the maximum or minimum of a function $f \colon \mathbb{R} \to \mathbb{R}$. To solve this problem, we examine the **critical points** of f, i.e., those points x for which $f'(x) = 0$. A critical point is not necessarily a maximum or minimum, or even a local maximum or minimum, but critical points are the only candidates for maxima or minima if f is everywhere differentiable. Similarly, for a function $f \colon \mathbb{R}^2 \to \mathbb{R}$ we consider points $(x, y) \in \mathbb{R}^2$ for which

$$(*) \qquad D_1 f(x, y) = D_2 f(x, y) = 0.$$

This is the same as saying that the curves

$$t \mapsto f(x + t, y)$$
$$t \mapsto f(x, y + t)$$

have derivative 0 at 0. We might try to get more information by considering the condition

$$0 = (f \circ c)'(0)$$

for every curve $c \colon (-\varepsilon, \varepsilon) \to \mathbb{R}^2$ with $c(0) = (x, y)$, but it turns out that these conditions follow from (∗), because of the chain rule.

To find maxima and minima for

$$J(f) = \int_a^b F(t, f(t), f'(t)) \, dt$$

we wish to proceed in an analogous way, by considering curves *in the set of all functions* $f \colon [a, b] \to \mathbb{R}$. This can be done by considering a "variation" of f, that is, a function

$$\alpha \colon (-\varepsilon, \varepsilon) \times [a, b] \to \mathbb{R}$$

such that

$$\alpha(0, t) = f(t).$$

The functions $t \mapsto \alpha(u, t)$ are then a family of functions on $(-\varepsilon, \varepsilon)$ which pass through f for $u = 0$. We will denote this function by $\bar{\alpha}(u)$. Thus $\bar{\alpha}$ is a function from $(-\varepsilon, \varepsilon)$ to the set of functions $f \colon [a, b] \to \mathbb{R}$. If each $\bar{\alpha}(u)$ satisfies $\bar{\alpha}(u)(a) = a'$, $\bar{\alpha}(u)(b) = b'$, in other words if

$$\alpha(u, a) = a'$$
$$\alpha(u, b) = b'$$

for all $u \in (-\varepsilon, \varepsilon)$, then we call α a variation of f **keeping endpoints fixed**.

For a variation α we now compute

$$\frac{dJ(\bar{\alpha}(u))}{du}\bigg|_{u=0} = \frac{d}{du}\bigg|_{u=0} \int_a^b F\left(t, \alpha(u, t), \frac{\partial \alpha}{\partial t}(u, t)\right) dt$$

$$= \int_a^b \left[\frac{d}{du}\bigg|_{u=0} F\left(t, \alpha(u,t), \frac{\partial \alpha}{\partial t}(u,t)\right) \right] dt$$

$$= \int_a^b \left[\frac{\partial \alpha}{\partial u}(0,t) \frac{\partial F}{\partial x}(t, f(t), f'(t)) \right.$$

$$\left. + \frac{\partial^2 \alpha}{\partial u \partial t}(0,t) \frac{\partial F}{\partial y}(t, f(t), f'(t)) \right] dt.$$

Since $\partial^2 \alpha / \partial u \partial t = \partial^2 \alpha / \partial t \partial u$, we can apply integration by parts to the second term in the integrand, thus obtaining

$$(*) \qquad \frac{dJ(\bar\alpha(u))}{du}\bigg|_{u=0} = \int_a^b \frac{\partial \alpha}{\partial u}(0,t) \left[\frac{\partial F}{\partial x}(t, f(t), f'(t)) \right.$$

$$\left. - \frac{d}{dt}\left(\frac{\partial F}{\partial y}(t, f(t), f'(t))\right) \right] dt$$

$$+ \frac{\partial \alpha}{\partial u}(0,t) \frac{\partial F}{\partial y}(t, f(t), f'(t)) \bigg|_a^b.$$

For variations α keeping endpoints fixed, the second term is 0, and we obtain

$$(**) \qquad \frac{dJ(\bar\alpha(u))}{du}\bigg|_{u=0} = \int_a^b \frac{\partial \alpha}{\partial u}(0,t) \left[\frac{\partial F}{\partial x}(t, f(t), f'(t)) \right.$$

$$\left. - \frac{d}{dt}\left(\frac{\partial F}{\partial y}(t, f(t), f'(t))\right) \right] dt.$$

In classical treatments of the calculus of variations, the variations α were taken to be of the special form

$$\alpha(u,t) = f(t) + u\eta(t),$$

for some $\eta \colon [a,b] \to \mathbb{R}$ with $\eta(a) = \eta(b) = 0$. Then we obtain

$$\frac{dJ(\bar\alpha(u))}{du}\bigg|_{u=0} = \int_a^b \eta(t) \left[\frac{\partial F}{\partial x}(t, f(t), f'(t)) - \frac{d}{dt}\left(\frac{\partial F}{\partial y}(t, f(t), f'(t))\right) \right] dt.$$

The final result is, of course, essentially the same. The derivative $\frac{d}{du}\big|_{u=0} J(\bar\alpha(u))$ is called the "first variation" of J and is denoted classically by

$$\delta J = \int_a^b \eta \left[\frac{\partial F}{\partial x} - \frac{d}{dt}\frac{\partial F}{\partial y} \right] dt.$$

As is usual in classical notation, the arguments of functions are either put in indiscriminately or left out indiscriminately—in this case, not only are the arguments t and $(t, f(t), f'(t))$ omitted (resulting in the disappearance of the function f for which we are solving), but the dependence of δJ on α is not indicated (which can make things pretty confusing).

If f is to maximize or minimize J, then $\delta J(\alpha)$ must be 0 for every variation α of f keeping endpoints fixed. As in the case of 1-dimensional calculus, there is no reason to expect that the condition $\delta J(\alpha) = 0$ for all α will imply that f is even a local maximum or minimum for J, and we emphasize this by introducing a definition. We call f a **critical point** of J (or an **extremal** for J) if $\delta J(\alpha) = 0$ for all variations α of f keeping endpoints fixed. The particular form (∗∗) into which we have put δJ now allows us to deduce an important condition.

8. THEOREM (EULER'S EQUATION). The C^2 function f is a critical point of J if and only if f satisfies

$$\frac{\partial F}{\partial x}(t, f(t), f'(t)) - \frac{d}{dt}\left(\frac{\partial F}{\partial y}(t, f(t), f'(t))\right) = 0.$$

PROOF. Clearly f must make the integral in (∗∗) vanish for *every*

$$\eta(t) = \frac{\partial \alpha}{\partial u}(0, t)$$

which vanishes at a and b. So the theorem is a consequence of the following simple

8′. LEMMA. If a continuous function $g : [a, b] \to \mathbb{R}$ satisfies

$$\int_a^b \eta(t)g(t)\, dt = 0$$

for every C^∞ function η on $[a, b]$ with $\eta(a) = \eta(b) = 0$, then $g = 0$.

PROOF. Choose η to be ϕg where ϕ is positive on (a, b) and $\phi(a) = \phi(b) = 0$. ❖

As an example, consider the case where $F(t, x, y) = \sqrt{1 + y^2}$. The Euler equation is

$$0 = \frac{d}{dt}\left(\frac{f'(t)}{\sqrt{1 + [f'(t)]^2}}\right),$$

so

$$0 = \frac{\sqrt{1 + f'^2} \cdot f'' - f' \cdot \dfrac{f''}{\sqrt{1 + f'^2}}}{(-)},$$

hence

$$0 = (1 + f'^2) f'' - f' f'' = (1 - f' + f'^2) f'',$$

which implies that $f'' = 0$, so f is linear.

Notice that we would have obtained the same result if we had considered the case $F(t, x, y) = 1 + y^2$, for then the Euler equation is simply

$$0 = \frac{d}{dt}(2 f'(t)).$$

This is analogous to the situation in 1-dimensional calculus, where the critical points of $\sqrt{f}$ are the same as those of f, since

$$(\sqrt{f})' = \frac{f'}{2\sqrt{f}}.$$

For the case of the surface of revolution, where $F(t, x, y) = x\sqrt{1 + y^2}$, the Euler equation is

$$0 = \sqrt{1 + [f'(t)]^2} - \frac{d}{dt}\left(\frac{f(t) f'(t)}{\sqrt{1 + [f'(t)]^2}} \right);$$

this leads to the equation

$$1 + f'^2 - f f'' = 0,$$

which we will also write in the classical form

$$1 + \left(\frac{dy}{dx}\right)^2 - y \frac{d^2 y}{dx^2} = 0.$$

To solve this, we use one of the $\aleph_0$ standard tricks (leaving justification of the details to the reader). We let

$$p = y' = \frac{dy}{dx}.$$

Then

$$\frac{d^2 y}{dx^2} = \frac{dp}{dx} = \frac{dp}{dy} \cdot \frac{dy}{dx} = p \frac{dp}{dy},$$

so our equation becomes

$$1 + p^2 - yp\frac{dp}{dy} = 0,$$

$$\frac{p}{1+p^2}\,dp = \frac{1}{y}\,dy,$$

$$\frac{1}{2}\log(1+p^2) = \log y + \text{constant}$$

$$y = \text{constant} \cdot \sqrt{1+p^2}$$

$$p = \frac{dy}{dx} = \sqrt{cy^2 - 1}$$

$$\frac{dy}{\sqrt{cy^2 - 1}} = dx$$

and thus (see Problem 20 for the definition and properties of the "hyperbolic cosine" function cosh and its inverse)

$$\frac{\cosh^{-1} cy}{c} = x + k.$$

Replacing c by $1/c$, we write this as

(∗) $$y = c\cosh\left(\frac{x+k}{c}\right).$$

The graph of

$$\cosh x = \frac{e^x + e^{-x}}{2}$$

is shown below; it is symmetric about the y-axis, decreasing for $x \leq 0$, and increasing for $x \geq 0$.

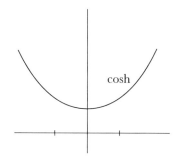

So our surface must look like the one drawn below. It is, by the way, not trivial to decide whether there *are* constants k and c which will make the graph of $(*)$ pass through (a, a') and (b, b'). Problem 21 investigates the special case where $a' = b'$.

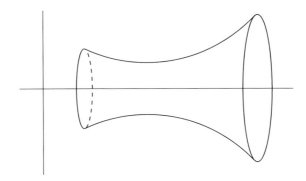

It is easy to generalize these considerations to the case where $f \colon [a, b] \to \mathbb{R}^n$ and

$$J(f) = \int_a^b F(t, f(t), f'(t)) \, dt \qquad \text{for } F \colon \mathbb{R} \times \mathbb{R}^n \times \mathbb{R}^n \to \mathbb{R}.$$

In this case we consider $\alpha \colon (-\varepsilon, \varepsilon) \times [a, b] \to \mathbb{R}^n$ with $\bar\alpha(0) = f$, and compute that

$$(***) \quad \frac{dJ(\bar\alpha(u))}{du}\bigg|_{u=0} = \int_a^b \sum_{l=1}^n \frac{\partial \alpha^l}{\partial u}(0, t) \left[\frac{\partial F}{\partial x^l}(t, f(t), f'(t)) \right.$$

$$\left. - \frac{d}{dt}\left(\frac{\partial F}{\partial y^l}(t, f(t), f'(t)) \right) \right] dt$$

$$+ \sum_{l=1}^n \frac{\partial \alpha^l}{\partial u}(0, t) \frac{\partial F}{\partial y^l}(t, f(t), f'(t)) \bigg|_a^b.$$

Thus, any critical point f of J must satisfy the n equations

$$\frac{\partial F}{\partial x^l}(t, f(t), f'(t)) - \frac{d}{dt}\left(\frac{\partial F}{\partial y^l}(t, f(t), f'(t)) \right) = 0.$$

We are now going to apply these results to the problem of finding shortest paths in a manifold M. If $\gamma \colon [a, b] \to M$ is a piecewise smooth curve, with $\gamma(a) = p$ and $\gamma(b) = q$, we define a **variation of** γ to be a function

$$\alpha \colon (-\varepsilon, \varepsilon) \times [a, b] \to M$$

for some $\varepsilon > 0$, such that

(1) $\alpha(0,t) = \gamma(t)$,

(2) there is a partition $a = t_0 < t_1 < \cdots < t_N = b$ of $[a,b]$ so that α is C^∞ on each strip $(-\varepsilon, \varepsilon) \times [t_{i-1}, t_i]$.

We call α a variation of γ **keeping endpoints fixed** if

(3)
$$\begin{aligned} \alpha(u,a) &= p \\ \alpha(u,b) &= q \end{aligned} \qquad \text{for all } u \in (-\varepsilon, \varepsilon).$$

As before, we let $\bar{\alpha}(u)$ be the path $t \mapsto \alpha(u,t)$. We would like to find which paths γ satisfy

$$\frac{dL(\bar{\alpha}(u))}{du}\bigg|_{u=0} = 0$$

for all variations α keeping endpoints fixed. However, we will take a hint from our first example and first find the critical points for the "energy"

$$E(\gamma) = \frac{1}{2} \int_a^b \left\| \frac{d\gamma}{dt} \right\|^2 dt = \frac{1}{2} \int_a^b \left\langle \frac{d\gamma}{dt}, \frac{d\gamma}{dt} \right\rangle dt,$$

which has a much nicer integrand; afterwards we will consider the relation between the two integrals.

We can assume that each $\gamma|[t_{i-1}, t_i]$ lies in some coordinate system (x, U) (otherwise we just refine the partition). If (u,t) is the standard coordinate system in $(-\varepsilon, \varepsilon) \times [a,b]$ we write

$$\frac{\partial \alpha}{\partial u}(u,t) = \alpha_* \left(\frac{\partial}{\partial u}\bigg|_{(u,t)} \right)$$

$$\frac{\partial \alpha}{\partial t}(u,t) = \alpha_* \left(\frac{\partial}{\partial t}\bigg|_{(u,t)} \right).$$

Then $\partial\alpha/\partial t(u,t)$ is the tangent vector at time t to the curve $\bar{\alpha}(u)$. If we adopt the abbreviations

$$\alpha^i(u,t) = x^i(\alpha(u,t)), \qquad \gamma^i(t) = x^i(\gamma(t)) = \alpha^i(0,t),$$

then

$$\frac{\partial\alpha}{\partial t}(u,t) = \sum_{i=1}^n \frac{\partial\alpha^i}{\partial t}(u,t) \cdot \frac{\partial}{\partial x^i}\bigg|_{\alpha(u,t)}, \qquad \frac{d\gamma}{dt} = \sum_{i=1}^n \frac{d\gamma^i}{dt} \cdot \frac{\partial}{\partial x^i}\bigg|_{\gamma(t)}.$$

So

$$E\big(\gamma\,|\,[t_{i-1},t_i]\big) = \frac{1}{2}\int_{t_{i-1}}^{t_i} \left\langle \frac{d\gamma}{dt}, \frac{d\gamma}{dt} \right\rangle dt$$

$$= \frac{1}{2}\int_{t_{i-1}}^{t_i} \sum_{i,j=1}^n g_{ij}(\gamma(t))\frac{d\gamma^i}{dt}\frac{d\gamma^j}{dt}\,dt.$$

If we use the coordinate system x to identify U with $\mathbb{R}^n$, and consider the g_{ij} as functions on $\mathbb{R}^n$, then we are considering

$$\int_{t_{i-1}}^{t_i} F(\gamma(t), \gamma'(t))\,dt$$

where

$$F(x,y) = \frac{1}{2}\sum_{i,j=1}^n g_{ij}(x) \cdot y^i y^j.$$

Then

$$\frac{\partial F}{\partial x^l}\left(\gamma(t), \frac{d\gamma}{dt}\right) = \frac{1}{2}\sum_{i,j=1}^n \frac{\partial g_{ij}}{\partial x^l}(\gamma(t))\frac{d\gamma^i}{dt}\frac{d\gamma^j}{dt},$$

and

$$\frac{\partial F}{\partial y^l}\left(\gamma(t), \frac{d\gamma}{dt}\right) = \sum_{r=1}^n g_{lr}(\gamma(t))\frac{d\gamma^r}{dt},$$

so

$$\frac{d}{dt}\left(\frac{\partial F}{\partial y^l}\left(\gamma(t), \frac{d\gamma}{dt}\right)\right) = \sum_{r=1}^n g_{lr}(\gamma(t))\frac{d^2\gamma^r}{dt^2} + \sum_{r,j=1}^n \frac{\partial g_{lr}}{\partial x^j}(\gamma(t))\frac{d\gamma^j}{dt}\frac{d\gamma^r}{dt}.$$

In order to obtain a symmetrical looking result, we note that a little index juggling gives

$$\sum_{r,j=1}^{n} \frac{\partial g_{lr}}{\partial x^j}\frac{d\gamma^j}{dt}\frac{d\gamma^r}{dt} = \sum_{i,j=1}^{n} \frac{\partial g_{il}}{\partial x^j}\frac{d\gamma^i}{dt}\frac{d\gamma^j}{dt} = \sum_{i,j=1}^{n} \frac{\partial g_{jl}}{\partial x^i}\frac{d\gamma^i}{dt}\frac{d\gamma^j}{dt}$$

so

$$\sum_{r,j=1}^{n} \frac{\partial g_{lr}}{\partial x^j}\frac{d\gamma^j}{dt}\frac{d\gamma^r}{dt} = \frac{1}{2}\sum_{i,j=1}^{n} \frac{\partial g_{il}}{\partial x^j}\frac{d\gamma^i}{dt}\frac{d\gamma^j}{dt} + \frac{1}{2}\sum_{i,j=1}^{n} \frac{\partial g_{jl}}{\partial x^i}\frac{d\gamma^i}{dt}\frac{d\gamma^j}{dt}.$$

From (∗∗∗) we now obtain

$$\frac{dE\left(\bar{\alpha}(u)|\,[t_{i-1},t_i]\right)}{du}\bigg|_{u=0}$$

$$= -\int_{t_{i-1}}^{t_i} \sum_{l=1}^{n} \frac{\partial \alpha^l}{\partial u}(0,t)\left[\sum_{r=1}^{n} g_{lr}(\gamma(t))\frac{d^2\gamma^r}{dt^2}\right.$$

$$+ \sum_{i,j=1}^{n} \frac{1}{2}\left(\frac{\partial g_{il}}{\partial x^j}(\gamma(t)) + \frac{\partial g_{jl}}{\partial x^i}(\gamma(t)) - \frac{\partial g_{ij}}{\partial x^l}(\gamma(t))\right)\frac{d\gamma^i}{dt}\frac{d\gamma^j}{dt}\bigg]\,dt$$

$$+ \sum_{l=1}^{n} \frac{\partial \alpha^l}{\partial u}(0,t)\sum_{r=1}^{n} g_{lr}(\gamma(t))\frac{d\gamma^r}{dt}\bigg|_{t_{i-1}}^{t_i}.$$

Remember that γ is only piecewise C^∞. Let

$$\frac{d\gamma}{dt}(t_i{}^+) = \text{right hand tangent vector of } \gamma \text{ at } t_i$$

$$\frac{d\gamma}{dt}(t_i{}^-) = \text{left hand tangent vector of } \gamma \text{ at } t_i.$$

Notice that the final sum in the above formula is simply

$$\left\langle \frac{\partial \alpha}{\partial u}(0,t_i), \frac{d\gamma}{dt}(t_i{}^-)\right\rangle - \left\langle \frac{\partial \alpha}{\partial u}(0,t_{i-1}), \frac{d\gamma}{dt}(t_{i-1}{}^+)\right\rangle.$$

To abbreviate the integral somewhat we introduce the symbols

$$\boxed{[ij,l] = \frac{1}{2}\left(\frac{\partial g_{il}}{\partial x^j} + \frac{\partial g_{jl}}{\partial x^i} - \frac{\partial g_{ij}}{\partial x^l}\right).}$$

These depend on the coordinate system, but the integral

$$-\int_{t_{i-1}}^{t_i} \sum_{l=1}^{n} \frac{\partial \alpha^l}{\partial u}(0,t) \left[\sum_{r=1}^{n} g_{lr}(\gamma(t)) \frac{d^2 \gamma^r}{dt^2} + \sum_{i,j=1}^{n} [ij,l](\gamma(t)) \frac{d\gamma^i}{dt} \frac{d\gamma^j}{dt} \right] dt,$$

which appears in our result, clearly cannot. Consequently, we will use the exact same expression for each $[t_{i-1}, t_i]$, even though different coordinate systems may actually be involved (and hence different g_{ij} and γ^i).

Now we just have to add up these results. Let

$$\Delta_{t_i} \frac{d\gamma}{dt} = \frac{d\gamma}{dt}(t_i^+) - \frac{d\gamma}{dt}(t_i^-) \qquad i = 1, \ldots, N-1$$

$$\Delta_{t_0} \frac{d\gamma}{dt} = \frac{d\gamma}{dt}(t_0^+)$$

$$\Delta_{t_N} \frac{d\gamma}{dt} = -\frac{d\gamma}{dt}(t_N^-).$$

Then we obtain the following formula (where there is a convention being used in the integral).

9. THEOREM (FIRST VARIATION FORMULA). For any variation α, we have

$$\left. \frac{dE(\bar\alpha(u))}{du} \right|_{u=0}$$

$$= -\int_{a}^{b} \sum_{l=1}^{n} \frac{\partial \alpha^l}{\partial u}(0,t) \left[\sum_{r=1}^{n} g_{lr}(\gamma(t)) \frac{d^2 \gamma^r}{dt^2} + \sum_{i,j=1}^{n} [ij,l](\gamma(t)) \frac{d\gamma^i}{dt} \frac{d\gamma^j}{dt} \right] dt$$

$$- \sum_{i=0}^{N} \left\langle \frac{\partial \alpha}{\partial u}(0,t_i), \Delta_{t_i} \frac{d\gamma}{dt} \right\rangle.$$

(In the case of a variation α leaving endpoints fixed, the sum can be written from 1 to $N-1$.)

This result is not very pretty, but there it is. It should be noted that $[ij,l]$ are *not* the components of a tensor. Nevertheless, later on we will have an invariant interpretation of the first variation formula. For the time being we present, with apologies, this coordinate dependent approach. From the first variation formula it is, of course, simple to obtain conditions for critical points of E.

10. COROLLARY. If $\gamma : [a,b] \to M$ is a C^∞ path, then γ is a critical point of E_a^b if and only if for every coordinate system (x, U) we have

$$\sum_{r=1}^{n} g_{lr}(\gamma(t)) \frac{d^2\gamma^r}{dt^2} + \sum_{i,j=1}^{n} [ij,l](\gamma(t)) \frac{d\gamma^i}{dt} \frac{d\gamma^j}{dt} = 0 \qquad \text{for } \gamma(t) \in U.$$

PROOF. Suppose γ is a critical point. Given t with $\gamma(t) \in U$, choose a partition of $[a,b]$ with $t \in (t_{i-1}, t_i)$ for some i, and such that $\gamma|[t_{i-1}, t_i]$ is in U. If α is a variation of γ keeping endpoints fixed, then in the first variation formula we can assume that the part of the integral from t_{i-1} to t_i is written in terms of (x, U). The final term in the formula vanishes since γ is C^∞. Now apply the method of proof in Lemma 8′, choosing all $\partial \alpha^l / \partial u(0, t)$ to be 0, except one, which is 0 outside of (t_{i-1}, t_i), but a positive function times the term in brackets on (t_{i-1}, t_i). ❖

In order to put the equations of Corollary 10 in a standard form we introduce another set of symbols

$$\Gamma_{ij}^k = \sum_{l=1}^{n} g^{kl}[ij,l] = \sum_{l=1}^{n} g^{kl} \frac{1}{2}\left(\frac{\partial g_{il}}{\partial x^j} + \frac{\partial g_{jl}}{\partial x^i} - \frac{\partial g_{ij}}{\partial x^l}\right).$$

Our equations can now be written

$$\frac{d^2\gamma^k}{dt^2} + \sum_{i,j=1}^{n} \Gamma_{ij}^k(\gamma(t)) \frac{d\gamma^i}{dt} \frac{d\gamma^j}{dt} = 0.$$

We know from the standard theorem about systems of second order differential equations (Problem 5-4), that for each $p \in M$ and each $v \in M_p$, there is a unique $\gamma : (-\varepsilon, \varepsilon) \to M$, for some $\varepsilon > 0$, such that γ satisfies

$$\gamma(0) = p$$

$$\frac{d\gamma}{dt}(0) = v$$

$$\frac{d^2\gamma^k}{dt^2} + \sum_{i,j=1}^{n} \Gamma_{ij}^k(\gamma(t)) \frac{d\gamma^i}{dt} \frac{d\gamma^j}{dt} = 0.$$

Moreover, this γ is C^∞ on $(-\varepsilon, \varepsilon)$. This last fact shows that if $\gamma_1 \colon [0, \varepsilon) \to M$ and $\gamma_2 \colon (-\varepsilon, 0] \to M$ are C^∞ functions satisfying this equation, and if moreover

$$\gamma_1(0) = \gamma_2(0)$$

$$\frac{d\gamma_1}{dt}(0^+) = \frac{d\gamma_2}{dt}(0^-),$$

then γ_1 and γ_2 together give a C^∞ function on $(-\varepsilon, \varepsilon)$. Naturally, we could replace 0 by any other t. We now have the more precise result,

11. COROLLARY. A piecewise C^∞ path $\gamma \colon [a, b] \to M$ is a critical point for E_a^b if and only if γ is actually C^∞ on $[a, b]$ and for every coordinate system (x, U) satisfies

$$\boxed{\frac{d^2\gamma^k}{dt^2} + \sum_{i,j=1}^{n} \Gamma_{ij}^k(\gamma(t))\frac{d\gamma^i}{dt}\frac{d\gamma^j}{dt} = 0} \qquad \text{for } \gamma(t) \in U.$$

PROOF. Let γ be a critical point. Choosing the same α^l as before (all α^l are 0 outside of (t_{i-1}, t_i)), we see that $\gamma \mid [t_{i-1}, t_i]$ satisfies the equation, because the final term in the first variation formula still vanishes. Now choose α so that

$$\frac{\partial\alpha}{\partial u}(0, t_i) = \Delta_{t_i}\frac{d\gamma}{dt}, \qquad i = 1, \ldots, N - 1.$$

We already know that the integral in the first variation formula vanishes. So we obtain

$$0 = -\sum_{i=1}^{N-1}\left\langle \Delta_{t_i}\frac{d\gamma}{dt}, \Delta_{t_i}\frac{d\gamma}{dt}\right\rangle,$$

which implies that all $\Delta_{t_i}\frac{d\gamma}{dt}$ are 0. By our previous remarks, this means that γ is actually C^∞ on all of $[a, b]$. ❖

As the simplest possible case, consider the Euclidean metric on $\mathbb{R}^n$,

$$\langle \ , \ \rangle = \sum_{i=1}^{n} dx^i \otimes dx^i.$$

Here $g_{ij} = \delta_{ij}$, so all $\partial g_{ij}/\partial x^k = 0$, and $\Gamma_{ij}^k = 0$. The critical points γ for the energy function satisfy

$$\frac{d^2\gamma^k}{dt^2} = 0.$$

Thus γ lies along a straight line, so γ is a critical point for the length function as well. The situation is now quite different from the first variational problem we considered, when we considered only curves of the form $t \mapsto (t, f(t))$. Any reparameterization of γ is also a critical point for length, since length is independent of parameterization (Problem 16). This shows that there are critical points for length which definitely aren't critical points for energy, since we have just seen that for γ to be a critical point for energy, the component functions of γ must be linear, and hence γ must be parameterized proportionally to arclength. This situation always prevails.

12. THEOREM. If $\gamma: [a, b] \to M$ is a critical point for E, then γ is parameterized proportionally to arclength.

PROOF. Observe first, from the definitions, that

$$\frac{\partial g_{ij}}{\partial x^l} = [il, j] + [jl, i].$$

Now we have

$$\frac{d}{dt} \left\| \frac{d\gamma}{dt} \right\|^2 = \frac{d}{dt} \left(\sum_{i,j=1}^{n} g_{ij}(\gamma(t)) \frac{d\gamma^i}{dt} \frac{d\gamma^j}{dt} \right)$$

$$= \sum_{i,j=1}^{n} \sum_{l=1}^{n} \frac{\partial g_{ij}}{\partial x^l}(\gamma(t)) \frac{d\gamma^l}{dt} \frac{d\gamma^i}{dt} \frac{d\gamma^j}{dt} + \sum_{r,j=1}^{n} g_{rj}(\gamma(t)) \frac{d^2\gamma^r}{dt^2} \frac{d\gamma^j}{dt}$$

$$+ \sum_{i,r=1}^{n} g_{ir}(\gamma(t)) \frac{d\gamma^i}{dt} \frac{d^2\gamma^r}{dt^2}.$$

Replacing $\partial g_{ij}/\partial x^l$ by the value given above, this can be written as

$$\frac{d}{dt} \left\| \frac{d\gamma}{dt} \right\|^2 = \sum_{j=1}^{n} \frac{d\gamma^j}{dt} \left(\sum_{r=1}^{n} g_{rj}(\gamma(t)) \frac{d^2\gamma^r}{dt^2} + \sum_{j,l=1}^{n} [il, j](\gamma(t)) \frac{d\gamma^i}{dt} \frac{d\gamma^l}{dt} \right)$$

$$+ \sum_{i=1}^{n} \frac{d\gamma^i}{dt} \left(\sum_{r=1}^{n} g_{ir}(\gamma(t)) \frac{d^2\gamma^r}{dt^2} + \sum_{j,l=1}^{n} [jl, i](\gamma(t)) \frac{d\gamma^j}{dt} \frac{d\gamma^l}{dt} \right).$$

Since γ is a critical point for E, both terms in parentheses are 0 (Corollary 10). Thus the length $\|d\gamma/dt\|$ is constant. ❖

The formula

(*)
$$\frac{\partial g_{ij}}{\partial x^k} = [ik, j] + [jk, i]$$

occurring in this proof will be used on several occasions later on. It will also be useful to know a formula for $\partial g^{ij}/\partial x^k$. To derive one, we first differentiate

$$\sum_{m=1}^{n} g_{lm} g^{mj} = \delta_l^j$$

to obtain

$$\sum_{m=1}^{n} g_{lm} \frac{\partial g^{mj}}{\partial y^k} = -\sum_{m=1}^{n} \frac{\partial g_{lm}}{\partial y^k} g^{mj}.$$

Thus we have

$$\frac{\partial g^{ij}}{\partial y^k} = \sum_{l,m} g^{il} g_{lm} \frac{\partial g^{mj}}{\partial y^k} = -\sum_{l,m} g^{il} g^{mj} \frac{\partial g_{lm}}{\partial y^k}$$

$$= -\sum_{l,m} g^{il} g^{mj} ([lk, m] + [mk, l]) \quad \text{by } (*)$$

$$= -\sum_{l} g^{il} \Gamma_{lk}^j - \sum_{m} g^{mj} \Gamma_{mk}^i,$$

or

(**)
$$\frac{\partial g^{ij}}{\partial y^k} = -\sum_{l=1}^{n} (g^{il} \Gamma_{lk}^j + g^{lj} \Gamma_{lk}^i).$$

We can find the equations for critical points of the length function L in exactly the same way as we treated the energy function. For the moment we consider only paths $\gamma : [a, b] \to M$ with $d\gamma/dt \neq 0$ everywhere. For the portion $\gamma | [t_{i-1}, t_i]$ of γ contained in a coordinate system (x, U), we have

$$L(\gamma | [t_{i-1}, t_i]) = \int_{t_{i-1}}^{t_i} \sqrt{\sum_{i,j=1}^{n} g_{ij}(\gamma(t)) \frac{d\gamma^i}{dt} \frac{d\gamma^j}{dt}} \, dt.$$

Considering our coordinate system as $\mathbb{R}^n$, we are now dealing with the case

$$F(x, y) = \sqrt{\sum_{i,j=1}^{n} g_{ij}(x) y^i y^j}.$$

We introduce the arclength function

$$s(t) = L_a^t(\gamma).$$

Then

$$\frac{ds}{dt} = \left\| \frac{d\gamma}{dt} \right\| = F\left(\gamma(t), \frac{d\gamma}{dt}\right).$$

So we have

$$\frac{\partial F}{\partial x^l}\left(\gamma(t), \frac{d\gamma}{dt}\right) = \frac{1}{2} \frac{\displaystyle\sum_{i,j=1}^{n} \frac{\partial g_{ij}}{\partial x^l}(\gamma(t))\frac{d\gamma^i}{dt}\frac{d\gamma^j}{dt}}{\dfrac{ds}{dt}}$$

$$\frac{\partial F}{\partial y^l}\left(\gamma(t), \frac{d\gamma}{dt}\right) = \frac{\displaystyle\sum_{r=1}^{n} g_{lr}(\gamma(t))\frac{d\gamma^r}{dt}}{\dfrac{ds}{dt}}.$$

After a little more calculation we finally obtain the equations for a critical point of L:

$$\frac{d^2\gamma^k}{dt^2} + \sum_{i,j=1}^{n} \Gamma_{ij}^k(\gamma(t))\frac{d\gamma^i}{dt}\frac{d\gamma^j}{dt} - \frac{d\gamma^k}{dt}\frac{\dfrac{d^2s}{dt^2}}{\dfrac{ds}{dt}} = 0.$$

It is clear from this that critical points of E are also critical points of L (since they satisfy $d^2s/dt^2 = 0$). Conversely, given a critical point γ for L with $d\gamma/dt \neq 0$ everywhere, the function

$$s: [a, b] \to [0, L_a^b(\gamma)]$$

is a diffeomorphism, and we can consider the reparameterized curve

$$\gamma \circ s^{-1}: [0, L_a^b(\gamma)] \to M.$$

This reparameterized curve is automatically also a critical point for L, so it must satisfy the same differential equation. Since it is now parameterized by arclength, the third term vanishes, so $\gamma \circ s^{-1}$ is a critical point for E.

There is only one detail which remains unsettled. Conceivably a critical point for L might have a kink, but be C^∞ because it has a zero tangent vector

pg. 119: Add the following at the end of the proof:

Smoothness of A follows from the fact that the function $A_{i_1 \cdots i_k}$ is $A(\partial/\partial x_{i_1}, \ldots, \partial/\partial x_{i_k})$.

pg. 131, Problem 9: Let F be a covariant functor from $\mathbf{V}, \ldots$.

pg. 133. Though there is considerable variation in terminology, what are here called "odd scalar densities" should probably simply be called "scalar densities"; what are called "even scalar densities" might best be called "signed scalar densities".

In part (c) of Problem 10, we should be considering the h of part (a), not the h of part (b)! Thus conclude that the bundle of signed scalar densities (*not* the scalar densities) is not trivial if M is not orientable.

pg. 134. Extending the changed terminology from pg. 133, we should probably speak of the bundle of "signed tensor densities of type $\binom{k}{l}$ and weight w" (though sometimes the term relative tensor is used instead, restricting densities to those of weight 1), when the transformation rule involves $(\det A)^w$, omitting the modifier "signed" when it involves $|\det A|^w$.

pg. 143. The hypothesis of Theorem 3 should be changed so that it reads:

Let $x \in U$ and let α_1, α_2 be two maps on some open interval I such that $\alpha_1(I), \alpha_2(I) \subset U$,

$$\alpha_i{}'(t) = f(\alpha_i(t)) \qquad i = 1, 2$$

and

$$\alpha_1(t_0) = \alpha_2(t_0) \qquad \text{for some } t_0 \in I.$$

And the first sentence of the proof should begin:

pg. 177. Problem 17, part (d) should be deleted.

(d) Let $f : M \to N$, and suppose that $f_{*p} = 0$. For $X_p, Y_p \in M_p$ and ....

pg. 198. In Problem 5, we must also assume that each $\Delta_i \oplus \Delta_j$ is integrable.

pg. 226. In the commutative diagram, the lower right entry should be "l-forms on N".

pg. 233. The reference "pg. V:375" refers to pg. 375 of Volume V.

pg. 237. In Problem 26, replace parts (b) and (c) with:

(b) Determine the ith component of $v_1 \times \cdots \times v_{n-1}$ in terms of the $(n-1) \times (n-1)$ submatrices of the matrix

$$\begin{pmatrix} v_1 \\ \vdots \\ v_n \end{pmatrix}.$$

In particular, for $\mathbb{R}^3$, show that

$$v \times w = (v^2 w^3 - v^3 w^2, v^3 w^1 - v^1 w^3, v^1 w^2 - v^2 w^1).$$

pg. 292. In Problem 20, the condition $U_i \cap U_j \neq \emptyset$ should be $U_i \cap U_{i+1} \neq \emptyset$.

pg. 408. Problem 16 (b) should read: "For any Lie group G, show that ...".

pp. 408–410. For consistency with standard usage, Aut should be replaced with Aut, and then replace End with End. In part (g) of Problem 19, add the hypothesis that H is a connected Lie subgroup.

pg. 411. The display in Problem 21, part (c) should read:

$$(-1)^{km}[\omega \wedge [\eta \wedge \lambda]] + (-1)^{kl}[\eta \wedge [\lambda \wedge \omega]] + (-1)^{lm}[\lambda \wedge [\omega \wedge \eta]] = 0.$$

CORRECTIONS FOR VOLUME I

pg. 3, line 3−: change $\bar{d}_i(x,y) < 1$ to $\bar{d}_i(x,y) \le 1$.

pg. 14: relabel the lower left part of the central figure as

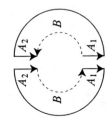

pg. 19: replace the next-to-last paragraph with the following:

The set of points in a manifold-with-boundary that do not have a neighborhood homeomorphic to $\mathbb{R}^n$ (but only one homeomorphic to $\mathbb{H}^n$) is called the **boundary** of M and is denoted by ∂M. Equivalently, $x \in \partial M$ if and only if there is a neighborhood V of x and a homeomorphism $\phi : V \to \mathbb{H}^n$ such that $\phi(x) = 0$. If M is actually a manifold, then $\partial M = \emptyset$, and ∂M itself is always a manifold (without boundary).

pg. 22: Replace the top left figure with

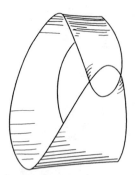

pg. 43: Replace the last line and displayed equation with the following:
Since rank $f = k$ in a neighborhood of p, the lower rectangle in the matrix

$$\left(\frac{\partial(v^i \circ f)}{\partial x^j} \right) = $$

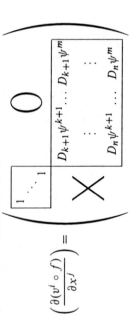

pg. 60, Problem 30: Change part (f) and add part (g):

(f) If M is a connected manifold, there is a proper map $f : M \to \mathbb{R}$; the function f can be made C^∞ if M is a C^∞ manifold.
(g) The same is true if M has at most countably many components.

pg. 61, Problem 32: For clarity, restate part (c) as follows:

(c) This is false if $f : M_1 \to \mathbb{R}$ is replaced with $f : M_1 \to N$ for a disconnected manifold N.

pg. 70: Replace the last two lines of page 70 and the first two lines of page 71 with the following:

theorem of topology). If there were a way to map $T(M, i)$, fibre by fibre, homeomorphically onto $M \times \mathbb{R}^2$, then each v_p would correspond to $(p, \mathbf{v}(p)) \in \mathbb{R}^2$ for some $\mathbf{v}(p) \in \mathbb{R}^2$, and we could continuously pick $\mathbf{w}(p) \in \mathbb{R}^2$, corresponding to a dashed vector, by using the criterion that $\mathbf{w}(p)$ should make a positive angle with $\mathbf{v}(p)$.

pg. 78: the third display should read:

$$0 = \ell(0) = \ell(fh) = f(p)\ell(h) + h(p)\ell(f) = 0 + \ell(f).$$

pg. 103, Problem 29(d). Add the hypothesis that M is orientable.

pg. 117: After the next to last display, $\bar{A}(X_1, \ldots, X_k)(p) = A(p)(X_1(p), \ldots, X_k(p))$, add:

If A is C^∞, then $\bar{A}$ is C^∞, in the sense that $\bar{A}(X_1, \ldots, X_k)$ is a C^∞ function for all C^∞ vector fields $X_1, \ldots, X_k$.

pg. 118: Add the following to the statement of the theorem: If $\mathcal{A}$ is C^∞, then A is also.

there, as in the figure below. In this case it would not be possible to reparame-

terize γ by arclength. Problem 37 shows that this situation cannot arise.

Henceforth we will call a critical point of E a **geodesic** on M (for the Riemannian metric $\langle\ ,\ \rangle$). This name comes from the science of geodesy, which is concerned with the measurement of the earth's surface, including surveying and the measurement of degrees of latitude and longitude. A geodesic on the earth's surface is a segment of a great circle, which is the shortest path between two points. Before we can say whether this is true for geodesics in general, which are so far merely known to be critical points for length, we must initiate a local study of geodesics.

The most elementary properties of geodesics depend only on facts about differential equations. Observe that the equations for a geodesic,

$$\frac{d^2\gamma^k}{dt^2} + \sum_{i,j=1}^{n} \Gamma^k_{ij} \frac{d\gamma^i}{dt} \frac{d\gamma^j}{dt} = 0,$$

have an important homogeneity property: if γ is a geodesic, then $t \mapsto \gamma(ct)$ is also clearly a geodesic. This feature of the equation allows us to improve the result given by the basic existence and uniqueness theorems.

13. THEOREM. Let $p \in M$. Then there is a neighborhood U of p and a number $\varepsilon > 0$ such that for every $q \in U$ and every tangent vector $v \in M_q$ with $\|v\| < \varepsilon$ there is a unique geodesic

$$\gamma_v \colon (-2, 2) \to M$$

satisfying

$$\gamma_v(0) = q, \qquad \frac{d\gamma_v}{dt}(0) = v.$$

PROOF. The fundamental existence and uniqueness theorem says that there is a neighborhood U of p and $\varepsilon_1, \varepsilon_2 > 0$ so that for $q \in U$ and $v \in M_q$ with $\|v\| < \varepsilon_1$ there is a unique geodesic

$$\gamma_v \colon (-2\varepsilon_2, 2\varepsilon_2) \to M$$

with the required initial conditions.

Choose $\varepsilon < \varepsilon_1 \varepsilon_2$: Then if $|v| < \varepsilon$ and $|t| < 2$ we have

$$\|v/\varepsilon_2\| < \varepsilon_1 \quad \text{and} \quad |\varepsilon_2 t| < 2\varepsilon_2.$$

So we can define $\gamma_v(t)$ to be $\gamma_{v/\varepsilon_2}(\varepsilon_2 t)$. ❖

If $v \in M_q$ is a vector for which there is a geodesic

$$\gamma \colon [0, 1] \to M$$

satisfying

$$\gamma(0) = q, \quad \frac{d\gamma}{dt}(0) = v,$$

then we define the **exponential** of v to be

$$\exp(v) = \exp_q(v) = \gamma(1).$$

(The reason for this terminology will be explained in the next chapter.) The geodesic γ can thus be described as

$$\gamma(t) = \exp_q(tv).$$

Since M_q is an n-dimensional vector space, there is a natural way to give it a C^∞ structure. If $\mathcal{O} \subset M_q$ is the set of all vectors $v \in M_q$ for which $\exp_q(v)$ is defined, then the map

$$\exp_q \colon \mathcal{O} \to M$$

is C^∞, since the solutions of the differential equations for geodesics have a C^∞ flow. Identifying the tangent space $(M_q)_v$ at $v \in M_q$ with M_q itself, we have an induced map

$$(\exp_q)_{v*} \colon M_q \to M_{\exp_q(v)}.$$

In particular, we claim that the map

$$(\exp_q)_{0*} \colon M_q \to M_q \quad \text{is the identity.}$$

In fact, to obtain a curve c in the manifold M_q with $dc/dt(0) = v \in M_q = (M_q)_0$, we can let $c(t) = tv$. Then $\exp_q \circ c(t) = \exp_q(tv)$, the geodesic with tangent vector v at time 0, so

$$(\exp_q)_{0*}(v) = \left. \frac{d}{dt} \right|_{t=0} \exp_q(c(t)) = v.$$

Before proving the next result, we recall some facts about the manifold TM. If (x, U) is a coordinate system on M, then for $q \in U$ we can express every vector $v \in M_q$ uniquely as

$$v = \sum_{i=1}^{n} a^i \left. \frac{\partial}{\partial x^i} \right|_q .$$

We will denote a^i by $\dot{x}^i(v)$, so that

$$v = \sum_{i=1}^{n} \dot{x}^i(v) \left. \frac{\partial}{\partial x^i} \right|_{\pi(v)} ,$$

where $\pi : TM \to M$ is the projection. Then

$$(x^1 \circ \pi, \dots, x^n \circ \pi, \dot{x}^1, \dots, \dot{x}^n) = (\bar{x}^1, \dots, \bar{x}^n, \dot{x}^1, \dots, \dot{x}^n)$$

is a coordinate system on $\pi^{-1}(U)$. For $v \in M_q$, $q \in U$ we therefore have tangent vectors

$$\left. \frac{\partial}{\partial \bar{x}^i} \right|_v , \quad \left. \frac{\partial}{\partial \dot{x}^i} \right|_v \in (TM)_v;$$

the vectors $\partial/\partial \dot{x}^i \big|_v$ are all in the tangent space of the submanifold $M_q \subset TM$, while the vectors $\partial/\partial \bar{x}^i \big|_v$ span a complimentary subspace.

14. THEOREM. For every $p \in M$ there is a neighborhood W and a number $\varepsilon > 0$ such that

(1) Any two points of W are joined by a unique geodesic in M of length $< \varepsilon$.

(2) Let $v(q, q')$ denote the unique vector $v \in M_q$ of length $< \varepsilon$ such that $\exp_q(v) = q'$. Then $(q, q') \mapsto v(q, q')$ is a C^∞ function from $W \times W \to TM$.

(3) For each $q \in W$, the map $\exp_q$ maps the open ε-ball in M_q diffeomorphically onto an open set $U_q \supset W$.

PROOF. Theorem 13 says that the vector $0 \in M_p$ has a neighborhood V in the manifold TM such that exp is defined on V. Define the C^∞ function $F\colon V \to M \times M$ by

$$F(v) = (\pi(v), \exp(v)).$$

Let (x, U) be a coordinate system around p. We will use the coordinate system

$$(\bar{x}^1, \ldots, \bar{x}^n, \dot{x}^1, \ldots, \dot{x}^n),$$

described above, for $\pi^{-1}(U)$. If $\pi_i\colon M \times M \to M$ is projection on the ith factor, then

$$(x^1 \circ \pi_1, \ldots, x^n \circ \pi_1, x^1 \circ \pi_2, \ldots, x^n \circ \pi_2) = (x_1{}^1, \ldots, x_1{}^n, x_2{}^1, \ldots, x_2{}^n)$$

is a coordinate system on $U \times U$. Now, using the fact that

$$(\exp_p)_{0*}\colon M_p \to M_p$$

is the identity, it is not hard to see that at $0 \in M_p$ we have

$$F_*\left(\frac{\partial}{\partial \bar{x}^i}\Big|_0\right) = \frac{\partial}{\partial x_1{}^i}\Big|_{(p,p)} + \frac{\partial}{\partial x_2{}^i}\Big|_{(p,p)}$$

$$F_*\left(\frac{\partial}{\partial \dot{x}^i}\Big|_0\right) = \frac{\partial}{\partial x_2{}^i}\Big|_{(p,p)}.$$

Consequently, F_* is one-one at $0 \in M_p$, so F maps some neighborhood V' of 0 diffeomorphically onto some neighborhood of $(p, p) \in M \times M$. We may assume that V' consists of all vectors $v \in M_q$ with q in some neighborhood U' of p and $\|v\| < \varepsilon$. Choose W to be a smaller neighborhood of p for which $F(V') \supset W \times W$. ❖

Given a W as in the theorem, and $q \in W$, consider the geodesics through q of the form $t \mapsto \exp_q(tv)$ for $\|v\| < \varepsilon$. These fill out U_q. The close analysis of geodesics depends on the following.

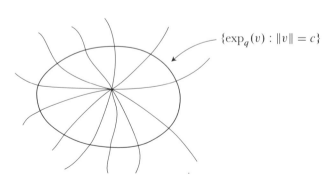

$$\{\exp_q(v) : \|v\| = c\}$$

15. LEMMA (GAUSS' LEMMA). In U_q, the geodesics through q are perpendicular to the hypersurfaces

$$\{\exp_q(v) : \|v\| = \text{constant} < \varepsilon\}.$$

FIRST PROOF. Let $v \colon \mathbb{R} \to M_q$ be a smooth curve with $\|v(t)\| = $ a constant $k < \varepsilon$ for all t, and define

$$\alpha(u,t) = \exp_q(u \cdot v(t)) \qquad -1 < u < 1.$$

We are claiming that for every such α we have

$$\left\langle \frac{\partial \alpha}{\partial u}(u,t), \frac{\partial \alpha}{\partial t}(u,t) \right\rangle = 0 \qquad \text{for all } (u,t).$$

A calculation precisely like that in the proof of Theorem 12 proves the following equation, in which the arguments (u,t) and $\alpha(u,t)$ are omitted, for convenience:

$$(1) \quad \frac{\partial}{\partial u} \left\langle \frac{\partial \alpha}{\partial u}, \frac{\partial \alpha}{\partial t} \right\rangle = \sum_{j=1}^{n} \frac{\partial \alpha^j}{\partial t} \left(\sum_{r=1}^{n} g_{rj} \frac{\partial^2 \alpha^r}{\partial u^2} + \sum_{i,l=1}^{n} [il,j] \frac{\partial \alpha^i}{\partial u} \frac{\partial \alpha^l}{\partial u} \right)$$

$$+ \sum_{i=1}^{n} \frac{\partial \alpha^i}{\partial u} \left(\sum_{r=1}^{n} g_{ir} \frac{\partial^2 \alpha^r}{\partial u \partial t} + \sum_{j,l=1}^{n} [jl,i] \frac{\partial \alpha^j}{\partial u} \frac{\partial \alpha^l}{\partial t} \right).$$

The first term on the right is 0 since each curve $u \mapsto \alpha(u,t)$ is a geodesic. Similarly, we obtain

$$(2) \quad \frac{\partial}{\partial t} \left\langle \frac{\partial \alpha}{\partial u}, \frac{\partial \alpha}{\partial u} \right\rangle = 2 \sum_{i=1}^{n} \frac{\partial \alpha^i}{\partial u} \left(\sum_{r=1}^{n} g_{ir} \frac{\partial^2 \alpha^r}{\partial u \partial t} + \sum_{j,l=1}^{n} [jl,i] \frac{\partial \alpha^j}{\partial u} \frac{\partial \alpha^l}{\partial t} \right),$$

which is just twice the second term on the right of (1). But $\partial \alpha / \partial u(u,t)$ is just the tangent vector at time u to the geodesic $u \mapsto \exp_q(u \cdot v(t))$, where $\|v(t)\| = k$; so $\|\partial \alpha / \partial u\| = k$. Thus the second term on the right of (2) is also 0. So

$$\left\langle \frac{\partial \alpha}{\partial u}, \frac{\partial \alpha}{\partial t} \right\rangle \qquad \text{is independent of } u.$$

But $\alpha(0,t) = \exp_q(0) = q$, so $\partial \alpha / \partial t(0,t) = 0$. It follows that

$$\left\langle \frac{\partial \alpha}{\partial u}, \frac{\partial \alpha}{\partial t} \right\rangle = 0 \qquad \text{for all } (u,t).$$

SECOND PROOF. Let $v \colon \mathbb{R} \to M_q$ be any smooth curve with $\|v(t)\| = $ a constant $k < \varepsilon$ for all t, and define

$$\beta(u,t) = \exp_q(t \cdot v(u)) \qquad \text{(note carefully the roles played by } t \text{ and } u).$$

Then β is a variation of the geodesic $\gamma(t) = \exp_q(t \cdot v(0))$, defined on $[0,1]$. By

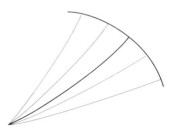

the first variation formula, we have

$$\left. \frac{dE(\bar{\beta}(u))}{du} \right|_{u=0} = -\left\langle \frac{\partial \beta}{\partial u}(0,1), \frac{d\gamma}{dt}(1) \right\rangle - \left\langle \frac{\partial \beta}{\partial u}(0,0), \frac{d\gamma}{dt}(0) \right\rangle$$

$$= -\left\langle \frac{\partial \beta}{\partial u}(0,1), \frac{d\gamma}{dt}(1) \right\rangle,$$

the integral vanishing since γ is a geodesic. But each curve $\bar{\beta}(u)$ has energy

$$E(\bar{\beta}(u)) = \int_0^1 \left\| \frac{d\bar{\beta}(u)(t)}{dt} \right\|^2 dt = \int_0^1 k^2 \, dt = k^2,$$

so

$$0 = \left. \frac{dE(\bar{\beta}(u))}{du} \right|_{u=0} = -\left\langle \frac{\partial \beta}{\partial u}(0,1), \frac{d\gamma}{dt}(1) \right\rangle. \quad ❖$$

16. COROLLARY. Let $c \colon [a,b] \to U_q - \{q\}$ be a piecewise smooth curve,

$$c(t) = \exp_q(u(t) \cdot v(t)),$$

for $0 < u(t) < \varepsilon$ and $\|v(t)\| = 1$. Then

$$L_a^b c \geq |u(b) - u(a)|,$$

with equality if and only if u is monotonic and v is constant, so that c is a radial geodesic joining two concentric spherical shells around q.

PROOF. If $\alpha(u,t) = \exp_q(u \cdot v(t))$, then $c(t) = \alpha(u(t),t)$ and

$$\frac{dc}{dt} = \frac{\partial \alpha}{\partial u} u'(t) + \frac{\partial \alpha}{\partial t}.$$

Since

$$\left\langle \frac{\partial \alpha}{\partial u}, \frac{\partial \alpha}{\partial t} \right\rangle = 0, \qquad \left\| \frac{\partial \alpha}{\partial u} \right\| = 1,$$

we have

$$\left\| \frac{dc}{dt} \right\|^2 = |u'(t)|^2 + \left\| \frac{\partial \alpha}{\partial t} \right\|^2 \geq |u'(t)|^2,$$

with equality if and only if $\partial \alpha / \partial t = 0$, and hence $v'(t) = 0$. Thus

$$\int_0^b \left\| \frac{dc}{dt} \right\| dt \geq \int_0^b |u'(t)| \, dt \geq |u(b) - u(a)|,$$

with equality if and only if u is monotonic and v is constant. ❖

17. COROLLARY. Let W and ε be as in Theorem 15, let $\gamma \colon [0,1] \to M$ be the geodesic of length $< \varepsilon$ joining $q, q' \in W$, and let $c \colon [0,1] \to M$ be any piecewise C^∞ path from q to q'. Then

$$L(\gamma) \leq L(c),$$

with equality holding if and only if c is a reparameterization of γ.

PROOF. We can assume that $q' = \exp_q(rv) \in U_q - \{q\}$ (otherwise break c up into smaller pieces). For $\delta > 0$, the path c must contain a segment which joins the spherical shell of radius δ to the spherical shell of radius r, and lies between them. By Corollary 16, the length of this segment has length $\geq r - \delta$. So the length of c is $\geq r$, and clearly c must be a reparameterization of γ for equality to hold. ❖

We thus see that *sufficiently small pieces* of geodesics are minimal paths for arc-length. We can use Corollary 17 to determine the geodesics on a few simple surfaces, without any computations, if we first introduce a notion which will play a crucial role later. If $(M, \langle \ , \ \rangle)$ and $(M', \langle \ , \ \rangle')$ are C^∞ manifolds with Riemannian metrics, then a one-one C^∞ function $f: M \to M'$ is called an **isometry** of M into M' if $f^*\langle \ , \ \rangle' = \langle \ , \ \rangle$. For example, reflection through a plane $E^2 \subset \mathbb{R}^{n+1}$ is an isometry $I: S^n \to S^n$. It is clear that if $c: [0, 1] \to M$ is a C^∞ curve, then the length of c with respect to $\langle \ , \ \rangle$ is the length of $f \circ c$ with respect to $\langle \ , \ \rangle'$; and if c is a geodesic, then $f \circ c$ is likewise a geodesic.

For the isometry $I: S^n \to S^n$ mentioned above, the fixed point set is the great circle $C = S^n \cap E^2$. Let $p, q \in C$ be two points with a unique geodesic C' of minimal length between them. Then $I(C')$ is a geodesic of the same length as C' between $I(p) = p$ and $I(q) = q$. So $C' = I(C')$, which implies that $C' \subset C$, so that C is a geodesic. Since there is a great circle through any point of S^n in any given direction, these are all the geodesics.

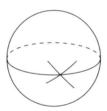

Notice that a portion of a great circle which is larger than a semi-circle is definitely not of minimal length, *even among nearby paths*. Antipodal points on

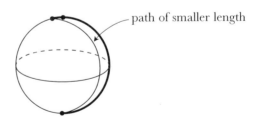

path of smaller length

the sphere have a continuum of geodesics of minimal length between them. All other pairs of points have a unique geodesic of minimal length between them but an infinite family of non-minimal geodesics, depending on how many times the geodesic goes around the sphere and in which direction it starts.

The geodesics on a right circular cylinder Z are the generating lines, the

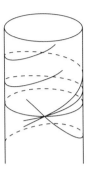

circles cut by planes perpendicular to the generating lines, and the helices on Z. In fact, if L is a generating line of Z, then we can set up an isometry $I : Z - L \to \mathbb{R}^2$ by rolling Z onto $\mathbb{R}^2$. The geodesics on Z are just the images

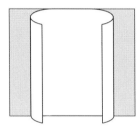

under I^{-1} of the straight lines in $\mathbb{R}^2$. Two points on Z have infinitely many geodesics between them.

We are now in a position to wind up our discussion of Riemannian metrics on M by establishing an important connection between the Riemannian metric $\langle\ ,\ \rangle$ and the metric $d : M \times M \to \mathbb{R}$ it determines,

$$d(p,q) = \inf\{L(\gamma) : \gamma \text{ a piecewise smooth curve from } p \text{ to } q\}.$$

Notice that on both the sphere and the infinite cylinder every geodesic γ defined on an interval $[a,b]$ can be extended to a geodesic defined on all of $\mathbb{R}$. This is false on a cylinder of bounded height, a bounded portion of $\mathbb{R}^n$, or $\mathbb{R}^n - \{0\}$. In general, a manifold M with a Riemannian metric $\langle\ ,\ \rangle$ is called **geodesically complete** if every geodesic $\gamma : [a,b] \to M$ can be extended to a geodesic from $\mathbb{R}$ to M.

18. THEOREM (HOPF-RINOW-DE RHAM). If $\langle\ ,\ \rangle$ is a Riemannian metric on M, then M is geodesically complete if and only if M is complete in the metric d determined by $\langle\ ,\ \rangle$. Moreover, any two points in a geodesically complete manifold can be joined by a geodesic of minimal length.

PROOF. Suppose M is geodesically complete. Given $p, q \in M$ with $d(p,q) = r > 0$, choose U_p as in Theorem 14. Let $S \subset U_p$ be the spherical shell of radius $\delta < \varepsilon$. There is a point

$$p_0 = \exp_p \delta v, \qquad \|v\| = 1$$

on S such that $d(p_0, q) \le d(s, q)$ for all $s \in S$. We claim that

(*) $$\exp_p(rv) = q;$$

this will show that the geodesic $\gamma(t) = \exp_p(tv)$ is a geodesic of minimal length between p and q. To prove this result, we will prove that

(**) $$d(\gamma(t), q) = r - t \qquad t \in [\delta, r].$$

First of all, since every curve from p to q must intersect S, we clearly have

$$d(p, q) = \min_{s \in S}\big(d(p, s) + d(s, q)\big) = \delta + d(p_0, q).$$

So $d(p_0, q) = r - \delta$. This proves that (**) holds for $t = \delta$.

Now let $t_0 \in [\delta, r]$ be the least upper bound of all t for which (**) holds. Then (**) holds for t_0 also, by continuity. Suppose $t_0 < r$. Let S' be a spherical shell

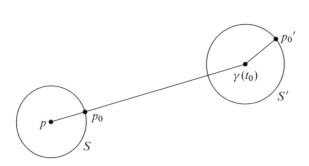

of radius δ' around $\gamma(t_0)$ and let $p_0' \in S'$ be a point closest to q. Then

$$d(\gamma(t_0), q) = \min_{s \in S'}\big(d(\gamma(t_0), s) + d(s, q)\big) = \delta' + d(p_0', q),$$

so

$(***)$ $$d(p_0', q) = (r - t_0) - \delta'.$$

Hence

$$d(p, p_0') \geq d(p, q) - d(p_0', q) = t_0 + \delta'.$$

But the path c obtained by following γ from p to $\gamma(t_0)$ and then the minimal geodesic from $\gamma(t_0)$ to p_0' has length precisely $t_0 + \delta'$. So c is a path of minimal length, and must therefore be a geodesic, which means that it coincides with γ. Hence

$$\gamma(t_0 + \delta') = p_0'.$$

Hence $(***)$ gives

$$d(\gamma(t_0 + \delta'), q) = r - (t_0 + \delta'),$$

showing that $(**)$ holds for $t_0 + \delta'$. This contradicts the choice of t_0, so it must be that $t_0 = r$. In other words, $(**)$ holds for $t = r$, which proves $(*)$.

From this result, it follows easily that M is complete with the metric d. In fact, if $A \subset M$ has diameter D, and $p \in A$, then the map $\exp_p : M_p \to M$ maps the closed disc of radius D in M_p onto a compact set containing A. In other words, bounded subsets of M have compact closure. From this it is clear that Cauchy sequences converge.

Conversely, suppose M is complete as a metric space. Given any geodesic $\gamma : (a, b) \to M$, choose $t_n \to b$. Clearly $\gamma(t_n)$ is a Cauchy sequence in M, so it converges to some point $p \in M$. Using Theorem 14, it is not difficult to show that γ can be extended past b. Consequently, by a least upper bound argument, any geodesic can be extended to $\mathbb{R}$. ❖

As a particular consequence of Theorem 18, note that there is always a minimal geodesic joining any two points of a compact manifold.

ADDENDUM
TUBULAR NEIGHBORHOODS

Let $M^n \subset N^{n+k}$ be a submanifold of N, with $i\colon M \to N$ the inclusion map, so that for every $p \in M$ we have $i_*(M_p) \subset N_p$. If $\langle \ , \ \rangle$ is a Riemannian metric for N, then we can define $M_p^{\perp} \subset N_p$ as

$$M_p{}^{\perp} = \{v \in N_p : \langle v, i_*w \rangle = 0 \text{ for all } w \in M_p\}.$$

Let

$$E = \bigcup_{p \in M} M_p{}^{\perp} \quad \text{and} \quad \varpi\colon E \to M \quad \text{take } M_p{}^{\perp} \text{ to } p.$$

It is not hard to see that $\nu = \varpi\colon E \to M$ is a k-plane bundle over M, the **normal bundle of M in N**.

For example, the normal bundle ν of $S^{n-1} \subset \mathbb{R}^n$ is the trivial 1-plane bundle, for ν has a section consisting of unit outward normal vectors. On the other hand

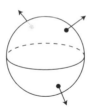

if M is the Möbius strip and $S^1 \subset M$ is a circle around the center, then it is not hard to see that the normal bundle ν will be isomorphic to the (non-trivial) bundle $M \to S^1$. If we consider $S^1 \subset M \subset \mathbb{P}^2$, then the normal bundle

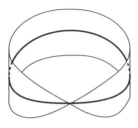

of S^1 in $\mathbb{P}^2$ is exactly the same as the normal bundle of S^1 in M, so it too is non-trivial.

Our aim is to prove that for compact M the normal bundle of M in N is always equivalent to a bundle $\pi\colon U \to M$ for which U is an open neighborhood of M in N, and for which the 0-section $s\colon M \to U$ is just the inclusion of M into U. In the case where N is the total space of a bundle over M, this open neighborhood can be taken to be the whole total space. But in general the neighborhood cannot be all of N. For example, as an appropriate neighborhood of $S^1 \subset \mathbb{R}^2$ we can choose $\mathbb{R}^2 - \{0\}$.

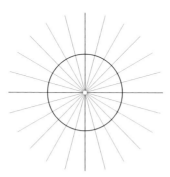

A bundle $\pi\colon U \to M$ with U an open neighborhood of M in N, for which the 0-section $s\colon M \to U$ is the inclusion of M in U, is called a **tubular neighborhood of M in N**. Before proving the existence of tubular neighborhoods, we add some remarks and a Lemma.

If $\pi\colon U \to M$ is a tubular neighborhood, then clearly

$$\pi \circ s = \text{ identity of } M,$$
$$s \circ \pi \quad \text{ is smoothly homotopic to the identity of } U,$$

so π is a deformation retraction, and $H^k(U) \approx H^k(M)$; thus M has the same de Rham cohomology as an open neighborhood. Moreover, if we choose a Riemannian metric $\langle\ ,\ \rangle$ for $\pi\colon U \to M$ and define $D = \{e \in U : \langle e, e \rangle \leq 1\}$, then D is a submanifold-with-boundary of U, and the map $\pi|D\colon D \to M$ is also a deformation retraction. So M also has the same de Rham cohomology as a closed neighborhood.

19. LEMMA. Let X be a compact metric space and $X_0 \subset X$ a closed subset. Let $f\colon X \to Y$ be a local homeomorphism such that $f|X_0$ is one-one. Then there is a neighborhood U of X_0 such that $f|U$ is one-one.

PROOF. Let $C \subset X \times X$ be

$$\{(x, y) \in X \times X : x \neq y \text{ and } f(x) = f(y)\}.$$

Then C is closed, for if (x_n, y_n) is a sequence in C with $x_n \to x$ and $y_n \to y$, then $f(x) = \lim f(x_n) = \lim f(y_n) = f(y)$, and also $x \neq y$ since f is locally one-one.

If $g \colon C \to \mathbb{R}$ is $g(x, y) = d(x, X_0) + d(y, X_0)$, then $g > 0$ on C. Since C is compact, there is $\varepsilon > 0$ such that $g \geq 2\varepsilon$ on C. Then f is one-one on the ε-neighborhood of X_0. ❖

20. THEOREM. Let $M \subset N$ be a compact submanifold of N. Then M has a tubular neighborhood $\pi \colon U \to M$ in N, which is equivalent to the normal bundle of M in N.

PROOF. Choose a Riemannian metric $\langle \ , \ \rangle$ for N, with the corresponding norm $\| \ \|$, and metric $d \colon N \times N \to \mathbb{R}$. Let

$$E = \{v : v \in N_p \text{ and } v \in M_p^\perp, \text{ for some } p \in M\}$$
$$E_\varepsilon = \{v \in E : \|v\| < \varepsilon\}$$
$$U_\varepsilon = \{q \in N : d(q, M) < \varepsilon\}.$$

It follows easily from Theorem 13, and compactness of M, that exp is defined on E_ε for sufficiently small $\varepsilon > 0$. We claim that for sufficiently small ε, the map exp is a diffeomorphism from E_ε onto U_ε. This will clearly prove the theorem.

Let $V \subset E$ be the set of a non-critical points for exp. Then $V \supset M$ (considered as a subset of E via the 0-section), and $V_1 = \overline{V \cap E_1}$ is compact; since exp is one-one on $M \subset V_1$, it follows from Lemma 19 that for sufficiently small ε the map exp is a diffeomorphism on E_ε.

It is clear also that $\exp(E_\varepsilon) \subset U_\varepsilon$. To prove that exp is onto U_ε, choose any $q \in U_\varepsilon$, and a point $p \in M$ closest to q. If $\gamma \colon [0, 1] \to N$ is the geodesic of length $< \varepsilon$ with $\gamma(0) = p$ and $\gamma(1) = q$, it is easy to see that γ is perpendicular to M at p (compare the second proof of Gauss' Lemma). This means that $q = \exp_p d\gamma/dt(0)$ where $d\gamma/dt(0) \in E_\varepsilon$. ❖

One of the interesting features of Theorem 20 is that all the paraphernalia of Riemannian metrics and geodesics are used in its proof, while they do not even appear in the statement. Theorem 20 will be needed only in Chapter 11, where we will also need the following modification.

21. THEOREM. Let N be a manifold-with-boundary, with compact boundary ∂N. Then ∂N has (arbitrarily small) open [and closed] neighborhoods for which there are deformation retractions onto ∂N.

PROOF. Exactly the same as the proof of Theorem 20, using only inward pointing normal vectors. ❖

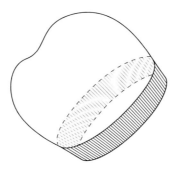

PROBLEMS

1. Let V be a vector space over a field F of characteristic $\neq 2$, and let $h\colon V \times V \to F$ be symmetric and bilinear.

(a) Define $q\colon V \to F$ by $q(v) = h(v, v)$. Show that if $\phi_1, \ldots, \phi_n$ is a basis for V^*, then

$$q = \sum_{i,j=1}^{n} a_{ij} v^*_i \cdot v^*_j$$

for some a_{ij}.

(b) Show that

$$q(-v) = q(v)$$

$$h(u, v) = \tfrac{1}{2}\big[q(u + v) - q(u) - q(v)\big].$$

(c) Suppose $q\colon V \to F$ satisfies $q(-v) = v$, and that $h(u, v) = q(u+v) - q(u) - q(v)$ is bilinear. Show that

$$q(u+v+w) - q(u) - q(v+w) = q(u+v) - q(u) - q(v) - q(u+w) - q(u) - q(w).$$

Conclude that $q(0) = 0$, and $q(2u) = 4q(u)$. Then show that $q(v) = h(v, v)$.

2. Let $\langle\ ,\ \rangle$ be a Euclidean metric for V^*. Suppose $\phi_i, \psi_i \in V^*$ satisfy $\phi_1 \wedge \cdots \wedge \phi_k = \psi_1 \wedge \cdots \wedge \psi_k \neq 0$, and let W_ϕ and W_ψ be the subspaces of V^* spanned by the ϕ_i and ψ_i.

(a) Show that $\omega \in W_\phi$ if and only if $\omega \wedge \phi_1 \wedge \cdots \wedge \phi_k = 0$. Conclude that $W_\phi = W_\psi$.

(b) Let $\sigma_1, \ldots, \sigma_k$ be an orthonormal basis of $W_\phi = W_\psi$. If $\phi_i = \sum_j a_{ji}\sigma_j$, show that the signed k-dimensional volume of the parallelepiped spanned by $\phi_1, \ldots, \phi_k$ is $\det(a_{ij})$. (The sign is $+$ if $\phi_1, \ldots, \phi_k$ has the same orientation as $\sigma_1, \ldots, \sigma_k$, and $-$ otherwise.)

(c) Using Problem 7-9, show that this volume is the same for $\psi_1, \ldots, \psi_k$.

(d) Conversely, if $W_\phi = W_\psi$, and the signed volumes of the parallelepipeds are the same, show that $\phi_1 \wedge \cdots \wedge \phi_k = \psi_1 \wedge \cdots \wedge \psi_k$.

If we identify V with V^{**}, so that we have a wedge product $v_1 \wedge \cdots \wedge v_k$ of vectors $v_i \in V$, then we have a geometric condition for equality with $w_1 \wedge \cdots \wedge w_k$. In *Leçons sur la Géométrie des Espaces de Riemann*, É. Cartan uses this condition to *define* $\Omega^k(V^*)$ as formal sums of equivalence classes of k vectors; he deduces geometrically the corresponding conditions on the coordinates of v_i, w_i.

3. Let V be an n-dimensional vector space, and $\langle\ ,\ \rangle$ an inner product on V which is not necessarily positive definite. A basis $v_1, \ldots, v_n$ for V is called **orthonormal** if $\langle v_i, v_j \rangle = \pm\delta_{ij}$.

(a) If $V \neq \{0\}$, then there is a vector $v \in V$ with $\langle v, v \rangle \neq 0$.

(b) For $W \subset V$, let $W^{\perp} = \{v \in V : \langle v, w \rangle = 0 \text{ for all } w \in W\}$. Prove that $\dim W^{\perp} \geq n - \dim W$. *Hint:* If $\{w_i\}$ is a basis for W, consider the linear functionals $\lambda_i : V \to \mathbb{R}$ defined by $\lambda_i(v) = \langle v, w_i \rangle$.

(c) If $\langle \ , \ \rangle$ is non-degenerate on W, then $V = W \oplus W^{\perp}$, and $\langle \ , \ \rangle$ is also non-degenerate on $W^{\perp}$.

(d) V has an orthonormal basis. Thus, there is an isomorphism $f : \mathbb{R}^n \to V$ with $f^*\langle \ , \ \rangle = \langle \ , \ \rangle_r$ for some r (the inner product $\langle \ , \ \rangle_r$ is defined on page 301).

(e) The **index** of $\langle \ , \ \rangle$ is the largest dimension of a subspace $W \subset V$ such that $\langle \ , \ \rangle | W$ is negative definite. Show that the index is $n - r$, thus showing that r is unique ("Sylvester's Law of Inertia").

4. Let $\langle \ , \ \rangle$ be a (possibly non-positive definite) inner product on V, and let $v_1, \ldots, v_n$ be an orthonormal basis (see Problem 3). Define an inner product $\langle \ , \ \rangle^k$ on $\Omega^k(V)$ by requiring that

$$v^*_{i_1} \wedge \cdots \wedge v^*_{i_k} \qquad 1 \leq i_1 < \cdots < i_k \leq n$$

be an orthonormal basis, with

$$\left\langle v^*_{i_1} \wedge \cdots \wedge v^*_{i_k}, v^*_{j_1} \wedge \cdots \wedge v^*_{j_k} \right\rangle^k = \det\left(\langle v_{i_\alpha}, v_{j_\beta} \rangle \right).$$

(a) Show that $\langle \ , \ \rangle^k$ is independent of the basis $v_1, \ldots, v_k$. (Use Problem 7-16.)

(b) Show that

$$\langle \phi_1 \wedge \cdots \wedge \phi_k, \psi_1 \wedge \cdots \wedge \psi_k \rangle^k = \det\left(\langle \phi_i, \psi_j \rangle^* \right) = \det\left(\langle \phi_i, \psi_i \rangle^1 \right).$$

(c) If $\langle \ , \ \rangle$ has index i, then

$$\langle v^*_1 \wedge \cdots \wedge v^*_n, v^*_1 \wedge \cdots \wedge v^*_n \rangle^n = (-1)^i.$$

(d) For those who know about $\otimes$ and Λ^k. Using the isomorphisms $\bigotimes^k V^* \approx \left(\bigotimes^k V \right)^*$ and $\Lambda^k(V^*) \approx (\Lambda^k V)^*$, define inner products on $\bigotimes^k V$ and $\Lambda^k V$ by using the isomorphism $V \to V^*$ given by the inner product on V. Show that these inner products agree with the ones defined above.

5. Recall the definition of $v_1 \times \cdots \times v_{n-1}$ in Problem 7-26.

(a) Show that $\langle v_1 \times \cdots \times v_{n-1}, v_i \rangle = 0$.

(b) Show that $|v_1 \times \cdots \times v_{n-1}| = \sqrt{\det(g_{ij})}$, where $g_{ij} = \langle v_i, v_j \rangle$. *Hint:* Apply the result on page 308 to a certain $(n-1)$-dimensional subspace of $\mathbb{R}^n$.

6. Let $\xi = \pi : E \to B$ be a vector bundle. An **indefinite metric** on ξ is a continuous choice of a non-positive definite inner product $\langle\ ,\ \rangle_p$ on each $\pi^{-1}(p)$. Show that the index of $\langle\ ,\ \rangle_p$ is constant on each component of B.

7. This problem requires a little knowledge of simple-connectedness and covering spaces.

(a) There is no way of continuously choosing a 1-dimensional subspace of $S^2{}_p$, for each $p \in S^2$. (Consider the space consisting of the two unit vectors in each subspace.)

(b) There is no Riemannian metric of index 1 on S^2.

8. Let $\langle\ ,\ \rangle$ and $\langle\ ,\ \rangle'$ be two Riemannian metrics on a vector bundle $\xi = \pi : E \to B$. Let S be the set of $e \in E$ with $\langle e, e \rangle = 1$, and define S' similarly. Show that S is homeomorphic to S'. If ξ is a smooth bundle over a manifold M, show that S is diffeomorphic to S'.

9. Show by a computation that if the functions g_{ij} and g'_{ij} are related by

$$g'_{\alpha\beta} = \sum_{i,j} g_{ij} \frac{\partial x^i}{\partial x'^\alpha} \frac{\partial x^j}{\partial x'^\beta},$$

with $\det(g_{ij}) \neq 0$, and the functions g^{ij}, g'^{ij} are defined by

$$\sum_{k=1}^{n} g^{ik} g_{kj} = \delta^i_j, \qquad \sum_{k=1}^{n} g'^{ik} g'_{kj} = \delta^i_j,$$

then

$$g'^{\alpha\beta} = \sum_{i,j} g^{ij} \frac{\partial x'^\alpha}{\partial x^i} \frac{\partial x'^\beta}{\partial x^j}.$$

This, of course, is the classical way of defining the tensor [having the components] g^{ij}.

10. (a) Let $\langle\ ,\ \rangle$ be a Riemannian metric on M, and A a tensor of type $\binom{1}{1}$, so that $A(p) : M_p \to M_p$. Define a tensor B of type $\binom{2}{0}$ by

$$B(p)(v_1, v_2) = \langle A(p)(v_1), v_2 \rangle.$$

If the expression for A in a coordinate system is

$$A = \sum_{i,j=1}^{n} A^j_i \, dx^i \otimes \frac{\partial}{\partial x^j},$$

show that $B = \sum_{i,k} B_{ik} \, dx^i \otimes dx^k$, where

$$B_{ik} = \sum_{j=1}^{n} A_i^j g_{jk}.$$

(b) Similarly, define a tensor C of type $\binom{0}{2}$ by

$$C(p)(\lambda_1, \lambda_2) = \langle A(p)^*(\lambda_1), \lambda_2 \rangle.$$

Show that if C has components C^{kj}, then

$$C^{kj} = \sum_{i=1}^{n} g^{ki} A_i^j.$$

The tensors B and C are said to be obtained from A by "raising and lowering indices".

11. (a) Let $X_1, \ldots, X_n$ be linearly independent vector fields on a manifold M with a Riemannian metric $\langle \ , \ \rangle$. Show that the Gram-Schmidt process can be applied to the vector fields all at once, so that we obtain n everywhere orthonormal vector fields $Y_1, \ldots, Y_n$.
(b) For the case of a non-positive definite metric, find $Y_1, \ldots, Y_n$ with $\langle Y_i, Y_j \rangle = \pm \delta_{ij}$ in a neighborhood of any point.

12. (a) If $f : [a,b] \to \mathbb{R}$ is positive, show that the area of the surface obtained by revolving the graph of f around the x-axis is

$$\int_a^b 2\pi f \sqrt{1 + (f')^2}.$$

(b) Compute the area of S^2.

13. Let $M \subset \mathbb{R}^n$ be an $(n-1)$-dimensional submanifold with orientation μ. The **outward unit normal** $v(p)$ at $p \in M$ is defined to be that vector in $\mathbb{R}^n{}_p$ of length 1 such that $v(p), (v_1)_p, \ldots, (v_{n-1})_p$ is positively oriented in $\mathbb{R}^n{}_p$ when $(v_1)_p, \ldots, (v_{n-1})_p$ is positively oriented in M_p.

(a) If $M = \partial N$ for an n-dimensional manifold-with-boundary $N \subset \mathbb{R}^n$, then $v(p)$ is outward pointing in the sense of Chapter 8.
(b) Let dV_{n-1} be the volume element of M determined by the Riemannian metric it acquires as a submanifold of $\mathbb{R}^n$. Show that if we consider $v(p)$ as an element of $\mathbb{R}^n$, then

$$dV_{n-1}(p)\big((v_1)_p, \ldots, (v_{n-1})_p\big) = \det \begin{pmatrix} v(p) \\ v_1 \\ \vdots \\ v_{n-1} \end{pmatrix}.$$

Conclude that $d V_{n-1}(p)$ is the restriction to M_p of

$$\sum_{i=1}^{n}(-1)^{i-1}v^i(p)\,dx^1(p)\wedge\cdots\wedge\widehat{dx^i(p)}\wedge\cdots\wedge dx^n(p).$$

(c) Note that $v_1\times\cdots\times v_{n-1}=\alpha\,v(p)$ for some $\alpha\in\mathbb{R}$ (by Problem 5). Show that for $w\in\mathbb{R}^n$ we have

$$\langle w,v(p)\rangle\cdot\langle v_1\times\cdots\times v_{n-1},v(p)\rangle=\langle w,v_1\times\cdots\times v_{n-1}\rangle.$$

Conclude that

$v^i(p)\cdot d V_{n-1}(p)=$ restriction to M_p of

$$(-1)^{i-1}dx^1(p)\wedge\cdots\wedge\widehat{dx^i(p)}\wedge\cdots\wedge dx^n(p).$$

(d) Let $M\subset\mathbb{R}^n$ be a compact n-dimensional manifold-with-boundary, with v the outward unit normal on ∂M. Denote the volume element of M by $d V_n$, and that of ∂M by $d V_{n-1}$. Let $X=\sum_i a^i\,\partial/\partial x^i$ be a vector field on M. Prove the *Divergence Theorem*:

$$\int_M \operatorname{div} X\,d V_n=\int_{\partial M}\langle X,v\rangle\,d V_{n-1}$$

(the function div X is defined in Problem 7-27). *Hint*: Consider the form ω on M defined by

$$\omega=\sum_{i=1}^{n}(-1)^{i-1}a^i\,dx^1\wedge\cdots\wedge\widehat{dx^i}\wedge\cdots\wedge dx^n.$$

(e) Let $M\subset\mathbb{R}^3$ be a compact 2-dimensional manifold-with-boundary, with orientation μ, and outward unit normal v. Let T be the vector field on ∂M consisting of positively oriented unit vectors. Denote the volume element of M by dA, and that of ∂M by ds. Let X be a vector field on M. Prove (the original) *Stokes' Theorem*:

$$\int_M\langle\nabla\times X,v\rangle\,dA=\int_{\partial M}\langle X,T\rangle\,ds$$

($\nabla\times X$ is defined in Problem 7-27).

14. (a) Let V_n be the volume of the unit ball in $\mathbb{R}^n$. Show that

$$V_n=\int_{-1}^{1}(1-x^2)^{(n-1)/2}V_{n-1}\,dx.$$

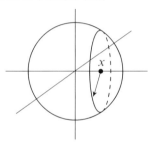

(b) If $I_n = \int_{-1}^{1} (1 - x^2)^{(n-1)/2}\, dx$, show that

$$I_n = \frac{n-1}{n} I_{n-2}.$$

(c) Using $V_1 = 2$, $V_2 = \pi$, show that

$$V_n = \begin{cases} \dfrac{\pi^{n/2}}{(n/2)!} & n \text{ even} \\[3mm] \dfrac{2^{(n+1)/2}\pi^{(n-1)/2}}{1 \cdot 3 \cdot 5 \cdots n} & n \text{ odd.} \end{cases}$$

(In terms of the Γ function, this can be written $\dfrac{\pi^{n/2}}{\Gamma(1 + n/2)}$.)

(d) Let A_{n-1} be the $(n-1)$-volume of S^{n-1}. Using the method of proof in Corollary 8-8, but reversing the order of integration, show that

$$V_n = \int_0^1 r^{n-1} A_{n-1}\, dr = \frac{1}{n} A_{n-1}.$$

(e) Obtain this same result by applying the Divergence Theorem (Problem 13), with $X(p) = p_p$.

15. (a) Let $c: [0, 1] \to \mathbb{R}^n$ be a differentiable curve, where $\mathbb{R}^n$ has the usual Riemannian metric $\langle\ ,\ \rangle = \sum_i dx^i \otimes dx^i$. Show that

$$L(c) = \int_0^1 \sqrt{\sum_{i=1}^n [(c^i)'(t)]^2}\ dt.$$

(b) For the special case $c: [0, 1] \to \mathbb{R}^2$ given by $c(t) = (t, f(t))$, show that this length,

$$\int_0^1 \sqrt{1 + [f'(t)]^2}\ dt,$$

is the least upper bound of the lengths of inscribed polygonal curves.

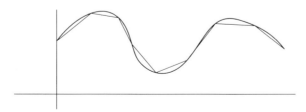

Hint: If the inscribed polygonal curve is determined by the points $(t_i, c(t_i))$ for a partition $0 = t_0 < \cdots < t_n = 1$ of $[0, 1]$, then we have

$$|c(t_i) - c(t_{i-1})| = \sqrt{(t_i - t_{i-1})^2 + \big(f(t_i) - f(t_{i-1})\big)^2}$$

$$= \sqrt{(t_i - t_{i-1})^2 + f'(\xi_i)(t_i - t_{i-1})^2}$$

for some $\xi_i \in [t_{i-1}, t_i]$.

(c) Prove the same result in the general case. *Hint*: Use the results of Problem 8-1, and uniform continuity of $\sqrt{}$ on a compact set.

It is natural to suppose that the area of a surface is, similarly, the least upper bound of the areas of inscribed polygonal surfaces, but as H. Schwarz first observed, this least upper bound is infinite for a bounded portion of a cylinder! To illustrate Schwarz's example I have plagiarized the following picture from a book called *Математический Анализ на Многообразиях*, written by someone called M. Спивак.

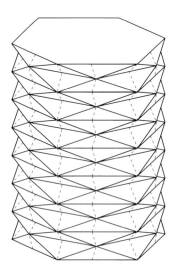
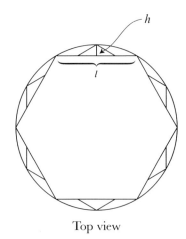

Top view

To increase the number of triangles, we maintain the hexagonal arrangement, but move the planes of the hexagons closer together, so that the triangles are more nearly in a plane parallel to the bases of the cylinder. In this way, we can increase the number of triangles indefinitely, while the area of each approaches $hl/2$.

The topic of surface area for non-differentiable surfaces is a complex one, which we will not go into here.

16. Let $c \colon [0, 1] \to M$ be a curve in a manifold M with a Riemannian metric $\langle \, , \, \rangle$. If $p \colon [0, 1] \to [0, 1]$ is a diffeomorphism, show that

$$L(c) = L(c \circ p).$$

17. Show that the metric d on M may be defined using C^∞, instead of piecewise C^∞ curves. (Show how to round off corners of a piecewise C^∞ path so that the length increases by less than any given $\varepsilon > 0$; remember that the formula for length involves only first derivatives.)

18. (a) Let $B \subset M$ be homeomorphic to the ball $\{p \in \mathbb{R}^n : |p| \leq 1\}$ and let $S \subset M$ be the subset corresponding to $\{p \in \mathbb{R}^n : |p| = 1\}$. Show that $M - S$ is disconnected, by showing that $M - B$ and $B - S$ are disjoint open subsets of $M - S$.

(b) If $p \in B - S$ and $q \in M - B$, show that $d(p, q) \geq \min\limits_{q' \in S} d(p, q')$. Use this fact and Lemma 7' to complete the proof of Theorem 7. (In the theory of infinite dimensional manifolds, these details become quite important, for $M - S$ does *not* have to be disconnected, and Theorem 7 is false.)

19. (a) By applying integration by parts to the equation on pages 318–319, show that

$$\frac{dJ(\bar{\alpha}(u))}{du}\bigg|_{u=0} = \int_a^b \frac{\partial^2 \alpha}{\partial u \partial t}(0, t)\left[\frac{\partial F}{\partial y}(t, f(t), f'(t)) \right.$$
$$\left. - \int_a^t \frac{\partial F}{\partial x}(t, f(t), f'(t))\, dt\right] dt;$$

this result makes sense even if f is only C^1.

(b) *Du Bois Reymond's Lemma.* If a continuous function g on $[a, b]$ satisfies

$$\int_a^b \eta'(t) g(t)\, dt = 0$$

for all C^∞ functions η on $[a, b]$ with $\eta(a) = \eta(b) = 0$, then g is a constant. *Hint*: The constant c must be

$$c = \frac{1}{b - a} \int_a^b g(u)\, du.$$

We clearly have

$$\int_a^b \eta'(t)[g(t) - c]\, dt = 0,$$

so we need to find a suitable η with $\eta'(t) = g(t) - c$.

(c) Conclude that if the C^1 function f is a critical point of J, then f still satisfies the Euler equations (which are not *a priori* meaningful if f is not C^2).

20. The **hyperbolic sine**, **hyperbolic cosine**, and **hyperbolic tangent** functions sinh, cosh, and tanh are defined by

$$\sinh x = \frac{e^x - e^{-x}}{2}, \qquad \cosh x = \frac{e^x + e^{-x}}{2}, \qquad \tanh x = \frac{\sinh x}{\cosh x}.$$

(a) Graph sinh, cosh, and tanh.
(b) Show that

$$\cosh^2 - \sinh^2 = 1$$
$$\tanh^2 + 1/\cosh^2 = 1$$
$$\sinh(x + y) = \sinh x \cosh y + \cosh x \sinh y$$
$$\cosh(x + y) = \cosh x \cosh y + \sinh x \sinh y$$
$$\sinh' = \cosh$$
$$\cosh' = \sinh.$$

(c) For those who know about complex power series:

$$\sinh x = \frac{\sin ix}{i}, \qquad \cosh x = \cos ix.$$

(d) The inverse functions of sinh and tanh are denoted by $\sinh^{-1}$ and $\tanh^{-1}$, respectively, while $\cosh^{-1}$ denotes the inverse of $\cosh \,|\, [0, \infty)$. Show that

$$\sinh(\cosh^{-1} x) = \sqrt{x^2 - 1}$$
$$(\sinh^{-1})'(x) = \frac{1}{\sqrt{1 + x^2}}$$
$$\cosh(\sinh^{-1} x) = \sqrt{1 + x^2}$$
$$\cosh(\tanh^{-1} x) = \frac{1}{\sqrt{1 - x^2}}$$
$$(\cosh^{-1})'(x) = \frac{1}{\sqrt{x^2 - 1}}.$$

21. Consider the problem of finding a surface of revolution joining two circles of radius 1, situated, for convenience, at a and $-a$. We are looking for a function

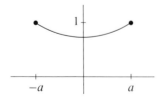
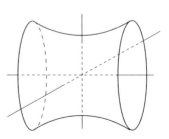

of the form

$$f(x) = c \cosh \frac{x}{c}$$

where c is supposed to satisfy

$$c \cosh \frac{a}{c} = 1 \quad (c > 0).$$

(a) There is a unique $y_0 > 0$ with $\tanh y_0 = 1/y_0$. Examine the sign of $1/y - \tanh y$ for $y > 0$.

(b) Examine the sign of $\cosh y - y \sinh y$ for $y > 0$.

(c) Let

$$A_a(c) = c \cosh \frac{a}{c} \quad c > 0.$$

Show that A_a has a minimum at a/y_0, find the value of A_a there, and sketch the graph.

(d) There exists c with $c \cosh a/c = 1$ if and only if $a \leq y_0/\cosh y_0$. If $a = y_0/\cosh y_0$, then there is a unique such c, namely $c = a/y_0 = 1/\cosh y_0$. If $a < y_0/\cosh y_0$, then there are two such c, with $c_1 < a/y_0 < c_2$. It turns out that the surface for c_2 has smaller area.

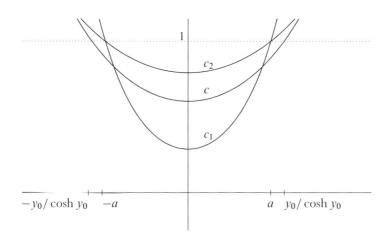

(e) Using Problem 20(d), show that

$$\frac{y_0}{\cosh y_0} = \sqrt{y_0^2 - 1}.$$

$[y_0 \sim 1.2, \text{ so } \sqrt{y_0^2 - 1} \sim .67.]$

[These phenomena can be pictured more easily if we use the notion of an envelope—c.f. Volume III, pp. 176ff. The envelope of the 1-parameter family of curves

$$f_c(x) = c \cosh \frac{x}{c}$$

is determined by solving the equations

$$0 = \frac{\partial f_c(x)}{\partial c} = \cosh \frac{x}{c} - \frac{x}{c} \sinh \frac{x}{c}.$$

We obtain

$$\frac{x}{c} = \pm y_0, \qquad y = c \cosh \frac{x}{c} = c \cosh y_0,$$

so the envelope consists of the straight lines

$$y = \pm \frac{\cosh y_0}{y_0} x.$$

The unique member of the family through $(y_0/\cosh y_0, 1)$ is tangent to the envelope at that point. For $a < y_0/\cosh y_0$, the graph of f_{c_1} is tangent to

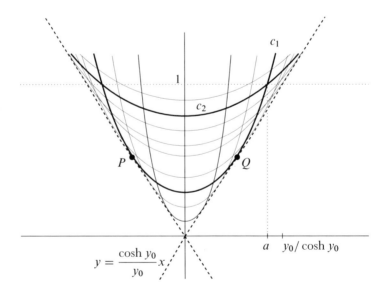

the envelope at points $P, Q \in (-a, a)$, but the graph of f_{c_2} is tangent to the envelope at points outside $[-a, a]$. The point Q is called **conjugate** to P along the extremal f_{c_1}, and it is shown in the calculus of variations that the existence of this conjugate point implies that the portion of f_{c_1} from P to $(a, 1)$ does *not* give a local minimum for $\int f\sqrt{1+(f')^2}$. (Compare with the discussion of conjugate points of a geodesic in Volume IV, Chapter 8, and note the remark on pg. V. 396.)]

22. All of our illustrations of calculus of variations problems involved an F which does not involve t, so that the Euler equations are actually

$$\frac{\partial F}{\partial x}(f(t), f'(t)) - \frac{d}{dt}\left(\frac{\partial F}{\partial y}(f(t), f'(t))\right) = 0.$$

(a) Show that for any f and $F: \mathbb{R}^2 \to \mathbb{R}$ we have

$$\frac{d}{dt}\left(F - f'\frac{\partial F}{\partial y}\right) = f'\left[\frac{\partial F}{\partial x} - \frac{d}{dt}\frac{\partial F}{\partial y}\right],$$

and conclude that the extremals for our problem satisfy

$$F - f'\frac{\partial F}{\partial y} = 0.$$

(b) Apply this to $F(x, y) = x\sqrt{1 + y^2}$ to obtain directly the equation $dy/dx = \sqrt{cy^2 - 1}$ which we eventually obtained in our solution to the problem.

23. (a) Let x and x' be two coordinate systems, with corresponding g_{ij} and g'_{ij} for the expression of a Riemannian metric. Show that

$$\frac{\partial g'_{\alpha\beta}}{\partial x'^\gamma} = \sum_{i,j,k=1}^n \frac{\partial g_{ij}}{\partial x^k}\frac{\partial x^k}{\partial x'^\gamma}\frac{\partial x^i}{\partial x'^\alpha}\frac{\partial x^j}{\partial x'^\beta}$$

$$+ \sum_{i,j=1}^n g_{ij}\left(\frac{\partial x^i}{\partial x'^\alpha}\frac{\partial^2 x^j}{\partial x'^\beta\partial x'^\gamma} + \frac{\partial x^j}{\partial x'^\beta}\frac{\partial^2 x^i}{\partial x'^\alpha\partial x'^\gamma}\right).$$

(b) For the corresponding $[ij, k]$ and $[\alpha\beta, \gamma]'$, show that

$$[\alpha\beta, \gamma]' = \sum_{i,j,k=1}^n [ij,k]\frac{\partial x^i}{\partial x'^\alpha}\frac{\partial x^j}{\partial x'^\beta}\frac{\partial x^k}{\partial x'^\gamma} + \sum_{i,j=1}^n \frac{\partial x^i}{\partial x'^\gamma}\frac{\partial^2 x^j}{\partial x'^\alpha\partial x'^\beta},$$

so that $[ij, k]$ are not the components of a tensor.
(c) Also show that

$$\Gamma'^\gamma_{\alpha\beta} = \sum_{i,j,k=1}^n \Gamma^k_{ij}\frac{\partial x^i}{\partial x'^\alpha}\frac{\partial x^j}{\partial x'^\beta}\frac{\partial x'^\gamma}{\partial x^k} + \sum_{l=1}^n \frac{\partial^2 x^l}{\partial x'^\alpha\partial x'^\beta}\frac{\partial x'^\gamma}{\partial x^l}.$$

24. Show that any C^∞ structure on $\mathbb{R}$ is diffeomorphic to the usual C^∞ structure. (Consider the arclength function on a geodesic for some Riemannian metric on $\mathbb{R}$.)

25. Let $\langle \ , \ \rangle = \sum_i dx^i \otimes dx^i$ be the usual Riemannian metric on $\mathbb{R}^n$, and let $\sum_{i,j} g_{ij} du^i \otimes du^j$ be another metric, where $u^1, \ldots, u^n$ again denotes the standard coordinate system on $\mathbb{R}^n$. Suppose we are told that there is a diffeomorphism $f\colon \mathbb{R}^n \to \mathbb{R}^n$ such that $\sum_{i,j} g_{ij} \, du^i \otimes du^j = f^* \langle \ , \ \rangle$. How can we go about finding f?

(a) Let $\partial f/\partial u^i = e_i\colon \mathbb{R}^n \to \mathbb{R}^n$. If we consider e_i as a vector field on $\mathbb{R}^n$, show that $f_*(\partial/\partial u^i) = e_i$.

(b) Show that $g_{ij} = \langle e_i, e_j \rangle$.

(c) To solve for f it is, in theory at least, sufficient to solve for the e_i, and to solve for these we want to find differential equations

$$\frac{\partial e_i}{\partial u_j} = \sum_{r=1}^{n} A^r_{ij} e_r$$

satisfied by the e_i's. Show that we must have

$$B_{ij,k} \overset{\text{def}}{=} \sum_{r=1}^{n} g_{kr} A^r_{ij} = \sum_{l=1}^{n} \frac{\partial^2 f^l}{\partial u^i \partial u^j} \cdot \frac{\partial f^l}{\partial u^k}.$$

(d) Show that

$$\frac{\partial g_{ij}}{\partial u^k} = \sum_{l=1}^{n} \frac{\partial^2 f^l}{\partial u^i \partial u^k} \frac{\partial f^l}{\partial u^j} + \frac{\partial^2 f^l}{\partial u^j \partial u^k} \frac{\partial f^l}{\partial u^i}$$

$$= \sum_{r=1}^{n} g_{jr} A^r_{ik} + g_{ir} A^r_{kj}$$

$$= B_{ik,j} + B_{kj,i}.$$

(e) By cyclically permuting i, j, k, deduce that

$$B_{ij,k} = [ij, k],$$

so that $A^r_{ij} = \Gamma^r_{ij}$. In *Leçons sur la Géométrie des Espaces de Riemann*, É. Cartan uses this approach to motivate the introduction of the Γ^k_{ij}.

(f) Deduce the result $A^r_{ij} = \Gamma^r_{ij}$ directly from our equations for a geodesic. (Note that the curves obtained by setting all but one f^i constant are geodesics, since they correspond to lines parallel to the x^i-axis.)

26. If $(V', \langle \ , \ \rangle')$ and $(V'', \langle \ , \ \rangle'')$ are two vector spaces with inner products, we define $\langle \ , \ \rangle$ on $V = V' \oplus V''$ by

$$\langle v' \oplus v'', w' \oplus w'' \rangle = \langle v', w' \rangle' + \langle v'', w'' \rangle''.$$

(a) Show that $\langle \ , \ \rangle$ is an inner product.
(b) Given Riemannian metrics on M and N, it follows that there is a natural way to put a Riemannian metric on $M \times N$. Describe the geodesics on $M \times N$ for this metric.

27. (a) Let $\gamma \colon [a,b] \to M$ be a geodesic, and let $p \colon [\alpha, \beta] \to [a,b]$ be a diffeomorphism. Show that $c = \gamma \circ p$ satisfies

$$\frac{d^2 c^k}{dt^2} + \sum_{i,j=1}^{n} \Gamma_{ij}^k (c(t)) \frac{dc^i}{dt} \frac{dc^j}{dt} = \frac{dc^k}{dt} \frac{p''(t)}{p'(t)}.$$

(b) Conversely, if c satisfies this equation, then γ is a geodesic.
(c) If c satisfies

$$\frac{d^2 c^k}{dt^2} + \sum_{i,j=1}^{n} \Gamma_{ij}^k (c(t)) \frac{dc^i}{dt} \frac{dc^j}{dt} = \frac{dc^k}{dt} \mu(t) \qquad \text{for } \mu \colon \mathbb{R} \to \mathbb{R},$$

then c is a reparameterization of a geodesic. (The equation $p''(t) = p'(t)\mu(t)$ can be solved explicitly: $p(t) = \int^t e^{M(s)} \, ds$, where $M'(s) = \mu(s)$.)

28. Let c be a curve in M with $dc/dt \neq 0$ everywhere, and consider the hypersurfaces

$$\{\exp_{c(t)} v : \|v\| = \text{constant, where } v \in M_{c(t)} \text{ with } \langle v, dc/dt \rangle = 0\}.$$

Show that for $v \in M_{c(t)}$ with $\langle v, dc/dt \rangle = 0$, the geodesic $u \mapsto \exp_{c(t)} u \cdot v$ is perpendicular to these hypersurfaces. (Gauss' Lemma is the "special case" where c is constant.)

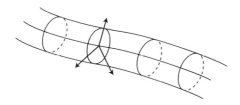

29. Let $\gamma\colon [a, b] \to M$ be a geodesic with $\gamma(a) = p$, and suppose that $\exp_p$ is a diffeomorphism on a neighborhood $\mathcal{O} \subset M_p$ of $\{t\gamma'(0) : 0 \leq t \leq 1\}$. Show that γ is a curve of minimal length between p and $q = \gamma(b)$, among all curves in $\exp(\mathcal{O})$. (Gauss' Lemma still works on $\exp(\mathcal{O})$.)

30. If $\langle \; , \; \rangle$ is a Riemannian metric on M and $d\colon M \times M \to \mathbb{R}$ is the corresponding metric, then a curve $\gamma\colon [a, b] \to M$ with $d(\gamma(a), \gamma(b)) = L(\gamma)$ is a geodesic.

31. *Schwarz's inequality* for continuous functions states that

$$\left(\int_a^b fg \right)^2 \leq \left(\int_a^b f^2 \right) \left(\int_a^b g^2 \right),$$

with equality if and only if f and g are linearly dependent (over $\mathbb{R}$).

(a) Prove Schwarz's inequality by imitating the proof of Theorem 1(2).
(b) For any curve γ show that

$$[L_a^b(\gamma)]^2 \leq (b - a) E_a^b(\gamma),$$

with equality if and only if γ is parameterized proportionally to arclength.
(c) Let $\gamma\colon [a, b] \to M$ be a geodesic with $L_a^b(\gamma) = d(\gamma(a), \gamma(b))$. If $c(a) = \gamma(a)$ and $c(b) = \gamma(b)$, show that

$$E(\gamma) = \frac{L(\gamma)^2}{b - a} \leq \frac{L(c)^2}{b - a} \leq E(c).$$

Conclude that $E(\gamma) < E(c)$ unless c is also a geodesic with

$$L_a^b(c) = d(c(a), c(b)).$$

In particular, sufficiently small pieces of a geodesic minimize energy.

32. Let p be a point of a manifold M with a Riemannian metric $\langle \; , \; \rangle$. Choose a basis $v_1, \ldots, v_n$ of M_p, so that we have a "rectangular" coordinate system χ on M_p given by $\sum_i a^i v_i \mapsto (a^1, \ldots, a^n)$; let x be the coordinate system $\chi \circ \exp^{-1}$, defined in a neighborhood U of p.

(a) Show that in this coordinate system we have $\Gamma_{ij}^k(p) = 0$. *Hint*: Recall the equations for a geodesic, and note that a geodesic γ through p is just $\exp$ composed with a straight line through 0 in M_p, so that each γ^k is linear.

(b) Let $r : U \to \mathbb{R}$ be $r(q) = d(p, q)$, so that $r \circ \gamma = \sum_k (\gamma^k)^2$. Show that

$$\frac{d^2\big((r \circ \gamma)^2\big)}{dt^2} = 2\left[\sum_k \left(\frac{d\gamma^k}{dt}\right)^2 - \sum_{i,j,k} \gamma^2 \Gamma_{ij}^k \frac{d\gamma^i}{dt} \frac{d\gamma^j}{dt}\right].$$

(c) Note that

$$\sum_{i,j} \frac{d\gamma^i}{dt} \frac{d\gamma^j}{dt} \leq n^2 \sum_k \left(\frac{d\gamma^k}{dt}\right)^2.$$

Using part (a), conclude that if $\|\gamma'(0)\|$ is sufficiently small, then

$$\frac{d^2 (r \circ \gamma)^2}{dt^2} > 0,$$

so that $d(r \circ \gamma)^2 / dt$ is strictly increasing in a neighborhood of 0.

(d) Let $B_\varepsilon = \{v \in M_p : \|v\| \leq \varepsilon\}$ and $S_\varepsilon = \{v \in M_p : \|v\| = \varepsilon\}$. Show that the following is true for all sufficiently small $\varepsilon > 0$: if γ is a geodesic such that $\gamma(0) \in \exp(S_\varepsilon)$ and such that $\gamma'(0)$ is tangent to $\exp(S_\varepsilon)$, then there is $\delta > 0$ (depending on γ) such that $\gamma(t) \notin \exp(B_\varepsilon)$ for $0 \neq t \in (-\delta, \delta)$. *Hint:* If $\gamma'(0)$ is

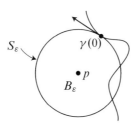

tangent to $\exp(S_\varepsilon)$, then $d(r \circ \gamma)/dt = 0$.

(e) Let q and q' be two points with $r(q), r(q') < \varepsilon$ and let γ be the unique geodesic of length $< 2\varepsilon$ joining them. Show that for sufficiently small ε the maximum of $r \circ \gamma$ occurs at either q or q'.

(f) A set $U \subset M$ is **geodesically convex** if every pair $q, q' \in U$ has a unique geodesic of minimum length between them, and this geodesic lies completely in U. Show that $\exp(\{v \in M_p : \|v\| < \varepsilon\})$ is geodesically convex for sufficiently small $\varepsilon > 0$.

(g) Let $f : U \to \mathbb{R}^n$ be a diffeomorphism of a neighborhood U of $0 \in \mathbb{R}^n$ into $\mathbb{R}^n$. Show that for sufficiently small ε, the image of the open ε-ball is convex.

33. (a) There is an everywhere differentiable curve $c(t) = (t, f(t))$ in $\mathbb{R}^2$ such that
$$\lim_{h \to 0} \frac{\text{length of } c \,|\, [0, h]}{|c(h) - c(0)|} \neq 1.$$

Hint: Make c look something like the following picture.

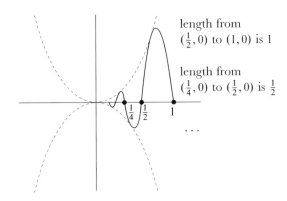

length from $(\frac{1}{2}, 0)$ to $(1, 0)$ is 1

length from $(\frac{1}{4}, 0)$ to $(\frac{1}{2}, 0)$ is $\frac{1}{2}$

(b) Consider the situation in Corollary 15, except that $c(t) = q$ if and only if $t = a$, and suppose $u'(t) > 0$ for t near 0. If c is C^1, then $v(t)$ approaches a limit as $t \to 0$ (even though $v(0)$ is undefined). Show that if c is C^1, then there is some $K > 0$ such that for all t near 0 we have
$$\left\| \frac{\partial \alpha}{\partial t}(u, t) \right\| \leq K u \left\| \frac{\partial \alpha}{\partial t}(1, t) \right\| \qquad 0 \leq u \leq 1.$$

Hint: In M_q we clearly have
$$\left\| \frac{d(u \cdot v(t))}{dt} \right\| = |u| \cdot \left\| \frac{dv(t)}{dt} \right\|.$$

Since $\exp_q$ is locally a diffeomorphism there are $0 < K_1 < K_2$ such that
$$K_1 \|v\| \leq \|\exp_{q*} v\| \leq K_2 \|v\|$$

for all tangent vectors v at points near q.

(c) Conclude that
$$\lim_{h \to 0} \frac{L(c \,|\, [0, h])}{d(p, c(h))} = \lim_{h \to 0} \frac{\displaystyle\int_0^h \sqrt{u'(t)^2 + \left\| \frac{\partial \alpha}{\partial t}(u(t), t) \right\|^2} \, dt}{u(h)} = 1.$$

(d) If c is C^1, show that $L(c)$ is the least upper bound of inscribed piecewise geodesic curves.

34. (a) Using the methods of Problem 33, show that if c is the straight line joining $v, w \in M_p$, then

$$\lim_{v,w \to 0} \frac{L(\exp \circ c)}{L(c)} = 1.$$

(b) Similarly, if $\gamma_{v,w}$ is the unique geodesic joining $\exp(v)$ and $\exp(w)$, and $\gamma_{v,w} = \exp \circ c_{v,w}$, then

$$\lim_{v,w \to 0} \frac{L(\gamma_{v,w})}{L(c_{v,w})} = 1.$$

(c) Conclude that

$$\lim_{v,w \to 0} \frac{d(\exp v, \exp w)}{\|v - w\|} = 1.$$

35. Let $f: M \to N$ be an isometry. Show that f is an isometry of the metric space structures determined on M and N by their respective Riemannian metrics.

36. Let M be a manifold with Riemannian metric $\langle \ , \ \rangle$ and corresponding metric d. Let $f: M \to M$ be a map of M onto itself which preserves the metric d.

(a) If γ is a geodesic, then $f \circ \gamma$ is a geodesic.

(b) Define $f': M_p \to M_{f(p)}$ as follows: For γ a geodesic with $\gamma(0) = p$, let

$$f'(\gamma'(0)) = \left. \frac{df(\gamma(t))}{dt} \right|_{t=0}.$$

Show that $\|f'(X)\| = \|X\|$, and that $f'(cX) = cf'(X)$.

(c) Given $X, Y \in M_p$, use Problem 34 to show that

$$\frac{2\langle X, Y \rangle}{\|X\| \cdot \|Y\|} = \frac{\|X\|^2 + \|Y\|^2}{\|X\| \cdot \|Y\|} - \frac{\|tX - tY\|^2}{\|tX\| \cdot \|tY\|}$$

$$= \frac{\|X\|^2 + \|Y\|^2}{\|X\| \cdot \|Y\|} - \lim_{t \to 0} \frac{[d(\exp tX, \exp tY)]^2}{\|tX\| \cdot \|tY\|}.$$

Conclude that $\langle X, Y \rangle_p = \langle f'(X), f'(Y) \rangle_{f(p)}$, and then that $f'(X + Y) = f'(X) + f'(Y)$.

(d) Part (c) shows that $f' \colon M_p \to M_{f(p)}$ is a diffeomorphism. Use this to show that f is itself a diffeomorphism, and hence an isometry.

37. (a) For $v, w \in \mathbb{R}^n$ with $w \neq 0$, show that

$$\lim_{t \to 0} \frac{\|v + tw\| - \|v\|}{t} = \frac{\langle v, w \rangle}{\|v\|}.$$

The same result then holds in any vector space with a Euclidean metric $\langle \ , \ \rangle$. *Hint*: If $v \colon \mathbb{R}^n \to \mathbb{R}$ is the norm, then the limit is $Dv(v)(w)$. Alternately, one can use the equation $\langle u, v \rangle = \|u\| \cdot \|v\| \cdot \cos\theta$ where θ is the angle between u and v.

(b) Conclude that if w is linearly independent of v, then

$$\lim_{t \to 0} \frac{\|v + tw\| - \|v\| - \|tw\|}{t} \neq 0.$$

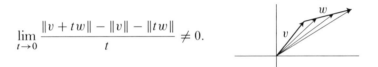

(c) Let $\gamma \colon [0, 1] \to M$ be a piecewise C^1 critical point for length, and suppose that $\gamma'(t_0{}^+) \neq \gamma'(t_0{}^-)$ for some $t_0 \in (0, 1)$. Choose $t_1 < t_0$ and consider the variation α for which $\bar{\alpha}(u)$ is obtained by following γ up to t_1, then the unique geodesic from $\gamma(t_1)$ to $\gamma(t_0 + u)$, and finally the rest of γ. Show that if t_1 is

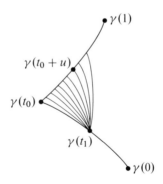

close enough to t_0, then $dL(\bar{\alpha}(u))/du\big|_{u=0} \neq 0$, a contradiction. Thus, critical paths for length cannot have kinks.

38. Consider a cylinder $Z \subset \mathbb{R}^3$ of radius r. Find the metric d induced by the Riemannian metric it acquires as a subset of $\mathbb{R}^3$.

39. Consider a cone C (without the vertex), and let L be a generating line. Unfolding $C - L$ onto $\mathbb{R}^2$ produces a map $f : C - L \to \mathbb{R}^2$ which is a local

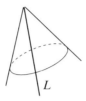

isometry, but which is usually not one-one. Investigate the geodesics on a cone (the number of geodesics between two points depends on the angle of the cone, and some geodesics may come back to their initial point).

40. Let $g : S^n \to \mathbb{P}^n$ be the map $g(p) = [p] = \{p, -p\}$.

(a) Show that there is a unique Riemannian metric $\langle\!\langle \ , \ \rangle\!\rangle$ on $\mathbb{P}^n$ such that $g^*\langle\!\langle \ , \ \rangle\!\rangle$ is the usual Riemannian metric on S^n (the one that makes the inclusion of S^n into $\mathbb{R}^{n+1}$ an isometry).

(b) Show that every geodesic $\gamma : \mathbb{R} \to \mathbb{P}^n$ is closed (that is, there is a number a such that $\gamma(t + a) = \gamma(t)$ for all t), and that every two geodesics intersect exactly once.

(c) Show that there are isometries of $\mathbb{P}^n$ onto itself taking any tangent vector at one point to any tangent vector at any other point.

These results show that $\mathbb{P}^n$ provides a model for "elliptical" non-Euclidean geometry. The sum of the angles in any triangle is $> \pi$.

41. The **Poincaré upper half-plane** $\mathcal{H}^2$ is the manifold $\{(x, y) \in \mathbb{R}^2 : y > 0\}$ with the Riemannian metric

$$\langle \ , \ \rangle = \frac{dx \otimes dx + dy \otimes dy}{y^2}.$$

(a) Compute that

$$\Gamma^2_{22} = \Gamma^1_{12} = \Gamma^1_{21} = -\frac{1}{y}, \quad \Gamma^2_{11} = \frac{1}{y}; \qquad \text{all other } \Gamma^k_{ij} = 0.$$

(b) Let C be a semi-circle in $\mathcal{H}^2$ with center at $(0, c)$ and radius R. Considering it as a curve $t \mapsto (t, \gamma(t))$, show that

$$\frac{d^2\gamma(t))}{dt^2} = \frac{-\gamma'(t)}{t - c} - \frac{\gamma'(t)^2}{\gamma(t)}.$$

(c) Using Problem 27, show that all the geodesics in $\mathcal{H}^2$ are the (suitably parameterized) semi-circles with center on the x-axis, together with the straight lines parallel to the y-axis.

(d) Show that these geodesics have infinite length in either direction, so that the upper half-plane is complete.

(e) Show that if γ is a geodesic and $p \notin \gamma$, then there are infinitely many geodesics through p which do not intersect γ.

(f) For those who know a little about conformal mapping (compare with Problem IV.7-6). Consider the upper half-plane as a subset of the complex numbers $\mathbb{C}$. Show that the maps

$$f(z) = \frac{az + b}{cz + d} \qquad a, b, c, d \in \mathbb{R}, \quad ad - bc > 0$$

are isometries, and that we can take any tangent vector at one point to any tangent vector at any other point by some f_*. Conclude that if length $AB =$ length $A'B'$ and length $AC =$ length $A'C'$ and the angle between the tangent vectors of β and γ at A equals the angle between the tangent vectors of β' and γ' at A', then length $BC =$ length $B'C'$ and the angles at B and B' and at C and C' are equal ("side-angle-side"). These results show that the Poincaré

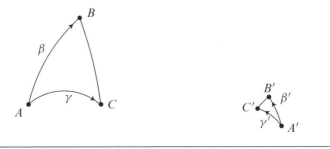

upper half-plane is a model for Lobachevskian non-Euclidean geometry. The sum of the angles in any triangle is $< \pi$.

42. Let M be a Riemannian manifold such that every two points of M can be joined by a unique geodesic of minimal length. Does it necessarily follow that the Riemannian manifold M is complete?

43. Let M be a manifold with a Riemannian metric $\langle\ ,\ \rangle$, and choose a fixed point $p \in M$. Suppose that every geodesic $\gamma: [a, b] \to M$ with initial value $\gamma(a) = p$ can be extended to all of $\mathbb{R}$. Show that the Riemannian manifold M is geodesically complete.

44. Let p be a point in a complete *non-compact* Riemannian manifold M. Prove that there is a geodesic $\gamma: [0, \infty) \to M$ with the initial value $\gamma(0) = p$, having the property that γ is a minimal geodesic between any two of its points.

45. Let M and N be geodesically complete Riemannian manifolds, and give $M \times N$ the Riemannian metric described in Problem 26. Show that the Riemannian manifold $M \times N$ is also complete.

46. This problem presupposes knowledge of covering spaces. Let $g: M \to N$ be a covering space, where N is a C^∞ manifold. Then there is a unique C^∞ structure on M which makes g an immersion. If $\langle\ ,\ \rangle$ is a Riemannian metric on N, then $g^*\langle\ ,\ \rangle$ is a Riemannian metric on M, and $(M, g^*\langle\ ,\ \rangle)$ is complete if and only if $(N, \langle\ ,\ \rangle)$ is complete.

47. (a) If $M^n \subset N^{n+k}$ is a submanifold of N, show that the normal bundle ν is indeed a k-plane bundle.
(b) Using the notion of Whitney sum $\oplus$ introduced in Problem 3-52, show that

$$\nu \oplus TM \simeq (TN)|M.$$

48. (a) Show that the normal bundles ν_1, ν_2 of $M^n \subset N^{n+k}$ defined for two different Riemannian metrics are equivalent.
(b) If $\xi = \pi: E \to M$ is a smooth k-plane bundle over M^n, show that the normal bundle of $M \subset E$ is equivalent to ξ.

49. (a) Given an exact sequence of bundle maps

$$0 \longrightarrow E_1 \overset{\tilde{f}}{\longrightarrow} E_2 \overset{\tilde{g}}{\longrightarrow} E_3 \longrightarrow 0$$

as in Problem 3-28, where the bundles are over a smooth manifold M [or, more generally, over a paracompact space], show that $E_2 \simeq E_1 \oplus E_3$.
(b) If $\xi = \pi: E \to M$ is a smooth bundle, conclude that $TE \simeq \pi^*(\xi) \oplus \pi^*(TM)$.

50. (a) Let M be a non-orientable manifold. According to Problem 3-22 there is $S^1 \subset M$ so that $(TM)|S^1$ is not orientable (the Problem deals with the case where $(TM)|S^1$ is always trivial, but the same conclusions will hold if each $(TM)|S^1$ is orientable; in fact, it is not hard to show that a bundle over S^1 is trivial if and only if it is orientable). Using Problem 47, show that the normal bundle ν of $S^1 \subset M$ is not orientable.

(b) Use Problem 3-29 to conclude that there is a neighborhood of some $S^1 \subset M$ which is not orientable. (Thus, any non-orientable manifold contains a "fairly small" non-orientable open submanifold.)

CHAPTER 10

LIE GROUPS

This chapter uses, and illuminates, many of the results and concepts of the preceding chapters. It will also play an important role in later Volumes, where we are concerned with geometric problems, because in the study of these problems the groups of automorphisms of various structures play a central role, and these groups can be studied by the methods now at our disposal.

A **topological group** is a space G which also has a group structure (the product of $a, b \in G$ being denoted by ab) such that the maps

$$(a, b) \mapsto ab \qquad \text{from } G \times G \text{ to } G$$
$$a \mapsto a^{-1} \qquad \text{from } G \text{ to } G$$

are continuous. It clearly suffices to assume instead that the single map $(a, b) \mapsto ab^{-1}$ is continuous. We will mainly be interested in a very special kind of topological group. A **Lie group** is a group G which is also a manifold with a C^∞ structure such that

$$(x, y) \mapsto xy$$
$$x \mapsto x^{-1}$$

are C^∞ functions. It clearly suffices to assume that the map $(x, y) \mapsto xy^{-1}$ is C^∞. As a matter of fact (Problem 1), it even suffices to assume that the map $(x, y) \mapsto xy$ is C^∞.

The simplest example of a Lie group is $\mathbb{R}^n$, with the operation $+$. The circle S^1 is also a Lie group. One way to put a group structure on S^1 is to consider it as the quotient group $\mathbb{R}/\mathbb{Z}$, where $\mathbb{Z} \subset \mathbb{R}$ denotes the subgroup of integers. The functions $x \mapsto \cos 2\pi x$ and $x \mapsto \sin 2\pi x$ are C^∞ functions on $\mathbb{R}/\mathbb{Z}$, and at each point at least one of them is a coordinate system. Thus the map

$$(x, y) \ \mapsto \ x - y \ \mapsto \ xy^{-1}$$
$$\cap \qquad\qquad \cap \qquad\quad \cap$$
$$\mathbb{R} \times \mathbb{R} \ \longrightarrow \ \mathbb{R} \ \longrightarrow \ S^1 = \mathbb{R}/\mathbb{Z},$$

which can be expressed in coordinates as one of the two maps

$$(x, y) \mapsto \cos 2\pi(x - y) = \cos 2\pi x \cos 2\pi y + \sin 2\pi x \sin 2\pi y$$
$$(x, y) \mapsto \sin 2\pi(x - y) = \sin 2\pi x \cos 2\pi y - \cos 2\pi x \sin 2\pi y,$$

is C^∞; consequently the map $(x, y) \mapsto xy^{-1}$ from $S^1 \times S^1$ to S^1 is also C^∞.

If G and H are Lie groups, then $G \times H$, with the product C^∞ structure, and the direct product group structure, is easily seen to be a Lie group. In particular, the torus $S^1 \times S^1$ is a Lie group. The torus may also be described as the quotient group

$$\mathbb{R} \times \mathbb{R}/(\mathbb{Z} \times \mathbb{Z});$$

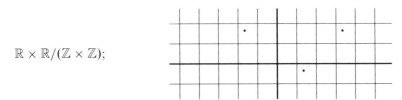

the pairs (a, b) and (a', b') represent the same element of $S^1 \times S^1$ if and only if $a' - a \in \mathbb{Z}$ and $b' - b \in \mathbb{Z}$.

Many important Lie groups are matrix groups. The **general linear group** GL$(n, \mathbb{R})$ is the group of all non-singular real $n \times n$ matrices, considered as a subset of $\mathbb{R}^{n^2}$. Since the function det: $\mathbb{R}^{n^2} \to \mathbb{R}$ is continuous (it is a polynomial map), the set GL$(n, \mathbb{R}) = \det^{-1}(\mathbb{R} - \{0\})$ is open, and hence can be given the C^∞ structure which makes it an open submanifold of $\mathbb{R}^{n^2}$. Multiplication of matrices is C^∞, since the entries of AB are polynomials in the entries of A and B. Smoothness of the inverse map follows similarly from Cramer's Rule:

$$(A^{-1})_{ji} = \det A^{ij} / \det A,$$

where A^{ij} is the matrix obtained from A by deleting row i and column j.

One of the most important examples of a Lie group is the **orthogonal group** O(n), consisting of all $A \in$ GL$(n, \mathbb{R})$ with $A \cdot A^{t} = I$, where A^{t} is the transpose of A. This condition is equivalent to the condition that the rows [and columns] of A are orthonormal, which is equivalent to the condition that, with respect to the usual basis of $\mathbb{R}^n$, the matrix A represents a linear transformation which is an "isometry", i.e., is norm preserving, and thus inner product preserving. Problem 2-33 presents a proof that O(n) is a closed submanifold of GL$(n, \mathbb{R})$, of dimension $n(n - 1)/2$. To show that O(n) is a Lie group we must show that the map $(x, y) \mapsto xy^{-1}$ which is C^∞ on GL$(n, \mathbb{R})$, is also C^∞ as a map from O$(n) \times$O(n) to O(n). By Proposition 2-11, it suffices to show that it is continuous; but this is true because the inclusion of O$(n) \to$ GL$(n, \mathbb{R})$ is a homeomorphism (since O(n) is a submanifold of GL$(n, \mathbb{R})$). Later in the chapter we will have another way of proving that O(n) is a Lie group, and in particular, a manifold.

The argument in the previous paragraph shows, generally, that if $H \subset G$ is a subgroup of G and also a submanifold of G, then H is a Lie group. (This gives another proof that S^1 is a Lie group, for $S^1 \subset \mathbb{R}^2$ can be considered as the group of complex numbers of norm 1. Similarly, S^3 is the Lie group of quaternions of norm 1. It is know that these are the only spheres which admit a Lie group structure.) It is possible for a subgroup H of G to be Lie group with respect to a C^∞ structure that makes it merely an immersed submanifold. For example, if $L \subset \mathbb{R} \times \mathbb{R}$ is a subgroup consisting of all (x, cx) for c irrational, then the

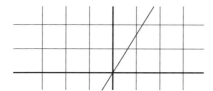

image of L in $S^1 \times S^1 = \mathbb{R} \times \mathbb{R}/(\mathbb{Z} \times \mathbb{Z})$ is a dense subgroup. We define a **Lie subgroup** H of G to be a subset H of G which is a subgroup of G, and also a Lie group for some C^∞ structure which makes the inclusion map $i \colon H \to G$ an immersion. As we have seen, a subgroup which is an (imbedded) submanifold is always a Lie subgroup. It even turns out, after some work (Problem 18), that a subgroup which is an immersed submanifold is always a Lie subgroup, but we will not need this fact.

The group $O(n)$ is disconnected; the two components consist of all $A \in O(n)$ with $\det A = +1$ and $\det A = -1$, respectively. Clearly $SO(n) = \{A \in O(n) : \det A = 1\}$, the component containing the identity I, is a subgroup. This is not accidental.

1. PROPOSITION. If G is a topological group, then the component K containing the identity $e \in G$ is a closed normal subgroup of G. If G is a Lie group, then K is an open Lie subgroup.

PROOF. If $a \in K$, then $a^{-1}K$ is connected, since $b \mapsto a^{-1}b$ is a homeomorphism of K to itself. Since $e = a^{-1}a \in a^{-1}K$, we have $a^{-1}K \subset K$. Since this is true for all $a \in K$, we have $K^{-1}K \subset K$, which proves that K is a subgroup.

For any $b \in G$, it follows similarly that bKb^{-1} is connected. Since $e \in bKb^{-1}$, we have $bKb^{-1} \subset K$, so K is normal. Moreover, K is closed since components are always closed.

If G is a Lie group, then K is also open, since G is locally connected, so K is a submanifold and a subgroup of G. Hence K is a Lie subgroup. ❖

The group $SO(2)$ is just S^1, which we have already seen is a Lie group. As a final example of a Lie group, we mention $E(n)$, the group of all **Euclidean**

motions, i.e., isometries of $\mathbb{R}^n$. A little argument shows (Problem 5) that every element of $E(n)$ can be written uniquely as $A \cdot \tau$ where $A \subset O(n)$, and τ is a translation,

$$\tau(x) = \tau_a(x) = x + a.$$

We can give $E(n)$ the C^∞ structure which makes it diffeomorphic to $O(n) \times \mathbb{R}^n$. Now $E(n)$ is not the direct product $O(n) \times \mathbb{R}^n$ as a group, since translations and orthogonal transformations do not generally commute. In fact,

$$A\tau_a A^{-1}(x) = A(A^{-1}x + a) = x + A(a) = \tau_{A(a)}(x),$$

so

$$A\tau_a A^{-1} = \tau_{A(a)}, \qquad A\tau_a = \tau_{A(a)} A.$$

Consequently,

$$A\tau_a(B\tau_b)^{-1} = A\tau_a\tau_b B^{-1} = A\tau_{a-b}B^{-1}$$
$$= AB^{-1}\tau_{B(a-b)},$$

which shows that $E(n)$ is a Lie group. Clearly the component of $e \in E(n)$ is the subgroup of all $A\tau$ with $A \in SO(n)$.

For any Lie group G, if $a \in G$ we define the left and right translations, $L_a \colon G \to G$ and $R_a \colon G \to G$, by

$$L_a(b) = ab$$
$$R_a(b) = ba.$$

Notice that L_a and R_a are both diffeomorphisms, with inverses $L_{a^{-1}}$ and $R_{a^{-1}}$, respectively. Consequently, the maps

$$L_{a*} \colon G_b \to G_{ab}$$
$$R_{a*} \colon G_b \to G_{ba}$$

are isomorphisms. A vector field X on G is called **left invariant** if

$$L_{a*}X = X \qquad \text{for all } a \in G.$$

Recall this means that

$$L_{a*}X_b = X_{ab} \qquad \text{for all } a, b \in G.$$

It is easy to see that this is true if we merely have

$$L_{a*}X_e = X_a \qquad \text{for all } a \in G.$$

Consequently, given $X_e \in G_e$, there is a unique left invariant vector field X on G which has the value X_e at e.

2. PROPOSITION. Every left invariant vector field X on a Lie group G is C^∞.

PROOF. It suffices to prove that X is C^∞ in a neighborhood of e, since the diffeomorphism L_a then takes X to the C^∞ vector field $L_{a*}X$ around a (Problem 5-1). Let (x, U) be a coordinate system around e. Choose a neighborhood V of e so that $a, b \in V$ implies $ab^{-1} \in U$. Then for $a \in V$ we have

$$X x^i(a) = L_{a*} X_e(x^i)$$
$$= X_e(x^i \circ L_a).$$

Since the map $(a, b) \mapsto ab$ is C^∞ on $V \times V$ we can write

$$x^i(ab) = x^i L_a(b) = f^i(x^1(a), \ldots, x^n(a), x^1(b), \ldots, x^n(b))$$

for some C^∞ function f^i on $x(V) \times x(V)$. Then

$$X x^i(a) = X_e(x^i \circ L_a)$$

$$= \sum_{j=1}^{n} c^j \frac{\partial(x^i \circ L_a)}{\partial x^j}\bigg|_e \qquad \text{where } X_e = \sum_{j=1}^{n} c^j \frac{\partial}{\partial x^j}\bigg|_e$$

$$= \sum_{j=1}^{n} c^j D_{n+j} f^i(x(a), x(e)),$$

which shows that $X x^i$ is C^∞. This implies that X is C^∞. ❖

3. COROLLARY. A Lie group G always has a trivial tangent bundle (and is consequently orientable).

PROOF. Choose a basis $X_{1e}, \ldots, X_{ne}$ for G_e. Let $X_1, \ldots, X_n$ be the left invariant vector fields with these values at e. Then $X_1, \ldots, X_n$ are clearly everywhere linearly independent, so we can define an equivalence

$$f : TG \to G \times \mathbb{R}^n$$

by

$$f\left(\sum_{j=1}^{n} c^i X_i(a)\right) = (a, c^1, \ldots, c^n). ❖$$

A left invariant vector field X is just one that is L_a-related to itself for all a. Consequently, Proposition 6-3 shows that $[X, Y]$ is left invariant if X and Y are. Henceforth we will use X, Y, etc., to denote elements of G_e, and $\widetilde{X}$, $\widetilde{Y}$, etc., to denote the left invariant vector fields with $\widetilde{X}(e) = X$, $\widetilde{Y}(e) = Y$, etc. We can then define an operation $[\ ,\ ]$ on G_e by

$$[X, Y] = [\widetilde{X}, \widetilde{Y}](e).$$

The vector space G_e, together with this $[\ ,\ ]$ operation, is called the **Lie algebra** of G, and will be denoted by $\mathcal{L}(G)$. (Sometimes the Lie algebra of G is defined instead to be the set of left invariant vector fields.) We will also use the more customary notation $\mathfrak{g}$ (a German Fraktur g) for $\mathcal{L}(G)$. This notation requires some conventions for particular groups; we write

$$\mathfrak{gl}(n, \mathbb{R}) \qquad \text{for the Lie algebra of } \mathrm{GL}(n, \mathbb{R})$$
$$\mathfrak{o}(n) \qquad \text{for the Lie algebra of } \mathrm{O}(n).$$

In general, a **Lie algebra** is a finite dimensional vector space V, with a bilinear operation $[\ ,\ ]$ satisfying

$$[X, X] = 0$$
$$[[X, Y], Z] + [[Y, Z], X] + [[Z, X], Y] = 0 \qquad \text{"Jacobi identity"}$$

for all $X, Y, Z \in V$.

Since the $[\ ,\ ]$ operation is assumed alternating, it is also skew-symmetric, $[X, Y] = -[Y, X]$. Consequently, we call a Lie algebra **abelian** or **commutative** if $[X, Y] = 0$ for all X, Y.

The Lie algebra of $\mathbb{R}^n$ is isomorphic as a vector space to $\mathbb{R}^n$. Clearly $\mathcal{L}(\mathbb{R}^n)$ is abelian, since the vector fields $\partial/\partial x^i$ are left invariant and $[\partial/\partial x^i, \partial/\partial x^j] = 0$. The Lie algebra $\mathcal{L}(S^1)$ of S^1 is 1-dimensional, and consequently must be abelian. If V_i are Lie algebras with bracket operations $[\ ,\ ]_i$ for $i = 1, 2$, then we can define an operation $[\ ,\ ]$ on the direct sum $V = V_1 \oplus V_2 \ (= V_1 \times V_2$ as a set) by

$$[(X_1, X_2), (Y_1, Y_2)] = \bigl([X_1, Y_1]_1, [X_2, Y_2]_2\bigr).$$

It is easy to check that this makes V into a Lie algebra, and that $\mathcal{L}(G \times H)$ is isomorphic to $\mathcal{L}(G) \times \mathcal{L}(H)$ with this bracket operation. Consequently, the Lie algebra $\mathcal{L}(S^1 \times \cdots \times S^1)$ is also abelian.

The structure of $\mathfrak{gl}(n, \mathbb{R})$ is more complicated. Since $\mathrm{GL}(n, \mathbb{R})$ is an open submanifold of $\mathbb{R}^{n^2}$, the tangent space of $\mathrm{GL}(n, \mathbb{R})$ at the identity I can be identified with $\mathbb{R}^{n^2}$. If we use the standard coordinates x^{ij} on $\mathbb{R}^{n^2}$, then an $n \times n$ (possibly singular) matrix $M = (M_{ij})$ can be identified with

$$M_I = \sum_{i,j} M_{ij} \frac{\partial}{\partial x^{ij}}\bigg|_I .$$

Let $\tilde{M}$ be the left invariant vector field on $GL(n, \mathbb{R})$ corresponding to M. We compute the function $\tilde{M}x^{kl}$ on $GL(n, \mathbb{R})$ as follows. For every $A \in GL(n, \mathbb{R})$,

$$\tilde{M}x^{kl}(A) = \tilde{M}_A(x^{kl}) = L_{A*}M_I(x^{kl}) = M_I(x^{kl} \circ L_A).$$

Now the function $x^{kl} \circ L_A \colon GL(n, \mathbb{R}) \to GL(n, \mathbb{R})$ is the linear function

$$(x^{kl} \circ L_A)(B) = x^{kl}(AB) = \sum_{\alpha=1}^{n} A_{k\alpha} B_{\alpha l},$$

with (constant) partial derivatives

$$\frac{\partial}{\partial x^{ij}}(x^{kl} \circ L_A) = \begin{cases} A_{ki} & j = l \\ 0 & j \neq l. \end{cases}$$

So

$$\tilde{M}x^{kl}(A) = M_I(x^{kl} \circ L_A) = \sum_{i,j} M_{ij} \frac{\partial}{\partial x^{ij}}(x^{kl} \circ L_A)$$

$$= \sum_{i=1}^{n} M_{il} A_{ki} = \sum_{\alpha=1}^{n} M_{\alpha l} A_{k\alpha}.$$

Thus,

$$\frac{\partial}{\partial x^{ij}} \tilde{M}x^{kl} = \begin{cases} M_{jl} & k = i \\ 0 & k \neq i. \end{cases}$$

So if N is another $n \times n$ matrix, we have

$$N_I(\tilde{M}x^{kl}) = \sum_{i,j} N_{ij} \frac{\partial}{\partial x^{ij}}(\tilde{M}x^{kl})$$

$$= \sum_{j=1}^{n} N_{kj} M_{jl} = (NM)_{kl}.$$

From this we see that

$$[\tilde{M}, \tilde{N}]_I = \sum_{k,l}(MN - NM)_{kl} \frac{\partial}{\partial x^{kl}}\bigg|_I ;$$

thus, if we identify $\mathfrak{gl}(n, \mathbb{R})$ with $\mathbb{R}^{n^2}$, the bracket operation is just

$$[M, N] = MN - NM.$$

Notice that in any ring, if we define $[a, b] = ab - ba$, then $[\ ,\ ]$ satisfies the Jacobi identity.

Since $O(n)$ is a submanifold of $GL(n, \mathbb{R})$ we can consider $O(n)_I$ as a subspace of $GL(n, \mathbb{R})_I$, and thus identify $\mathfrak{o}(n)$ with a certain subspace of $\mathbb{R}^{n^2}$. This subspace may be determined as follows. If $A \colon (-\varepsilon, \varepsilon) \to O(n)$ is a curve with $A(0) = I$, and we denote $(A(t))_{ij}$ by $A_{ij}(t)$, then

$$\sum_{k=1}^{n} A_{ik}(t) A_{jk}(t) = \delta_{ij};$$

differentiating gives

$$A_{ik}{}'(0)\delta_{jk} + \delta_{ik} A_{jk}{}'(0) = 0,$$

which shows that

$$A_{ij}{}'(0) = -A_{ji}{}'(0).$$

Thus $O(n)_I \subset \mathbb{R}^{n^2}$ can contain only matrices M which are skew-symmetric,

$$M = \begin{pmatrix} 0 & M_{12} & M_{13} & \dots & M_{1n} \\ -M_{12} & 0 & & & \\ -M_{13} & & 0 & & \\ \vdots & & & \ddots & \\ -M_{1n} & & & & 0 \end{pmatrix}.$$

This subspace has dimension $n(n-1)/2$, which is exactly the dimension of $O(n)$, so $O(n)_I$ must consist exactly of skew-symmetric matrices. If we did not know the dimension of $O(n)$, we could use the following line of reasoning. For each i, j with $i < j$, we can define a curve $A \colon \mathbb{R} \to O(n)$ by

$$A(t) = \begin{pmatrix} 1 & & & & & \\ & \ddots & & & & \\ & & \cos t & & \sin t & \\ & & & \ddots & & \\ & & -\sin t & & \cos t & \\ & & & & & \ddots \\ & & & & & & 1 \end{pmatrix} \begin{matrix} \\ \\ i \\ \\ j \\ \\ \end{matrix} \qquad \text{(rotation in the } (i, j)\text{-plane)}$$

with $\sin t$ and $-\sin t$ at (i, j) and (j, i), 1's on the diagonal except at (i, i) and (j, j), and 0's elsewhere. Then the set of all $A'(0)$ span the skew-symmetric matrices. Hence $O(n)_I$ must consist exactly of skew-symmetric matrices, and $O(n)$ must have dimension $n(n-1)/2$.

We do not need any new calculations to determine the bracket operation in $\mathfrak{o}(n)$. In fact, consider a Lie subgroup H of any Lie group G, and let $i: H \to G$ be the inclusion. Since $i_*: H_e \to G_e$ is an isomorphism into, we can identify H_e with a subspace of G_e. Any $X \in H_e$ can be extended to a left invariant vector field $\tilde{X}$ on H and a left invariant vector field $\tilde{\tilde{X}}$ on G. For each $a \in H \subset G$, we have left translations

$$L_a: H \to H, \qquad L_a: G \to G$$

and

$$L_a \circ i = i \circ L_a.$$

So

$$i_* \tilde{X}(a) = i_* L_{a*} X = L_{a*}(i_* X) = \tilde{\tilde{X}}(a).$$

In other words, $\tilde{X}$ and $\tilde{\tilde{X}}$ are i-related. Consequently, if $Y \in H_e$, then $[\tilde{X}, \tilde{Y}]$ and $[\tilde{\tilde{X}}, \tilde{\tilde{Y}}]$ are i-related, which means that

$$[\tilde{\tilde{X}}, \tilde{\tilde{Y}}](e) = i_*([\tilde{X}, \tilde{Y}](e)).$$

Thus, $H_e \subset G_e = \mathfrak{g}$ is a **subalgebra** of $\mathfrak{g}$, that is, H_e is a subspace of $\mathfrak{g}$ which is closed under the $[\ ,\ ]$ operation; moreover, H_e with this induced $[\ ,\ ]$ operation is just $\mathfrak{h} = \mathcal{L}(H)$.

This correspondence between Lie subgroups of G and subalgebras of $\mathfrak{g}$ turns out to work in the other direction also.

4. THEOREM. Let G be a Lie group, and $\mathfrak{h}$ a subalgebra of $\mathfrak{g}$. Then there is a unique connected Lie subgroup H of G whose Lie algebra is $\mathfrak{h}$.

PROOF. For $a \in G$, let Δ_a be the subspace of G_a consisting of all $\tilde{X}(a)$ for $X \in \mathfrak{h}$. The fact that $\mathfrak{h}$ is a subalgebra of $\mathfrak{g}$ implies that Δ is an integrable distribution. Let H be the maximal integral manifold of Δ containing e. If $b \in G$, then clearly $L_{b*}(\Delta_a) = \Delta_{ba}$, so L_{b*} leaves the distribution Δ invariant. It follows immediately that L_b permutes the various maximal integral manifolds of Δ among themselves. In particular, if $b \in H$, then $L_{b^{-1}}$ takes H to the maximal integral manifold containing $L_{b^{-1}}(b) = e$, so $L_{b^{-1}}(H) = H$. This implies that H is a subgroup of G. To prove that it is a Lie subgroup we just need to show that $(a, b) \mapsto ab^{-1}$ is C^∞. Now this map is clearly C^∞ as a map into G. Using Theorem 6-7, it follows that it is C^∞ as a map into H.

The proof of uniqueness is left to the reader. ❖

There is a very difficult theorem of Ado which states that every Lie algebra is isomorphic to a subalgebra of $GL(N, \mathbb{R})$ for some N. It then follows from Theorem 4 that *every Lie algebra is isomorphic to the Lie algebra of some Lie group.* Later on we will be able to obtain a "local" version of this result. We will soon see to what extent the Lie algebra of G determines G.

We continue the study of Lie groups along the same route used in the study of groups. Having considered subgroups of Lie groups (and subalgebras of their Lie algebras), we next consider, more generally, homomorphisms between Lie groups. If $\phi \colon G \to H$ is a C^∞ homomorphism, then $\phi_{*e} \colon G_e \to H_e$. For any $a \in G$ we clearly have

$$\phi \circ L_a = L_{\phi(a)} \circ \phi,$$

so if $X \in G_e$, and $\widetilde{X} = \widetilde{\phi_{*e} X}$ is the left invariant vector field on H with value $\phi_{*e} X$ at e, then

$$\phi_{*a} \widetilde{X}(a) = \phi_{*a} L_{a*} X = L_{\phi(a)*} \phi_{*e} X$$
$$= \widetilde{X}(\phi(a)).$$

Thus $\widetilde{X}$ and $\widetilde{X}$ are ϕ-related. Consequently, the map $\phi_{*e} \colon \mathfrak{g} \to \mathfrak{h}$ is a **Lie algebra homomorphism**, that is,

$$\phi_{*e}(aX + bY) = a\phi_{*e}X + b\phi_{*e}Y$$
$$\phi_{*e}[X, Y] = [\phi_{*e}X, \phi_{*e}Y].$$

Usually, we will denote ϕ_{*e} simply by $\phi_* \colon \mathfrak{g} \to \mathfrak{h}$.

For example, suppose that $G = H = \mathbb{R}$. There are an enormous number of homomorphisms $\phi \colon \mathbb{R} \to \mathbb{R}$, because $\mathbb{R}$ is a vector space of uncountable dimension over $\mathbb{Q}$, and every linear transformation is a group homomorphism. But if ϕ is C^∞, then the condition

$$\phi(s + t) = \phi(s) + \phi(t)$$

implies that

$$\frac{d\phi(t + s)}{ds} = \frac{d\phi(s)}{ds};$$

evaluating at $s = 0$ gives

$$\phi'(t) = \phi'(0),$$

which means that $\phi(t) = ct$ for some $c \; (= \phi'(0))$. It is not hard to see that even a continuous ϕ must be of this form (one first shows that ϕ is of this form on the rational numbers). We can identify $\mathcal{L}(\mathbb{R})$ with $\mathbb{R}$. Clearly the map $\phi_*\colon \mathbb{R} \to \mathbb{R}$ is just multiplication by c.

Now suppose that $G = \mathbb{R}$, but $H = S^1 = \mathbb{R}/\mathbb{Z}$. A neighborhood of the identity $e \in S^1$ can be identified with a neighborhood of $0 \in \mathbb{R}$, giving rise to an identification of $\mathcal{L}(S^1)$ with $\mathbb{R}$. The continuous homomorphisms $\phi\colon \mathbb{R} \to S^1$ are clearly of the form

$$\mathbb{R} \xrightarrow{\times c} \mathbb{R} \longrightarrow \mathbb{R}/\mathbb{Z};$$

once again, $\phi_*\colon \mathbb{R} \to \mathbb{R}$ is multiplication by c.

Notice that the only continuous homomorphism $\phi\colon S^1 \to \mathbb{R}$ is the 0 map (since $\{0\}$ is the only compact subgroup of $\mathbb{R}$). Consequently, a Lie algebra homomorphism $\mathfrak{g} \to \mathfrak{h}$ may not come from any C^∞ homomorphism $\phi\colon G \to H$. However, we do have a local result.

5. THEOREM. Let G and H be Lie groups, and $\Phi\colon \mathfrak{g} \to \mathfrak{h}$ a Lie algebra homomorphism. Then there is a neighborhood U of $e \in G$ and a C^∞ map $\phi\colon U \to H$ such that

$$\phi(ab) = \phi(a)\phi(b) \qquad \text{when } a, b, ab \in U,$$

and such that for every $X \in \mathfrak{g}$ we have

$$\phi_{*e} X = \Phi(X).$$

Moreover, if there are two C^∞ homomorphisms $\phi, \psi\colon G \to H$ with $\phi_{*e} = \psi_{*e} = \Phi$, and G is connected, then $\phi = \psi$.

PROOF. Let $\mathfrak{k}$ (German Fraktur k) be the subset $\mathfrak{k} \subset \mathfrak{g} \times \mathfrak{h}$ of all $(X, \Phi(X))$, for $X \in \mathfrak{g}$. Since Φ is a homomorphism, $\mathfrak{k}$ is a subalgebra of $\mathfrak{g} \times \mathfrak{h} = \mathcal{L}(G \times H)$. By Theorem 4, there is a unique connected Lie subgroup K of $G \times H$ whose Lie algebra is $\mathfrak{k}$. If $\pi_1\colon G \times H \to G$ is projection on the first factor, and $\omega = \pi_1|K$, then $\omega\colon K \to G$ is a C^∞ homomorphism. For $X \in \mathfrak{g}$ we have

$$\omega_*(X, \Phi(X)) = X,$$

so $\omega_*\colon K_{(e,e)} \to G_e$ is an isomorphism. Consequently, there is an open neighborhood V of $(e, e) \in K$ such that ω takes V diffeomorphically onto an open neighborhood U of $e \in G$. If $\pi_2\colon G \times H \to H$ is projection on the second factor, we can define

$$\phi = \pi_2 \circ \omega^{-1} \quad \text{on} \quad U.$$

The first condition on ϕ is obvious. As for the second, if $X \in \mathfrak{g}$, then

$$\omega_*(X, \Phi(X)) = X,$$

so

$$\phi_* X = \pi_{2*}(X, \Phi(X)) = \Phi(X).$$

Given $\phi, \psi : G \to H$, define the one-one map $\theta : G \to G \times H$ by

$$\theta(a) = (a, \psi(a)).$$

The image G' of θ is a Lie subgroup of $G \times H$ and for $X \in \mathfrak{g}$ we clearly have

$$\theta_* X = (X, \Phi(X)),$$

so $\mathcal{L}(G') = \mathfrak{k}$. Thus $G' = K$, which implies that $\psi(a) = \phi(a)$ for all $a \in G$. ❖

6. COROLLARY. If two Lie groups G and H have isomorphic Lie algebras, then they are locally isomorphic.

PROOF. Given an isomorphism $\Phi : \mathfrak{g} \to \mathfrak{h}$, let ϕ be the map given by Theorem 5. Since $\phi_{*e} = \Phi$ is an isomorphism, ϕ is a diffeomorphism in a neighborhood of $e \in G$. ❖

Remark: For those who know about simply-connected spaces it is fairly easy (Problem 8) to conclude that two simply-connected Lie groups with isomorphic Lie algebras are actually isomorphic, and that all connected Lie groups with a given Lie algebra are covered by the same simply-connected Lie group.

7. COROLLARY. A connected Lie group G with an abelian Lie algebra is itself abelian.

PROOF. By Corollary 6, G is locally isomorphic to $\mathbb{R}^n$, so $ab = ba$ for a, b in a neighborhood of e. It follows that G is abelian, since (Problem 4) any neighborhood of e generates G. ❖

8. COROLLARY. For every $X \in G_e$, there is a unique C^∞ homomorphism $\phi : \mathbb{R} \to G$ such that

$$\left. \frac{d\phi}{dt} \right|_{t=0} = X.$$

FIRST PROOF. Define $\Phi\colon \mathbb{R} \to \mathcal{L}(G)$ by

$$\Phi(\alpha) = \alpha X.$$

Clearly Φ is a Lie algebra homomorphism. By Theorem 5, on some neighborhood $(-\varepsilon, \varepsilon)$ of $0 \in \mathbb{R}$ there is a map $\phi\colon (-\varepsilon, \varepsilon) \to G$ with

$$\phi(s + t) = \phi(s)\phi(t) \qquad |s|, |t|, |s + t| < \varepsilon$$

and

$$\left. \frac{d\phi}{dt} \right|_{t=0} = \phi_* \left(\left. \frac{d}{dt} \right|_{t=0} \right) = X.$$

To extend ϕ to $\mathbb{R}$ we write every t with $|t| \geq \varepsilon$ uniquely as

$$t = k(\varepsilon/2) + r \qquad k \text{ an integer}, |r| < \varepsilon/2$$

and define

$$\phi(t) = \begin{cases} \phi(\varepsilon/2) \cdots \phi(\varepsilon/2) \cdot \phi(r) & [\phi(\varepsilon/2) \text{ appears } k \text{ times}] & k \geq 0 \\ \phi(-\varepsilon/2) \cdots \phi(-\varepsilon/2) \cdot \phi(r) & [\phi(-\varepsilon/2) \text{ appears } -k \text{ times}] & k < 0. \end{cases}$$

Uniqueness also follows from Theorem 5.

SECOND (DIRECT) PROOF. If $f\colon G \to \mathbb{R}$ is C^∞, and $\phi\colon \mathbb{R} \to G$ is a C^∞ homomorphism, then

$$\begin{aligned} \frac{d\phi}{dt}(f) &= \lim_{h \to 0} \frac{f(\phi(t + h)) - f(\phi(t))}{h} \\ &= \lim_{h \to 0} \frac{f(\phi(t)\phi(h)) - f(\phi(t))}{h} \\ &= \left. \frac{d}{du} \right|_{u=0} f \circ L_{\phi(t)} \circ \phi \\ &= L_{\phi(t)*} \left. \frac{d\phi}{du} \right|_{u=0} (f) \\ &= L_{\phi(t)*} X(f) = \widetilde{X}(\phi(t))(f). \end{aligned}$$

Thus ϕ must be an integral curve of $\widetilde{X}$, which proves uniqueness. Conversely, if $\phi\colon \mathbb{R} \to G$ is an integral curve of $\widetilde{X}$, then

$$t \mapsto \phi(s) \cdot \phi(t)$$

is an integral curve of $\widetilde{X}$ which passes through $\phi(s)$ at time $t = 0$. The same is clearly true for

$$t \mapsto \phi(s + t),$$

so ϕ is a homomorphism. We know that integral curves of $\widetilde{X}$ exist locally; they can be extended to all of $\mathbb{R}$ using the method of the first proof. ❖

A homomorphism $\phi\colon \mathbb{R} \to G$ is called a 1-**parameter subgroup** of G. We thus see that there is a unique 1-parameter subgroup ϕ of G with given tangent vector $d\phi/dt(0) \in G_e$. We have already examined the 1-parameter subgroups of $\mathbb{R}$. More interesting things happen when we take G to be $\mathbb{R} - \{0\}$, with multiplication as the group operation. Then all C^∞ homomorphisms $\phi\colon \mathbb{R} \to \mathbb{R} - \{0\}$, with

$$\phi(s + t) = \phi(s)\phi(t),$$

must satisfy

$$\phi'(t) = \phi'(0)\phi(t)$$
$$\phi(0) = 1.$$

The solutions of this equation are

$$\phi(t) = e^{\phi'(0)t}.$$

Notice that $\mathbb{R} - \{0\}$ is just $\mathrm{GL}(1, \mathbb{R})$. All C^∞ homomorphisms $\phi\colon \mathbb{R} \to \mathrm{GL}(n, \mathbb{R})$ must satisfy the analogous differential equation

$$(*) \qquad \begin{aligned} \phi'(t) &= \phi'(0) \cdot \phi(t), \\ \phi(0) &= I, \end{aligned}$$

where $\cdot$ now denotes matrix multiplication. The solutions of these equations can be written formally in the same way

$$(**) \qquad \phi(t) = \exp(t\phi'(0)),$$

where exponentiation of matrices is defined by

$$\exp(A) = I + \frac{A}{1!} + \frac{A^2}{2!} + \frac{A^3}{3!} + \cdots .$$

This follows from the facts in Problem 5-6, some of which will be briefly recapitulated here.

If $A = (a_{ij})$ and $|A| = \max |a_{ij}|$, then clearly

$$|A + B| \leq |A| + |B|$$
$$|AB| \leq n|A| \cdot |B|;$$

hence $|A|^k \leq n^{k-1}|A|^k \leq n^k|A|^k$. Consequently,

$$\left| \frac{A^N}{N!} + \cdots + \frac{A^{N+K}}{(N+K)!} \right| \leq \frac{(n|A|)^N}{N!} + \cdots + \frac{(n|A|)^{N+K}}{(N+K)!} \to 0 \quad \text{as } N \to \infty,$$

so the series for $\exp(A)$ converges (the $(i, j)^{\text{th}}$ entry of the partial sums converge), and convergence is absolute and uniform in any bounded set. Moreover (see Problem 5-6), if $AB = BA$, then

$$\exp(A + B) = (\exp A)(\exp B).$$

Hence, if $\phi(t)$ is defined by (**), then

$$\phi'(t) = \lim_{h \to 0} \frac{\exp(t\phi'(0) + h\phi'(0)) - \exp(t\phi'(0))}{h}$$

$$= \lim_{h \to 0} \frac{[\exp(h\phi'(0)) - I]}{h} \exp(t\phi'(0))$$

$$= \lim_{h \to 0} \frac{\dfrac{h\phi'(0)}{1!} + \dfrac{h^2\phi'(0)^2}{2!} + \cdots}{h} \exp(t\phi'(0))$$

$$= \phi'(0)\phi(t),$$

so ϕ does satisfy (*).

For any Lie group G, we now define the "exponential map"

$$\exp \colon \mathfrak{g} \to G$$

as follows. Given $X \in \mathfrak{g}$, let $\phi \colon \mathbb{R} \to G$ be the unique C^∞ homomorphism with $d\phi/dt(0) = X$. Then

$$\exp(X) = \phi(1).$$

We clearly have

$$\exp(t_1 + t_2)X = (\exp t_1 X)(\exp t_2 X)$$
$$\exp(-tX) = (\exp tX)^{-1}.$$

9. PROPOSITION. The map $\exp \colon G_e \to G$ is C^∞ (note that $G_e \approx \mathbb{R}^n$ has a natural C^∞ structure), and 0 is a regular point, so that $\exp$ takes a neighborhood of $0 \in G_e$ diffeomorphically onto a neighborhood of $e \in G$. If $\psi \colon G \to H$ is any C^∞ homomorphism, then

$$\exp \circ \psi_* = \psi \circ \exp.$$

$$\begin{array}{ccc} G_e & \xrightarrow{\ \psi_*\ } & H_e \\ {\scriptstyle \exp}\downarrow & & \downarrow{\scriptstyle \exp} \\ G & \xrightarrow{\ \psi\ } & H \end{array}$$

PROOF. The tangent space $(G_e \times G)_{(X,a)}$ of the C^∞ manifold $G_e \times G$ at the point (X, a) can be identified with $G_e \oplus G_a$. We define a vector field Y on $G_e \times G$ by

$$Y_{(X,a)} = 0 \oplus \widetilde{X}(a).$$

$(S^1)_e \times S^1$

Then Y has a flow $\alpha \colon \mathbb{R} \times (G_e \times G) \to G_e \times G$, which we know is C^∞. Since

$$\exp X = \text{projection on } G \text{ of } \alpha(1, 0 \oplus X),$$

it follows that exp is C^∞.

If we identify a vector $v \in (G_e)_0$ with G_e, then the curve $c(t) = tv$ in G_e has tangent vector v at 0. So

$$\exp_{*0}(v) = \left.\frac{d\,\exp(c(t))}{dt}\right|_{t=0} = \left.\frac{d}{dt}\right|_{t=0} \exp(tv)$$

$$= v.$$

So $\exp_{*0}$ is the identity, and hence one-one. Therefore exp is a diffeomorphism in a neighborhood of 0.

Given $\psi \colon G \to H$, and $X \in G_e$, let $\phi \colon \mathbb{R} \to G$ be a homomorphism with

$$\left.\frac{d\phi}{dt}\right|_{t=0} = X.$$

Then $\psi \circ \phi \colon \mathbb{R} \to H$ is a homomorphism with

$$\left.\frac{d(\psi \circ \phi)}{dt}\right|_{t=0} = \psi_* X.$$

Consequently,

$$\exp(\psi_* X) = \psi \circ \phi(1) = \psi(\exp X). \; \diamond$$

10. COROLLARY. Every one-one C^∞ homomorphism $\phi \colon G \to H$ is an immersion (so $\phi(G)$ is a Lie subgroup of H).

PROOF. If $\phi_{*p}(\widetilde{X}(p)) = 0$ for some non-zero $X \in \mathfrak{g}$, then also $\phi_{*e}(X) = 0$. But then

$$e = \exp \phi_{*e}(tX) = \phi(\exp(tX)),$$

contradicting the fact that ϕ is one-one. $\diamond$

11. COROLLARY. Every continuous homomorphism $\phi \colon \mathbb{R} \to G$ is C^∞.

PROOF. Let U be a star-shaped open neighborhood of $0 \in G_e$ on which $\exp$ is one-one. For any $a \in \exp\left(\frac{1}{2}U\right)$, if $a = \exp(X/2)$ for $X \in U$, then
$$a = \exp(X/2) = [\exp(X/4)]^2, \qquad \exp X/4 \in \exp\left(\tfrac{1}{2}U\right).$$
So a has a square root in $\exp\left(\frac{1}{2}U\right)$. Moreover, if $a = b^2$ for $b \in \exp\left(\frac{1}{2}U\right)$, then $b = \exp(Y/2)$ for $Y \in U$, so
$$\exp(X/2) = a = b^2 = [\exp(Y/2)]^2 = \exp Y.$$
Since $X/2, Y \in U$ it follows that $X/2 = Y$, so $X/4 = Y/2$. This shows that every $a \in \exp\left(\frac{1}{2}U\right)$ has a unique square root *in the set* $\exp\left(\frac{1}{2}U\right)$.

Now choose $\varepsilon > 0$ so that $\phi(t) \in \exp\left(\frac{1}{2}U\right)$ for $|t| \leq \varepsilon$. Let $\phi(\varepsilon) = \exp X$, $X \in \exp\left(\frac{1}{2}U\right)$. Since
$$[\phi(\varepsilon/2)]^2 = \phi(\varepsilon) = [\exp X/2]^2,$$
it follows from the above that $\phi(\varepsilon/2) = \exp(X/2)$. By induction we have
$$\phi(\varepsilon/2^n) = \exp(X/2^n).$$

Hence
$$\phi(m/2^n \cdot \varepsilon) = \phi(\varepsilon/2^n)^m = [\exp(X/2^n)]^m = \exp(m/2^n \cdot X).$$

By continuity,
$$\phi(s\varepsilon) = \exp sX \qquad \text{for all } s \in [-1, 1]. \; \diamondsuit$$

12. COROLLARY. Every continuous homomorphism $\phi \colon G \to H$ is C^∞.

PROOF. Choose a basis $X_1, \ldots, X_n$ for G_e. The map $t \mapsto \phi(\exp t X_i)$ is a continuous homomorphism of $\mathbb{R}$ to H, so there is $Y_i \in H_e$ such that
$$\phi(\exp t X_i) = \exp t Y_i.$$
Thus,
$$(*) \qquad \phi\big((\exp t_1 X_1) \cdots (\exp t_n X_n)\big) = (\exp t_1 Y_1) \cdots (\exp t_n Y_n).$$
Now the map $\psi \colon \mathbb{R}^n \to G$ given by
$$\psi(t_1, \ldots, t_n) = (\exp t_1 X_1) \cdots (\exp t_n X_n)$$
is C^∞ and clearly
$$\psi_*\left(\left.\frac{\partial}{\partial x^i}\right|_0\right) = X_i,$$
so ψ is a diffeomorphism of a neighborhood U of $0 \in \mathbb{R}^n$ onto a neighborhood V of $e \in G$. Then on V,
$$\phi = (\phi \circ \psi) \circ \psi^{-1},$$
and $(*)$ shows that $\phi \circ \psi$ is C^∞. So ϕ is C^∞ at e, and thus everywhere. $\diamondsuit$

13. COROLLARY. If G and G' are Lie groups which are isomorphic as topological groups, then they are isomorphic as Lie groups, that is, there is a diffeomorphism between them which is also a group isomorphism.

PROOF. Apply Corollary 11 to the continuous isomorphism and its inverse. ❖

The properties of the particular exponential map

$$\exp\colon \mathbb{R}^{n^2}\ (= \mathfrak{gl}(n,\mathbb{R}))\ \rightarrow\ \mathrm{GL}(n,\mathbb{R})$$

may now be used to show that $O(n)$ is a Lie group. It is easy to see that

$$\exp(M^{\mathbf{t}}) = (\exp M)^{\mathbf{t}}.$$

Moreover, since $\exp(M + N) = (\exp M)(\exp N)$ when $MN = NM$, we have

$$(\exp M)(\exp -M) = I.$$

So if M is skew-symmetric, $M = -M^{\mathbf{t}}$, then

$$(\exp M)(\exp M)^{\mathbf{t}} = I,$$

i.e., $\exp M \in O(n)$. Conversely, any $A \in O(n)$ sufficiently close to I can be
written $A = \exp M$ for some M. Let $A^{\mathbf{t}} = \exp N$. Then $I = A \cdot A^{\mathbf{t}} =
(\exp M)(\exp N)$, so $\exp N = (\exp M)^{-1} = \exp(-M)$. For sufficiently small M
and N this implies that $N = -M$. So $\exp M^{\mathbf{t}} = A^{\mathbf{t}} = \exp(-M)$; hence
$M^{\mathbf{t}} = -M$. It follows that a neighborhood of I in $O(n)$ is an $n(n-1)/2$
dimensional submanifold of $\mathrm{GL}(n,\mathbb{R})$. Since $O(n)$ is a subgroup, $O(n)$ is itself
a submanifold of $\mathrm{GL}(n,\mathbb{R})$.

Just as in $\mathrm{GL}(n,\mathbb{R})$, the equation $\exp(X + Y) = \exp X \exp Y$ holds whenever
$[X, Y] = 0$ (Problem 13). In general, $[X, Y]$ measures, up to first order, the
extent to which this equation fails to hold. In the following Theorem, and in
its proof, to indicate that a function $c\colon \mathbb{R} \rightarrow G_e$ has the property that $c(t)/t^3$ is
bounded for small t, we will denote it by $O(t^3)$. Thus $O(t^3)$ will denote different
functions at different times.

14. THEOREM. If G is a Lie group and $X, Y \in G_e$, then

(1) $\exp tX \exp tY = \exp\left\{t(X + Y) + \dfrac{t^2}{2}[X, Y] + O(t^3)\right\}$

(2) $\exp(-tX)\exp(-tY)\exp tX \exp tY = \exp\{t^2[X, Y] + O(t^3)\}$

(3) $\exp tX \exp tY \exp(-tX) = \exp\{tY + t^2[X, Y] + O(t^3)\}.$

PROOF. We have

(i) $\widetilde{X}f(a) = \widetilde{X}_a(f) = L_{a*}X(f) = X(f \circ L_a) = \dfrac{d}{du}\bigg|_{u=0} f(a \cdot \exp uX).$

Similarly,

(ii) $$\widetilde{Y}f(a) = \dfrac{d}{du}\bigg|_{u=0} f(a \cdot \exp uY).$$

For fixed s, let

$$\phi(t) = f(\exp sX \exp tY).$$

Then

(iii) $\phi'(t) = \dfrac{d}{dt} f(\exp sX \exp tY) = \dfrac{d}{du}\bigg|_{u=0} f(\exp sX \exp tY \exp uY)$

$= (\widetilde{Y}f)(\exp sX \exp tY)$ by (ii).

Applying (iii) to $\widetilde{Y}f$ instead of f gives

(iv) $$\phi''(t) = [\widetilde{Y}(\widetilde{Y}f)](\exp sX \exp tY).$$

Now Taylor's Theorem says that

$$\phi(t) = \phi(0) + \phi'(0)t + \dfrac{\phi''(0)}{2!}t^2 + O(t^3).$$

Suppose that $f(e) = 0$. Then we have

(v) $f(\exp sX \exp tY) = f(\exp sX) + t(\widetilde{Y}f)(\exp sX)$

$$+ \dfrac{t^2}{2}[\widetilde{Y}(\widetilde{Y}f)](\exp sX) + O(t^3).$$

Similarly, for any F,

$$\dfrac{d}{ds} F(\exp sX) = (\widetilde{X}F)(\exp sX)$$

$$\dfrac{d^2}{ds^2} F(\exp sX) = [\widetilde{X}(\widetilde{X}F)](\exp sX)$$

$$F(\exp sX) = F(e) + s(\widetilde{X}F)(e) + \dfrac{s^2}{2}[\widetilde{X}(\widetilde{X}F)](e) + O(s^3).$$

Substituting in (v) for $F = f$, $F = \widetilde{Y}f$, and $F = \widetilde{Y}(\widetilde{Y}f)$ gives

(vi) $f(\exp sX \exp tY) = s(\widetilde{X}f)(e) + t(\widetilde{Y}f)(e)$

$$+ \frac{s^2}{2}[\widetilde{X}(\widetilde{X}f)](e) + \frac{t^2}{2}[\widetilde{Y}(\widetilde{Y}f)](e) + st\widetilde{X}(\widetilde{Y}f)(e)$$

$$+ O(s^3) + O(t^3) + O(s^2t) + O(st^2).$$

In particular,

(vii) $f(\exp tX \exp tY) = t[(\widetilde{X} + \widetilde{Y})f](e)$

$$+ t^2\left[\left(\frac{\widetilde{X}\widetilde{X}}{2} + \widetilde{X}\widetilde{Y} + \frac{\widetilde{Y}\widetilde{Y}}{2}\right)f\right](e) + O(t^3).$$

Now for small t we can write

$$\exp tX \exp tY = \exp Z(t)$$

for some C^∞ function Z with values in G_e. Applying Taylor's formula to Z gives

$$Z(t) = tZ_1 + t^2Z_2 + O(t^3),$$

for some $Z_1, Z_2 \in G_e$. If $f(e) = 0$, then clearly $f(A(t) + O(t^3)) = f(A(t)) + O(t^3)$, so by (vi) we have

(viii) $f(\exp Z(t)) = f\left(\exp(tZ_1 + t^2Z_2)\right) + O(t^3)$

$$= t(\widetilde{Z}_1 f)(e) + t^2(\widetilde{Z}_2 f)(e)$$

$$+ \frac{t^2}{2}[\widetilde{Z}_1(\widetilde{Z}_1 f)](e) + O(t^3).$$

Since we can take the f's to be coordinate functions, comparison of (vii) and (viii) gives

$$\widetilde{X} + \widetilde{Y} = \widetilde{Z}_1$$

$$\frac{\widetilde{Z}_1\widetilde{Z}_1}{2} + \widetilde{Z}_2 = \frac{\widetilde{X}\widetilde{X}}{2} + \widetilde{X}\widetilde{Y} + \frac{\widetilde{Y}\widetilde{Y}}{2},$$

which gives

$$Z_1 = X + Y, \qquad Z_2 = \frac{1}{2}[X, Y],$$

thus proving (1).

Equation (2) follows immediately from (1).

To prove (3), again choose f with $f(e) = 0$. Then similar calculations give

(ix) $\quad f(\exp tX \exp tY \exp(-tX))$

$$= t[(\tilde{X} + \tilde{Y} - \tilde{X})f](e) + t^2 \left[\left(\frac{\tilde{X}\tilde{X}}{2} + \frac{\tilde{Y}\tilde{Y}}{2} + \frac{\tilde{X}\tilde{X}}{2} + \tilde{X}\tilde{Y} - \tilde{X}\tilde{X} - \tilde{Y}\tilde{X} \right) \right](e)$$

$$+ O(t^3).$$

If we write

$$\exp tX \exp tY \exp(-tX) = \exp(tS_1 + t^2 S_2 + O(t^3)),$$

then we also have

(x) $\qquad f(\exp tX \exp tY \exp(-tX)) = f\left(\exp(tS_1 + t^2 S_2)\right) + O(t^3)$

$$= t(\tilde{S}_1 f)(e) + t^2(\tilde{S}_2 f)(e)$$

$$+ \frac{t^2}{2}[\tilde{S}_1(\tilde{S}_1 f)](e) + O(t^3).$$

Comparing (ix) and (x) gives the desired result. ❖

Notice that formula (2) is a special case of Theorem 5-16 (compare also with Problems 5-16 and 5-18).

The work involved in proving Theorem 14 is justified by its role in the following beautiful theorem.

15. THEOREM. If G is a Lie group and $H \subset G$ is a closed subset which is also a subgroup (algebraically), then H is a Lie subgroup of G. More precisely, there is a C^∞ structure on H, *with the relative topology*, that makes it a Lie subgroup of G.

PROOF. We attempt to reconstruct the Lie algebra of H as follows. Let $\mathfrak{h} \subset G_e$ be the set of all $X \in G_e$ such that $\exp tX \in H$ for all t.

Assertion 1. Let $X_i \in G_e$ with $X_i \to X$ and let $t_i \to 0$ with each $t_i \neq 0$. Suppose $\exp t_i X_i \in H$ for all i. Then $X \in \mathfrak{h}$.

Proof. We can assume $t_i > 0$, since $\exp(-t_i X_i) = (\exp t_i X_i)^{-1} \in H$. For $t > 0$, let

$$k_i(t) = \text{ largest integer } \leq \frac{t}{t_i}.$$

Then

$$\frac{t}{t_i} - 1 < k_i(t) \leq \frac{t}{t_i},$$

so

$$t_i k_i(t) \to t.$$

Now

$$\exp\big(k_i(t) t_i X_i\big) = \big[\exp(t_i X_i)\big]^{k_i(t)} \in H,$$
$$k_i(t) t_i X_i \to t X.$$

Thus $\exp tX \in H$, since H is closed and exp is continuous. We clearly also have $\exp tX \in H$ for $t < 0$, so $X \in \mathfrak{h}$. **Q.E.D.**

We now claim that $\mathfrak{h} \subset G_e$ is a vector subspace. Clearly $X \in \mathfrak{h}$ implies $sX \in \mathfrak{h}$ for all $s \in \mathbb{R}$. If $X, Y \in \mathfrak{h}$, we can write by (1) of Theorem 14

$$\exp tX \exp tY = \exp\{t(X + Y) + t Z(t)\}$$

where $Z(t) \to 0$ as $t \to 0$. Choose positive $t_i \to 0$ and let $X_i = X + Y + Z(t_i)$. Then *Assertion 1* implies that $X + Y \in \mathfrak{h}$. Alternatively, we can write, for fixed t,

$$\left(\exp \frac{t}{n} X \exp \frac{t}{n} Y\right)^n = \exp\left\{t(X + Y) + \frac{t^2}{2n}[X, Y] + O(1/n^2)\right\};$$

taking limits as $n \to \infty$ gives $\exp t(X + Y) \in H$.

[Similarly, using (2) of Theorem 14 we see that $[X, Y] \in \mathfrak{h}$, so that $\mathfrak{h}$ is a subalgebra, but we will not even use this fact.]

Now let U be an open neighborhood of $0 \in G_e$ on which exp is a diffeomorphism. Then $\exp(\mathfrak{h} \cap U)$ is a submanifold of G. It clearly suffices to show that if U is small enough, then

$$H \cap \exp(U) = \exp(\mathfrak{h} \cap U).$$

Choose a subspace $\mathfrak{h}' \subset G_e$ complementary to $\mathfrak{h}$, so that $G_e = \mathfrak{h} \oplus \mathfrak{h}'$.

Assertion 2. The map $\phi : G_e \to G$ defined by

$$\phi(X + X') = \exp X \exp X' \qquad\qquad X \in \mathfrak{h},\ X' \in \mathfrak{h}'$$

is a diffeomorphism in some neighborhood of 0.

Proof. Choose a basis $X_1, \ldots, X_k, \ldots, X_n$ of G_e with $X_1, \ldots, X_k$ a basis for $\mathfrak{h}$. Then ϕ is given by

$$\phi\left(\sum_{i=1}^{n} a_i X_i\right) = \exp\left(\sum_{i=1}^{k} a_i X_i\right) \exp\left(\sum_{i=k+1}^{n} a_i X_i\right).$$

Since the map $\sum_{i=1}^{n} a_i X_i \mapsto (a_1, \ldots, a_n)$ is a diffeomorphism of G_e onto $\mathbb{R}^n$, it suffices to show that

$$\psi(a_1, \ldots, a_n) = \exp\left(\sum_{i=1}^{k} a_i X_i\right) \exp\left(\sum_{i=k+1}^{n} a_i X_i\right)$$

is a diffeomorphism in a neighborhood of $0 \in \mathbb{R}^n$. This is clear, since

$$\psi_*\left(\left.\frac{\partial}{\partial x^i}\right|_0\right) = X_i. \qquad \textbf{Q.E.D.}$$

Assertion 3. There is a neighborhood V' of 0 in $\mathfrak{h}'$ such that $\exp X' \notin H$ if $0 \neq X' \in V'$.

Proof. Choose an inner product on $\mathfrak{h}'$ and let $K \subset \mathfrak{h}'$ be the compact set of all $X' \in \mathfrak{h}'$ with $1 \leq |X'| \leq 2$. If the assertion were false, there would be $X_i' \in \mathfrak{h}'$ with $X_i' \to 0$ and $\exp X_i' \in H$. Choose integers n_i with

$$n_i X_i' \in K.$$

Choosing a subsequence if necessary, we can assume $X_i' \to X' \in K$. Since

$$1/n_i \to 0, \qquad \exp(1/n_i)(n_i X_i') \in H,$$

it follows from *Assertion 1* that $X' \in \mathfrak{h}$, a contradiction. **Q.E.D.**

We can now complete the proof of the theorem. Choose a neighborhood $U = W \times W'$ of G_e on which $\exp$ is a diffeomorphism, with

$$W \text{ a neighborhood of } 0 \in \mathfrak{h}$$
$$W' \text{ a neighborhood of } 0 \in \mathfrak{h}'$$

such that W' is contained in V' of *Assertion 3*, and ϕ of *Assertion 2* is a diffeomorphism on $W \times W'$. Clearly

$$\exp(\mathfrak{h} \cap U) \subset H \cap \exp(U).$$

To prove the reverse inclusion, let $a \in H \cap \exp(U)$. Then

$$a = \exp X \exp X' \qquad X \in W, \ X' \in W'.$$

Since $a, \exp X \in H$ we obtain $\exp X' \in H$, so $0 = X'$, and $a \in \exp(\mathfrak{h} \cap U)$. ❖

Up to now, we have concentrated on the left invariant vector fields, but many properties of Lie groups are better expressed in terms of forms. A form ω is called **left invariant** if $L_a{}^*\omega = \omega$ for all $a \in G$. This means that

$$\omega(b) = L_a{}^*[\omega(ab)].$$

Clearly, a left invariant k-form ω is determined by its value $\omega(e) \in \Omega^k(G_e)$. Hence, if $\omega^1, \ldots, \omega^n$ are left invariant 1-forms such that $\omega^1(e), \ldots, \omega^n(e)$ span $G_e{}^*$, then every left invariant k-form is

$$\sum_{i_1 < \cdots < i_k} a_{i_1 \ldots i_k}\, \omega^{i_1} \wedge \cdots \wedge \omega^{i_k} = \sum_I A_I \omega^I$$

for certain *constants* a_I. If $\omega^1(e), \ldots, \omega^n(e)$ is the dual basis to $X_1, \ldots, X_n \in G_e$, then any C^∞ vector field X can be written

$$X = \sum_{j=1}^n f^j \widetilde{X}_j \qquad \text{for } C^\infty \text{ functions } f^j.$$

Then

$$\omega^i(X) = f^i,$$

so ω^i is C^∞. It follows that any left invariant form is C^∞.

If ω is left invariant, then for $a \in G$ we have

$$L_a{}^*d\omega = d(L_a{}^*\omega) = d\omega,$$

so $d\omega$ is also left invariant. The formula on page 215 implies that for a left invariant 1-form ω and left invariant vector fields $\widetilde{X}$ and $\widetilde{Y}$ we have

$$d\omega(\widetilde{X}, \widetilde{Y}) = \widetilde{X}(\omega(\widetilde{Y})) - \widetilde{Y}(\omega(\widetilde{X})) - \omega([\widetilde{X}, \widetilde{Y}])$$
$$= -\omega([\widetilde{X}, \widetilde{Y}]).$$

Hence

$$(*) \qquad\qquad d\omega(e)(X, Y) = -\omega(e)([X, Y]),$$

the bracket being the operation in $\mathfrak{g}$.

The interplay between left invariant and right invariant vector fields is the subject of Problem 11. Here we consider the case of forms.

16. PROPOSITION. Let $\psi : G \to G$ be $\psi(a) = a^{-1}$.

(1) A form ω is left invariant if and only if $\psi^*\omega$ is right invariant.

(2) If $\omega_e \in \Omega^k(G_e)$, then $\psi^*\omega_e = (-1)^k\omega_e$.

(3) If ω is left and right invariant, then $d\omega = 0$.

(4) If G is abelian, then $\mathfrak{g}$ is abelian (converse of Corollary 7).

PROOF. (1) Clearly

$$\psi \circ R_b = L_{b^{-1}} \circ \psi,$$

so

$$R_b{}^*\psi^* = \psi^*L_{b^{-1}}{}^*.$$

If ω is left invariant, then

$$R_b{}^*(\psi^*\omega) = \psi^*L_{b^{-1}}{}^*\omega = \psi^*\omega,$$

so $\psi^*\omega$ is right invariant. The converse is similar.

(2) It clearly suffices to prove this for $k = 1$. So it is enough to show that $\psi_{*e}(X) = -X$ for $X \in G_e$. Now X is the tangent vector at $t = 0$ of the curve $t \mapsto \exp tX$. So $\psi_{*e}X$ is the tangent vector at $t = 0$ of $t \mapsto (\exp tX)^{-1} = \exp(-tX)$; this tangent vector is just $-X$.

(3) If ω is a left and right invariant k-form, then

$$\psi^*(\omega_e) = (-1)^k\omega_e.$$

Since $\psi^*\omega$ and ω are both left invariant, we have

$$\psi^*\omega = (-1)^k\omega.$$

The form $d\omega$ is also left and right invariant, so

$$\psi^*(d\omega) = (-1)^{k+1}d\omega.$$

But

$$\psi^*(d\omega) = d(\psi^*\omega) = d((-1)^k\omega) = (-1)^k d\omega.$$

So $d\omega = 0$.

(4) If G is abelian, then all left invariant 1-forms ω are also right invariant. So $d\omega = 0$ for all left invariant 1-forms. It follows from $(*)$ that $[X, Y] = 0$ for all $X, Y \in \mathfrak{g}$.

Alternate proof of (4). By Theorem 14, if G is abelian, then for $X, Y \in G_e$ we have

$$\frac{t^2}{2}[X, Y] + O(t^3) = \frac{t^2}{2}[Y, X] + O(t^3).$$

Hence

$$\tfrac{1}{2}[X, Y] + O(t^3)/t^2 = \tfrac{1}{2}[Y, X] + O(t^3)/t^2.$$

Letting $t \to 0$, we obtain $[X, Y] = [Y, X]$. ❖

Since $d\omega$ is left invariant for any left invariant ω, it follows that for a basis ω^1, $\ldots, \omega^n$ of invariant 1-forms we can express each $d\omega^k$ in terms of the $\omega^i \wedge \omega^j$. First choose $X_1, \ldots, X_n \in G_e$ dual to $\omega^1(e), \ldots, \omega^n(e)$. There are constants C_{ij}^k such that

$$[X_i, X_j] = \sum_{k=1}^{n} C_{ij}^k X_k;$$

clearly we also have

$$[\widetilde{X}_i, \widetilde{X}_j] = \sum_{k=1}^{n} C_{ij}^k \widetilde{X}_k.$$

The numbers C_{ij}^k are called the **constants of structure** of G (with respect to the basis $X_1, \ldots, X_n$ of $\mathfrak{g}$). From skew-symmetry of $[\ ,\ ]$ and the Jacobi identity we obtain

(1) $C_{ij}^k = -C_{ji}^k$

(2) $\displaystyle\sum_{h=1}^{n} (C_{ij}^h C_{hk}^l + C_{jk}^h C_{hi}^l + C_{ki}^h C_{hj}^l) = 0.$

From $(*)$ on page 394 we obtain

$$d\omega^k = -\sum_{i<j} C_{ij}^k\, \omega^i \wedge \omega^j = -\frac{1}{2}\sum_{i,j} C_{ij}^k\, \omega^i \wedge \omega^j.$$

It turns out that (2) is exactly what we obtain from the relation $d^2\omega^k = 0$. Condition (2) is thus an integrability condition. In fact, we can prove (Problem 30) that if C_{ij}^k are constants satisfying (1) and (2), then we can find everywhere linearly independent 1-forms $\omega^1, \ldots, \omega^n$ in a neighborhood of $0 \in \mathbb{R}^n$ such that

$$d\omega^k = -\frac{1}{2}\sum_{i,j} C_{ij}^k\, \omega^i \wedge \omega^j.$$

Moreover, the existence of such ω^i implies (Problem 29) that we can define a multiplication $(a, b) \mapsto ab$ in a neighborhood of 0 which is a group as far as it can be and which has the ω^i as left invariant 1-forms. From this latter fact and (a suitable local version of) Theorem 5 we could immediately deduce the following Theorem, for which we supply an independent proof.

17. THEOREM. Let G be a Lie group with a basis of left invariant 1-forms $\omega^1, \ldots, \omega^n$ and constants of structure C_{ij}^k. Let M^n be a differentiable manifold and let $\theta^1, \ldots, \theta^n$ be everywhere linearly independent 1-forms on M satisfying

$$d\theta^k = -\sum_{i<j} C_{ij}^k \theta^i \wedge \theta^j.$$

Then for every $p \in M$ there is a neighborhood U and a diffeomorphism $f: U \to G$ such that

$$\theta^i = f^* \omega^i.$$

PROOF. Let $\pi_1: M \times G \to M$ and $\pi_2: M \times G \to G$ be the projections. Let

$$\bar{\theta}^k = \pi_1{}^* \theta^k, \qquad \bar{\omega}^k = \pi_2{}^* \omega^k.$$

Then

$$d(\bar{\theta}^k - \bar{\omega}^k) = -\sum_{i<j} C_{ij}^k \left([\bar{\theta}^i \wedge \bar{\theta}^j] - [\bar{\omega}^i \wedge \bar{\omega}^j]\right)$$

$$= -\sum_{i<j} C_{ij}^k [\bar{\theta}^i \wedge (\bar{\theta}^j - \bar{\omega}^j) + (\bar{\theta}^i - \bar{\omega}^i) \wedge \bar{\omega}^j].$$

By Proposition 7-14, $M \times G$ is foliated by n-dimensional manifolds whose tangent spaces at each point are annihilated by all $\bar{\theta}^k - \bar{\omega}^k$. Choose $a \in G$ and let Γ be the folium through (p, a). Now $\bar{\theta}^1, \ldots, \bar{\theta}^n, \bar{\omega}^1, \ldots, \bar{\omega}^n$ are linearly independent everywhere; so on $\Gamma_{(p,a)}$, which is the set of vectors in $(M \times G)_{(p,a)}$ where $\bar{\theta}^k - \bar{\omega}^k = 0$, the sets $\bar{\theta}^1, \ldots, \bar{\theta}^n$ and $\bar{\omega}^1, \ldots, \bar{\omega}^n$ are each linearly independent. Hence $\pi_1: \Gamma \to M$ and $\pi_2: \Gamma \to G$ are each diffeomorphisms in some neighborhood of (p, a). This means that Γ contains the graph of a diffeomorphism f from a neighborhood U of p to a neighborhood of a.

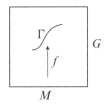

Let $\bar{f}\colon U \to M \times G$ be the map

$$\bar{f}(q) = (q, f(q)) \subset \Gamma.$$

Since $\bar{\theta}^k - \bar{\omega}^k = 0$ on Γ, we have

$$\begin{aligned}
0 = \bar{f}^*(\bar{\theta}^k - \bar{\omega}^k) &= \bar{f}^*\pi_1{}^*\theta^k - \bar{f}^*\pi_2{}^*\omega^k \\
&= (\pi_1 \circ \bar{f})^*\theta^k - (\pi_2 \circ \bar{f})^*\omega^k \\
&= \theta^k - f^*\omega^k. \ \diamondsuit
\end{aligned}$$

It is also possible to say by how much any two such maps differ:

18. THEOREM. Let M be a connected manifold, let G be a Lie group, and let $f_1, f_2\colon M \to G$ be two C^∞ maps such that

$$f_1{}^*(\omega) = f_2{}^*(\omega)$$

for all left invariant 1-forms ω. Then f_1 and f_2 differ by a left translation, that is, there is a (unique) $a \in G$ such that

$$f_2 = L_a \circ f_1.$$

PEDESTRIAN PROOF. Case 1. $M = \mathbb{R}$ and the two maps $\gamma_1, \gamma_2\colon \mathbb{R} \to G$ satisfy $\gamma_1(0) = \gamma_2(0)$. We must show that $\gamma_1 = \gamma_2$. For every left invariant 1-form ω we have

$$\begin{aligned}
\omega(\gamma_2(t))\left(\frac{d\gamma_2}{dt}\right) &= \gamma_2{}^*\omega\left(\frac{d}{dt}\Big|_t\right) = \gamma_1{}^*\omega\left(\frac{d}{dt}\Big|_t\right) \\
&= \omega(\gamma_1(t))\left(\frac{d\gamma_1}{dt}\right) \\
&= \left[(L_{\gamma_2(t)\gamma_1(t)^{-1}})^*\omega(\gamma_2(t))\right]\left(\frac{d\gamma_1}{dt}\right) \\
&= \omega(\gamma_2(t))\left(\left[L_{\gamma_2(t)\gamma_1(t)^{-1}}\right]_*\frac{d\gamma_1}{dt}\right).
\end{aligned}$$

It follows that

$$\frac{d\gamma_2}{dt} = \left[L_{\gamma_2(t)\gamma_1(t)^{-1}}\right]_*\frac{d\gamma_1}{dt}.$$

If we regard γ_1 as given, and write this equation out in a coordinate system, then it becomes an ordinary differential equation for γ_2 (of the type considered in the Addendum to Chapter 5), so it has a unique solution with the initial condition $\gamma_2(0) = \gamma_1(0)$. But this solution is clearly $\gamma_2 = \gamma_1$.

Case 2. $M = \mathbb{R}$, but the maps γ_1, γ_2 are arbitrary. Choose $a \in G$ so that

$$\gamma_2(0) = a \cdot \gamma_1(0).$$

If ω is a left invariant 1-form, then

$$(L_a \circ \gamma_1)^*(\omega) = \gamma_1^*(L_a^*\omega) = \gamma_1^*(\omega) = \gamma_2^*(\omega).$$

Since $L_a \circ \gamma_1(0) = \gamma_2(0)$, it follows from *Case 1* that $L_a \circ \gamma_1 = \gamma_2$.

Case 3. General case. Let $p_0 \in M$. Choose $a \in G$ so that

$$f_2(p_0) = a \cdot f_1(p_0).$$

For any $p \in M$ there is a C^∞ curve $c \colon \mathbb{R} \to M$ with $c(0) = p_0$ and $c(1) = p$. Let $\gamma_i = f_i \circ c$. Then

$$\gamma_2^*(\omega) = c^* f_2^*(\omega) = c^* f_1^*(\omega) = \gamma_1^*(\omega).$$

By *Case 2*, we have

$$\gamma_2(t) = a \cdot \gamma_1(t) \qquad \text{for all } t,$$

in particular for $t = 1$, so $f_2(p) = a \cdot f_1(p)$.

ELEGANT PROOF. Let $\pi_i \colon G \times G \to G$ be projection on the ith factor. Choose a basis $\omega^1, \ldots, \omega^n$ for the left invariant 1-forms. For $(a, b) \in G \times G$, let

$$\Delta_{(a,b)} = \bigcap_{i=1}^n \ker(\pi_1^*\omega^i - \pi_2^*\omega^i).$$

Then Δ is an integrable distribution on $G \times G$. In fact, if $\Delta(G) \subset G \times G$ is the diagonal subgroup $\{(a, a) : a \in G\}$, then the maximal integral manifolds of Δ are the left cosets of $\Delta(G)$. Now define $h \colon M \to G \times G$ by

$$h(p) = (f_1(p), f_2(p)).$$

By assumption,

$$h^*(\pi_1^*\omega^i - \pi_2^*\omega^i) = f_1^*\omega^i - f_2^*\omega^i = 0.$$

Since M is connected, it follows that $h(M)$ is contained in some left coset of $\Delta(G)$. In other words, there are $a, b \in G$ with

$$af_1(p) = bf_2(p) \qquad \text{for all } p \in M. \ \clubsuit$$

19. COROLLARY. If G is a connected Lie group and $f : G \to G$ is a C^∞ map preserving left invariant forms, then $f = L_a$ for a unique $a \in G$.

While left invariant 1-forms play a fundamental role in the study of G, the left invariant n-forms are also very important. Clearly, all left invariant n-forms are a constant multiple of any non-zero one. If σ^n is a left invariant n-form, then σ^n determines an orientation on G, and if $f : G \to \mathbb{R}$ is a C^∞ function with compact support, we can define

$$\int_G f\sigma^n.$$

Since σ^n is usually kept fixed in any discussion, this is often abbreviated to

$$\int_G f \quad \text{or} \quad \int_G f(a)\,da.$$

The latter notation has advantages in certain cases. For example, left invariance of σ^n implies that

$$\int_G f(a)\,da = \int_G f(ba)\,da,$$

in other words,

$$\int_G f\sigma^n = \int_G g\sigma^n, \qquad \text{where } g(a) = f(ba);$$

[note that L_b is an orientation preserving diffeomorphism, so

$$\int_G f\sigma^n = \int_G L_b{}^*(f\sigma^n) = \int_G (f \circ L_b)L_b{}^*\sigma^n = \int_G (f \circ L_b)\sigma^n,$$

which proves the formula]. We can, of course, also consider right invariant n-forms. These generally turn out to be quite different from the left invariant n-forms (see the example in Problem 25). But in one case they coincide.

20. PROPOSITION. If G is compact and connected and ω is a left invariant n-form, then ω is also right invariant.

PROOF. Suppose $\omega \neq 0$. For each $a \in G$, the form $R_a{}^*\omega$ is left invariant, so there is a unique real number $f(a)$ with

$$R_a{}^*\omega = f(a)\omega.$$

Since $R_a{}^* \circ R_b{}^* = (R_{ab})^*$, we have

$$f(ab) = f(ba) = f(a) \cdot f(b).$$

So $f(G) \subset \mathbb{R}$ is a compact connected subgroup of $\mathbb{R} - \{0\}$. Hence $f(G) = \{1\}$. ❖

We can also consider Riemannian metrics on G. In the case of a compact group G there is always a Riemannian metric on G which is both left and right invariant. In fact, if $\langle \ , \ \rangle$ is any Riemannian metric we can choose a bi-invariant n-form σ^n and define a bi-invariant $\langle\!\langle \ , \ \rangle\!\rangle$ on G by

$$\langle\!\langle V, W \rangle\!\rangle = \int_{G \times G} \langle L_{a*} R_{b*}(V), L_{a*} R_{b*}(W) \rangle \, da \, db.$$

We are finally ready to account for some terminology from Chapter 9.

21. PROPOSITION. Let G be a Lie group with a bi-invariant metric.

(1) For any $a \in G$, the map $I_a \colon G \to G$ given by $I_a(b) = ab^{-1}a$ is an isometry which reverses geodesics through a, i.e., if γ is a geodesic and $\gamma(0) = a$, then $I_a(\gamma(t)) = \gamma(-t)$.

(2) The geodesics γ with $\gamma(0) = e$ are precisely the 1-parameter subgroups of G, i.e., the maps $t \mapsto \exp(tX)$ for some $X \in \mathfrak{g}$.

PROOF. (1) Since

$$I_e(b) = b^{-1},$$

the map $I_{e*} \colon G_e \to G_e$ is just multiplication by -1 (see the proof of Proposition 16(2)), so it is an isometry on G_e. Since

$$I_e = R_{a^{-1}} I_e L_{a^{-1}}$$

for any $a \in G$, the map $I_{e*} \colon G_a \to G_{a^{-1}}$ is also an isometry. Clearly I_e reverses geodesics through e.

Since

$$I_a = R_a I_e R_a^{-1},$$

it is clear that I_a is an isometry reversing geodesic through a.

(2) Let $\gamma \colon \mathbb{R} \to G$ be a geodesic with $\gamma(0) = e$. For fixed t, let

$$\bar{\gamma}(u) = \gamma(t + u).$$

Then $\bar{\gamma}$ is a geodesic and $\bar{\gamma}(0) = \gamma(t)$. So

$$\begin{aligned}
I_{\gamma(t)} I_e(\gamma(u)) = I_{\gamma(t)}(\gamma(-u)) &= I_{\gamma(t)}(\bar{\gamma}(-u - t)) \\
&= \bar{\gamma}(t + u) = \gamma(u + 2t).
\end{aligned}$$

But also

$$I_{\gamma(t)} I_e(b) = \gamma(t) b \gamma(t),$$

so

$$\gamma(t)\gamma(u)\gamma(t) = \gamma(u + 2t).$$

It follows by induction that

$$\gamma(nt) = \gamma(t)^n \qquad \text{for any integer } n.$$

If $t' = n't$ and $t'' = n''t$ for integers n' and n'', then

$$\gamma(t' + t'') = \gamma(t)^{n'+n''} = \gamma(t')\gamma(t''),$$

so γ is a homomorphism on $\mathbb{Q}$. By continuity, γ is a 1-parameter subgroup.

These are the only geodesics, since there are 1-parameter subgroups with any tangent vector at $t = 0$, and geodesics through e are determined by their tangent vectors at $t = 0$. ❖

We conclude this chapter by introducing some neat formalism which allows us to write the expression for $d\omega^k$ in an invariant way that does not use the constants of structure of G. If V is a d-dimensional vector space, we define a *V*-**valued** k-**form** on M to be a function ω such that each $\omega(p)$ is an alternating map

$$\omega(p)\colon \underbrace{M_p \times \cdots \times M_p}_{k \text{ times}} \to V.$$

If $v_1, \ldots, v_d$ is a basis for V, then there are ordinary k-forms $\omega^1, \ldots, \omega^d$ such that for $X_1, \ldots, X_k \in M_p$ we have

$$\omega(p)(X_1, \ldots, X_k) = \sum_{i=1}^{d} \omega^i(p)(X_1, \ldots, X_k)v_i;$$

we will write simply

$$\omega = \sum_{i=1}^{d} \omega^i \cdot v_i.$$

For any V-valued k-form ω we define a V-valued $(k+1)$-form $d\omega$ by

$$d\omega = \sum_{i=1}^{d} d\omega^i \cdot v_i;$$

a simple calculation shows that this definition does not depend on the choice of basis $v_1, \ldots v_d$ for V.

Similarly, suppose $\rho: U \times V \to W$ is a bilinear map, where U and V have bases $u_1, \ldots, u_c$ and $v_1, \ldots, v_d$, respectively. If ω is a U-valued k-form

$$\omega = \sum_{i=1}^{c} \omega^i \cdot u_i$$

and η is a V-valued l-form

$$\eta = \sum_{j=1}^{d} \eta^j \cdot v_j,$$

then

$$\sum_{i=1}^{c} \sum_{j=1}^{d} \omega^i \wedge \eta^j \cdot \rho(u_i, v_j)$$

is a W-valued $(k+l)$-form; a calculation shows that this does not depend on the choice of bases $u_1, \ldots, u_c$ or $v_1, \ldots, v_d$. We will denote this W-valued $(k+l)$-form by $\rho(\omega \wedge \eta)$.

These concepts have a natural place in the study of a Lie group G. Although there is no natural way to choose a basis of left invariant 1-forms on G, there *is* a natural $\mathfrak{g}$-valued 1-form on G, namely the form ω defined by

$$(*) \qquad\qquad \omega(a)(\tilde{X}(a)) = X \in \mathfrak{g}.$$

Using the bilinear map $[\ ,\ ]: \mathfrak{g} \times \mathfrak{g} \to \mathfrak{g}$, we have, for any $\mathfrak{g}$-valued k-form η and any $\mathfrak{g}$-valued l-form λ on G, a new $\mathfrak{g}$-valued $(k+l)$-form $[\eta \wedge \lambda]$ on G.

Now suppose that $X_1, \ldots, X_n \in G_e = \mathfrak{g}$ is a basis, and that $\omega^1, \ldots, \omega^n$ is a dual basis of left invariant 1-forms. The form ω defined by $(*)$ can clearly be written

$$\omega = \sum_{k=1}^{n} \omega^k \cdot X_k.$$

Then

$$(1) \qquad\qquad d\omega = \sum_{k=1}^{n} d\omega^k \cdot X_k$$

$$= \sum_{k=1}^{n} \left(\sum_{i<j} C_{ij}^k \omega^i \wedge \omega^j \right) \cdot X_k.$$

On the other hand,

$$[X_i, X_j] = \sum_{k=1}^{n} C_{ij}^k X_k,$$

so

(2)
$$[\omega \wedge \omega] = \sum_{k=1}^{n} \left(\sum_{i=1}^{n} \sum_{j=1}^{n} C_{ij}^k \, \omega^i \wedge \omega^j \cdot X_k \right).$$

Comparing (1) and (2), we obtain the **equations of structure** of G:

$$\boxed{d\omega = -\frac{1}{2}[\omega \wedge \omega].}$$

The equations of structure of a Lie group will play an important role in Volume III. For the present we merely wish to point out that the terms $d\omega$ and $[\omega \wedge \omega]$ appearing in this equation can also be defined in an invariant way. For the term $d\omega$ we just modify the formula in Theorem 7-13: If U is a vector field on G and f is a $\mathfrak{g}$-valued function on G, then (Problem 20) we can define a $\mathfrak{g}$-valued function $U(f)$ on G. On the other hand, $\omega(U)$ is a $\mathfrak{g}$-valued function on G. For vector fields U and V we can then define

$$d\omega(U, V) = U(\omega(V)) - V(\omega(U)) - \omega([U, V]).$$

Recall that the value at $a \in G$ of the right side depends only on the values U_a and V_a of U and V at a. If we choose $U = \widetilde{X}$, $V = \widetilde{Y}$ for some $X, Y \in G_e$, then

$$
\begin{aligned}
d\omega(a)(\widetilde{X}_a, \widetilde{Y}_a) &= 0 - 0 - \omega(a)([\widetilde{X}, \widetilde{Y}]_a) \\
&= -\omega(e)([\widetilde{X}, \widetilde{Y}]_e) && \text{since } [\widetilde{X}, \widetilde{Y}] \text{ is left invariant} \\
&= -\omega(e)([X, Y]) && \text{by definition of } [\ ,\ ] \text{ in } G_e \\
&= -[X, Y] && \left. \begin{aligned} & \\ & \end{aligned} \right\} \text{ by definition of } \omega. \\
&= -[\omega(a)(\widetilde{X}_a), \omega(a)(\widetilde{Y}_a)]
\end{aligned}
$$

It follows that for any vector fields U and V we have

$$\boxed{d\omega(U, V) = -[\omega(U), \omega(V)].}$$

Problem 20 gives an invariant definition of $\rho(\omega \wedge \eta)$ and shows that this equation is equivalent to the equations of structure.

WARNING: In some books the equation which we have just deduced appears as $d\omega(U,V) = -\frac{1}{2}[\omega(U),\omega(V)]$. The appearance of the factor $\frac{1}{2}$ here has *nothing* to do with the $\frac{1}{2}$ in the other form of the structure equations. It comes about because some books do not use the factor $(k+l)!/k!l!$ in the definition of $\wedge$. This makes their $\lambda \wedge \eta$ equal to $\frac{1}{2}$ of ours for 1-forms λ and η. Then the definition of $d(\sum \omega_i \, dx^i)$ as $\sum d\omega_i \wedge dx^i$ makes their $d\omega$ equal to $\frac{1}{2}$ of ours for 1-forms ω.

PROBLEMS

1. Let G be a group which is also a C^∞ manifold, and suppose that $(x, y) \mapsto xy$ is C^∞.

(a) Find f^{-1} when $f: G \times G \to G \times G$ is $f(x, y) = (x, xy)$.
(b) Show that (e, e) is a regular point of f.
(c) Conclude that G is a Lie group.

2. Let G be a topological group, and $H \subset G$ a subgroup. Show that the closure $\overline{H}$ of H is also a subgroup.

3. Let G be a topological group and $H \subset G$ a subgroup.

(a) If H is open, then so is every coset gH.
(b) If H is open, then H is closed.

4. Let G be a connected topological group, and U a neighborhood of $e \in G$. Let U^n denote all products $a_1 \cdots a_n$ for $a_i \in U$.

(a) Show that U^{n+1} is a neighborhood of U^n.
(b) Conclude that $\bigcup_n U^n = G$. (Use Problem 3.)
(c) If G is locally compact and connected, then G is σ-compact.

5. Let $f: \mathbb{R}^n \to \mathbb{R}^n$ be distance preserving, with $f(0) = 0$.

(a) Show that f takes straight lines to straight lines.
(b) Show that f takes planes to planes.
(c) Show that f is a linear transformation, and hence an element of $O(n)$.
(d) Show that any element of $E(n)$ can be written $A \cdot \tau$ for $A \in O(n)$ and τ a translation.

6. Show that the tangent bundle TG of a Lie group G can always be made into a Lie group.

7. We have computed that for $M \in \mathfrak{gl}(n, \mathbb{R})$ we have

$$\widetilde{M} = \sum_{k,l} \widetilde{M} x^{kl} \cdot \frac{\partial}{\partial x^{kl}}, \qquad \text{where } \widetilde{M} x^{kl}(A) = \sum_{\alpha=1}^{n} M_{\alpha l} A_{k\alpha}.$$

(a) Show that this means that

$$\widetilde{M}(A) = A \cdot M \in \mathbb{R}^{n^2} = GL(n, \mathbb{R})_A.$$

(It is actually clear *a priori* that $\widetilde{M}$ defined in this way is left invariant, for $L_{A*} = L_A$ since L_A is linear.)
(b) Find the right invariant vector field with value M at I.

8. Let G and H be topological groups and $\phi \colon U \to H$ a map on a connected open neighborhood U of $e \in G$ such that $\phi(ab) = \phi(a)\phi(b)$ when $a, b, ab \in U$.

(a) For each $c \in G$, consider pairs (V, ψ), where $V \subset G$ is an open neighborhood of c with $V \cdot V^{-1} \subset U$, and where $\psi \colon V \to H$ satisfies $\psi(a) \cdot \psi(b)^{-1} = \phi(ab^{-1})$ for $a, b \in V$. Define $(V_1, \psi_1) \underset{c}{\sim} (V_2, \psi_2)$ if $\psi_1 = \psi_2$ on some smaller neighborhood of c. Show that the set of all $\underset{c}{\sim}$ equivalence classes, for all $c \in G$, can be made into a covering space of G.

(b) Conclude that if G is simply-connected, then ϕ can be extended uniquely to a homomorphism of G into H.

9. In Theorem 5, show that ϕ and ψ are equal even if they are defined only on a neighborhood U of $e \in G$, provided that U is connected.

10. Show that Corollary 7 is false if G is not assumed connected.

11. If G is a group, we define the **opposite group** G° to be the same set with the multiplication $\bullet$ defined by $a \bullet b = b \cdot a$. If $\mathfrak{g}$ is a Lie algebra, with operation $[\ ,\ ]$, we define the **opposite Lie algebra** $\mathfrak{g}^{\circ}$ to be the same set with the operation $[X, Y]^{\circ} = -[X, Y]$.

(a) G° is a group, and if $\psi \colon G \to G$ is $a \mapsto a^{-1}$, then ψ is an isomorphism from G to G°.

(b) $\mathfrak{g}^{\circ}$ is a Lie algebra, and $X \mapsto -X$ is an isomorphism of $\mathfrak{g}$ onto $\mathfrak{g}^{\circ}$.

(c) $\mathcal{L}(G^{\circ})$ is isomorphic to $[\mathcal{L}(G)]^{\circ} = \mathfrak{g}^{\circ}$.

(d) Let $[\ ,\ ]$ be the operation on G_e obtained by using right invariant vector fields instead of left invariant ones. Then $(\mathfrak{g}, [\ ,\ ])$ is isomorphic to $\mathcal{L}(G^{\circ})$, and hence to $\mathfrak{g}^{\circ}$.

(e) Use this to give another proof that $\mathfrak{g}$ is abelian when G is abelian.

12. (a) Show that

$$\exp \begin{pmatrix} 0 & a \\ -a & 0 \end{pmatrix} = \begin{pmatrix} \cos a & \sin a \\ -\sin a & \cos a \end{pmatrix}.$$

(b) Use the matrices A and B below to show that $\exp(A + B)$ is not generally equal to $(\exp A)(\exp B)$.

$$A = \begin{pmatrix} 0 & 1 \\ 0 & 0 \end{pmatrix} \qquad B = \begin{pmatrix} 0 & 0 \\ -1 & 0 \end{pmatrix}$$

13. Let $X, Y \in G_e$ with $[X, Y] = 0$.

(a) Use Lemma 5-13 to show that $(\exp sX)(\exp tY) = (\exp tY)(\exp sX)$.

(b) More generally, use Theorem 5 to show that $\exp$ is a homomorphism on the subspace of G_e spanned by X and Y. In particular, $\exp(X + Y) = (\exp X)(\exp Y)$.

14. Problem 13 implies that $\exp t(X + Y) = (\exp tX)(\exp tY)$ if $[X, Y] = 0$. A more general result holds. Let X and Y be vector fields on a C^∞ manifold M with corresponding local 1-parameter families of local diffeomorphisms $\{\phi_t\}$, $\{\psi_s\}$. Suppose that $[X, Y] = 0$, and let $\eta_t = \phi_t \circ \psi_t = \psi_t \circ \phi_t$.

(a) Show that

$$\frac{d\eta_t(p)}{dt} = X(\eta_t(p)) + \phi_{t*}(Y(\psi_t(p))).$$

(b) Using Corollary 5-12, show that

$$\frac{d\eta_t(p)}{dt} = X(\eta_t(p)) + Y(\eta_t(p)).$$

In other words, $\{\eta_t\}$ is generated by $X + Y$.

15. (a) If M is a diagonal matrix with complex entries, show that

$$\det \exp M = e^{\operatorname{trace} M}.$$

(b) Show that the same equation holds for all diagonalizable M with complex entries.

(c) Conclude that it holds for all M with complex entries. (The diagonalizable matrices are dense; compare Problem 7-15.)

(d) Using Proposition 9, show that for the homomorphism $\det \colon \mathrm{GL}(n, \mathbb{R}) \to \mathbb{R} - \{0\}$, the map $\det_* \colon \mathfrak{gl}(n, \mathbb{R}) \to \mathcal{L}(\mathbb{R} - \{0\}) = \mathbb{R}$ is just $M \mapsto \operatorname{trace} M$.

(e) Use this fact to give a fancy proof that $\operatorname{trace} MN = \operatorname{trace} NM$. (Look at $\operatorname{trace}(MN - NM) = \operatorname{trace}[M, N]$.)

(f) Prove the result in part (d) directly, without using (c). (Since $\det_*$ and trace are homomorphisms, it suffices to look at matrices with only one non-zero entry.)

(g) Now use this result and Proposition 9 to give a fancy proof of (c).

16. (a) Let U be a neighborhood of the identity $(1, 0)$ of S^1 (considered as a subset of $\mathbb{R}^2$). Show that no matter how small U is, there are elements $a \in U$ which have square roots outside U in addition to their square root in U.

(b) Show that for each $n \geq 1$, there is a neighborhood U of $e \in G$ such that every element in U has a unique n^{th} root in U.

(c) For $G = S^1$, show that there is no neighborhood U which has this property for all n.

17. (a) Let (x, V) be a coordinate system around $e \in G$ with $x^i(e) = 0$. Let

$$x^i(ab) = f^i(x^1(a), \ldots, x^n(a), x^1(b), \ldots, x^n(b))$$

for C^∞ functions f^i. Show that

$$D_j f^i(0) = D_{n+j} f^i(0) = \delta^i_j.$$

(b) If $\alpha, \beta : (-\varepsilon, \varepsilon) \to G$ are differentiable, show that

$$(\alpha \cdot \beta)'(0) = \alpha'(0) + \beta'(0).$$

(c) Also deduce this result from Theorem 14(1). (Not even the full strength of (1) is needed; it suffices to know that $\exp tX \exp tY = \exp\{t(X+Y) + O(t)\}$. The argument of part (a) is essentially equivalent to the initial part of the deduction of (1).)

18. Let G be a Lie group, and let $H \subset G$ be a subgroup of G (algebraically), such that every $a \in H$ can be joined to e by a C^∞ path lying in H. Let $\mathfrak{h} \subset G_e$ be the set of tangent vectors to all C^∞ paths lying in H.

(a) Show that $\mathfrak{h}$ is a subalgebra of G_e. (Use Theorem 14.)

(b) Let $K \subset G$ be the connected Lie subgroup of G with Lie algebra $\mathfrak{h}$. Show that $H \subset K$. *Hint*: Join any $a \in H$ to e by a C^∞ curve c, and show that the tangent vectors of c lie in the distribution constructed in the proof of Theorem 4.

(c) Let $c_1, \dots, c_k$ be curves in H with $\{c_i'(0)\}$ a basis for $\mathfrak{h}$. By considering the map $f(t^1, \dots, t^k) = c_1(t^1) \cdots c_k(t^k)$, show that $K \subset H$. Thus, H is a Lie subgroup of G. It is even true that $H \subset G$ is a Lie subgroup if H is path connected (by not necessarily C^∞ paths); see Yamabe, *On an arcwise connected subgroup of a Lie group*, Osaka Math. J. 2 (1950), 13–14.

(d) If $H \subset G$ is a subgroup and an immersed submanifold, then H is a Lie subgroup.

19. For $a \in G$, consider the map $b \mapsto aba^{-1} = L_a R_a^{-1}(b)$. The map

$$(L_a R_a^{-1})_* : \mathfrak{g} \to \mathfrak{g}$$

is denoted by $\mathrm{Ad}(a)$; usually $\mathrm{Ad}(a)(X)$ is denoted simply by $\mathrm{Ad}(a)X$.

(a) $\mathrm{Ad}(ab) = \mathrm{Ad}(a) \circ \mathrm{Ad}(b)$. Thus we have a homomorphism $\mathrm{Ad} : G \to \mathit{Aut}(\mathfrak{g})$, where $\mathit{Aut}(\mathfrak{g})$, the automorphism group of $\mathfrak{g}$, is the set of all non-singular linear transformations of the vector space $\mathfrak{g}$ onto itself (thus, isomorphic to $\mathrm{GL}(n, \mathbb{R})$ if $\mathfrak{g}$ has dimension n). The map Ad is called the **adjoint representation**.

(b) Show that

$$\exp(\mathrm{Ad}(a)X) = a(\exp X)a^{-1}.$$

Hint: This follows immediately from one of our propositions.

(c) For $A \in GL(n, \mathbb{R})$ and $M \in \mathfrak{gl}(n, \mathbb{R})$ show that

$$Ad(A)M = AMA^{-1}.$$

(It suffices to show this for M in a neighborhood of 0.)

(d) Show that
$$Ad(\exp tX)Y = Y + t[X, Y] + O(t^2).$$

(e) Since $Ad \colon G \to \mathfrak{g}$, we have the map

$$Ad_{*e} \colon \mathfrak{g} \ (= G_e) \ \to \quad \begin{array}{l} \text{tangent space of } Aut(\mathfrak{g}) \text{ at the} \\ \text{identity map } 1_{\mathfrak{g}} \text{ of } \mathfrak{g} \text{ to itself.} \end{array}$$

This tangent space is isomorphic to $End(\mathfrak{g})$, where $End(\mathfrak{g})$ is the vector space of all linear transformations of $\mathfrak{g}$ into itself: If c is a curve in $Aut(\mathfrak{g})$ with $c(0) = 1_{\mathfrak{g}}$, then to regard $c'(0)$ as an element of $Aut(\mathfrak{g})$, we let it operate on $Y \in \mathfrak{g}$ by

$$c'(0)(Y) = \frac{d}{dt}\Big|_{t=0} c(Y).$$

(Compare with the case $\mathfrak{g} = \mathbb{R}^n$, $Aut(\mathfrak{g}) = GL(n, \mathbb{R})$, $End(\mathfrak{g}) = n \times n$ matrices.) Use (d) to show that

$$Ad_{*e}(X)(Y) = [X, Y].$$

(A proof may also be given using the fact that $[\widetilde{X}, \widetilde{Y}] = L_{\widetilde{X}}\widetilde{Y}$.) The map $Y \mapsto [X, Y]$ is denoted by $ad\, X \in End(\mathfrak{g})$.

(f) Conclude that

$$Ad(\exp X) = \exp(ad\, X) = 1_{\mathfrak{g}} + ad\, X + \frac{(ad\, X)^2}{2!} + \cdots.$$

(g) Let G be a connected Lie group and $H \subset G$ a Lie subgroup. Show that H is a normal subgroup of G if and only if $\mathfrak{h} = \mathcal{L}(H)$ is an **ideal** of $\mathfrak{g} = \mathcal{L}(G)$, that is, if and only if $[X, Y] \in \mathfrak{h}$ for all $X \in \mathfrak{g}$, $Y \in \mathfrak{h}$.

20. (a) Let $f \colon M \to V$, where V is a finite dimensional vector space, with basis $v_1, \ldots, v_d$. For $X_p \in M_p$, define $X_p(f) \in V$ by

$$X(f) = \sum_{i=1}^{d} X_p(f^i) \cdot v_i,$$

where $f = \sum_{i=1}^{d} f^i \cdot v_i$ for $f^i \colon M \to \mathbb{R}$. Show that this definition is independent of the choice of basis $v_1, \ldots, v_d$ for V.

(b) If ω is a V-valued k-form, show that $d\omega$ may be defined invariantly by the formula in Theorem 7-13 (using the definition in part (a)).

(c) For $\rho\colon U \times V \to W$, show that $\rho(\omega \wedge \eta)$ may be defined invariantly by

$$\rho(\omega \wedge \eta)(X_1, \ldots, X_k, X_{k+1}, \ldots, X_{k+l})$$

$$= \frac{1}{k!\,l!} \sum_{\sigma \in S_{k+l}} \operatorname{sgn} \sigma \cdot \rho\big(\omega(X_{\sigma(1)}, \ldots, X_{\sigma(k)}), \eta(X_{\sigma(k+1)}, \ldots, X_{\sigma(k+l)})\big).$$

Conclude, in particular, that

$$[\omega \wedge \omega](X, Y) = 2[\omega(X), \omega(Y)].$$

(d) Deduce the structure equations from (b) and (c).

21. (a) If ω is a U-valued k-form and η is a V-valued l-form, and $\rho\colon U \times V \to W$, then

$$d(\rho(\omega \wedge \eta)) = \rho(d\omega \wedge \eta) + (-1)^k \rho(\omega \wedge d\eta).$$

(b) For a $\mathfrak{g}$-valued k-form ω and l-form η we have

$$[\omega \wedge \eta] = (-1)^{kl+1}[\eta \wedge \omega].$$

(c) Moreover, if λ is a $\mathfrak{g}$-valued m-form, then

$$(-1)^{km}[\omega \wedge [\eta \wedge \lambda]] + (-1)^{kl}[\eta \wedge [\lambda \wedge \omega]] + (-1)^{lm}[\lambda \wedge [\eta \wedge \omega]] = 0.$$

22. Let $G \subset \mathrm{GL}(n, \mathbb{R})$ be a Lie subgroup. The inclusion map $G \to \mathrm{GL}(n, \mathbb{R}) \to \mathbb{R}^{n^2}$ will be denoted by P (for "point"). Then dP is an $\mathbb{R}^{n^2}$-valued 1-form (it corresponds to the identity map of the tangent space of G into itself). We can also consider dP as a matrix of 1-forms; it is just the matrix (dx^{ij}), where each dx^{ij} is restricted to the tangent bundle of G. We also have the $\mathbb{R}^{n^2}$-valued 1-form (or matrix of 1-forms) $P^{-1} \cdot dP$, where $\cdot$ denotes matrix multiplication, and P^{-1} denotes the map $A \mapsto A^{-1}$ on G.

(a) $P^{-1} \cdot dP = \rho(P^{-1} \wedge dP)$, where $\rho\colon \mathbb{R}^{n^2} \times \mathbb{R}^{n^2} \to \mathbb{R}^{n^2}$ is matrix multiplication.
(b) $L_A^* dP = A \cdot dP$. (Use $f^* d = df^*$.)
(c) $P^{-1} \cdot dP$ is left invariant; and $(dP) \cdot P^{-1}$ is right invariant.
(d) $P^{-1} \cdot dP$ is the natural $\mathfrak{g}$-valued 1-form ω on G. (It suffices to check that $P^{-1} \cdot dP = \omega$ at I.)
(e) Using $dP = P \cdot \omega$, show that $0 = dP \cdot \omega + P \cdot d\omega$, where the matrix of 2-forms $P \cdot d\omega$ is computed by formally multiplying the matrices of 1-forms dP and ω. Deduce that

$$d\omega + \omega \cdot \omega = 0.$$

If ω is the matrix of 1-forms $\omega = (\omega^{ij})$, this says that

$$d\omega^{ij} = -\sum_k \omega^{ik} \wedge \omega^{kj}.$$

Check that these equations are equivalent to the equations of structure (use the form $d\omega(X, Y) = -[\omega(X), \omega(Y)]$.)

23. Let $G \subset GL(2, \mathbb{R})$ consist of all matrices $\begin{pmatrix} a & b \\ 0 & 1 \end{pmatrix}$ with $a \neq 0$. For convenience, denote the coordinates x^{11} and x^{12} on $GL(2, \mathbb{R})$ by x and y.

(a) Show that for the natural $\mathfrak{g}$-valued form ω on G we have

$$\omega = \frac{1}{x} \begin{pmatrix} dx & dy \\ 0 & 0 \end{pmatrix},$$

so that dx/x and dy/x are left invariant 1-forms on G, and a left invariant 2-form is $(dx \wedge dy)/x^2$.

(b) Find the structure constants for these forms.

(c) Show that

$$(dP) \cdot P^{-1} = \frac{1}{x} \begin{pmatrix} dx & -y\,dx + x\,dy \\ 0 & 0 \end{pmatrix}$$

and find the right invariant 2-forms.

24. (a) Show that the natural $\mathfrak{gl}(n, \mathbb{R})$-valued 1-form ω on $GL(n, \mathbb{R})$ is given by

$$\omega^{ij} = \frac{1}{\det(x^{\alpha\beta})} \sum_{k=1}^{n} y^{ik}\,dx^{kj},$$

where

$$(y^{\alpha\beta}) = \det(x^{\alpha\beta}) \cdot (x^{\alpha\beta})^{-1}.$$

(b) Show that both the left and right invariant n^2-forms are multiples of

$$\frac{1}{\left(\det(x^{\alpha\beta})\right)^n}(dx^{11} \wedge \cdots \wedge dx^{n1}) \wedge \cdots \wedge (dx^{1n} \wedge \cdots \wedge dx^{nn}).$$

25. The **special linear group** $SL(n, \mathbb{R}) \subset GL(n, \mathbb{R})$ is the set of all matrices of determinant 1.

(a) Using Problem 15, show that its Lie algebra $\mathfrak{sl}(n, \mathbb{R})$ consists of all matrices with trace $= 0$.

(b) For the case of $SL(2, \mathbb{R})$, show that

$$P^{-1} \cdot dP = \begin{pmatrix} v\,dx - y\,du & v\,dy - y\,dv \\ -u\,dx + x\,du & -u\,dy + x\,dv \end{pmatrix},$$

where we use x, y, u, v for x^{11}, x^{12}, x^{21}, x^{22}. Check that the trace is 0 by differentiating the equation $xv - yu = 1$.

(c) Show that a left invariant 3-form is

$$v\,dx \wedge du \wedge dy - y\,dx \wedge du \wedge dv.$$

26. For $M, N \in \mathfrak{o}(n) = \mathcal{L}(O(n)) = \{M : M = -M^{\mathbf{t}}\}$, define

$$\langle N, M \rangle = -\operatorname{trace} M \cdot N^{\mathbf{t}}.$$

(a) $\langle\ ,\ \rangle$ is a positive definite inner product on $\mathfrak{o}(n)$.

(b) If $A \in O(n)$, then

$$\langle \mathrm{Ad}(A)M, \mathrm{Ad}(A)N \rangle = \langle M, N \rangle.$$

$(\mathrm{Ad}(A)$ is defined in Problem 19.)

(c) The left invariant metric on $O(n)$ with value $\langle\ ,\ \rangle$ at $O(n)_I$ is also right invariant.

27. (a) If G is a compact Lie group, then $\exp\colon \mathfrak{g} \to G$ is onto. *Hint:* Use Proposition 21.

(b) Let $A \in SL(2, \mathbb{R})$. Recall that A satisfies its characteristic polynomial, so $A^2 - (\operatorname{trace} A)A + I = 0$. Conclude that trace $A^2 \geq -2$.

(c) Show that the following element of $SL(2, \mathbb{R})$ is not A^2 for any A. Conclude that it is not in the image of exp.

$$\begin{pmatrix} -2 & 0 \\ 0 & -1/2 \end{pmatrix}$$

(d) $SL(2, \mathbb{R})$ does not have a bi-invariant metric.

28. Let x be a coordinate system around e in a Lie group G, let $\pi_j\colon G \times G \to G$ be the projections, and let (y, z) be the coordinate system around (e, e) given by $y^i = x^i \circ \pi_1$, $z^i = x^i \circ \pi_2$. Define $\phi^i\colon G \times G \to \mathbb{R}$ by

$$\phi^i(a, b) = x^i(ab),$$

and let X_i be the left invariant vector field on G with

$$X_i(e) = \frac{\partial}{\partial x^i}\bigg|_e.$$

(a) Show that

$$X_i = \sum_{j=1}^n \psi_i^j \frac{\partial}{\partial x^j},$$

where

$$\psi_i^j(a) = \frac{\partial \phi^j}{\partial z^i}(a, e).$$

(b) Using $L_a L_b = L_{ab}$, show that

$$[L_{a*} X_i(b)](x^l) = [X_i(ab)](x^l).$$

Deduce that

$$X_i(b)(x^l \circ L_a) = \psi_i^l(ab),$$

and then that

$$\sum_{j=1}^n \psi_i^j(b) \cdot \frac{\partial \phi^l}{\partial z^j}(a, b) = \psi_i^l(ab).$$

Letting $\tilde{\psi} = (\tilde{\psi}_j^i)$ be the inverse matrix of $\psi = (\psi_i^j)$, we can write

$$\frac{\partial \phi^l}{\partial z^j}(a, b) = \sum_{i=1}^n \psi_i^l(ab) \cdot \tilde{\psi}_j^i(b).$$

This equation (or any of numerous things equivalent to it) is known as *Lie's first fundamental theorem*. The associativity of G is implicitly contained in it, since we used the fact that $L_a L_b = L_{ab}$.

(c) Prove the *converse of Lie's first fundamental theorem*, which states the following. Let $\phi = (\phi^1, \dots, \phi^n)$ be a differentiable function in a neighborhood of $0 \in \mathbb{R}^{2n}$ [with standard coordinate system $y^1, \dots, y^n, z^1, \dots, z^n$] such that

$$\phi(a, 0) = a \qquad \text{for } a \in \mathbb{R}^n.$$

Suppose there are differentiable functions ψ_j^i in a neighborhood of $0 \in \mathbb{R}^n$ [with standard coordinate system $x^1, \dots, x^n$] such that

$$\psi_j^i(0) = \delta_j^i$$

$$(*) \qquad \frac{\partial \phi^l}{\partial z^j}(a, b) = \sum_{i=1}^n \psi_i^l(\phi(a, b)) \cdot \tilde{\psi}_j^i(b) \qquad \begin{array}{l}\text{for } (a, b) \text{ in a neighborhood} \\ \text{of } 0 \in \mathbb{R}^{2n}.\end{array}$$

Then $(a, b) \mapsto \phi(a, b)$ is a **local Lie group structure** on a neighborhood of $0 \in \mathbb{R}^n$ (it is associative and has inverses for points close enough to 0, which serves as the identity); the corresponding left invariant vector fields are

$$X_i = \sum_{j=1}^{n} \psi_i^j \frac{\partial}{\partial x^j}.$$

[To prove associativity, note that

$$\frac{\partial \phi^l(\phi(a, b), z)}{\partial z^j} = \sum_{i=1}^{n} \psi_i^l(\phi(\phi(a, b), z)) \cdot \tilde{\psi}_j^i(z) \qquad \text{by } (*),$$

and then show that $\phi(a, \phi(b, z))$ satisfies the same equation.]

29. *Lie's second fundamental theorem* states that the left invariant vector fields X_i of a Lie group G satisfy

$$[X_i, X_j] = \sum_{k=1}^{n} C_{ij}^k X_k$$

for certain *constants* C_{ij}^k—in other words, the bracket of two left invariant vector fields is left invariant. The aim of this problem is to prove the *converse of Lie's second fundamental theorem*, which states the following: A Lie algebra of vector fields on a neighborhood of $0 \in \mathbb{R}^n$, which is of dimension n over $\mathbb{R}$ and contains a basis for $\mathbb{R}^n{}_0$, is the set of left invariant vector fields for some local Lie group structure on a neighborhood of $0 \in \mathbb{R}^n$.

(a) Choose $X_1, \ldots, X_n$ in the Lie algebra so that $X_i(0) = \partial/\partial x^i|_0$ and set

$$X_i = \sum_{j=1}^{n} \psi_i^j \frac{\partial}{\partial x^j}.$$

If

$$\omega^i = \sum_{j=1}^{n} \tilde{\psi}_j^i \, dx^j,$$

then the ω^i are the dual forms, and consequently

$$d\omega^k = -\sum_{i<j} C_{ij}^k \, \omega^i \wedge \omega^j \qquad C_{ij}^k \text{ constants.}$$

(b) Let $\pi_j : \mathbb{R}^n \times \mathbb{R}^n \to \mathbb{R}^n$ be the projections. Then

$$\pi_2{}^* \omega^j - \pi_1{}^* \omega^j = \sum_{l=1}^{n} (\tilde{\psi}_l^j \circ \pi_2) \left[d(x^l \circ \pi_2) - \sum_{i=1}^{n} (\psi_i^l \circ \pi_2) \cdot \pi_1{}^* \omega^i \right].$$

Consequently, the ideal generated by the forms $d(x^l \circ \pi_2) - \sum_{i=1}^{n}(\psi_i^l \circ \pi_2) \cdot \pi_1{}^* \omega^i$ is the same as the ideal $\mathcal{I}$ generated by the forms $\pi_2{}^* \omega^j - \pi_1{}^* \omega^j$. Using the fact that the C_{jk}^i are constants, show that $d(\mathcal{I}) \subset \mathcal{I}$. Hence $\mathbb{R}^n \times \mathbb{R}^n$ is foliated by n-dimensional manifolds on which the forms $d(x^l \circ \pi_2) - \sum_{i=1}^{n}(\psi_i^l \circ \pi_2) \cdot \pi_1{}^* \omega^i$ all vanish.

(c) Conclude, as in the proof of Theorem 17, that for fixed a, there is a function $\Phi_a : \mathbb{R}^n \to \mathbb{R}^n$ satisfying $\Phi_a(0) = a$ and

$$d\Phi_a^l(b) = \sum_{i=1}^{n} \psi_i^l(\Phi_a(b)) \cdot \omega^i(b),$$

or equivalently,

$$\frac{\partial \Phi_a^l}{\partial x^j}(b) = \sum_{i=1}^{n} \psi_i^l(\Phi_a(b)) \cdot \tilde{\psi}_j^i(b).$$

Now set $\phi(a, b) = \Phi_a(b)$, and use the converse of Lie's first fundamental theorem.

30. *Lie's third fundamental theorem* states that the C_{jk}^i satisfy equations (1) and (2) on page 396, i.e., that the left invariant vector fields form a Lie algebra under $[\ ,\ ]$. The aim of this problem is to prove the *converse of Lie's third fundamental theorem*, which states that any n-dimensional Lie algebra is the Lie algebra for some local Lie group in a neighborhood of $0 \in \mathbb{R}^n$.

Let C_{ij}^k be constants satisfying equations (1) and (2) on page 396. We would like to find vector fields $X_1, \ldots, X_n$ on a neighborhood of $0 \in \mathbb{R}^n$ such that $[X_i, X_j] = \sum_{k=1}^{n} C_{ij}^k X_k$. Equivalently, we want to find forms ω^i with

$$d\omega^k = -\sum_{i<j} C_{ij}^k \, \omega^i \wedge \omega^j.$$

Then the result will follow from the converse of Lie's second fundamental theorem.

(a) Let h_r^k be functions on $\mathbb{R} \times \mathbb{R}^n$ such that

$$\frac{\partial h_r^k}{\partial t} = \delta_r^k - \sum_{i,j} C_{ij}^k x^i h_r^j$$

$$h_r^k(0, x) = 0.$$

These are equations "depending on the parameters x" (see Problem 5-5(b)). Note that $h_r^k(t, 0) = \delta_r^k t$, so that $h_r^k(1, 0) = \delta_r^k$. Let σ^k be the 1-form on $\mathbb{R} \times \mathbb{R}^n$ defined by

$$\sigma^k = \sum_r h_r^k \, dx^r,$$

and write
$$d\sigma^k = \lambda^k + (dt \wedge \alpha^k),$$

where λ^k and α^k do not involve dt. Show that

$$\lambda^k = \sum_{i<j} \left(\frac{\partial h_j^k}{\partial x^i} - \frac{\partial h_i^k}{\partial x^j} \right) dx^i \wedge dx^j$$

$$\alpha^k = dx^k - \sum_{i,j} C_{ij}^k x^i \sigma^j.$$

(b) Show that

$$d\lambda^k = dt \wedge \left(-\sum_{i,j} C_{ij}^k \, dx^i \wedge \sigma^j - \sum_{i,j} C_{ij}^k \, x^k \lambda^j \right).$$

(c) Let

$$\theta^k = \lambda^k + \frac{1}{2} \sum_{i,j} C_{ij}^k \, \sigma^i \wedge \sigma^j.$$

Show that

$$d\theta^k = dt \wedge \left(-\sum_{i,j} C_{ij}^k \, x^i \lambda^j - \sum_{i,j} \sum_{r,s} C_{ij}^k C_{rs}^i \, x^r \sigma^s \wedge \sigma^j \right)$$
$$+ \text{ terms not involving } dt.$$

Using

$$\sum_{i,j} \sum_{r,s} C_{ij}^k C_{rs}^i \, \sigma^s \wedge \sigma^j = \frac{1}{2} \sum_{i,j} \sum_{r,s} (C_{ij}^k C_{rs}^i - C_{is}^k C_{rj}^i) \, \sigma^s \wedge \sigma^j$$

$$= \frac{1}{2} \sum_{i,j} \sum_{r,s} (C_{ij}^k C_{rs}^i + C_{is}^k C_{jr}^i) \, \sigma^s \wedge \sigma^j,$$

and equation (2) on page 396, show that

$$d\theta^k = dt \wedge \left(-\sum_{i,j} C_{ij}^k \, x^i \lambda^j + \frac{1}{2} \sum_{i,j} \sum_{r,s} C_{ir}^k C_{sj}^i \, x^r \, \sigma^s \wedge \sigma^j \right)$$
$$+ \text{ terms not involving } dt.$$

Finally deduce that

$$d\theta^k = dt \wedge -\sum_{j,l} C_{jl}^k \, x^j \theta^l + \text{ terms not involving } dt.$$

(d) We can write

$$\theta^k = \sum_{i<j} g^k_{ij} \, dx^i \wedge dx^j,$$

where $g^k_{ij}(0, x) = 0$ (Why?). Using (c), show that

$$\frac{\partial g^k_{ij}}{\partial t} = -\sum_{r,s} C^k_{rs} \, x^r \, g^s_{ij}.$$

Conclude that $\theta^k = 0$.

(e) We now have

$$\lambda^k = -\frac{1}{2} \sum_{i,j} C^k_{ij} \, \sigma^i \wedge \sigma^j,$$

$$d\sigma^k = -\frac{1}{2} \sum_{i,j} C^k_{ij} \, \sigma^i \wedge \sigma^j + (dt \wedge \alpha^k).$$

Show that the forms $\omega^k(x) = \sigma^k(1, x)$ satisfy

$$d\omega^k = -\frac{1}{2} \sum_{i,j} C^k_{ij} \, \omega^i \wedge \omega^j.$$

CHAPTER 11

EXCURSION IN THE REALM
OF ALGEBRAIC TOPOLOGY

This chapter explores further properties of the de Rham cohomology vector spaces of a manifold. Our main results will be restatements, in terms of the de Rham cohomology, of fundamental properties of the ordinary cohomology which is studied in algebraic topology. Because we deal only with manifolds, many of the proofs become significantly easier. On the other hand, we will be using some of the main tools of algebraic topology, thus retaining much of the flavor of that subject. Along the way we will deduce all sorts of interesting consequences, including a theorem about the possibility of imbedding n-manifolds in $\mathbb{R}^{n+1}$.

Let M be a manifold with $M = U \cup V$ for open sets $U, V \subset M$. Before examining the cohomology of M we will simply look at the vector space $C^k(M)$ of k-forms on M. Let

$$i_U : U \to M \qquad\qquad i_V : V \to M$$
$$j_U : U \cap V \to U \qquad\qquad j_V : U \cap V \to V$$

be the inclusions. Then we have two linear maps α and β,

$$C^k(M) \xrightarrow{\ \alpha = i_U{}^* \oplus i_V{}^*\ } C^k(U) \oplus C^k(V) \xrightarrow{\ \beta = j_U{}^* - j_V{}^*\ } C^k(U \cap V)$$

defined by

$$\alpha(\omega) = (i_U{}^*(\omega), i_V{}^*(\omega)) \qquad \beta(\lambda_1, \lambda_2) = j_U{}^*(\lambda_1) - j_V{}^*(\lambda_2).$$

Here $i_U{}^*(\omega)$ is just the restriction of ω to U, etc. Clearly $\beta \circ \alpha = 0$. In other words, image $\alpha \subset \ker \beta$. Moreover, the converse holds: $\ker \beta \subset \text{image} \, \alpha$. For, if $\beta(\lambda_1, \lambda_2) = 0$, then $\lambda_1 = \lambda_2$ on $U \cap V$, so we can define ω on M to be λ_1 on U and λ_2 on V, and then $\alpha(\omega) = (\lambda_1, \lambda_2)$. The equation image $\alpha = \ker \beta$ is expressed by saying that the above diagram is **exact** at the middle vector space. We can extend this diagram by putting the vector space containing only 0 at the ends; the arrows at either end of the following sequence are the only possible linear maps.

1. LEMMA. The sequence

$$0 \to C^k(M) \xrightarrow{\alpha} C^k(U) \oplus C^k(V) \xrightarrow{\beta} C^k(U \cap V) \to 0$$

is exact at all places.

PROOF. It is clear that α is one-one. This is equivalent to exactness at $C^k(M)$, since the image of the first map is $\{0\} \subset C^k(M)$. Similarly, exactness at $C^k(U \cap V)$ is equivalent to β being onto. To prove that β is onto, let $\{\phi_U, \phi_V\}$ be a partition of unity subordinate to $\{U, V\}$. Then $\omega \in C^k(U \cap V)$ is

$$\omega = \beta(\phi_V \omega, -\phi_U \omega),$$

where $\phi_V \omega$ denotes the form equal to $\phi_V \omega$ on $U \cap V$, and equal to 0 on $U - (U \cap V)$. ❖

By putting in the maps d, we can expand our diagram as follows,

so that the rows are all exact. It is easy to check that this diagram **commutes**, that is, any two compositions from one vector space to another are equal:

$$(d \oplus d) \circ \alpha = \alpha \circ d$$

$$d \circ \beta = \beta \circ (d \oplus d)$$

Our first main theorem depends only on the simple algebraic structure inherent in this diagram. To isolate this purely algebraic structure, we make the following definitions. A **complex** C is a sequence of vector spaces C^k, $k = 0, 1, 2, \ldots$, together with a sequence of linear maps

$$d^k : C^k \to C^{k+1}$$

satisfying $d^{k+1} \circ d^k = 0$, or briefly, $d^2 = 0$. A **map** $\alpha : C_1 \to C_2$ between complexes is a sequence of linear maps

$$\alpha^k : C_1{}^k \to C_2{}^k$$

such that the following diagram commutes for all k.

$$
\begin{array}{ccc}
C_1{}^k & \xrightarrow{\ \alpha^k\ } & C_2{}^k \\
\downarrow{\scriptstyle d_1{}^k} & & \downarrow{\scriptstyle d_2{}^k} \\
C_1{}^{k+1} & \xrightarrow{\ \alpha^{k+1}\ } & C_2{}^{k+1}
\end{array}
$$

The most important examples of complexes are obtained by choosing $C^k = C^k(M)$ for some manifold M, with d^k the operator d on k-forms. Another example, implicit in our discussion, is the **direct sum** $C = C_1 \oplus C_2$ of two complexes, defined by

$$C^k = C_1{}^k \oplus C_2{}^k, \qquad d^k = d_1{}^k \oplus d_2{}^k.$$

For any complex C we can define the **cohomology vector spaces** of C by

$$H^k(C) = \frac{\ker d^k}{\operatorname{image} d^{k-1}}.$$

Naturally, if $C = \{C^k(M)\}$, then $H^k(C)$ is just $H^k(M)$. If $\alpha : C_1 \to C_2$ is a map between complexes, then we have a map, also denoted by α,

$$\alpha : H^k(C_1) \to H^k(C_2).$$

To define α we note that every element of $H^k(C_1)$ is determined by some $x \in C_1{}^k$ with $d_1{}^k(x) = 0$. Commutativity of the above diagram shows that $d_2{}^k(\alpha^k(x)) = \alpha^{k+1} d_1{}^k(x) = 0$, so $\alpha^k(x)$ determines an element of $H^k(C_2)$, which we define to be α(the class determined by x). This map is well-defined, for if we change x to $x + d_1{}^{k-1}(y)$ for some $y \in C_1{}^{k-1}$, then $\alpha^k(x)$ is changed to

$$\alpha^k(x + d_1{}^{k-1}(y)) = \alpha^k(x) + \alpha^k(d_1{}^{k-1}(y))$$
$$= \alpha^k(x) + d_2{}^{k-1}(\alpha^{k-1}(y)),$$

which determines the same element of $H^k(C_2)$. When $C_1{}^k = C^k(M)$, $C_2{}^k = C^k(N)$, and $\alpha \colon C^k(M) \to C^k(N)$ is f^* for $f \colon N \to M$, then this map is just $f^* \colon H^k(M) \to H^k(N)$.

Now suppose that we have an exact sequence of complexes

$$0 \to C_1 \xrightarrow{\alpha} C_2 \xrightarrow{\beta} C_3 \to 0,$$

which really means a vast commutative diagram in which all rows are exact.

What does this imply about the maps $\alpha \colon H^k(C_1) \to H^k(C_2)$ and $\beta \colon H^k(C_2) \to H^k(C_3)$? The nicest thing that could happen would be for the following diagram to be exact:

$$0 \to H^k(C_1) \xrightarrow{\alpha} H^k(C_2) \xrightarrow{\beta} H^k(C_3) \to 0.$$

This is *not* true. For example, if U and V are overlapping portions of S^2 for which there is a deformation retraction of $U \cap V$ into S^1, then we have an exact sequence

$$0 \to C^k(S^2) \to C^k(U) \oplus C^k(V) \to C^k(U \cap V) \to 0,$$

but *not* an exact sequence

$$0 \longrightarrow H^1(S^2) \longrightarrow H^1(U) \oplus H^1(V) \longrightarrow H^1(U \cap V) \longrightarrow 0.$$

$$\wr\wr \qquad\qquad\qquad \wr\wr \qquad\qquad\qquad \wr\wr$$

$$0 \qquad\qquad\qquad\qquad 0 \qquad\qquad\qquad\qquad \mathbb{R}$$

Nevertheless, something very nice is true:

2. THEOREM. If $0 \to C_1 \xrightarrow{\alpha} C_2 \xrightarrow{\beta} C_3 \to 0$ is a short exact sequence of complexes, then there are linear maps

$$\delta^k : H^k(C_3) \to H^{k+1}(C_1)$$

so that the following infinitely long sequence is exact (everywhere):

$$0 \longrightarrow H^0(C_1) \xrightarrow{\alpha} H^0(C_2) \xrightarrow{\beta} H^0(C_3) \xrightarrow{\delta} H^1(C_1) \longrightarrow \cdots$$

$$\cdots \longrightarrow H^k(C_1) \xrightarrow{\alpha} H^k(C_2) \xrightarrow{\beta} H^k(C_3) \xrightarrow{\delta} H^{k+1}(C_1) \longrightarrow \cdots$$

PROOF. Throughout the proof, diagram (∗) should be kept at hand. Let $x \in C_3{}^k$ with $d_3{}^k(x) = 0$. By exactness of the middle row of (∗), there is $y \in C_2{}^k$ with $\beta^k(y) = x$. Then

$$0 = d_3{}^k(x) = d_3{}^k \beta^k(y) = \beta^{k+1} d_2{}^k(y).$$

So $d_2{}^k(y) \in \ker \beta^{k+1} = \text{image } \alpha^{k+1}$; thus $d_2{}^k(y) = \alpha^{k+1}(z)$ for some (unique) $z \in C_1{}^{k+1}$. Moreover,

$$\alpha^{k+1} d_1{}^{k+1}(z) = d_2{}^{k+1} \alpha^{k+1}(z) = d_2{}^{k+1} d_2{}^k(y) = 0.$$

Since α^{k+1} is one-one, this implies that $d_1{}^{k+1}(z) = 0$, so z determines an element of $H^{k+1}(C_1)$; this element is defined to be δ^k of the element of $H^k(C_3)$ determined by x.

In order to prove that δ^k is well-defined, we must check that the result does not depend on the choice of $x \in C_3{}^k$ representing the element of $H^k(C_3)$. So we have to show that we obtain $0 \in H^{k+1}(C_1)$ if we start with an element of the form $d_3{}^{k-1}(x')$ for $x' \in C_3{}^{k-1}$. In this case, let $x' = \beta^{k-1}(y')$. Then

$$x = d_3{}^{k-1}(x') = d_3{}^{k-1} \beta^{k-1}(y') = \beta^k d_2{}^{k-1}(y'),$$

so we choose $d_2{}^{k-1}(y')$ as y. This means that $d_2{}^k(y) = 0$, and hence $z = 0$.

It is also necessary to check that our definition is independent of the choice of y with $\beta^k(y) = x$; this is left to the reader.

The proof that the sequence is exact consists of 6 similar diagram chases. We will supply the proof that $\ker \alpha \subset \operatorname{image} \delta$. Let $x \in C_1{}^k$ satisfy $d_1{}^k(x) = 0$, and suppose that $\alpha^k(x) \in C_2{}^k$ represents $0 \in H^k(C_2)$. This means that $\alpha^k(x) = d_2{}^{k-1}(y)$ for some $y \in C_2{}^{k-1}$. Now

$$d_3{}^{k-1}\beta^{k-1}(y) = \beta^k d_2{}^{k-1}(y) = \beta^k \alpha^k(x) = 0.$$

So $\beta^{k-1}(y)$ represents an element of $H^{k-1}(C_3)$. Moreover, the definition of δ immediately shows that the image of this element under δ is precisely the class represented by x. ❖

It is a worthwhile exercise to check that the main step in the proof of Theorem 8-16 is precisely the proof that $\ker \alpha \subset \operatorname{image} \delta$, together with the first part of the proof that δ is well-defined. All of Theorem 8-16 can be derived directly from the following corollary of Lemma 1 and Theorem 2.

3. THEOREM (THE MAYER-VIETORIS SEQUENCE). If $M = U \cup V$, where U and V are open, then we have an exact sequence (eventually ending in 0's):

$$0 \to H^0(M) \to \cdots \to H^k(M) \to H^k(U) \oplus H^k(V) \to H^k(U \cap V) \overset{\delta}{\to} H^{k+1}(M) \to \cdots .$$

As several of the Problems show, the cohomology of nearly everything can be computed by a suitable application of the Mayer-Vietoris sequence. As a simple example, we consider the torus $T = S^1 \times S^1$, and the open sets U and V illustrated below. Since there is a deformation retraction of U and V

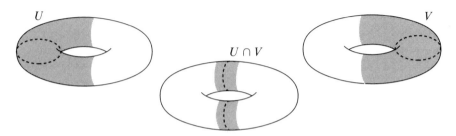

onto circles, and a deformation retraction of $U \cap V$ onto 2 circles, the Mayer-Vietoris sequence is

$$\longrightarrow H^0(T) \longrightarrow H^0(U) \oplus H^0(V) \longrightarrow H^0(U \cap V) \longrightarrow H^1(T) \longrightarrow H^1(U) \oplus H^1(V) \longrightarrow$$

$$\wr\wr \qquad \wr\wr$$
$$\mathbb{R} \qquad \mathbb{R}$$

$$\longrightarrow H^1(U \cap V) \longrightarrow H^2(T) \longrightarrow 0.$$

$$\wr\wr \qquad \wr\wr$$
$$\mathbb{R} \oplus \mathbb{R} \qquad \mathbb{R}$$

The map $H^1(U \cap V) \to H^2(T)$ is not 0 (it is onto $H^2(T)$), so its kernel is 1-dimensional. Thus the image of the map $H^1(U) \oplus H^1(V) \to H^1(U \cap V)$ is 1-dimensional. So the kernel of *this* map is 1-dimensional, and consequently the map $H^1(T) \to H^1(U) \oplus H^1(V)$ has a 1-dimensional image. Similar reasoning shows that this map also has a 1-dimensional kernel. It follows that dim $H^1(T) = 2$. The reasoning used here can fortunately be systematized.

4. PROPOSITION. If the sequence

$$0 \to V_1 \xrightarrow{\alpha} V_2 \to \cdots \to V_{k-1} \to V_k \to 0$$

is exact, then

$$0 = \dim V_1 - \dim V_2 + \dim V_3 - \cdots + (-1)^{k-1} \dim V_k.$$

PROOF. By induction on k. For $k = 1$ we have the sequence

$$0 \to V_1 \to 0.$$

Exactness means that $\{0\} \subset V_1$ is the kernel of the map $V_1 \to 0$, which implies that $V_1 = 0$.

Assume the theorem for $k - 1$. Since the map $V_2 \to V_3$ has kernel $\alpha(V_1)$, it induces a map $V_2/\alpha(V_1) \to V_3$. Moreover, this map is one-one. So we have an exact sequence of $k - 1$ vector spaces

$$0 \to V_2/\alpha(V_1) \to V_3 \to \cdots \to V_{k-1} \to V_k \to 0;$$

hence

$$0 = \dim V_2/\alpha(V_1) - \dim V_3 + \cdots$$
$$= -\dim V_1 + \dim V_2 - \dim V_3 + \cdots,$$

which proves the theorem for k. ❖

Rather than compute the cohomology of other manifolds, we will use the Mayer-Vietoris sequence to relate the dimensions of $H^k(M)$ to an entirely different set of numbers, arising from a "triangulation" of M, a new structure which we will now define.

The **standard n-simplex** Δ_n is defined as the set

$$\Delta_n = \left\{ x \in \mathbb{R}^{n+1} : 0 \le x^i \le 1 \text{ and } \sum_{i=1}^{n+1} x^i = 1 \right\}.$$

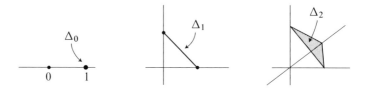

(In Problem 8-5, Δ_n is defined to be a different, although homeomorphic, set.) The subset of Δ_n obtained by setting $n - k$ of the coordinates x^i equal to 0 is homeomorphic to Δ_k, and is called a **k-face** of Δ_n. If $\Delta \subset M$ is a diffeomorphic image of some Δ_m, then the image of a k-face of Δ_m is called a **k-face** of Δ. Now by a **triangulation** of a compact n-manifold M we mean a finite collection $\{\sigma^n_i\}$ of diffeomorphic images of Δ_n which cover M and which satisfy the following condition:

> If $\sigma^n_i \cap \sigma^n_j \neq \emptyset$, then for some k the intersection $\sigma^n_i \cap \sigma^n_j$ is a k-face of both σ^n_i and σ^n_j.

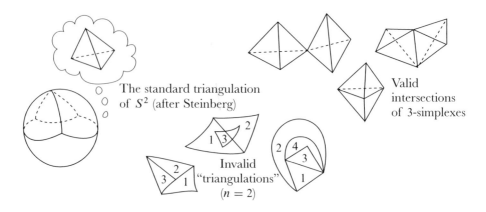

It is a difficult theorem that every C^∞ manifold has a triangulation; for a proof see Munkres, *Elementary Differential Topology*, or Whitney, *Geometric Integration*

Theory. Assuming that our manifold M has a triangulation $\{\sigma^n{}_i\}$ we will call each $\sigma^n{}_i$ an *n*-**simplex of the triangulation**; any k-face of any $\sigma^n{}_i$ will be called a k-**simplex of the triangulation**. We let α_k be the number of these k-simplexes.

Now let U be the disjoint union of open balls, one within each n-simplex $\sigma^n{}_i$, and let V_{n-1} be the complement of the set consisting of the centers of these balls, so that V_{n-1} is a neighborhood of the union of all $(n-1)$-simplexes of M. Then

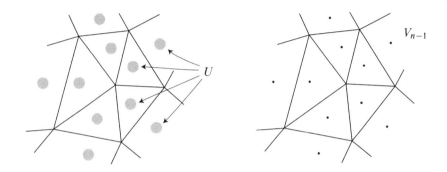

$M = U \cup V_{n-1}$ where $U \cap V_{n-1}$ has the same cohomology as a disjoint union of α_n copies of S^{n-1}. Consider first the case where $n > 2$. The Mayer-Vietoris sequence breaks into pieces:

(1) $\qquad 0 \longrightarrow H^0(M) \longrightarrow H^0(U) \oplus H^0(V_{n-1}) \longrightarrow H^0(U \cap V_{n-1}) \longrightarrow H^1(M)$

$$\longrightarrow \underset{\overset{\|}{0}}{H^1(U) \oplus H^1(V_{n-1})} \longrightarrow \underset{\overset{\|}{0}}{H^1(U \cap V_{n-1})}$$

(2) $\qquad$ For $1 < k < n-1$,

$$H^{k-1}(U \cap V_{n\ 1}) \to \underset{\overset{\|}{0}}{H^k(M)} \to \underset{\overset{\|}{0}}{H^k(U) \oplus H^k(V_{n-1})} \to \underset{\overset{\|}{0}}{H^k(U \cap V_{n-1})}$$

(3) $\qquad H^{n-2}(U \cap V_{n-1}) \longrightarrow \underset{\overset{\|}{0}}{H^{n-1}(M)} \longrightarrow \underset{\overset{\|}{0}}{H^{n-1}(U) \oplus H^{n-1}(V_{n-1})} \longrightarrow$

$$\longrightarrow H^{n-1}(U \cap V_{n-1}) \longrightarrow H^n(M) \longrightarrow \underset{\overset{\|}{0}}{H^n(U) \oplus H^n(V_{n-1})}$$

Applying Proposition 4 to these pieces yields

$$\dim H^k(V_{n-1}) = \dim H^k(M) \qquad\qquad 0 \le k \le n-2$$
$$\dim H^{n-1}(V_{n-1}) = \dim H^{n-1}(M) - \dim H^n(M) + \alpha_n.$$

For the case $n = 2$ we easily obtain the same result without splitting up the sequence. We now introduce the **Euler characteristic** $\chi(M)$ of M, defined by

$$\chi(M) = \dim H^0(M) - \dim H^1(M) + \dim H^2(M) - \cdots + (-1)^n \dim H^n(M).$$

This makes sense for any manifold in which all $H^k(M)$ are finite dimensional; we anticipate here a later result that $H^k(M)$ is finite dimensional whenever M is compact. The above equations then imply that

$$\chi(V_{n-1}) = \sum_{k=0}^{n-1}(-1)^k \dim H^k(V_{n-1})$$

$$= \sum_{k=0}^{n-2}(-1)^k \dim H^k(M)$$
$$\qquad + (-1)^{n-1}[\dim H^{n-1}(M) - \dim H^n(M) + \alpha_n]$$
$$= \chi(M) - (-1)^n \alpha_n,$$

or

$$\chi(M) = \chi(V_{n-1}) + (-1)^n \alpha_n.$$

5. THEOREM. For any triangulation of a compact manifold M we have

$$\chi(M) = \alpha_0 - \alpha_1 + \alpha_2 - \cdots + (-1)^n \alpha_n.$$

PROOF. In the manifold V_{n-1} we define a new open set U which consists of a disjoint union of sets diffeomorphic to $\mathbb{R}^n$, one for each $(n-1)$-face, joining the balls of the old U.

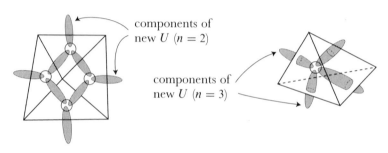

components of
new U $(n = 2)$

components of
new U $(n = 3)$

We will let V_{n-2} be the complement of arcs, in the new U, joining the centers of the balls in the old U.

V_{n-2} is the
complement of

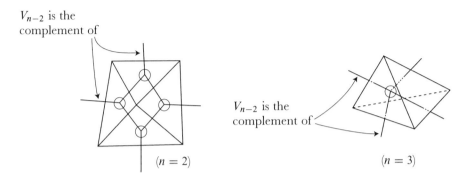

V_{n-2} is the
complement of

$(n = 2)$ $(n = 3)$

An argument precisely like that which proves the equation

$$\chi(M) = \chi(V_{n-1}) + (-1)^n \alpha_n$$

also shows that

$$\chi(V_{n-1}) = \chi(V_{n-2}) + (-1)^{n-1} \alpha_{n-1}.$$

Similarly, we introduce $V_{n-3}, \ldots, V_0$; the last of these is a disjoint union of α_0 sets each of which is smoothly contractible to a point. Hence $\chi(V_0) = \alpha_0$, while in all other cases we have

$$\chi(V_k) = \chi(V_{k-1}) + (-1)^k \alpha_k.$$

Combining these equations, we have

$$\begin{aligned}
\chi(M) &= \chi(V_{n-1}) + (-1)^n \alpha_n \\
&= \chi(V_{n-2}) + [(-1)^{n-1} \alpha_{n-1} + (-1)^n \alpha_n] \\
&\;\;\vdots \\
&= \chi(V_0) + [(-1)^1 \alpha_1 + \cdots + (-1)^n \alpha_n] \\
&= \alpha_0 - \alpha_1 + \cdots + (-1)^n \alpha_n. \;\; \clubsuit
\end{aligned}$$

6. COROLLARY (DESCARTES-EULER). If a convex polyhedron has V vertices, E edges, and F faces, then

$$V - E + F = 2.$$

If we turn from H^k to H_c^k we encounter a very different situation. If $U \subset M$ is open, a form ω with compact support $\subset M$ may not restrict to a form with compact support $\subset U$; the inclusion map of U into M is not proper. On the

other hand, if ω is a form with compact support $\subset U$, then ω can be extended to M by letting it be 0 outside U; we will denote this extended form by

$$i_U{'}(\omega).$$

If $C_c^k(M)$ denotes the vector space of k-forms with compact support on M, we can define a new sequence.

7. **LEMMA.** The sequence

$$0 \to C_c^k(U \cap V) \xrightarrow{\; j_U{'} \oplus -j_V{'} \;} C_c^k(U) \oplus C_c^k(V) \xrightarrow{\; i_U{'} + i_V{'} \;} C_c^k(M) \to 0$$

is exact.

PROOF. It is clear that $j_U{'} \oplus -j_V{'}$ is one-one; in fact, each map $j_U{'}$ and $j_V{'}$ is one-one.

To prove that $i_U{'} + i_V{'}$ is onto, let ω be a k-form with compact support on M, and let $\{\phi_U, \phi_V\}$ be a partition of unity for the cover $\{U, V\}$. Then

$$\omega = \phi_U \omega + \phi_V \omega$$

is clearly the image of $(\phi_U \omega, \phi_V \omega) \in C_c^k(U) \oplus C_c^k(V)$.

It is clear that image $(j_U{'} \oplus -j_V{'}) \subset \ker(i_U{'} + i_V{'})$. To prove the converse, suppose that

$$(\lambda_1, \lambda_2) \in C_c^k(U) \oplus C_c^k(V) \quad \text{satisfies} \quad i_U{'}(\lambda_1) + i_V{'}(\lambda_2) = 0.$$

This means that $\lambda_1 = -\lambda_2$. Since support $\lambda_1 \subset U$ and support $\lambda_2 \subset U$, this shows that support $\lambda_1 \subset U \cap V$ and support $\lambda_2 \subset U \cap V$. So (λ_1, λ_2) is the image of $\lambda_1 \in C_c^k(U \cap V)$. ❖

8. THEOREM (MAYER-VIETORIS FOR COMPACT SUPPORTS). If the manifold $M = U \cup V$ for U, V open in M, then there is a long exact sequence

$$\cdots \to H_c^k(U \cap V) \to H_c^k(U) \oplus H_c^k(V) \to H_c^k(M) \xrightarrow{\delta} H_c^{k+1}(U \cap V) \to \cdots .$$

PROOF. Apply Theorem 2 to the short exact sequence of complexes given by the Lemma. ❖

This sequence is much harder to work with than the Mayer-Vietoris sequence. For example, suppose we want to find H_c^k for $\mathbb{R}^n - \{0\}$, which is diffeomorphic to $S^{n-1} \times \mathbb{R}$. If we write $S^n = U \cup V$ in the usual way, so that $U \cap V$ is diffeomorphic to $S^{n-1} \times \mathbb{R}$, then $S^n \times \mathbb{R} = (U \times \mathbb{R}) \cup (V \times \mathbb{R})$, where $(U \times \mathbb{R}) \cap (V \times \mathbb{R})$ is diffeomorphic to $S^{n-1} \times \mathbb{R}^2$. The only way to use induction is to find H_c^k for all $S^n \times \mathbb{R}^m$, starting with $S^1 \times \mathbb{R}^m$. The details will be left to the reader; we will merely record one further result, for later use, and then proceed to yet another application of Theorem 2.

9. COROLLARY. If $M = U \cup V$ for U, V open in M, then there is a dual long exact sequence

$$\cdots \to H_c^{k+1}(U \cap V)^* \to H_c^k(M)^* \to [H_c^k(U) \oplus H_c^k(V)]^* \to H_c^k(U \cap V)^* \to \cdots .$$

PROOF. We just have to show that if the sequence of linear maps

$$W_1 \xrightarrow{\alpha} W_2 \xrightarrow{\beta} W_3$$

is exact at W_2, then so is the sequence of dual maps and spaces

$$W_3{}^* \xrightarrow{\beta^*} W_2{}^* \xrightarrow{\alpha^*} W_1{}^* .$$

For any $\lambda \in W_3{}^*$ we have

$$\alpha^* \beta^*(\lambda) = \alpha^*(\lambda \circ \beta) = \lambda \circ (\beta \circ \alpha) = \lambda \circ 0 = 0.$$

So $\alpha^* \circ \beta^* = 0$.

Now suppose $\lambda \in W_2{}^*$ satisfies $\alpha^*(\lambda) = 0$. Then $\lambda \circ \alpha = 0$. We claim that

there is $\bar{\lambda} \colon W_3 \to \mathbb{R}$ with $\lambda = \beta^*(\bar{\lambda})$, i.e., $\lambda = \beta \circ \bar{\lambda}$. Given a $w \in W_3$ which is of the form $\beta(w')$, we define

$$\bar{\lambda}(w) = \lambda(\beta').$$

This makes sense, for if $\beta(w') = \beta(w'')$, then $w - w'' = \alpha(z)$ for some z, so $\lambda(w) - \lambda(w'') = \lambda\alpha(z) = 0$. This defines $\bar{\lambda}$ on $\beta(W_2) \subset W_3$. Now choose $W \subset W_3$ with $W_3 = \beta(W_2) \oplus W$, and define $\bar{\lambda}$ to be 0 on W. ❖

We now consider a rather different situation. Let $N \subset M$ be a *compact* sub-manifold of M. Then $M - N$ is also a manifold. We therefore have the sequence

$$C_c^k(M - N) \xrightarrow{e} C_c^k(M) \xrightarrow{i^*} C^k(N),$$

where e is "extension". This sequence is *not* exact at $C_c^k(M)$: the kernel of i^* contains all $\omega \in C_c^k(M)$ which are 0 on N, while the image of e contains all $\omega \in C_c^k(M)$ which are 0 in a neighborhood of N.

To circumvent this difficulty, we will have to use a technical device. We appeal first to a result from the Addendum to Chapter 9. There is a compact neighborhood V of N and a map $\pi \colon V \to N$ such that V is a manifold-with-boundary, and if $j \colon N \to V$ is the inclusion, then $\pi \circ j$ is the identity of N, while $j \circ \pi$ is smoothly homotopic to the identity of V. We now construct a sequence of such neighborhoods $V = V_1 \supset V_2 \supset V_3 \supset \cdots$ with $\bigcap_i V_i = N$.

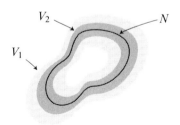

Now consider two forms $\omega_i \in C^k(V_i)$, $\omega_j \in C^k(V_j)$. We will call ω_i and ω_j *equivalent* if there is $l > i, j$ such that

$$\omega_i | V_l = \omega_j | V_l.$$

It is clear that we can make the set of all equivalence classes into a vector space $\mathcal{G}^k(N)$, the "germs of k-forms in a neighborhood of M". Moreover, it is easy to define $d \colon \mathcal{G}^k(N) \to \mathcal{G}^{k+1}(N)$, so that we obtain a complex $\mathcal{G}$. Finally, we define a map of complexes

$$C_c^k(M) \xrightarrow{i^*} \mathcal{G}^k(N)$$

in the obvious way: $\omega \mapsto$ the equivalence class of any $\omega | V_i$.

10. LEMMA. The sequence

$$0 \to C_c^k(M - N) \xrightarrow{e} C_c^k(M) \xrightarrow{i^*} \mathcal{G}^k(N) \to 0$$

is exact.

PROOF. Clearly e is one-one.

If $\omega \in C_c^k(M - N)$, then $\omega = 0$ in some neighborhood U of N. Since N is compact and $\bigcap_i V_i = N$, there is some i such that $V_i \subset U$, and consequently $\omega = 0$ on V_i. This means that $i^*e(\omega) = 0$. Conversely, suppose $\lambda \in C_c^k(M)$ satisfies $i^*(\lambda) = 0$. By definition of $\mathcal{G}^k(N)$, this means that $\lambda|V_i = 0$ for some i. Hence $\lambda|M - N$ has compact support $\subset M - N$, and $\lambda = e(\lambda|M - N)$.

Finally, any element of $\mathcal{G}^k(N)$ is represented by a form η on some V_i. Let $f : M \to [0,1]$ be a C^∞ function which is 1 on V_{i+1}, having support $f \subset$ interior V_i. Then $f\eta \in C_c^k(M)$, and $f\eta$ represents the same element of $\mathcal{G}^k(N)$ as η; consequently this element is $i^*(f\eta)$. ❖

11. LEMMA. The cohomology vector spaces $H^k(\mathcal{G})$ of the complex $\{\mathcal{G}^k(N)\}$ are isomorphic to $H^k(N)$ for all k.

PROOF. This follows easily from the fact that $j^* : H^k(V_i) \to H^k(N)$ is an isomorphism for each V_i. Details are left to the reader. ❖

12. THEOREM (THE EXACT SEQUENCE OF A PAIR). If $N \subset M$ is a compact submanifold of M, then there is an exact sequence

$$\cdots \to H_c^k(M - N) \to H_c^k(M) \to H^k(N) \xrightarrow{\delta} H_c^{k+1}(M - N) \to \cdots .$$

PROOF. Apply Theorem 2 to the exact sequence of complexes given by Lemma 10, and then use Lemma 11. ❖

In the proof of this theorem, the de Rham cohomology of the manifold-with-boundary V_i entered only as an intermediary (and we could have replaced the V_i by their interiors). But in the next theorem, which we will need later, it is the object of primary interest.

13. THEOREM. Let M be a manifold-with-boundary, with compact boundary ∂M. Then there is an exact sequence

$$\cdots \to H_c^k(M - \partial M) \to H_c^k(M) \to H^k(\partial M) \xrightarrow{\delta} H_c^{k+1}(M - \partial M) \to \cdots .$$

PROOF. Just like the proof of Theorem 12, using tubular neighborhoods V_i of ∂M in M. ❖

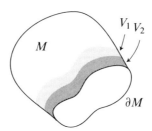

As a simple application of Theorem 13, we can rederive $H_c^k(\mathbb{R}^n)$ from a knowledge of $H^k(S^{n-1})$, by choosing M to be the closed ball B in $\mathbb{R}^n$, with $H_c^k(B) \approx H^k(B) = 0$ for $k \neq 0$. The reader may use Theorem 12 to compute $H_c^k(S^n \times \mathbb{R}^m)$, by considering the pair $(S^n \times \mathbb{R}^m, \{p\} \times \mathbb{R}^m)$. Then Theorem 13 may be used to compute the cohomology of $S^n \times S^{m-1} = \partial(S^n \times$ closed ball in $\mathbb{R}^m)$. For our next application we will seek bigger game.

Let $M \subset \mathbb{R}^{n+1}$ be a compact n-dimensional submanifold of $\mathbb{R}^{n+1}$ (a compact "hypersurface" of $\mathbb{R}^{n+1}$). Using Theorem 8-17, the sequence of the pair $(\mathbb{R}^{n+1}, M)$ gives

$$H_c^n(\mathbb{R}^{n+1}) \longrightarrow H^n(M) \overset{\delta}{\longrightarrow} H_c^{n+1}(\mathbb{R}^{n+1} - M) \longrightarrow H_c^{n+1}(\mathbb{R}^{n+1}) \longrightarrow H^{n+1}(M).$$
$$\| \qquad\qquad\qquad\qquad\qquad\qquad\qquad\qquad\qquad\quad \wr\wr \qquad\qquad\qquad \|$$
$$0 \qquad\qquad\qquad\qquad\qquad\qquad\qquad\qquad\qquad\quad \mathbb{R} \qquad\qquad\qquad 0$$

It follows that

$(*)$ $\qquad$ number of components of $\mathbb{R}^{n+1} - M = \dim H^n(M) + 1$.

But we also know (Problem 8-25) that

$(**)$ $\qquad$ number of components of $\mathbb{R}^{n+1} - M \geq 2$.

14. THEOREM. If $M \subset \mathbb{R}^{n+1}$ is a compact hypersurface, then M is orientable, and $\mathbb{R}^{n+1} - M$ has exactly 2 components. Moreover, M is the boundary of each component.

PROOF. From $(*)$ and $(**)$ we obtain

$$\dim H^n(M) + 1 \geq 2.$$

Since dim $H^n(M)$ is either 0 or 1, we conclude that dim $H^n(M) = 1$, so M is orientable; then (∗) shows that $\mathbb{R}^{n+1} - M$ has exactly two components. The proof in Problem 8-25 shows that every point of M is arbitrarily close to points in different components of $\mathbb{R}^{n+1} - M$, so every point of M is in the boundary of each of the two components. ❖

15. COROLLARY (GENERALIZED $[C^\infty]$ JORDAN CURVE THEOREM). If $M \subset \mathbb{R}^{n+1}$ is a submanifold homeomorphic to S^n, then $\mathbb{R}^{n+1} - M$ has two components, and M is the boundary of each.

16. COROLLARY. Neither the projective plane nor the Klein bottle can be imbedded in $\mathbb{R}^3$.

Our next main result will combine some of the theorems we already have. However, there are a number of technicalities involved, which we will have to dispose of first.

Consider a bounded open set $U \subset \mathbb{R}^n$ which is star-shaped with respect to 0. Then U can be described as

$$U = \{tx : x \in S^{n-1} \text{ and } 0 \le t < \rho(x)\}$$

for a certain function $\rho\colon S^{n-1} \to \mathbb{R}$. We will call ρ the **radial function** of U.

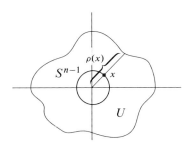

If ρ is C^∞, then we can prove that U is diffeomorphic to the open ball B of radius 1 in $\mathbb{R}^n$. The basic idea of the proof is to take $tx \in B$ to $\rho(x)t \cdot x \in U$. This produces difficulties at 0, so a modification is necessary.

17. LEMMA. If the radial function ρ of a star-shaped open set $U \subset \mathbb{R}^n$ is C^∞, then U is diffeomorphic to the open ball B of radius 1 in $\mathbb{R}^n$.

PROOF. We can assume, without loss of generality, that $\rho \geq 1$ on S^{n-1}. Let $f : [0, 1] \to [0, 1]$ be a C^∞ function with

$f = 0$ in a neighborhood of 0

$f' \geq 0$

$f(1) = 1.$

Define $h : B \to U$ by

$$h(tx) = [t + (\rho(x) - 1) f(t)]x, \qquad x \in S^{n-1}, \quad 0 \leq t < 1.$$

Clearly h is a one-one map of B onto U. It is the identity in a neighborhood of 0, so it is C^∞, with a non-zero Jacobian, at 0. At any other point the same conclusion follows from the fact that $t \mapsto t + (\rho(x) - 1) f(t)$ is a C^∞ function with strictly positive derivative. ❖

In general, the function ρ need not be C^∞; it might not even be continuous.

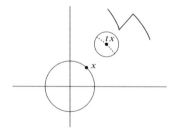

However, the discontinuities of ρ can be of a certain form only.

18. LEMMA. At each point $x \in S^{n-1}$, the radial function ρ of a star-shaped open set $U \subset \mathbb{R}^n$ is "lower semi-continuous": for every $\varepsilon > 0$ there is a neighborhood W of x in S^{n-1} such that $\rho(y) > \rho(x) - \varepsilon$ for all $y \in W$.

PROOF. Choose $tx \in U$ with $\rho(x) - t < \varepsilon$. Since U is open, there is an open

ball B with $x \in B \subset U$. There is clearly a neighborhood W of x with the property that for $y \in W$ the point ty is in B, and hence in U. This means that for $y \in W$ we have $\rho(y) \geq t > \rho(x) - \varepsilon$. ❖

Even when ρ is discontinuous, it looks as if U should be diffeomorphic to $\mathbb{R}^n$. Proving this turns out to be quite a feat, and we will be content with proving the following.

19. LEMMA. If U is an open star-shaped set in $\mathbb{R}^n$, then $H^k(U) \approx H^k(\mathbb{R}^n)$ and $H_c^k(U) \approx H_c^k(\mathbb{R}^n)$ for all k.

PROOF. The proof for H^k is clear, since U is smoothly contractible to a point. We also know that $H_c^n(U) \approx \mathbb{R} \approx H_c^n(\mathbb{R}^n)$. By Theorem 8-17, we just have to show that $H_c^k(U) = 0$ for $0 \le k < n$.

Let ω be a closed k-form with compact support $K \subset U$. We claim that there is a C^∞ function $\bar{\rho} \colon S^{n-1} \to \mathbb{R}$ such that $\bar{\rho} < \rho$ and

$$K \subset V = \{tx : x \in S^{n-1} \text{ and } 0 \le t < \bar{\rho}(x)\}.$$

This will prove the Lemma, for then V is diffeomorphic to $\mathbb{R}^n$, and consequently $\omega = d\eta$ where η has compact support contained in V, and hence in U.

For each $x \in S^{n-1}$, choose $t_x < \rho(x)$ such that all points in K of the form ux for $0 \le u \le \rho(x)$ actually have $u < t_x$. Since K is closed and ρ is lower semi-

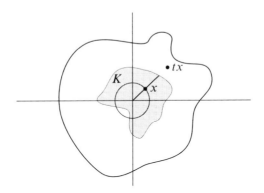

continuous, there is a neighborhood W_x of x in S^{n-1} such that t_x may also be used as t_y for all $y \in W$. Let $W_{x_1}, \ldots, W_{x_r}$ cover S^{n-1}, let $\phi_1, \ldots, \phi_r$ be a partition of unity subordinate to this cover, and define

$$\bar{\rho} = t_{x_1}\phi_1 + \cdots + t_{x_r}\phi_r.$$

Any point $x \in S^{n-1}$ is in a certain subcollection of the W_{x_i}, say $W_{x_1}, \ldots, W_{x_l}$ for convenience. Then $\rho_{l+1}(x), \ldots, \rho_r(x)$ are 0. Each $t_{x_1}, \ldots, t_{x_l}$ is $< \rho(x)$. Since $\phi_1(x) + \cdots + \phi_l(x) = 1$, it follows that $\bar{\rho}(x) < \rho(x)$. Similarly, $K \subset V$. ❖

We can apply this last Lemma in the following way. Let M be a compact manifold, and choose a Riemannian metric for M. According to Problem 9-32, every point has a neighborhood U which is geodesically convex; we can also choose U so that for any $p \in U$ the map $\exp_p$ takes an open subset of M_p diffeomorphically onto U. Let $\{U_1, \ldots, U_r\}$ be a finite cover by such open sets. If any $V = U_{i_1} \cap \cdots \cap U_{i_l}$ is non-empty, then V is clearly geodesically convex. If $p \in V$, then $\exp_p$ establishes a diffeomorphism of V with an open star-shaped set in M_p. It follows from Lemma 19 that V has the same H^k and H_c^k as $\mathbb{R}^n$. In general, a manifold M will be called of **finite type** if there is a finite cover $\{U_1, \ldots, U_r\}$ such that each non-empty intersection has the same H^k and H_c^k as $\mathbb{R}^n$; such a cover will be called **nice**.

It is fairly clear that if we consider $\mathbb{N} = \{1, 2, 3, \ldots\}$ as a subset of $\mathbb{R}^2$, then $M = \mathbb{R}^2 - \mathbb{N}$ is not of finite type. To prove this rigorously, we first use the Mayer-Vietoris sequence for $\mathbb{R}^2 = M \cup V$, where V is a disjoint union of balls around $1, 2, 3, \ldots$. We obtain

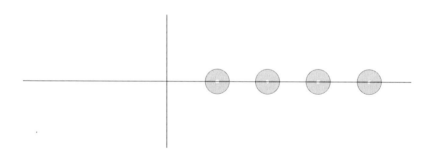

$$H^1(\mathbb{R}^2) \longrightarrow H^1(M) \oplus H^1(V) \longrightarrow H^1(M \cap V) \longrightarrow H^2(\mathbb{R}^2),$$
$$\| \qquad\qquad\qquad \| \qquad\qquad\qquad\qquad\qquad\qquad\qquad \|$$
$$0 \qquad\qquad\qquad 0 \qquad\qquad\qquad\qquad\qquad\qquad\qquad\qquad 0$$

where $M \cap V$ has the same H^1 as a disjoint union of infinitely many copies of S^1; this shows that $H^1(M)$ is infinite dimensional (see Problem 7 for more information about the cohomology of M). On the other hand,

20. PROPOSITION. If M has finite type, then $H^k(M)$ and $H_c^k(M)$ are finite dimensional for all k.

PROOF. By induction on the number of open sets r in a nice cover. It is clear for $r = 1$. Suppose it is true for a certain r, and consider a nice cover $\{U_1, \ldots, U_r, U\}$ of M. Then the theorem is true for $V = U_1 \cup \cdots \cup U_r$ and for U. It is also true for $U \cap V$, since this has the nice cover $\{U \cap U_1, \ldots, U \cap U_r\}$. Now consider the Mayer-Vietoris sequence

$$\cdots \to H^{k-1}(U \cap V) \xrightarrow{\delta} H^k(M) \xrightarrow{\alpha} H^k(U) \oplus H^k(V) \to \cdots.$$

The map α maps $H^k(M)$ onto a finite dimensional vector space, and the kernel of α is also finite dimensional. So $H^k(M)$ must be finite dimensional.

The proof for $H_c^k(M)$ is similar. ❖

For any manifold M we can define (see Problem 8-31) the **cup product** map

$$H^k(M) \times H^l(M) \xrightarrow{\cup} H^{k+l}(M)$$

by

$$([\omega], [\eta]) \mapsto [\omega \wedge \eta].$$

We can also define

$$H^k(M) \times H_c^l(M) \xrightarrow{\cup} H_c^{k+l}(M)$$

by the same formula, since $\omega \wedge \eta$ has compact support if η does. Now suppose that M^n is connected and oriented, with orientation μ. There is then a unique element of $H_c^n(M)$ represented by any $\eta \in C_c^n(M)$ with

$$\int_{(M,\mu)} \eta = 1.$$

It is convenient to also use μ to denote both this element of $H_c^n(M)$ and the isomorphism $H_c^n(M) \to \mathbb{R}$ which takes this element to $1 \in \mathbb{R}$. Now every $\alpha \in H^k(M)$ determines an element of the dual space $H_c^{n-k}(M)^*$ by

$$\beta \mapsto \alpha \cup \beta \in H_c^n(M) \xrightarrow{\mu} \mathbb{R}.$$

We denote this element of $H_c^{n-k}(M)^*$ by $PD(\alpha)$, the "Poincaré dual" of α, so that we have a map

$$PD\colon H^k(M) \to H_c^{n-k}(M)^*, \qquad PD(\alpha)(\beta) = \mu(\alpha \cup \beta).$$

One of the fundamental theorems of manifold theory states that PD is always an isomorphism. We are all set up to prove this fact, but we shall restrict the theorem to manifolds of finite type, in order not to plague ourselves with additional technical details. As with most big theorems of algebraic topology, the main part of the proof is called a Lemma, and the theorem itself is a simple corollary.

21. LEMMA. If $M = U \cup V$ for open sets U and V and PD is an isomorphism for all k on U, V, and $U \cap V$, then PD is also an isomorphism for all k on M.

PROOF. Let $l = n - k$. Consider the following diagram, in which the top row is the Mayer-Vietoris sequence, and the bottom row is the dual of the Mayer-Vietoris sequence for compact supports.

$$H^{k-1}(U) \oplus H^{k-1}(V) \longrightarrow H^{k-1}(U \cap V) \longrightarrow H^k(M) \longrightarrow H^k(U) \oplus H^k(V) \longrightarrow H^k(U \cap V)$$

$$\downarrow {PD \oplus PD} \qquad \downarrow {PD} \qquad \downarrow {PD} \qquad \downarrow {PD \oplus PD} \qquad \downarrow {PD}$$

$$[H_c^{l+1}(U) \oplus H_c^{l+1}(V)]^* \to H_c^{l+1}(U \cap V)^* \to H_c^l(M)^* \to [H_c^l(U) \oplus H_c^l(V)]^* \to H_c^l(U \cap V)^*$$

By assumption, all vertical maps, except possibly the middle one, are isomorphisms. It is not hard to check (Problem 8) that every square in this diagram commutes up to sign, so that by changing some of the vertical isomorphisms to their negatives, we obtain a commutative diagram. We now forget all about our manifold and use a purely algebraic result.

"THE FIVE LEMMA". Consider the following commutative diagram of vector spaces and linear maps. Suppose that the rows are exact, and that ϕ_1, ϕ_2, ϕ_4, ϕ_5 are isomorphisms. Then ϕ_3 is also an isomorphism.

$$\begin{array}{ccccccccc}
V_1 & \xrightarrow{\alpha_1} & V_2 & \xrightarrow{\alpha_2} & V_3 & \xrightarrow{\alpha_3} & V_4 & \xrightarrow{\alpha_4} & V_5 \\
\downarrow{\phi_1} & & \downarrow{\phi_2} & & \downarrow{\phi_3} & & \downarrow{\phi_4} & & \downarrow{\phi_5} \\
W_1 & \xrightarrow{\beta_1} & W_2 & \xrightarrow{\beta_2} & W_3 & \xrightarrow{\beta_3} & W_4 & \xrightarrow{\beta_4} & W_5
\end{array}$$

PROOF. Suppose $\phi_3(x) = 0$ for some $x \in V_3$. Then $\beta_3\phi_3(x) = 0$, so $\phi_4\alpha_3(x) = 0$. Hence $\alpha_3(x) = 0$, since ϕ_4 is an isomorphism. By exactness at V_3, there is $y \in V_2$ with $x = \alpha_2(y)$. Thus $0 = \phi_3(x) = \phi_3\alpha_2(y) = \beta_2\phi_2(y)$. Hence $\phi_2(y) = \beta_1(z)$ for some $z \in W_1$. Moreover, $z = \phi_1(w)$ for some $w \in V_1$. Then

$$\phi_2(y) = \beta_1(z) = \beta_1\phi_1(w) = \phi_2\alpha_1(w),$$

which implies that $y = \alpha_1(w)$. Hence

$$x = \alpha_2(y) = \alpha_2(\alpha_1(w)) = 0.$$

So ϕ_3 is one-one.

The proof that ϕ_3 is onto is similar, and is left to the reader. This proves the original Lemma. ❖

22. THEOREM (THE POINCARÉ DUALITY THEOREM). If M is a connected oriented n-manifold of finite type, then the map

$$PD: H^k(M) \to H_c^{n-k}(M)^*$$

is an isomorphism for all k.

PROOF. By induction on the number r of open sets in a nice cover of M. The theorem is clearly true for $r = 1$. Suppose it is true for a certain r, and consider a nice cover $\{U_1, \ldots, U_r, U\}$ of M. Let $V = U_1 \cup \cdots \cup U_r$. The theorem is true for U, V, and for $U \cap V$ (as in the proof of Proposition 19). By the Lemma, it is true for M. This completes the induction step. ❖

23. COROLLARY. If M is a connected oriented n-manifold of finite type, then $H^k(M)$ and $H_c^{n-k}(M)$ have the same dimension.

PROOF. Use the Theorem and Proposition 19, noting that V^* is isomorphic to V if V is finite dimensional. ❖

Even though the Poincaré Duality Theorem holds for manifolds which are not of finite type, Corollary 23 does not. In fact, Problem 7 shows that $H^1(\mathbb{R}^2 - \mathbb{N})$ and $H_c^1(\mathbb{R}^2 - \mathbb{N})$ have different (infinite) dimensions.

24. COROLLARY. If M is a compact connected orientable n-manifold, then $H^k(M)$ and $H^{n-k}(M)$ have the same dimension.

25. COROLLARY. If M is a compact orientable odd-dimensional manifold, then $\chi(M) = 0$.

PROOF. In the expression for $\chi(M)$, the terms $(-1)^k \dim H^k(M)$ and

$$(-1)^{n-k} \dim H^{n-k}(M) = (-1)^{k+1} \dim H^{n-k}(M)$$

cancel in pairs. ❖

A more involved use of Poincaré duality will eventually allow us to say much more about the Euler characteristic of any compact connected oriented manifold M^n. We begin by considering a smooth k-dimensional orientable vector bundle $\xi = \pi: E \to M$ over M. Orientations μ for M and ν for ξ give an orientation $\mu \oplus \nu$ for the $(n + k)$-manifold E, since E is locally a product. If $\{U_1, \ldots, U_r\}$ is a nice cover of M by geodesically convex sets so small that each bundle $\xi|U_i$ is trivial, then a slight modification of the proof for Lemma 19 shows that $\{\pi^{-1}(U_1), \ldots, \pi^{-1}(U_r)\}$ is a nice cover of E, so E is a manifold of finite type. Notice also that for the maps

$$M \xrightarrow[\pi]{\overset{s \; = \; 0\text{-section}}{\longleftarrow}} E$$

we have

$$\pi \circ s = \text{ identity of } M$$
$$s \circ \pi \quad \text{is smoothly homotopic to identity of } E,$$

so $\pi^* : H^l(M) \to H^l(E)$ is an isomorphism for all l. The Poincaré duality theorem shows that there is a unique class $U \in H_c^k(E)$ such that

$$\pi^* \mu \cup U = \mu \oplus \nu \in H_c^{n+k}(E).$$

This class U is called the **Thom class** of ξ. Our first goal will be to find a simpler property to characterize U.

Let $F_p = \pi^{-1}(p)$ be the fibre of ξ over any point $p \in M$, and let $j_p : F_p \to E$ be the inclusion map. Since j_p is proper, there is an element $j_p^* U \in H_c^k(F_p)$. On the other hand, the orientation ν for ξ determines an orientation ν_p for F_p, and hence an element $\nu_p \in H_c^k(F_p)$.

26. THEOREM. Let (M, μ) be a compact connected oriented manifold, and $\xi = \pi : E \to M$ an oriented k-plane bundle over M with orientation ν. Then the Thom class U is the unique element of $H_c^k(E)$ with the property that for all $p \in M$ we have $j_p^* U = \nu_p$. (This condition means that

$$\int_{(F_p, \nu_p)} j_p^* \omega = 1,$$

where U is the class of the closed form ω.)

PROOF. Pick some closed form $\omega \in C_c^k(E)$ representing U, and let $\eta \in C^n(M)$ be a form representing μ, so that $\int_{(M, \mu)} \eta = 1$. Our definition of U states that

(1) $$\int_E \pi^* \eta \wedge \omega = 1.$$

Let $A \subset M$ be an open set which is diffeomorphic to $\mathbb{R}^n$, so that A is smoothly contractible to any point $p \in A$. Also choose A so that there is an equivalence

$$f : \pi^{-1}(A) \to A \times \mathbb{R}^k.$$

This equivalence allows us to identify $\pi^{-1}(A)$ with $A \times F_p$. Under this identification, the map $j_p \colon F_p \to \pi^{-1}(A)$ corresponds to the map $e \mapsto (p, e)$ for $e \in F_p$, which we will continue to denote by j_p. We will also use $\pi_2 \colon A \times F_p \to F_p$ to denote projection on the second factor.

Let $\| \ \|$ be a norm on F_p. By choosing a smaller A if necessary, we can assume that there is some $K > 0$ such that, under the identification of $\pi^{-1}(A)$ with $A \times F_p$, the support of $\omega | \pi^{-1}(A)$ is contained in $\{(q, e) : q \in A, \|e\| < K\}$.

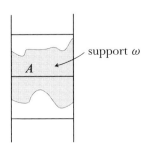

Using the fact that A is smoothly contractible to p, it is easy to see that there is a smooth homotopy $H \colon (A \times F_p) \times [0, 1] \to A \times F_p$ such that

$$H(e, 0) = e$$
$$H(e, 1) = (p, \pi_2(e)) = j_p(\pi_2(e));$$

we just pull the fibres along the smooth homotopy which makes A contractible

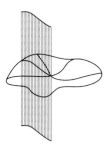

to e. For the H constructed in this way it follows that

$$H(e, t) \notin \text{support } \omega \text{ if } \|e\| \geq K.$$

Consequently, the form $H^* \omega$ on $(A \times F_p) \times [0, 1]$ has support contained in $\{(q, e, t) : \|e\| < K\}$. A glance at the definition of I (page 224) shows that the form $IH^*\omega$ on $A \times F_p$ has support contained in $\{(q,e) : \|e\| < K\}$. Theorem 7-14 shows that

$$(j_p \circ \pi_2)^*\omega - \omega = i_1^*(H^*\omega) - i_0^*(H^*\omega)$$
$$= d(IH^*\omega) + I(dH^*\omega)$$
$$= d(IH^*\omega).$$

Thus

$$(2) \qquad \pi_2^* j_p^*\omega - \omega = d\lambda, \qquad \text{support } \lambda \subset \{(q,e) : \|e\| < K\}.$$

So

$$(3) \qquad \int_{A \times F_p} \pi^*\eta \wedge \omega = \int_{A \times F_p} \pi^*\eta \wedge \pi_2^* j_p^*\omega - \int_{A \times F_p} \pi^*\eta \wedge d\lambda.$$

Now, on the one hand we have (Problem 8-17)

$$(4) \qquad \int_{A \times F_p} \pi^*\eta \wedge \pi_2^* j_p^*\omega = \int_A \pi^*\eta \cdot \int_{F_p} j_p^*\omega.$$

On the other hand, we claim that the last integral in (3) is 0. To prove this, it clearly suffices to prove that the integral is 0 over $A' \times F_p$ for any closed ball $A' \subset A$. Since

$$\pi^*\mu \wedge d\lambda = \pm d(\pi^*\mu \wedge d\lambda),$$

we have

$$(5) \qquad \int_{A' \times F_p} \pi^*\mu \wedge d\lambda = \pm \int_{A' \times F_p} d(\pi^*\mu \wedge \lambda) \qquad \begin{array}{l}\text{where } \pi^*\mu \wedge \lambda \text{ has} \\ \text{compact support on} \\ A' \times F_p \text{ by (2)}\end{array}$$

$$= \pm \int_{\partial A' \times F_p} \pi^*\mu \wedge \lambda \qquad \text{by Stokes' Theorem}$$

$$= 0,$$

because the form $\pi^*\mu \wedge \lambda$ is clearly 0 on $\partial A' \times F_p$ (since $\partial A'$ is $(n-1)$-dimensional).

Combining (3), (4), (5) we see that

$$\int_{A \times F_p} \pi^*\eta \wedge \omega = \int_A \pi^*\eta \cdot \int_{F_p} j_p^*\omega.$$

This shows that $\int_{F_p} j_p^* \omega$ is independent of p, for $p \in A$. Using connectedness, it is easy to see that it is independent of p for all $p \in M$, so we will denote it simply by $\int_F j^* \omega$. Thus

$$\int_{\pi^{-1}(A)} \pi^* \eta \wedge \omega = \int_A \pi^* \eta \cdot \int_F j^* \omega.$$

Comparing with equation (1), and utilizing partitions of unity, we conclude that

$$\int_F j^* \omega = 1,$$

which proves the first part of the theorem.

Now suppose we have another class $U' \in H_c^k(E)$. Since

$$H_c^k(E) \approx H^n(E) \approx H^n(M) \approx \mathbb{R},$$

it follows that $U' = cU$ for some $c \in \mathbb{R}$. Consequently,

$$j_p^* U' = j_p^* cU = c \cdot v_p.$$

Hence U' has the same property as U only if $c = 1$. ❖

The Thom class U of $\xi = \pi \colon E \to M$ can now be used to determine an element of $H^k(M)$. Let $s \colon M \to E$ be any section; there always is one (namely, the 0-section) and any two are clearly smoothly homotopic. We define the **Euler class** $\chi(\xi) \in H^k(M)$ of ξ by

$$\chi(\xi) = s^* U.$$

Notice that if ξ has a non-zero section $s \colon M \to E$, and $\omega \in C_c^k(E)$ represents U, then a suitable multiple $c \cdot s$ of s takes M to the complement of support ω. Hence, in this case

$$\chi(\xi) = (c \cdot s)^* U = 0.$$

The terminology "Euler class" is connected with the special case of the bundle TM, whose sections are, of course, vector fields on M. If X is a vector field on M which has an isolated 0 at some point p (that is, $X(p) = 0$, but $X(q) \neq 0$ for $q \neq p$ in a neighborhood of p), then, quite independently of our previous considerations, we can define an "index" of X at p. Consider first a vector field X on an open set $U \subset \mathbb{R}^n$ with an isolated zero at $0 \in U$. We can define a function $f_X : U - \{0\} \to S^{n-1}$ by $f_X(p) = X(p)/|X(p)|$. If $i : S^{n-1} \to U$ is $i(p) = \varepsilon p$, mapping S^{n-1} into U, then the map $f_X \circ i : S^{n-1} \to S^{n-1}$ has a certain degree; it is independent of ε, for small ε, since the maps $i_1, i_2 : S^{n-1} \to U$ corresponding to ε_1 and ε_2 will be smoothly homotopic. This degree is called the **index** of X at 0.

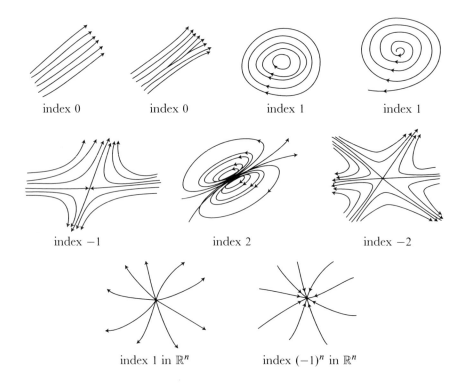

Now consider a diffeomorphism $h : U \to V \subset \mathbb{R}^n$ with $h(0) = 0$. Recall that $h_* X$ is the vector field on V with

$$(h_* X)(y) = h_*(X_{h^{-1}(y)}).$$

Clearly 0 is also an isolated zero of $h_* X$.

27. **LEMMA.** If $h : U \to V \subset \mathbb{R}^n$ is a diffeomorphism with $h(0) = 0$, and X has an isolated 0 at 0, then the index of $h_* X$ at 0 equals the index of X at 0.

PROOF. Suppose first that h is orientation preserving. Define

$$H: \mathbb{R}^n \times [0, 1] \to \mathbb{R}^n$$

by

$$H(x, t) = \begin{cases} h(tx) & 0 < t \le 1 \\ Dh(0)(x) & t = 0. \end{cases}$$

This is a smooth homotopy; to prove that it is smooth at 0 we use Lemma 3-2 (compare Problem 3-32). Each map $H_t = x \mapsto H(x, t)$ is clearly a diffeomorphism, $0 \le t \le 1$. Note that $H_1 \in SO(n)$, since h is orientation preserving. There is also a smooth homotopy $\{H_t\}$, $1 \le t \le 2$ with each $H_t \in SO(n)$ and $H_2 = $ identity, since $SO(n)$ is connected. So (see Problem 8-25), the map h is smoothly homotopic to the identity, via maps which are diffeomorphisms. This shows that $f_{h_* X}$ is smoothly homotopic to f_X on a sufficiently small region of $\mathbb{R}^n - \{0\}$. Hence the degree of $f_{h_* X} \circ i$ is the same as the degree of $f_X \circ i$.

To deal with non-orientation preserving h, it obviously suffices to check the theorem for $h(x) = (x^1, \ldots, x^{n-1}, -x^n)$. In this case

$$f_{h_* X} = h \circ f_X \circ h^{-1},$$

which shows that degree $f_{h_* X} \circ i = $ degree $f_X \circ i$. ❖

As a consequence of Lemma 27, we can now define the index of a vector field on a manifold. If X is a vector field on a manifold M, with an isolated zero at $p \in M$, we choose a coordinate system (x, U) with $x(p) = 0$, and define the **index** of X at p to be the index of $x_* X$ at 0.

28. THEOREM. Let M be a compact connected manifold with an orientation μ, which is, by definition, also an orientation for the tangent bundle $\xi = \pi: TM \to M$. Let $X: M \to TM$ be a vector field with only a finite number of zeros, and let σ be the sum of the indices of X at these zeros. Then

$$\chi(\xi) = \sigma \cdot \mu \in H^n(M).$$

PROOF. Let $p_1, \ldots, p_r$ be the zeros of X. Choose disjoint coordinate systems $(U_1, x_1), \ldots, (U_r, x_r)$ with $x_i(p_i) = 0$, and let

$$B_i = x_i^{-1}(\{p \in \mathbb{R}^n : |p| \le 1\}).$$

If $\omega \in C_c^n(E)$ is a closed form representing the Thom class U of ξ, then we are trying to prove that

$$\int_{(M, \mu)} X^*(\omega) = \sigma.$$

We can clearly suppose that $X(q) \notin$ support ω for $q \notin \bigcup_i B_i$. So

$$\int_M X^*(\omega) = \sum_{i=1}^r \int_{B_i} X^*(\omega);$$

thus it suffices to prove that

(∗)
$$\int_{B_i} X^*(\omega) = \text{index of } X \text{ at } p_i.$$

It will be convenient to drop the subscript i from now on.

We can assume that TM is trivial over B, so that $\pi^{-1}(B)$ can be identified with $B \times M_p$. Let j_p and π_2 have the same meaning as in the proof of Theorem 26. Also choose a norm $\| \; \|$ on M_p. We can assume that under the identification of $\pi^{-1}(B)$ with $B \times M_p$, the support of $\omega | \pi^{-1}(B)$ is contained in $\{(q, v) : q \in A, \|v\| \leq 1\}$. Recall from the proof of Theorem 26 that

$$\pi_2^* j_p^* \omega - \omega = d\lambda \qquad \text{support } \lambda \subset \{(q, v) : \|v\| \leq 1\}.$$

Since we can assume that $X(q) \notin$ support λ for $q \in \partial B$, we have

(1)
$$\int_B X^*(\omega) = \int_B X^* \pi_2^*(j_p^* \omega) - \int_B X^*(d\lambda)$$

$$= \int_B X^* \pi_2^*(j_p^* \omega) - \int_{\partial B} X^*(\lambda) \qquad \text{by Stokes' Theorem}$$

$$= \int_B X^* \pi_2^*(j_p^* \omega).$$

On the manifold M_p we have

$$j_p^* \omega = d\rho \qquad \begin{array}{l} \rho \text{ an } (n-1)\text{-form on } M_p \\ \text{(with non-compact support).} \end{array}$$

If $D \subset M_p$ is the unit disc (with respect to the norm $\| \; \|$) and S^{n-1} denotes $\partial D \subset M_p$, then

(2)
$$\int_{S^{n-1}} \rho = \int_{\partial D} \rho = \int_D d\rho$$

$$= \int_D j_p^* \omega$$

$$= 1, \qquad \begin{array}{l} \text{by Theorem 26, and the fact} \\ \text{that support } j_p^* \omega \subset D. \end{array}$$

Now, for $q \in B - \{p\}$, we can define

$$\bar{X}(q) = X(q)/|X(q)|,$$

and $\bar{X} \colon \partial B \to TM$ is smoothly homotopic to $X \colon \partial B \to TM$. So

(3)
$$\int_{B} X^* \pi_2{}^* (j_p{}^* \omega) = \int_{B} X^* \pi_2{}^* \, d\rho$$

$$= \int_{\partial B} X^* \pi_2{}^* \rho \qquad \text{by Stokes' Theorem}$$

$$= \int_{\partial B} \bar{X}^* \pi_2{}^* \rho$$

$$= \int_{\partial B} (\pi_2 \circ \bar{X})^* \rho.$$

From the definition of the index of a vector field, together with equation (2), it follows that

(4)
$$\int_{\partial B} (\pi_2 \circ \bar{X}^*)\rho = \text{ index of } X \text{ at } p.$$

Equations (1), (3), (4) together imply (∗). ❖

29. COROLLARY. If X and Y are two vector fields with only finitely many zeros on a compact orientable manifold, then the sum of the indices of X equals the sum of the indices of Y.

At the moment, we do not even know that there is a vector field on M with finitely many zeros, nor do we know what this constant sum of the indices is (although our terminology certainly suggests a good guess). To resolve these questions, we consider once again a triangulation of M. We can then find a vector field X with just one zero in each k-simplex of the triangulation. We begin by drawing the integral curves of X along the 1-simplexes, with a zero at each 0-simplex and at one point in each 1-simplex. We then extend this picture

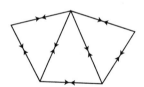

to include the integral curves of X on the 2-simplexes, producing a zero at one

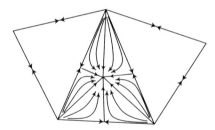

point in each of them. We then continue similarly until the n-simplexes are filled.

30. THEOREM (POINCARÉ-HOPF). The sum of the indices of this vector field (and hence of any vector field) on M is the Euler characteristic $\chi(M)$. Thus, for $\xi = \pi : TM \to M$ we have $\chi(\xi) = \chi(M) \cdot \mu$.

PROOF. At each 0-simplex of the triangulation, the vector field looks like

with index 1.

Now consider the vector field in a neighborhood of the place where it is zero on a 1-simplex. The vector field looks like a vector field on $\mathbb{R}^n = \mathbb{R}^1 \times \mathbb{R}^{n-1}$ which points directly inwards on $\mathbb{R}^1 \times \{0\}$ and directly outwards on $\{0\} \times \mathbb{R}^{n-1}$.

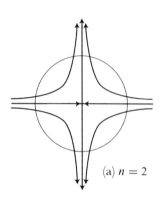

(a) $n = 2$

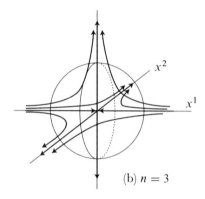

(b) $n = 3$

For $n = 2$, the index is clearly -1. To compute the index in general, we note that f_X takes the "north pole" $N = (0, \ldots, 0, 1)$ to itself and no other point goes to N. By Theorem 8-12 we just have to compute $\text{sign}_N \, f_X$. Now at N we can pick projection on $\mathbb{R}^{n-1} \times \{0\}$ as the coordinate system. Along the inverse image of the x^1-axis the vector field looks exactly like figure (a) above, where we already know the degree is -1, so f_{X*} takes the subspace of $S^{n-1}{}_N$ consisting of tangent vectors to this curve into the same subspace, in an orientation reversing way. Along the inverse image of the x^2-, $\ldots$, x^{n-1}-axes the vector field looks like

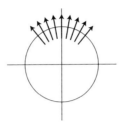

so f_{X*} takes the corresponding subspaces of $S^{n-1}{}_N$ into themselves in an orientation preserving way. Thus $\text{sign}_N \, f_X = -1$, which is therefore the index of the vector field.

In general, near a zero within a k-simplex, X looks like a vector field on $\mathbb{R}^n = \mathbb{R}^k \times \mathbb{R}^{n-k}$ which points directly inwards on $\mathbb{R}^k \times \{0\}$ and directly outwards on $\{0\} \times \mathbb{R}^{n-k}$. The same argument shows that the index is $(-1)^k$.

Consequently, the sum of the indices is

$$\alpha_0 - \alpha_1 + \alpha_2 - \cdots = \chi(M). \; \clubsuit$$

We end this chapter with one more observation, which we will need in the last chapter of Volume V! Let $\xi = \pi \colon E \to M$ be a smooth oriented k-plane bundle over a compact connected oriented n-manifold M, and let $\langle \, , \, \rangle$ be a Riemannian metric for ξ. Then we can form the "associated disc bundle" and "associated sphere bundle"

$$D = \{e : \langle e, e \rangle \le 1\}$$
$$S = \{e : \langle e, e \rangle = 1\}.$$

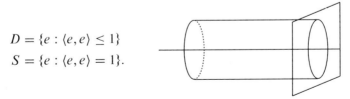

It is easy to see that D is a compact oriented $(n + k)$-manifold, with $\partial D = S$; moreover, the D constructed for any other Riemannian metric is diffeomorphic to this one. We let $\pi_0 \colon S \to M$ be $\pi | S$.

31. THEOREM. A class $\alpha \in H^k(M)$ satisfies $\pi_0{}^*(\alpha) = 0$ if and only if α is a multiple of $\chi(\xi)$.

PROOF. Consider the following picture. The top row is the exact sequence

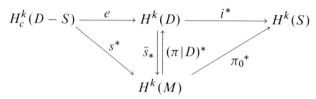

for (D, S) given by Theorem 13. The map $s: M \to D - S$ is the 0-section, while $\bar{s}: M \to D$ is the same 0-section. Note that everything commutes,

$$\pi_0{}^* = i^* \circ (\pi|D)^* \qquad \text{since } \pi_0 = (\pi|D) \circ i,$$

$$s^* = \bar{s}_* \circ e \qquad \begin{array}{l} \text{since extending a form to } D \\ \text{does not affect its value on } s(M), \end{array}$$

and that

$$\bar{s}^* \circ (\pi|D)^* = \text{ identity of } H^k(M),$$

since $(\pi|D) \circ \bar{s}$ is smoothly homotopic to the identity.

Now let $\alpha \in H^k(M)$ satisfy $\pi_0{}^*(\alpha) = 0$. Then $i^*(\pi|D)^*\alpha = 0$, so $(\pi|D)^*\alpha \in$ image e. Since $D - S$ is diffeomorphic to E, and every element of $H_c^k(D - S)$ is a multiple of the Thom class U of ξ, we conclude that

$$(\pi|D)^*\alpha = c \cdot e(U) \qquad \text{for some } c \in \mathbb{R}.$$

Hence

$$\alpha = \bar{s}^*(\pi|D)^*\alpha = c \cdot \bar{s}^*(e(U)) = c \cdot s^*U$$
$$= c \cdot \chi(\xi).$$

The proof of the converse is similar. ❖

PROBLEMS

1. Find $H^k(S^1 \times \cdots \times S^1)$ by induction on the number n of factors. [Answer: $\dim H^k = \binom{n}{k}$.]

2. (a) Use the Mayer-Vietoris sequence to determine $H^k(M - \{p\})$ in terms of $H^k(M)$, for a connected manifold M.

(b) If M and N are two connected n-manifolds, let $M \# N$ be obtained by joining M and N as shown below. Find the cohomology of $M \# N$ in terms of that of M and N.

(c) Find χ for the n-holed torus. [Answer: $2 - 2n$.]

3. (a) Find $H^k(\text{Möbius strip})$.

(b) Find $H^k(\mathbb{P}^2)$.

(c) Find $H^k(\mathbb{P}^n)$. (Use Problem 1-15(b); it is necessary to consider whether a neighborhood of $\mathbb{P}^{n-1}$ in $\mathbb{P}^n$ is orientable or not.) [Answer: $\dim H^k(\mathbb{P}^n) = 1$ if k even and $\leq n$, $= 0$ otherwise.]

(d) Find $H^k(\text{Klein bottle})$.

(e) Find the cohomology of $M \# (\text{Möbius strip})$ and $M \# (\text{Klein bottle})$ if M is the n-holed torus.

4. (a) The figure below is a triangulation of a rectangle. If we perform the indicated identifications of edges we do *not* obtain a triangulation of the torus. Why not?

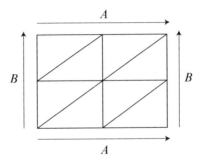

(b) The figure below does give a triangulation of the torus when sides are identified. Find α_0, α_1, α_2 for this triangulation; compare with Theorem 5 and Problem 1.

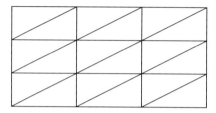

5. (a) For any triangulation of a compact 2-manifold M, show that

$$3\alpha_2 = 2\alpha_1$$

$$\alpha_1 = 3(\alpha_0 - \chi(M))$$

$$\frac{\alpha_0(\alpha_0 - 1)}{2} \geq \alpha_1$$

$$\alpha_0 \geq \frac{1}{2}\left(7 + \sqrt{49 - 24\chi(M)}\right).$$

(b) Show that for triangulations of S^2 and the torus $T^2 = S^1 \times S^1$ we have

$$S^2: \quad \alpha_0 \geq 4 \quad \alpha_1 \geq 6 \quad \alpha_2 \geq 4$$

$$T^2: \quad \alpha_0 \geq 7 \quad \alpha_1 \geq 21 \quad \alpha_2 \geq 14.$$

Find triangulations for which these inequalities are all equalities.

6. (a) Find $H_c^k(S^n \times \mathbb{R}^m)$ by induction on n, using the Mayer-Vietoris sequence for compact supports.
(b) Use the exact sequence of the pair $(S^n \times \mathbb{R}^m, \{p\} \times \mathbb{R}^m)$ to compute the same vector spaces.
(c) Compute $H^k(S^n \times S^{m-1})$, using Theorem 13.

7. (a) The vector space $H^1(\mathbb{R}^2 - \mathbb{N})$ may be described as the set of all sequences of real numbers. Using the exact sequence of the pair $(\mathbb{R}^2, \mathbb{N})$, show that $H_c^1(\mathbb{R}^2 - \mathbb{N})$ may be considered as the set of all real sequences $\{a_n\}$ such that $a_n = 0$ for all but finitely many n.
(b) Describe the map $PD: H^1(\mathbb{R}^2 - \mathbb{N}) \to H_c^1(\mathbb{R}^2 - \mathbb{N})^*$ in terms of these descriptions of $H^1(\mathbb{R}^2 - \mathbb{N})$ and $H_c^1(\mathbb{R}^2 - \mathbb{N})$, and show that it is an isomorphism.
(c) Clearly $H_c^1(\mathbb{R}^2 - \mathbb{N})$ has a countable basis. Show that $H^1(\mathbb{R}^2 - \mathbb{N})$ does not. *Hint:* If $v_i = \{a_i{}^j\} \in H^1(\mathbb{R}^2 - \mathbb{N})$, choose $(b_1, b_2) \in \mathbb{R}^2$ linearly independent of $(a_1{}^1, a_1{}^2)$; then choose $(b_3, b_4, b_5) \in \mathbb{R}^3$ linearly independent of both $(a_1{}^3, a_1{}^4, a_1{}^5)$ and $(a_2{}^3, a_2{}^4, a_2{}^5)$; etc.

8. Show that the squares in the diagram in the proof of Lemma 21 commute, except for the square

$$
\begin{array}{ccc}
H^{k-1}(U \cap V) & \longrightarrow & H^k(M) \\
\downarrow{\scriptstyle PD} & & \downarrow{\scriptstyle PD} \\
H_c^{l+1}(U \cap V)^* & \longrightarrow & H_c^l(M)^*
\end{array}
$$

which commutes up to the sign $(-1)^k$. (It will be necessary to recall how various maps are defined, which is a good exercise; the only slightly difficult maps are the ones involved in the above diagram.)

9. (a) Let $M = M_1 \cup M_2 \cup M_3 \cup \cdots$ be a disjoint union of oriented n-manifolds. Show that $H_c^k(M) \approx \bigoplus_i H_c^k(M_i)$, this "direct sum" consisting of all sequences $(\alpha_1, \alpha_2, \alpha_3, \dots)$ with $\alpha_i \in H_c^k(M_i)$ and all but finitely many $\alpha_i = 0 \in H_c^k(M_i)$.
(b) Show that $H^k(M) \approx \prod_i H^k(M_i)$, this "direct product" consisting of *all* sequences $(\alpha_1, \alpha_2, \alpha_3, \dots)$ with $\alpha_i \in H^k(M_i)$.
(c) Show that if the Poincaré duality theorem holds for each M_i, then it holds for M.
(d) The figure below shows a decomposition of a triangulated 2-manifold into three open sets U_0, U_1, and U_2. Use an analogous decomposition in n dimensions to prove that Poincaré duality holds for any triangulated manifold.

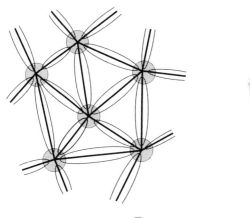

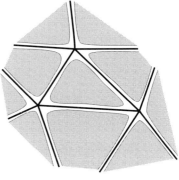

U_0 is union of shaded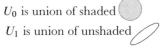
U_1 is union of unshaded

U_2 is union of shaded

10. Let $\xi = \pi \colon E \to M$ and $\xi' = \pi' \colon E' \to M$ be oriented k-plane bundles, over a compact oriented manifold M, and $(\tilde{f}, f)$ a bundle map from ξ' to ξ which is an isomorphism on each fibre.

(a) If $U \in H_c^k(E)$ and $U' \in H_c^k(E')$ are the Thom classes, then $\tilde{f}^*(U) = U'$.

(b) $f^*(\chi(\xi)) = \chi(\xi')$. (Using the notation of Problem 3-23, we have $f^*(\chi(\xi)) = \chi(f^*(\xi))$.)

11. (a) Let $\xi = \pi \colon E \to M$ be an oriented k-plane bundle over an oriented manifold M, with Thom class U. Using Poincaré duality, prove the Thom Isomorphism Theorem: The map $H^l(E) \to H_c^{l+k}(E)$ given by $\alpha \mapsto \alpha \cup U$ is an isomorphism for all l.

(b) Since we can also consider U as being in $H^k(E)$, we can form $U \cup U \in H_c^{2k}(E)$. Using anticommutativity of $\wedge$, show that this is 0 for k odd. Conclude that U represents $0 \in H^k(E)$, so that $\chi(\xi) = 0$. It follows, in particular, that $\chi(\xi) = 0$ when $\xi = \pi \colon TM \to M$ for M of odd dimension, providing another proof that $\chi(M) = 0$ in this case.

12. If a vector field X has an isolated singularity at $p \in M^n$, show that the index of $-X$ at p is $(-1)^n$ times the index of X at p. This provides another proof that $\chi(M) = 0$ for odd n.

13. (a) Let $p_1, \ldots, p_r \in M$. Using Problem 8-26, show that there is a subset $D \subset M$ diffeomorphic to the closed ball, such that all $p_i \in$ interior D.

(b) If M is compact, then there is a vector field X on M with only one singularity.

(c) It is a fact that a C^∞ map $f \colon S^{n-1} \to S^{n-1}$ of degree 0 is smoothly homotopic to a constant map. Using this, show that if $\chi(M) = 0$, then there is a nowhere 0 vector field on M.

(d) If M is connected and not compact, then there is a nowhere 0 vector field on M. (Begin with a triangulation to obtain a vector field with a discrete set of zeros. Join these by a ray going to infinity, enclose this ray in a cone, and push everything off to infinity.)

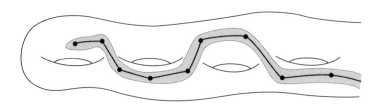

(e) If M is a connected manifold-with-boundary, with $\partial M \neq \emptyset$, then there is a nowhere zero vector field on M.

14. This Problem proves de Rham's Theorem. Basic knowledge of singular cohomology is required. We will denote the group of singular k-chains of X by $S_k(X)$. For a manifold M, we let $S_k^\infty(M)$ denote the C^∞ singular k-chains, and let $i\colon S_k^\infty(M) \to S_k(M)$ be the inclusion. It is not hard to show that there is a chain map $\tau\colon S_k(M) \to S_k^\infty(M)$ so that $\tau \circ i =$ identity of $S_k^\infty(M)$, while $i \circ \tau$ is chain homotopic to the identity of $S_k(M)$ [basically, τ is approximation by a C^∞ chain]. This means that we obtain the correct singular cohomology of M if we consider the complex $\mathrm{Hom}(S_k^\infty(M), \mathbb{R})$.

(a) If ω is a closed k-form on M, let $Rh(\omega) \in \mathrm{Hom}(S_k^\infty(M), \mathbb{R})$ be

$$Rh(\omega)(c) = \int_c \omega.$$

Show that Rh is a chain map from $\{C^k(M)\}$ to $\{\mathrm{Hom}(S_k^\infty(M), \mathbb{R})\}$. (*Hint:* Stokes' Theorem.) It follows that there is an induced map Rh from the de Rham cohomology of M to the singular cohomology of M.

(b) Show that Rh is an isomorphism on a smoothly contractible manifold (Lemmas 17, 18, and 19 will not be necessary for this.)

(c) Imitate the proof of Theorem 21, using the Mayer-Vietoris sequence for singular cohomology, to show that if Rh is an isomorphism for U, V, and $U \cap V$, then it is an isomorphism for $U \cup V$.

(d) Conclude that Rh is an isomorphism if M is of finite type. (Using the method of Problem 9, it follows that Rh is an isomorphism for any triangulated manifold.)

(e) Check that the cup product defined using $\wedge$ corresponds to the cup product defined in singular cohomology.

APPENDIX A

CHAPTER 1

Following the suggestions in this chapter, we will now define a **manifold** to be a topological space M such that

  (1) M is Hausdorff,
  (2) For each $x \in M$ there is a neighborhood U of x and an integer $n \geq 0$ such that U is homeomorphic to $\mathbb{R}^n$.

Condition (1) is necessary, for there is even a 1-dimensional "manifold" which is not Hausdorff. It consists of $\mathbb{R} \cup \{*\}$ where $* \notin \mathbb{R}$, with the following topology: A set U is open if and only if

  (1) $U \cap \mathbb{R}$ is open,
  (2) If $* \in U$, then $(U \cap \mathbb{R}) \cup \{0\}$ is a neighborhood of 0 (in $\mathbb{R}$).

Thus the neighborhoods of $*$ look just like neighborhoods of 0. This space may also be obtained by identifying all points except 0 in one copy of $\mathbb{R}$ with the corresponding point in another copy of $\mathbb{R}$. Although non-Hausdorff manifolds are important in certain cases, we will not consider them.

We have just seen that the Hausdorff property is not a "local property", but local compactness is, so every manifold is locally compact. Moreover, a Hausdorff locally compact space is regular, so every manifold is regular. (By the way, this argument does not work for "infinite dimensional" manifolds, which are locally like Banach spaces; these need not be regular even if they are Hausdorff.) On the other hand, there are manifolds which are not normal (Problem 6). Every manifold is also clearly locally connected, so every component is open, and thus a manifold itself. Before exhibiting non-metrizable manifolds, we first note that almost all "nice" properties of a manifold are equivalent.

THEOREM. The following properties are equivalent for any manifold M:

  (a) Each component of M is σ-compact.
  (b) Each component of M is second countable (has a countable base for the topology).
  (c) M is metrizable.
  (d) M is paracompact.

(In particular, a compact manifold is metrizable.)

459

FIRST PROOF. (a) $\Rightarrow$ (b) follows immediately from the simple proposition that a σ-compact locally second countable space is second countable.

(b) $\Rightarrow$ (c) follows from the Urysohn metrization theorem.

(c) $\Rightarrow$ (d) because any metric space is paracompact (Kelley, *General Topology*, pg. 160). The second proof does not rely on this difficult theorem.

(d) $\Rightarrow$ (a) is a consequence of the following.

LEMMA. A connected, locally compact, paracompact space is σ-compact.

Proof. There is a locally finite cover of the space by open sets with compact closure. If U_0 is one of these, then $\overline{U}_0$ can intersect only a finite number $U_1, \ldots, U_{n_1}$ of the others. Similarly $\overline{U}_0 \cup \overline{U}_1 \cup \cdots \cup \overline{U}_{n_1}$ intersects only $U_{n_1+1}, \ldots, U_{n_2}$; and so on. The union

$$\overline{U}_0 \cup \cdots \cup \overline{U}_{n_1} \cup \cdots \cup \overline{U}_{n_2} \cup \cdots = U_0 \cup \cdots \cup U_{n_1} \cup \cdots \cup U_{n_2} \cup \cdots$$

is clearly open. It is also closed, for if x is in the closure, then x must be in the closure of a finite union of these U_i, because x has a neighborhood which intersects only finitely many. Thus x is in the union.

Since the space is connected, it equals this countable union of compact sets. This proves the Lemma and the Theorem.

SECOND PROOF. (a) $\Rightarrow$ (b) $\Rightarrow$ (c) and (d) $\Rightarrow$ (a) as before.

(c) $\Rightarrow$ (a) is Theorem 1-2.

(a) $\Rightarrow$ (d). Let $M = C_1 \cup C_2 \cup \cdots$, where each C_i is compact. Clearly C_1 has an open neighborhood U_1 with compact closure. Then $\overline{U}_1 \cup C_2$ has an open neighborhood U_2 with compact closure. Continuing in this way, we obtain open sets U_i with $\overline{U}_i$ compact and $\overline{U}_i \subset U_{i+1}$, whose union contains all C_i, and hence is M. It is easy to show from this that M is paracompact. ❖

It turns out that there are even 1-manifolds which are not paracompact. The construction of these examples requires the ordinal numbers, which are briefly explained here. (Ordinal numbers will not be needed for a 2-dimensional example to come later.)

ORDINAL NUMBERS

Recall that an **ordering** $<$ on a set A is a relation such that

(1) $a < b$ and $b < c$ implies $a < c$ for all $a, b, c \in A$ (transitivity)

(2) For all $a, b \in A$, one and only one of the following holds:

      (i) $a = b$

      (ii) $a < b$ (trichotomy).

      (iii) $b < a$ (also written $a > b$)

An **ordered set** is just a pair $(A, <)$ where $<$ is an ordering on A. Two ordered sets $(A, <)$ and $(B, \prec)$ are **order isomorphic** if there is a one-one onto function $f : A \to B$ such that $a < b$ implies $f(a) \prec f(b)$; the map f itself is called an **order isomorphism**, and f^{-1} is easily seen to be an order isomorphism also.

An ordering $<$ on A is a **well-ordering** if every non-empty subset $B \subset A$ has a *first* element, that is, an element b such that $b \le b'$ for all $b' \in B$. Some well-ordered sets are illustrated below; in this scheme we do not list any of the $<$ relations which are consequences of the ones already listed.

$\emptyset$

$\{0\}$

$0 < 1$ $(A = \{0, 1\})$

$0 < 1 < 2$ $(A = \{0, 1, 2\})$

$0 < 1 < 2 < 3$ etc.

$\vdots$

$0 < 1 < 2 < 3 < \cdots$

$0 < 1 < 2 < \cdots < \omega$ (ω is some set $\ne 0, 1, 2, 3, \ldots$)

                                  ($\omega + 1$ is, for the present,

$0 < 1 < 2 < \cdots < \omega < \omega + 1$ just a set distinct from

                                  those already mentioned)

$\vdots$

$0 < 1 < 2 < \cdots < \omega < \omega + 1 < \omega + 2 < \cdots$

$0 < 1 < 2 < \cdots < \omega < \omega + 1 < \omega + 2 < \cdots < \omega \cdot 2$

$\vdots$

$0 < 1 < 2 < \cdots < \omega < \omega + 1 < \omega + 2 < \cdots < \omega \cdot 2 < \omega \cdot 2 + 1 < \cdots$

$0 < 1 < 2 < \cdots < \omega < \omega + 1 < \omega + 2 < \cdots < \omega \cdot 2 < \omega \cdot 2 + 1 < \cdots < \omega \cdot 3$

$\vdots$

$0 < 1 < 2 < \cdots < \omega < \cdots < \omega \cdot 2 < \cdots < \omega \cdot 3 < \cdots < \cdots$

$0 < 1 < 2 < \cdots < \omega < \cdots < \omega \cdot 2 < \cdots < \omega \cdot 3 < \cdots < \cdots < \omega^2.$

Any subset of a well-ordered set is, of course, also a well-ordered set with the same ordering. In particular, a subset B of a well-ordered set A is called an (**initial**) **segment** if $b \in B$ and $a < b$ imply $a \in B$. It is easy to see that if B is a segment of A, then either $B = A$ or else there is some $a \in A$ such that

$$B = \{a' \in A : a' < a\};$$

in fact, a is the first element of $A - B$. Notice that each set on our list is a segment of the succeeding ones. It is not hard to see that no two sets on our list are order isomorphic. For example,

$$0 < 1 < \cdots < \omega \quad \text{and} \quad 0 < 1 < \cdots < \omega < \omega + 1$$

are not order isomorphic because the second has both a last and a next to last element, while the first does not. But there is a much more general proposition which will settle all cases at once:

1. PROPOSITION. If $B \neq A$ is a segment of A, then B is not order isomorphic to A. In fact, the only order isomorphism from B to a *segment* of A is the identity.

PROOF. If $f : B \to B' \subset A$ is an order isomorphism and B' is a segment of A, then for the first element b of B (and hence of A) we clearly must have $f(b) = b$. Then $f(b')$ must be b', where b' is the second element. And so on; even for the "ω^{th}" element (the first one after the first, second, third, etc.)! The way we prove this rigorously is amazingly simple: If $f(b) \neq b$ for some $b \in B$, just consider the first element of $\{b \in B : f(b) \neq b\}$; an outright contradiction appears almost immediately. ❖

Proposition 1 has a companion, which makes the study of well-ordered sets simply delightful.

2. PROPOSITION. If $(A, <)$ and $(B, \prec)$ are well-ordered sets, then one is order isomorphic to a segment of the other.

PROOF. We match the first element of A with the first of B, the second with the second, ... , the "ω^{th}" with the "ω^{th}", etc., until we run out of one set. To do this rigorously, consider order isomorphisms from segments of A onto segments of B. It is easy to show that any two such order isomorphisms agree on the smaller of their two domains (just consider the smallest element where they don't). So all such order isomorphisms can be put together to give another, which is clearly the largest of all. If it is defined on all of A we are done. If it is not, then its range must be all of B (or we could easily extend it) and we are still done. ❖

Suppose we define a relation $<$ between well-ordered sets by stipulating that $(A, <) < (B, \prec)$ when $(A, <)$ is order isomorphic to a proper segment of $(B, \prec)$. Transitivity of $<$ is obvious, and Propositions 1 and 2 show that we almost have trichotomy. "Almost", because the condition "$(A, <) = (B, \prec)$" must be replaced by "$(A, <)$ order isomorphic to $(B, \prec)$". To obviate this difficulty we need only work with order isomorphism classes of well-ordered sets, instead of with the well-ordered sets themselves. These order isomorphism classes are called **ordinal numbers**. They are beautiful:*

3. PROPOSITION. $<$ is a well-ordering of the ordinal numbers.

PROOF. Given a non-empty set $\mathcal{A}$ of ordinal numbers, let $(A, <)$ be a well-ordered set representing one of its elements α. To produce a smallest element of $\mathcal{A}$ we can obviously ignore elements $\geq \alpha$. Every element $< \alpha$ is represented by an ordered set which is order isomorphic to some proper segment of A; each of these is the segment consisting of elements of A less that some $a \in A$. Consider the least of these a's. It determines a segment which represents some $\beta \in \mathcal{A}$. This β is the smallest element of $\mathcal{A}$. ❖

Notice that if α is an ordinal number, represented by a well-ordered set $(A, <)$, then the well-ordered set of all ordinals $\beta < \alpha$ has a particularly simple representation: it is order isomorphic to the set $(A, <)$! Roughly speaking: An ordinal number is order isomorphic to the set of all ordinals less than it.

If α is an ordinal number, we will denote by $\alpha + 1$ the smallest ordinal after α (if α is represented by the well-ordered set $(A, <)$, then $\alpha + 1$ is represented by a well-ordered set with just one more element, larger than all members of A). Notice that some ordinals are not of the form $\alpha + 1$ for any α; these are called **limit ordinals**, while those of the form $\alpha + 1$ are called **successor**

---

*Only one feature mars the beauty of the ordinal numbers as presented here. Each ordinal number is a horribly large set; it would be much nicer to choose one specific well-ordered set from each order isomorphism class, and define these specific sets to be the ordinal numbers. There is a particularly elegant way to do this, due to von Neumann, which can be found in the Appendix to Kelley, *General Topology*.

ordinals. We will also denote some ordinals by the symbols appearing before: $0, 1, 2, 3, \ldots, \omega, \omega + 1, \ldots$, etc.

Our list of well-ordered sets only begins to suggest the complexity which well-ordered sets can achieve. With a little thought, one can see how the symbols ω^3, ω^4, $\ldots$ would appear (symbols like $\omega^3 + \omega^2 \cdot 3 + \omega \cdot 4 + 6$ would be used somewhere between ω^3 and ω^4); after all these one would need

$$\omega^\omega, \omega^{\omega^\omega}, \ldots,$$

and after all these the symbol ε_0 pops up. After

$$\varepsilon_0{}^2, \varepsilon_0{}^3, \ldots, \varepsilon_0{}^\omega, \ldots, \varepsilon_0{}^{\omega^\omega}, \ldots, \varepsilon_0{}^{\omega^{\omega^\omega}}, \ldots,$$

one comes to

$$\varepsilon_1, \varepsilon_2, \ldots, \varepsilon_\omega, \ldots, \varepsilon_{\omega^\omega}, \ldots, \varepsilon_{\varepsilon_0}, \ldots, \varepsilon_{\varepsilon_{\varepsilon_0}}, \ldots;$$

and this is only the beginning!

All the well-ordered sets mentioned so far are *countable*. There are indeed an enormous number of countable well-ordered sets:

4. PROPOSITION. Let Ω be the collection of all countable ordinals (ordinals represented by a countable well-ordered set). Then Ω is uncountable.

PROOF. By Proposition 3, $(\Omega, <)$ is a well-ordered set. If it were countable, it would represent a countable ordinal $\alpha \in \Omega$. By the remark after Proposition 3, this would mean that Ω is order isomorphic to the collection of ordinals $< \alpha$, i.e., to a proper segment of itself, contradicting* Proposition 1. ❖

We have thus established the existence of an uncountable ordinal. Our specific example, represented by Ω, is clearly the first uncountable ordinal; any member of Ω is countable, and consequently has only countably many predecessors. (It is hopeless to try to "reach" Ω by continuing the listing of well-ordered sets begun above, for one would have to go uncountably far, and encounter sets with an uncountable number of degrees of complexity. A leap of faith is required.)

Although the countable ordinals exhibit uncountably many degrees of complexity, they are each simple in one way:

---

*By deleting the words countable and uncountable in this proof one obtains the "Burali-Forti Paradox": the set *Ord* of all ordinal numbers is well-ordered, so it represents an ordinal $\alpha \in Ord$, and hence is order isomorphic to an initial segment of *Ord*. For a resolution of this paradox, see Kelley's Appendix.

5. PROPOSITION. If $\alpha \in \Omega$ is a limit ordinal, then there is a sequence $\beta_1 < \beta_2 < \beta_3 < \cdots < \alpha$, such that every $\beta < \alpha$ satisfies $\beta < \beta_n$ for some n (we say that $\{\beta_n\}$ is "cofinal" in α).

PROOF. Since α is countable, *all* its members can be listed (in not-necessarily increasing order) $\gamma_1, \gamma_2, \gamma_3, \ldots$. Let $\beta_1 = \gamma_1$ and let β_{n+1} be the first γ in the list which comes after β_n. ❖

6. COROLLARY. If $\alpha \in \Omega$, then α is represented by some well-ordered subset of $\mathbb{R}$. However, no subset of $\mathbb{R}$ is order isomorphic to Ω.

PROOF. Suppose there were one, and hence a smallest, $\alpha \in \Omega$ not represented by some subset of $\mathbb{R}$. It cannot happen that $\alpha = \beta + 1$, for then β would be represented by a subset of $\mathbb{R}$, thus also by a subset of $(-\infty, 0)$ and α could by represented by a subset of $\mathbb{R}$. So by Proposition 5, there is a sequence $\beta_1 < \beta_2 < \beta_3 < \cdots < \alpha$ cofinal in α. Then β_i is represented by a subset of $(-\infty, i)$, and we can easily arrange that the subset representing β_i is a segment of the subset representing β_j for $i < j$. The union of all these sets would then represent α, a contradiction.

If a subset of $\mathbb{R}$ were order isomorphic to Ω, then there would be uncountably many disjoint intervals in $\mathbb{R}$, namely those between the points representing α and $\alpha + 1$ for all $\alpha \in \Omega$. This is impossible. ❖

The first example of a non-metrizable manifold is defined in terms of Ω. Consider $\Omega \times [0, 1)$, with the order $<$ defined as follows:

$$(\alpha, s) < (\beta, t) \qquad \text{if } \alpha < \beta \text{ or if } \alpha = \beta \text{ and } s < t.$$

This can be pictured as follows:

$(0,0) \qquad (1,0) \qquad (2,0) \qquad \cdots \qquad (\omega,0) \ (\omega+1,0) \ (\omega+2,0) \qquad (\omega\cdot2,0)$

The set $\Omega \times [0, 1)$ with the order topology (a subbase consists of sets of the form $\{x : x < x_0\}$ and $\{x : x > x_0\}$) is called the **closed long ray** (with "origin" $(0,0)$), and $L^+ = \Omega \times [0, 1) - \{(0,0)\}$ is the **(open) long ray**. The disjoint union of two copies of the closed long ray with their origins identified is the **long line** L. To distinguish L^+ and L, the names "half-long line" and "long line" may also be used. The Corollary to Proposition 5 implies easily that the long ray and the long line are 1-dimensional manifolds; aside from the line and the circle, there are no other connected 1-manifolds.

Quite a few new 2-manifolds can now be constructed:

$L^+ \times S^1$ (half-long cylinder), $L \times S^1$ (long cylinder),

$L^+ \times \mathbb{R}$ (half-long strip), $L \times \mathbb{R}$ (long strip),

$L \times L$ (big plane), $L \times L^+$ (big half-plane),

$L^+ \times L^+$ (big quadrant).

Identifying all points $((0,0),\theta)$ in the product of the closed long ray and S^1 produces another 2-manifold, which might be called the "big disc".

There is another way of producing a non-metrizable 2-manifold which does not use Ω at all. We begin with the open upper half-plane $\mathbb{R}_+^2 = \{(x,y) \in \mathbb{R}^2 : y > 0\}$ and another copy $\mathbb{R}^2 \times \{0\}$ of the plane; we will denote this set by $\mathbb{R}_0^2$, and denote the point $(x,y,0)$ by $(x,y)_0$. Define a map $f_0 \colon (\mathbb{R}_0^2)_+ \to \mathbb{R}_+^2$ by

$$f_0((x,y)_0) = (xy, y).$$

Consider the disjoint union of $\mathbb{R}_+^2$ and $\mathbb{R}_0^2$, with $p \in (\mathbb{R}_0^2)_+$ and $f_0(p) \in \mathbb{R}_+^2$ identified. This is a Hausdorff manifold; the following diagram shows two open sets homeomorphic to $\mathbb{R}^2$. The manifold itself is, in fact, homeomorphic

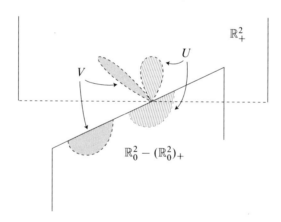

to $\mathbb{R}^2$; we could have thrown away $\mathbb{R}_+^2$ to begin with since it is identified by a homeomorphism with $(\mathbb{R}_0^2)_+$.

But consider now, for each $a \in \mathbb{R}$, another copy of $\mathbb{R}^2$, say $\mathbb{R}^2 \times \{a\}$, which we will denote by $\mathbb{R}_a^2$. Define $f_a \colon (\mathbb{R}_a^2)_+ \to \mathbb{R}_+^2$ by

$$f_a((x,y)_a) = (a + yx, y).$$

In the disjoint union of $\mathbb{R}^2_+$ and *all* $\mathbb{R}^2_a$, $a \in \mathbb{R}$ we wish to identify each $p \in (\mathbb{R}^2_a)_+$ with $f_a(p) \in \mathbb{R}^2_+$. We may dispense with $\mathbb{R}^2_+$ completely, and in the disjoint union of all $\mathbb{R}^2_a$ identify each $(x, y)_a$ and $(x', y')_b$ for which $y = y' > 0$ and $xy + a = x'y' + b$. The equivalence classes, of course, are a space homeomorphic to $\mathbb{R}^2_+$, so we will consider $\mathbb{R}^2_+$ a subset of the resulting space. This space is still a Hausdorff manifold, but it cannot be second countable, for it has an uncountable discrete subset, namely the set $\{(0, 0)_a\}$. This manifold, the **Prüfer manifold**, and related manifolds, have some very strange properties, developed in the problems.

PROBLEMS

1. (a) A well-ordered set cannot contain a decreasing infinite sequence $x_1 > x_2 > x_3 > \cdots$.
(b) If we denote $(\alpha + 1) + 1$ by $\alpha + 2$, $(\alpha + 2) + 1$ by $\alpha + 3$, etc., then any α equals $\beta + n$ for a unique limit ordinal β and integer $n \geq 0$. (Thus one can define *even* and *odd* ordinals.)

2. Let c be a "choice function", i.e., $c(A)$ is defined for each set $A \neq \emptyset$, and $c(A) \in A$ for all A. Given a set X, a well-ordering $<$ on a subset Y of X will be called "distinguished" if for all $y \in Y$,

$$y = c\big(Y - \{y' \in Y : y' < y\}\big).$$

(a) Show that of any two distinguished well-orderings, one is an extension of the other.
(b) Show that there is a well-ordering on X. (Zorn's Lemma may be deduced from this fact fairly easily.)
(c) Given two sets, show that one of them is equivalent to (can be put in one-one correspondence with) a subset of the other.
(d) Show that on any infinite set there is a well-ordering which represents a limit ordinal.
(e) From (d), and Problem 1, show that if X and Y are disjoint equivalent infinite sets, then $X \cup Y$ is equivalent to Y.

3. (a) L^+ and L are not metrizable.
(b) If $x_1 \leq x_2 \leq x_3 \leq \cdots$ is a sequence in L^+, then $\{x_n\}$ converges to some point. Consequently, any sequence has a convergent subsequence (but L^+ is not compact!).

(c) If $\{x_n\}$ and $\{y_n\}$ are sequences in L^+ with $x_n \leq y_n \leq x_{n+1}$ for all n, then both sequences converge to the same point.

(d) L^+ (and also L) are normal. (Use (c)).

(e) More generally, any order topology is normal (completely different proof).

(f) If $f: L^+ \to \mathbb{R}$ is continuous, and $r > s$, then one of the sets $f^{-1}((-\infty, s])$ and $f^{-1}([r, \infty))$ is countable.

(g) If $f: L^+ \to \mathbb{R}$ is continuous, then f is eventually constant.

4. (a) L^+ is not contractible. *Hint*: Given $H: L^+ \times [0,1] \to L^+$ with $H(x, 0) = x$ for all x, show that for every t we have $\{H(x, t)\} = L^+$.

(b) $\pi_1(L^+) = \pi_1(L) = 0$. Similarly for $L^+ \times \mathbb{R}$, $L \times \mathbb{R}$, $L \times L$, $L \times L^+$, $L^+ \times L^+$.

(c) $\pi_1(L^+ \times S^1) = \pi_1(L \times S^1) = \mathbb{Z}$.

5. (a) L^+ and L are not homeomorphic. *Hint*: Imitating Problem 1-19, define "paracompact ends".

(b) $L^+ \times \mathbb{R}$ and $L \times \mathbb{R}$ are not homeomorphic; $L^+ \times S^1$ and $L \times S^1$ are not homeomorphic.

(c) Of the 2-manifolds constructed from L^+ or L with $\pi_1 = 0$ and one paracompact end, only $L^+ \times \mathbb{R}$ has the homotopy type of L^+.

(d) The Stone-Čech compactifications of $L \times L$, $L^+ \times L$, $L^+ \times L^+$, and the big disc are all distinct. (Using Problem 3(g), one can explicitly construct these Stone-Čech compactifications.)

6. (a) Show that the Prüfer manifold P is Hausdorff.

(b) P does not have a countable dense subset.

(c) Let U be an open set in $\mathbb{R}^2_+$ which is the union of "wedges" centered at $(a, 0)$ for every irrational a. Show that U includes a whole rectangle of the form

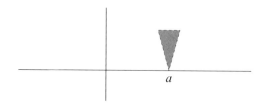

$(a, b) \times (0, \varepsilon)$. *Hint*: Let $A_n = \{a : $ the wedge centered at a has width $\geq 1/n\}$. Since $\mathbb{R} = \mathbb{Q} \cup \bigcup_n A_n$, some A_n is not nowhere dense.

(d) Let $C_1, C_2 \subset P$ be

$$C_1 = \{(0,0)_a : a \text{ irrational}\}$$
$$C_2 = \{(0,0)_a : a \text{ rational}\}.$$

Show that C_1 and C_2 are closed, but that they are not contained in disjoint open sets.

(e) Define $H \colon P \times [0,1] \to P$ by

$$H((x,y)_a, s) = \begin{cases} \left(x\sqrt{\dfrac{1-s+sy}{1+sy}}, \; y\sqrt{\dfrac{1+sy}{1-s+sy}} \right)_a & \text{if } y > 0 \\[3mm] \left(x\sqrt{1-s^2}, \, y\sqrt{1-s^2} \right)_a & \text{if } y \leq 0. \end{cases}$$

Show that H is well-defined and that $H(p, 1) \in \mathbb{R}^2_+ \cup \{(0,0)_a\}$ for all $p \in P$. Conclude that P is contractible.

(f) $P - \{(x,y)_a : y < 0\}$ is a manifold-with-boundary P', whose boundary is a disjoint union of uncountably many copies of $\mathbb{R}$.

(g) The disjoint union of two copies of P', with corresponding points on the boundary identified, is a manifold which is not metrizable, but which has a countable dense subset. Its fundamental group is uncountable.

7. It is known that every second countable contractible 2-manifold is S^2 or $\mathbb{R}^2$. Hence the result of constructing the Prüfer manifold using only copies $\mathbb{R}^2_a$ for rational a must be homeomorphic to $\mathbb{R}^2$. Describe a homeomorphism of this manifold onto $\mathbb{R}^2$.

8. Let M be a connected Hausdorff manifold which is not a point.

(a) If $A \subset M$ has cardinality $\mathfrak{c}$ (the cardinality of $\mathbb{R}$), then the closure $\bar{A}$ has cardinality $\mathfrak{c}$.

(b) If $C \subset M$ is closed and has cardinality $\mathfrak{c}$, then C has an open neighborhood with cardinality $\mathfrak{c}$.

(c) Let $p \in M$. There is a function $f \colon \Omega \to$ (set of subsets of M) such that $f(\alpha)$ has cardinality $\mathfrak{c}$ for all $\alpha \in \Omega$, and such that

$$f(0) = \{p\}$$

$f(\alpha)$ is an open neighborhood of the closure of $\bigcup_{\beta < \alpha} f(\beta)$.

(Consider functions defined on initial segments of Ω with these same properties, and apply Zorn's Lemma. Alternatively, one can require $f(\alpha)$ to be the result of applying the choice function to the set of all open neighborhoods of the closure

of $\bigcup_{\beta<\alpha} f(\beta)$ with cardinality $\mathfrak{c}$. Then there is a unique f with the required properties. This is an example of defining a function by "transfinite induction".)
(d) A function $f: \Omega \to$ (set of subsets of $[0, 1]$) with the properties of the function in part (c) is eventually constant.
(e) M has cardinality $\mathfrak{c}$. (Given $p' \in M$, consider an arc from p to p'.)

9. (a) A connected 1-manifold whose topology is the order topology for some order, is homeomorphic to either the real line, the long line, or the half-long line.
(b) Every 1-manifold M contains a maximal open submanifold N whose topology is the order topology for some order.
(c) If M is connected and $N \neq M$, then M is homeomorphic to S^1.

CHAPTER 2

The long ray L^+ can be given a C^∞ structure, and even a C^ω structure. To see this we need the result of Problem 9-24—any C^∞ [or C^ω] structure on a manifold M homeomorphic to $\mathbb{R}$ is diffeomorphic to $\mathbb{R}$ with the usual structure. This implies that it is also diffeomorphic to $(0, 1)$, and consequently that the structure on M can be extended if M is a proper subset of L^+. An easy application of Zorn's Lemma then shows that C^∞ and C^ω structures exist on L^+.

I do not know whether all C^∞ structures on L^+ are diffeomorphic. It is known that there are uncountably many inequivalent C^ω structures on L^+. If $p \in L^+$, and $L^+{}_p$ denotes all points $\leq p$, then $L^+ - L^+{}_p$ is clearly homeomorphic to L^+. If $\mathcal{O}$ is a C^ω structure for L^+, then it yields a C^ω structure for $L^+ - L^+{}_p$, and hence for L^+. These are all distinct, in other words, there is no C^ω map

$$f : L^+ - L^+{}_p \to L^+ - L^+{}_q \qquad q > p$$

with a C^ω inverse. In fact, we must have $f(q) > q$, and then it is easy to see

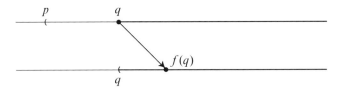

that we must also have $f(f(q)) > f(q)$, $f(f(f(q))) > f(f(q))$, etc. The increasing sequence $q, f(q), f(f(q)), \ldots$ has a limit point $x_0 \in L^+ - L^+{}_p$, and $f(x_0) = x_0$. Now f cannot be the identity on all points $> x_0$ (for then it would be the identity everywhere, since it is C^ω). So for some $q_1 > x_0$ we have $f(q_1) \neq q_1$; we can assume $f(q_1) > q_1$, since we can consider f^{-1} in the contrary case. Reasoning as before, we obtain $x_1 > x_0$ with $f(x_1) = x_1$. Continuing in this way, we obtain $x_0 < x_1 < x_2 < \cdots$ with $f(x_n) = x_n$. This sequence has a limit in $L^+ - L^+{}_p$, but this implies that $f(x) = x$ for all x, a contradiction.

A C^ω structure exists on the Prüfer manifold; this follows immediately from the fact that the maps f_a, used for identifying points in various $(\mathbb{R}^2_a)_+$ with points in $\mathbb{R}^2_+$, are all C^ω. I do not know whether every 2-manifold has a C^∞ structure.

Using the C^ω structure on L^+, we can get a C^ω structure on $L^+ \times L^+$. However, the method used for obtaining a C^ω structure on L^+ will not yield a *complex analytic structure* on $L^+ \times L^+$; the problem is that a complex analytic structure on $\mathbb{R}^2$ may be conformally equivalent to the disc, and hence extendable, but it may also be equivalent to the complex plane, and not extendable. In fact, it is a classical theorem of Rado that every Riemannian surface (2-manifold with a complex analytic structure) is second countable. On the other hand, a modification of the Prüfer manifold yields a non-metrizable manifold of complex dimension 2. References to these matters are to be found in

Calabi and Rosenlicht, *Complex Analytic Manifolds without Countable Base*, Proc. Amer. Math. Soc. **4** (1953), pp. 335–340.

H. Kneser, *Analytische Struktur und Abzählbarkeit*, Ann. Acad. Sic. Fennicae Series A, I 251/5 (1958), pp. 1–8.

PROBLEMS

10. Prove that for $q > p$ there is no non-constant C^ω map $f : L^+ - L^+{}_p \to L^+ - L^+{}_q$.

11. Let (Y, ρ) be a metric space and let $f : X \to Y$ be a continuous locally one-one map, where X is Hausdorff, connected, locally connected, and locally compact.

(a) Every two points $x, y \in X$ are contained in a compact connected $C \subset X$.

(b) Let $d(x, y)$ be the greatest lower bound of the diameters of $f(C)$ (in the ρ-metric) for all compact connected C containing x and y. Show that d is a metric on X which gives the same topology for X.

12. Of the various manifolds mentioned in the previous section, try to determine which can be immersed in which.

CHAPTER 6

Problem A-6(g) describes a non-paracompact 2-manifold in which two open half-planes are a dense set. We will now describe a 3-dimensional version with a twist.

Let $A = \{(x, y, z) \in \mathbb{R}^3 : y \neq 0\}$, and for each $a \in \mathbb{R}$ let $\mathbb{R}^3_a$ be a copy of $\mathbb{R}^3$, points in $\mathbb{R}^3_a$ being denoted by $(x, y, z)_a$. In the disjoint union of A and all $\mathbb{R}^3_a$, $a \in \mathbb{R}$ we identify

$$(x, y, z)_a \quad \text{for } y > 0 \quad \text{with} \quad (a + yx, y, z + a)$$
$$(x, y, z)_a \quad \text{for } y < 0 \quad \text{with} \quad (a + yx, y, z - a).$$

The equivalence classes form a 3-dimensional Hausdorff manifold M. On this manifold there is an obvious function "z", and the sets $z = $ constant form a foliation of M by a 2-dimensional manifold N. The remarkable fact about this 2-dimensional manifold N is that it is connected. For, the set of points $(x, y, c) \in A$ with $y > 0$ is identified with the set of points $(x, y, c - a)_a \in \mathbb{R}^3_a$ with $y > 0$. Now the folium containing $\{(x, y, c - a)_a\}$ contains the points $(x, y, c - a)_a$ with $y < 0$, and these are identified with the set of points $(x, y, c - 2a) \in A$ with $y < 0$. Since we can choose $a = c/2$, we see that all leaves of the foliation are the same as the leaf containing $\{(x, y, 0) : y < 0\} \subset A$.

This example is due to M. Kneser, *Beispiel einer dimensiönserhohenden analytischen Abbildung zwischen überäbzahlbaren Mannigfaltigkeiten*. Archiv. Math. **11** (1960), pp. 280–281.

CHAPTERS 7, 9, 10

1. We have seen that any paracompact C^∞ manifold has a Riemannian metric. The converse also holds, since a Riemannian metric determines an ordinary metric.

2. Problem A-11 implies that a manifold N immersed in a paracompact manifold M is paracompact, but a much easier proof is now available: Let $\langle\ ,\ \rangle$ be a Riemannian metric on M; if $f: N \to M$ is an immersion, then N has the Riemannian metric $f^*\langle\ ,\ \rangle$.

We can now dispense with the argument in the proof of Theorem 6-6 which was used to show that each folium of a distribution on a metrizable manifold is also metrizable, for the folium is a submanifold, and hence paracompact.

3. Since there is no Riemannian metric on a non-paracompact manifold M, the tangent bundle TM cannot be trivial. Thus the tangent bundle of the long line is not trivial, nor is the tangent bundle of the Prüfer manifold, even though the Prüfer manifold is contractible. (On the other hand, a basic result about bundles says that a bundle over a paracompact contractible space is trivial. Compare pg. V.272.)

4. The tangent bundle of the long line L is clearly orientable, so there *cannot* be a nowhere zero 1-form ω on L, for ω and the orientation would determine a nowhere zero vector field, contradicting the fact that the tangent bundle is not trivial. Thus, Theorem 7-9 fails for L. Notice also that if M is non-paracompact, then TM is definitely not equivalent to T^*M, since an equivalence would determine a Riemannian metric. So there are at least two inequivalent non-trivial bundles over M.

5. Although the results in the Addendum to Chapter 9 can be extended to closed, not necessarily compact, submanifolds, they cannot be extended to non-paracompact manifolds, as can be seen by considering the 0-dimensional submanifold $\{(0, 0)_a\}$ of the Prüfer manifold.

6. A Lie group is automatically paracompact, since its tangent bundle is trivial. More generally, a locally compact connected topological group is σ-compact (Problem 10-4).

7. It is not clear that a non-paracompact manifold cannot have an indefinite metric (a non-degenerate inner product on each tangent space). This will be proved in Volume II (Chapter 8, Addendum 1).

PROBLEM

13. Is there a nowhere zero 2-form on the various non-paracompact 2-manifolds which have been described?

NOTATION INDEX

INDEX

These books were typeset using Donald E. Knuth's TEX typesetting system, together with Berthold Horn's DVIPSONE PostScript driver. The figures were produced with Adobe Illustrator, and new or modified fonts were created using Fontographer.

The text font is 11 point Monotype Baskerville—though the em-dash has been modified—together with its italic. The elegant swashes of the italic *y* and *f* cause problems in words like *topology* and *apology*, so a special *gy* ligature was added; special *gg* and *gf* ligatures were also required.

Although a Baskerville bold face is unhistorical, bold type was useful in special circumstances—mainly for indicating **defined terms**. The bold face supplied by Monotype, even the "semi-bold", is obtrusively **extended**, so a non-extended version was created.

The somewhat bold appearance of chapter headings, in 16 point type, results from the linear scaling, as well as the fact that the upper case Baskerville letters are of somewhat heavier weight than the lower case. On the other hand, the tall initial letters beginning each chapter were designed specially, since simple scaling would have made them unpleasantly heavy.

A thicker set of numerals was constructed for use with the upper case lettering in chapter headings and statements of theorems, and special parentheses and other punctuation symbols were also required. Numerous other modifications of this sort, including additional kerns and alterations of set widths, were made for various purposes.

The mathematics fonts are a variation of the *MathTime* fonts, now based on the Monotype Times New Roman family, together with the Monotype Times N R Seven and Times Small Text families—presumably these three families are based on the original designs for Times New Roman, which was created in three essential sizes: 9, 7 and $5\frac{1}{2}$ point.

The italic fonts of these three families were used as the basis for creating the three separate "math italic" fonts—for use at ordinary size, in superscripts, and in second order superscripts. The proportions and weights for these were then used for the three sizes of the symbol font and the other mathematics fonts, including bold symbols, script letters, and additional special symbols, as well as for the extension font and its bold version.

The bold letters for mathematics come directly from the bold fonts of the Times families, while blackboard bold letters were made by hollowing out these bold letters. The Adobe Mathematical Pi 2 font was used for the ordinary sized German Fraktur letters, with suitably modified versions used for superscripts.

The covers, painted by the author in his spare moments, are loosely based on Samuel Taylor Coleridge's poem *The Rime of the Ancient Mariner*.